Examples in Parametric Inference with R

Ulhas Jayram Dixit

Examples in Parametric Inference with R

Ulhas Jayram Dixit
Department of Statistics
University of Bombay
Mumbai, Maharashtra
India

ISBN 978-981-10-9276-3   ISBN 978-981-10-0889-4   (eBook)
DOI 10.1007/978-981-10-0889-4

Printed on acid-free paper

This Springer imprint is published by Springer Nature
The registered company is Springer Science+Business Media Singapore Pte Ltd.

To
my parents late Shri Jayram Shankar
Dixit alias Appa

and

late Shrimati Kamal J. Dixit alias Yammi

Preface

This book originally grew out of my notes on the statistical inference courses at the Department of Statistics, University of Mumbai. I have experienced that reasonably good M.Sc. (Statistics) students are many a time not able to understand or solve problems from some available texts on statistical inference. These books are excellent in terms of content, but the presentation is highly sophisticated. For instance, proofs of various theorems are given in brief and a few examples are provided. To overcome this difficulty, I have solved many examples and, wherever necessary, a program in R is also given. Further, important proofs in this book are presented in such a manner that they are easy to understand.

Through this book, we expect students to know matrix algebra, calculus, probability theory, and distribution theory. This book will serve as an excellent tool for teaching statistical inference courses. The book consists of many solved and unsolved problems. Instructors can assign homework problems from the exercises and students will find the solved examples hugely beneficial in solving the exercise problems.

In "Prerequisite", we have discussed some basic concepts like distribution function and order statistics and illustrated them by using interesting examples. Chapter 1 deals with sufficiency and completeness. In this chapter, we have solved 37 examples. Chapter 2 deals with unbiased estimation. In the last 30 years of my teaching, I found that students were always confused about the relationship between sufficiency and unbiasedness. We have explained this relationship with various examples in this chapter. Chapter 3 is devoted to method of moments and maximum likelihood. In Chap. 4, we deal with lower bound for the variance of an unbiased estimator. Popular concepts like Cramer–Rao (1945, 1946) and Bhattacharya (1946, 1950) lower bound are discussed in detail. Chapter 4 also deals with Chapman and Robbins (1951) and Kiefer (1952) lower bound for the variance of an estimate but does not require regularity conditions. In Chap. 5, the concept of consistency is discussed in detail and illustrated by using different examples. In Chap. 6, Bayesian estimation is briefly discussed. Chapters 7 and 8 are significantly large chapters. Testing of hypothesis is studied in Chap. 7, whereas unbiased and other tests are studied in Chap. 8. We have given R programs in various chapters. No originality is claimed except perhaps in the presentation of the material.

It will prove difficult to thank all my friends who have contributed in some or other way to make this book a reality. I am thankful to Prof. R.B. Bapat for his valuable suggestions to improve upon the content and presentation of the book. I also thank Dr. T.V. Ramanathan for making some valuable suggestions. I am thankful to Shamim Ahmad, senior editor at Springer India for encouraging me to publish this book through Springer and making it easy to go through the process. I thank Prof. Seema C. for reading the book for language. I am equally thankful to Dr. Alok Dabade, Prof. Shailaja Kelkar, Dr. Mehdi Jabbari Nooghabi, Prof. S. Annapurna and Prof. Mandar Bhanushe for various academic discussions related to the book and drawing figures. I am also very thankful to my son Anand and daughter Vaidehi who helped me solve various problems. Further, I am thankful to my wife Dr. (Mrs.) Vaijayanti for the insightful discussions on our book.

We are grateful to Prof. Y.S. Sathe and Late Prof. M.N. Vartak for the diverse discussions which were helpful in understanding statistical inference. These discussions were particularly helpful in solving problems on UMVUE and testing of hypotheses. We are thankful to Prof. B.V. Dhandra, Dr. D.B. Jadhav, and Prof. D.T. Jadhav for providing their M.Phil. dissertations.

In spite of my best efforts, there might be some errors and misprints in the presentation. I owe these mistakes and request the readers to kindly bring them to my notice.

Ulhas Jayram Dixit

Contents

About the Author

Ulhas Jayram Dixit is Professor, at the Department of Statistics, University of Mumbai, India. He is the first Rothamsted International Fellow at Rothamsted Experimental Station in the UK, which is the world's oldest statistics department. Further, he received the Sesqui Centennial Excellence Award in research and teaching from the University of Mumbai in 2008. He is member of the New Zealand Statistical Association, the Indian Society for Probability and Statistics, Bombay Mathematical Colloquium, and the Indian Association for Productivity, Quality and Reliability. Editor of *Statistical Inference and Design of Experiment* (published by Narosa), Prof. Dixit has published over 40 papers in several international journals of repute. His topics of interest are outliers, measure theory, distribution theory, estimation, elements of stochastic process, non-parametric inference, stochastic process, linear models, queuing and information theory, multivariate analysis, financial mathematics, statistical methods, design of experiments, and testing of hypothesis. He received his Ph.D. degree from the University of Mumbai in 1989.

General distribution theory and that of order statistics are an inevitable part of learning theory of estimation and testing of hypothesis. Therefore, we briefly discuss these two topics with some interesting examples.

Distribution Function

Let $F(x)$ be a real-valued function of the variables x; we denote as

$$F(\infty) = \lim_{x \to \infty} F(x),$$

$$F(-\infty) = \lim_{x \to -\infty} F(x),$$

$$F(x_+) = \lim_{h \to 0^+} F(x+h) = F(x+0),$$

$$F(x_-) = \lim_{h \to 0^-} F(x+h) = F(x-0),$$

Definition 1 A function $F(x)$ of a random variable X is called a distribution function (df) if it satisfies the following three conditions:

1. $F(x)$ is non-decreasing, i.e. $F(x+h) \geq F(x)$ if $h > 0$
2. $F(x)$ is right continuous, i.e. $F(x_+) = F(x)$
3. $F(\infty) = 1$ and $F(-\infty) = 0$.

A point x is called a discontinuity point of the distribution function $F(x)$ if $F(x_+) = F(x) \neq F(x_-)$.

Further, if $F(x) = F(x_-)$ then x is called a continuity point of $F(x)$. The quantity $P(x) = F(x_+) - F(x_-) = F(x) - F(x_-)$ is called the jump of $F(x)$ at the point x. Hence, jump of a distribution function is positive at its discontinuity points and zero at its continuity points. An interval is called a continuity for the distribution function $F(x)$ if both its end points are continuity points of $F(x)$. A point x is called a point of jump of the df $F(x)$ if $F(x+\varepsilon) - F(x-\varepsilon) > 0$ for any $\varepsilon > 0$.

We discuss some following examples of distribution function.

Example 1 Let the random variable X follow binomial distribution with parameters $n = 3$ and $p = 0.5$. Then probability mass function (pmf) is given by,

$$P(X = x) = \begin{cases} \binom{3}{x}(0.5)^3 & ; \ x = 0, 1, 2, 3 \\ 0 & ; \ \text{otherwise.} \end{cases}$$

The df $F(x)$ is given by

$$F(x) = \begin{cases} 0 & ; \ x < 0 \\ \sum_{i=0}^{[x]} \binom{3}{x}(0.5)^i & ; \ 0 \leq x < 3 \\ 1 & ; \ x \geq 3. \end{cases}$$

The function $F(x)$ is a df with discontinuity points $(0, 1, 2, 3)$. For the discontinuity point 1, $F(1 - 0) = 0.125$ and $F(1) = F(1 + 0) = 0.5$, one can see $P[X = x]$ and $F(x)$ in Figs. 1 and 2, respectively.

Example 2 Let X be distributed as triangular distribution with probability density function (pdf):

$$f(x) = \begin{cases} x & ; \ 0 < x \leq 1 \\ 2 - x & ; \ 1 \leq x \leq 2 \\ 0 & ; \ \text{otherwise} \end{cases}$$

Then df of X is the df obtained as follows:

$$F(x) = \begin{cases} 0 & ; \ x \leq 0 \\ \int_0^x t\,dt = \frac{x^2}{2} & ; \ 0 \leq x \leq 1 \\ \int_0^1 t\,dt + \int_1^x (2 - t)\,dt = 2x - \frac{x^2}{2} - 1 & ; \ 1 \leq x \leq 2 \\ 1 & ; \ x \geq 2 \end{cases}$$

Fig. 1 Probability mass function

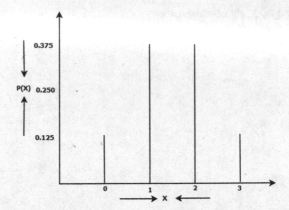

Fig. 2 Distribution function

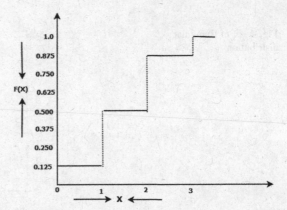

The df $F(x)$ is a continuous function as there are no jump points, i.e., discontinuity points. Figures 3 and 4 give the graph of $f(x)$ and $F(x)$, respectively.

Example 3 The random variable X has the following df

$$F(x) = \begin{cases} 0 & ; \ x < 0 \\ 1 - pe^{-\frac{x}{\theta}} & ; \ x \geq 0, \ 0 < p < 1, \ \theta > 0 \end{cases}$$

The function $F(x)$ is a df with one discontinuity point $x = 0$; since $F(0 - 0) = 0$ and $F(0) = F(0^+) = 1 - p$. Such a function is called a mixture df, i.e., mixture of a step function and a continuous function. We can see the graph of $F(x)$ in Fig. 5.

Every df has a countable set of discontinuity points, and it can be decomposed into two parts as a step function and a continuous function.

Now we consider the Jordan Decomposition Theorem to prove this fact.

Fig. 3 pdf of triangular distribution

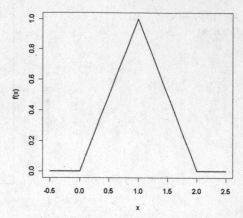

Fig. 4 df of triangular distribution

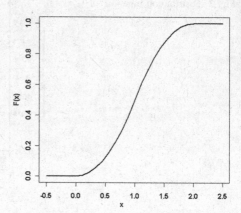

Fig. 5 Distribution Function

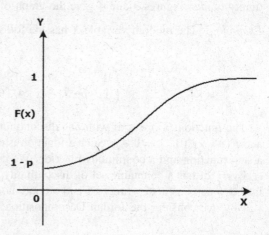

Theorem 1 Every df F has a countable set of discontinuity points. Moreover, $F(x) = \alpha F_d(x) + (1 - \alpha)F_c(x), \forall x, 0 \le \alpha \le 1$, where $F_d(x)$ is a step function and $F_c(x)$ is a continuous function. Further, decomposition is unique.

Proof First, we shall prove the first part

Let $(a, b]$ be a finite interval with at least n discontinuity points $x_1, x_2, \ldots, x_n$ such that $a < x_1 < x_2 < \cdots < x_n < b$,

Hence,

$$F(a) \le F(x_{1-}) < F(x_1) \le F(x_{2-}) < F(x_2) < \cdots \le F(x_{n-}) < F(x_n) \le F(b)$$

Let $p_k = P[X = x_k] = F(x_k) - F(x_{k-}), k = 1, 2, \ldots, n$

Now, $\sum\limits_{k=1}^{n} p_k = F(b) - F(a)$

$\Rightarrow$ The number of points with jump greater than ε will be less than or equal to $F(b) - F(a)$

$\Rightarrow n\varepsilon = F(b) - F(a)$.

$\Rightarrow n = \frac{F(b) - F(a)}{\varepsilon}$.

$\Rightarrow$ The number of points of discontinuity in the interval is finite.

Now, $\Re$ can be looked upon as a countable union of intervals of the type (a,b].

Therefore, the set of discontinuity points of a df F is countable.

Next, we shall prove the second part.

If $\alpha = 1$ then $F(x) = F_d(x) \Rightarrow X$ is a discrete random variable.

If $\alpha = 0$ then $F(x) = F_c(x) \Rightarrow X$ is a continuous random variable.

If $\alpha \in (0, 1)$ then $F(x) = \alpha F_d(x) + (1 - \alpha)F_c(x)$

Let D denote the points of discontinuity of the df F.

Let $\alpha = P(X \in D)$

Since the number points of discontinuity is countable, without loss of generality we assume them to be $x_1, x_2, \ldots$

Let $p(x_i) = P[X = x_i] = F(x_i) - F(x_{i-}); i = 1, 2, \ldots$

Let

$$F_d(x) = \sum_{x_i \le x} p(x_i) \left(\frac{1}{\alpha}\right) \tag{1}$$

Now,

$$F_d(-\infty) = \sum_{x_i \le -\infty} p(x_i) \frac{1}{\alpha} = 0$$

$$F_d(\infty) = \sum_{x_i \le \infty} p(x_i) \frac{1}{\alpha} = \frac{\alpha}{\alpha} = 1$$

Let $x < x'$

$$F_d(x') = \sum_{x_i \leq x'} p(x_i) \frac{1}{\alpha}$$

$$= \sum_{x_i \leq x} p(x_i) \frac{1}{\alpha} + \sum_{x < x_i \leq x'} p(x_i) \frac{1}{\alpha}$$

$$= F_d(x) + \sum_{x < x_i \leq x'} p(x_i) \frac{1}{\alpha}$$

$$F_d(x') - F_d(x) = \sum_{x < x_i \leq x'} p(x_i) \frac{1}{\alpha} \geq 0$$

$$\Rightarrow F_d(x') \geq F_d(x)$$

$F_d(x)$ is non-decreasing function of x.
Now

$$F_d(x+h) = \sum_{x_i < x+h} p(x_i) \frac{1}{\alpha}$$

$$= \sum_{x_i \leq x} p(x_i) \frac{1}{\alpha} + \sum_{x < x_i \leq x+h} p(x_i) \frac{1}{\alpha}$$

$$= F_d(x) + \sum_{x < x_i \leq x+h} p(x_i) \frac{1}{\alpha}$$

Taking limit as $h \to 0$

$$F_d(x+) = F_d(x).$$

$F_d(x)$ is right continuous $\forall x \Rightarrow F_d(x)$ is a df of a discrete random variable.
Now,

$$F(x) = \alpha F_d(x) + (1 - \alpha) F_c(x)$$

$$F_c(x) = \frac{1}{1 - \alpha} [F(x) - \alpha F_d(x)]$$

$$F_c(-\infty) = \frac{0}{1 - \alpha} = 0$$

$$F_c(\infty) = \frac{1 - \alpha}{1 - \alpha} = 1$$

Since $F(x)$ and $F_d(x)$ are right continuous at x, $F_c(x)$ is also right continuous.
Let $x < x'$

$$F_c(x') - F_c(x) = \frac{1}{1-\alpha}[F(x') - F(x) - \alpha(F_d(x') - F_d(x))]$$

$$= \frac{1}{1-\alpha}\left[F(x') - F(x) - \alpha\left\{\sum_{x_i < x'}\frac{p(x_i)}{\alpha} - \sum_{x_i \leq x}\frac{p(x_i)}{\alpha}\right\}\right]$$

$$= \frac{1}{1-\alpha}\left[F(x') - F(x) - \sum_{x < x_i \leq x'}p(x_i)\right] \geq 0 \qquad (2)$$

Since $\sum_{x < x_i \leq x'}p(x_i)$ = sum of the total jumps, in $F(x)$ between points x and x'
$\sum\limits_{x < x_i \leq x'}p(x_i) \leq$ Jump in $F(x)$ between x and x' i.e. $F(x') - F(x)$

$$\Rightarrow F_c(x') - F_c(x) \geq 0$$

Therefore, if $x' > x \Rightarrow F_c(x') \geq F_c(x)$
$\Rightarrow F_c(x)$ is a non-decreasing function of x.
$\Rightarrow F_c(x)$ is a df.
Now, we shall prove that $F_c(x)$ is left continuous function of x.
From (2),

$$F_c(x') - F_c(x) = \frac{1}{1-\alpha}\left[F(x') - F(x) - \sum_{x < x_i \leq x'}p(x_i)\right]$$

$$= \frac{1}{1-\alpha}\left[F(x') - F(x) - \sum_{x < x_i < x'}p(x_i) - P(X = x')\right]$$

$$F_c(x') - F_c(x) = \frac{1}{1-\alpha}\left[F(x') - F(x) - \alpha\left(\sum_{x < x_i \leq x'}\frac{p(x_i)}{\alpha}\right) - (F(x') - F(x'-))\right]$$

$$= \frac{1}{1-\alpha}[F(x'-) - F(x) - \alpha(F_d(x'-) - F_d(x))]$$

$$= \frac{1}{1-\alpha}[F(x'-) - \alpha F_d(x'-) - F(x) + \alpha F_d(x)]$$

$$= \frac{F(x'-) - \alpha F_d(x'-)}{1-\alpha} - \frac{[F(x) - \alpha F_d(x)]}{1-\alpha}$$

$$= F_c(x'-) - F_c(x)$$

$$\Rightarrow F_c(x') = F_c(x'-)$$

$\Rightarrow F_c(x)$ is left continuous at any point x.
But $F_c(x)$ is also right continuous at all points of x.
Hence, $F_c(x)$ is a df of a continuous random variable X.
Therefore $F(x) = \alpha F_d(x) + (1 - \alpha)F_c(x)$.
We shall now prove the third part, i.e., the decomposition is unique.
Suppose that the decomposition is not unique.

$$F(x) = \alpha F_{d_1}(x) + (1 - \alpha)F_{c_1}(x) = \alpha F_{d_2}(x) + (1 - \alpha)F_{c_2}(x)$$

$$\alpha[F_{d_1}(x) - F_{d_2}(x)] = (1 - \alpha)[F_{c_2}(x) - F_{c_1}(x)]$$

$\Rightarrow$ Step function = Continuous function
This cannot be true. Hence, our assumption is wrong and therefore, decomposition is unique.

Example 4 Decompose the following

$$F(x) = \begin{cases} 0 & ; x < -1 \\ \frac{x+1}{12} & ; -1 \leq x < 2 \\ \frac{3}{4} & ; 2 \leq x < 3 \\ 1 - \frac{3}{4x} & ; x \geq 3 \end{cases}$$

$F(-1_-) = 0$, $F(-1) = 0 \Rightarrow F(-1_-) = F(-1)$. Therefore $F(x)$ is continuous at $x = -1$.
Now $F(2_-) = \frac{1}{4}$, $F(2) = \frac{3}{4} \Rightarrow F(2_-) \neq F(2)$, which implies that $F(x)$ is discontinuous at $x = 2$. The jump at the point 2 is $P[X = 2] = F(2) - F(2_-) = \frac{3}{4} - \frac{1}{4} = \frac{1}{2}$. At the point 3 $F(3_-) = \frac{3}{4}$, $F(3) = \frac{3}{4} \Rightarrow F(3_-) = F(3)$ and therefore $F(x)$ is continuous at $x = 3$.
Let D be the set of discontinuity points $D = \{2\}$.

$$\alpha = P(X \in D) = P(X = 2) = \frac{1}{2}$$

$$F_d(x) = \sum_{x_i \leq x} \frac{p(x_i)}{\alpha} = 2\left(\frac{1}{2}\right) = 1$$

$$F_d(x) = \begin{cases} 0 & ; x < 2 \\ 1 & ; x \geq 2 \end{cases}$$

Therefore,

$$F(x) = \alpha F_d(x) + (1 - \alpha)F_c(x)$$

$$= \frac{1}{2}F_d(x) + \frac{1}{2}F_c(x)$$

where

$$F_c(x) = 2\left[F(x) - \frac{1}{2}F_d(x)\right]$$

$F_c(x) = 0, x < -1$.
For $-1 \leq x < 2$; $F_c(x) = 2\left(\frac{x+1}{12}\right) = \frac{x+1}{6}$
For $2 \leq x < 3$; $F_c(x) = 2\left[\frac{3}{4} - \frac{1}{2}\right] = \frac{1}{2}$
For $x \geq 3$; $F_c(x) = 2\left[1 - \frac{3}{4x} - \frac{1}{2}\right] = 1 - \frac{3}{2x}$
Hence

$$F_c(x) = \begin{cases} 0 & ; x < -1 \\ \frac{x+1}{6} & ; -1 \leq x < 2 \\ \frac{1}{2} & ; 2 \leq x < 3 \\ 1 - \frac{3}{2x} & ; x \geq 3 \end{cases}$$

Example 5 Decompose the following

$$F(x) = \begin{cases} 0 & ; x < 1 \\ \frac{1}{4} + \frac{x^2 - 1}{12} & ; 1 \leq x < 2 \\ \frac{3}{4} + \frac{x-2}{4} & ; 2 \leq x < 3 \\ 1 & ; x \geq 3 \end{cases}$$

$$F(1_-) = 0, \ F(1) = \frac{1}{4} \Rightarrow F(1_-) \neq F(1)$$

$F(x)$ is discontinuous at $x = 1$. $P(X = 1) = \frac{1}{4}$
$F(2_-) = \frac{1}{4} + \frac{1}{4} = \frac{1}{2}$, $F(2) = \frac{3}{4} \Rightarrow F(2_-) \neq F(2)$.
$F(x)$ is discontinuous at $x = 2 \Rightarrow P(X = 2) = \frac{1}{4}$
$F(3_-) = 1, \ F(3) = 1 \Rightarrow F(x)$ is continuous at $x = 3$.
The set of points of discontinuity is $D = \{1, 2\}$ and $\alpha = P(X \in D) = \frac{1}{2}$.

$$F_d(x) = \sum_{x_i \leq x} \frac{p(x_i)}{\alpha} = 2 \sum_{x_i \leq x} p(x_i)$$

Hence

$$F_d(x) = \begin{cases} 0 & ; \ x < 1 \\ \frac{1}{2} & ; \ 1 \le x < 2 \\ 1 & ; \ x \ge 2 \end{cases}$$

$$F(x) = \alpha F_d(x) + (1 - \alpha) F_c(x)$$

$$= \frac{1}{2} F_d(x) + \frac{1}{2} F_c(x)$$

$$F_c(x) = 2\left[F(x) - \frac{1}{2} F_d(x)\right]$$

For $x < 1$; $F_c(x) = 2(0) = 0$.

For $1 \le x < 2$; $F_c(x) = 2\left[\frac{1}{4} + \frac{x^2-1}{12} - \frac{1}{4}\right] = \frac{x^2-1}{6}$.

For $2 \le x < 3$; $F_c(x) = 2\left[\frac{3}{4} + \frac{x-2}{4} - \frac{1}{2}\right] = \frac{x-1}{2}$.

Hence

$$F_c(x) = \begin{cases} 0 & ; \ x < 1 \\ \frac{x^2-1}{6} & ; \ 1 \le x < 2 \\ \frac{x-1}{2} & ; \ 2 \le x < 3 \\ 1 & ; \ x \ge 3 \end{cases}$$

If X is discrete random variable then

X	1	2
$P(X = x)$	$\frac{1}{2}$	$\frac{1}{2}$

If X is continuous random variable then

$$f_c(x) = \begin{cases} \frac{x}{3} & ; \ 1 \le x < 2 \\ \frac{1}{2} & ; \ 2 \le x < 3 \\ 0 & ; \ \text{otherwise} \end{cases}$$

Example 6 Decompose $F(x)$ and find $E(X)$ and $V(X)$

$$F(x) = \begin{cases} 0 & ; \ x < 0 \\ \frac{1}{4} & ; \ 0 \le x < 1 \\ \frac{1}{2} + \frac{1}{2}(1 - e^{-(x-1)}) & ; \ x \ge 1 \end{cases}$$

$F(0_-) = 0$, $F(0) = \frac{1}{4} \Rightarrow F(0_-) \ne F(0) \Rightarrow F(x)$ is discontinuous at $x = 0$ and $P[X = 0] = \frac{1}{4}$.

$F(1_-) = \frac{1}{4}$, $F(1) = \frac{1}{2}$, $F(1_-) \neq F(1)$ ∴ $F(x)$ is not continuous at $x = 1$
therefore $P[X = 1] = \frac{1}{4}$.
$D = \{0, 1\}$ and $P(X = 1) = F(1) - F(1_-) = \frac{1}{4}$

$$\alpha = P(X = 0) + P(X = 1) = \frac{1}{4} + \frac{1}{4} = \frac{1}{2}$$

$$F_d(x) = \sum_{x_i \le x} \frac{p(x_i)}{\alpha} = 2 \sum_{x_i \le x} p(x_i)$$

$$F_d(x) = \begin{cases} 0 & ; x < 0 \\ \frac{1}{2} & ; 0 \le x < 1 \\ 1 & ; x \ge 1 \end{cases}$$

$$F(x) = \frac{1}{2} F_d(x) + \frac{1}{2} F_c(x)$$
$$= \frac{1}{2} [F_d(x) + F_c(x)]$$

$$F_c(x) = 2 \left[F(x) - \frac{1}{2} F_d(x) \right]$$

For $x < 0$; $F_c(x) = 2[0 - 0] = 0$. For $0 \le x < 1$; $F_c(x) = 2\left[\frac{1}{4} - \frac{1}{4}\right] = 0$.
For $x \ge 1$; $F_c(x) = 2\left[\frac{1}{2} + \frac{1}{2}(1 - e^{-(x-1)}) - \frac{1}{2}\right] = 1 - e^{-(x-1)}$.
Hence

$$F_c(x) = \begin{cases} 1 - e^{-(x-1)} & ; x \ge 1 \\ 0 & ; \text{otherwise} \end{cases}$$

Therefore,
If X is discrete random variable then

X	0	1
P(X = x)	$\frac{1}{2}$	$\frac{1}{2}$

If X is continuous random variable then

$$f(x) = e^{-(x-1)}; \ x \ge 1$$

Let $A = \{0, 1\}$ and $B = \{(1, \infty)\}$.
Further, let $\mathbf{I}_A$ and $\mathbf{I}_B$ be two indicator functions such that

$$\mathbf{I}_A(x) = \begin{cases} 1 & ; \; x \in A \\ 0 & ; \; \text{otherwise} \end{cases}$$

and

$$\mathbf{I}_B(x) = \begin{cases} 1 & ; \; x \in B \\ 0 & ; \; \text{otherwise} \end{cases}$$

Then the random variable X can be written as

$$X = \frac{1}{2}\mathbf{I}_A(x) + \frac{1}{2}\mathbf{I}_B(x)$$

$$EX = \frac{1}{2}E_{\mathbf{I}_A}X + \frac{1}{2}E_{\mathbf{I}_B}X$$

$$E_{\mathbf{I}_A}X = 0 \times \frac{1}{2} + 1 \times \frac{1}{2} = \frac{1}{2}$$

$$E_{\mathbf{I}_B}X = \int_1^\infty x e^{-(x-1)}dx = 2$$

$$EX = \frac{1}{4} + 1 = \frac{5}{4}$$

$$EX^2 = \frac{1}{2}E_{\mathbf{I}_A}X^2 + \frac{1}{2}E_{\mathbf{I}_B}X^2$$

$$E_{\mathbf{I}_A}X^2 = \frac{1}{2}$$

$$E_{\mathbf{I}_B}X^2 = \int_1^\infty x^2 e^{-(x-1)}dx = 5$$

$$EX^2 = \frac{1}{4} + \frac{5}{2} = \frac{11}{4}$$

$$V(X) = \frac{11}{4} - \frac{25}{16} = \frac{19}{16}$$

Example 7 Decompose $F(x)$ and find $F_d(x)$ and $F_c(x)$.

$$F(x) = \begin{cases} 0 & ; x < 0 \\ \frac{x}{4} & ; 0 \le x < 2 \\ \frac{3}{4} & ; 2 \le x < 3 \\ 1 & ; x \ge 3 \end{cases}$$

$F(0_-) = 0$, $F(0) = 0 \Rightarrow F(0_-) = F(0) \Rightarrow F(x)$ is continuous at $x = 0$.
$F(2_-) = \frac{1}{2}$, $F(2) = \frac{3}{4} \Rightarrow F(2_-) \ne F(2) \Rightarrow F(x)$ is not continuous at $x = 2$.
Therefore $P(X = 2) = F(2) - F(2_-) = \frac{1}{4}$.
$F(3_-) = \frac{3}{4}$, $F(3) = 1 \Rightarrow F(3_-) \ne F(3) \Rightarrow F(x)$ is not continuous at $x = 3$.
Therefore $P(X = 3) = F(3) - F(3_-) = \frac{1}{4}$.
The set of points of discontinuity is $D = \{2, 3\} \Rightarrow \alpha = P(X = 2) + P(X = 3) = \frac{1}{2}$.

$$F_d(x) = \sum_{x_i \le x} \frac{p(x_i)}{\alpha} = 2 \sum_{x_i \le x} p(x_i)$$

Hence

$$F_d(x) = \begin{cases} 0 & ; x < 2 \\ \frac{1}{2} & ; 2 \le x < 3 \\ 1 & ; x \ge 3 \end{cases}$$

$$F_c(x) = 2 \left[F(x) - \frac{1}{2} F_d(x) \right]$$

For $x < 0 \Rightarrow F_c(x) = 0$.
For $0 \le x < 2$; $F_c(x) = 2 \left[\frac{x}{4} - 0 \right] = \frac{x}{2}$.
For $2 \le x < 3$; $F_c(x) = 2 \left[\frac{3}{4} - \frac{1}{4} \right] = 1$.
For $x \ge 3$; $F_c(x) = 2 \left[1 - \frac{1}{2} \right] = 1$.
Hence,

$$F_c(x) = \begin{cases} 0 & ; x < 0 \\ \frac{x}{2} & ; 0 \le x < 2 \\ 1 & ; x \ge 2 \end{cases}$$

Therefore,
If X is discrete random variable then

X	2	3
$P(X = x)$	$\frac{1}{2}$	$\frac{1}{2}$

If X is continuous random variable then

$$f_c(x) = \begin{cases} \frac{1}{2} & ; \ 0 \le x < 2 \\ 0 & ; \ \text{otherwise} \end{cases}$$

Let $A = \{2, 3\}$ and $B = \{(0, 2)\}$. Further, let $\mathbf{I}_A$ and $\mathbf{I}_B$ be two indicator functions such that

$$\mathbf{I}_A(x) = \begin{cases} 1 & ; \ x \in A \\ 0 & ; \ \text{otherwise} \end{cases}$$

and

$$\mathbf{I}_B(x) = \begin{cases} 1 & ; \ x \in B \\ 0 & ; \ \text{otherwise} \end{cases}$$

Then the random variable X can be written as

$$X = \frac{1}{2}\mathbf{I}_A(x) + \frac{1}{2}\mathbf{I}_B(x)$$

$$EX = \frac{1}{2}E_{\mathbf{I}_A}X + \frac{1}{2}E_{\mathbf{I}_B}X$$

$$E_{\mathbf{I}_A}X = 1 + \frac{3}{2} = \frac{5}{2}$$

$$E_{\mathbf{I}_B}X = \int_0^2 \frac{x}{2}\,dx = 1$$

Therefore,

$$EX = \frac{5}{4} + \frac{1}{2} = \frac{7}{4}$$

Also,

$$EX^2 = \frac{1}{2}E_{\mathbf{I}_A}X^2 + \frac{1}{2}E_{\mathbf{I}_B}X^2$$

$$E_{\mathbf{I}_A}X^2 = 4 \times \frac{1}{2} + 9 \times \frac{1}{2} = \frac{13}{2}$$

$$E1_B X = \int_0^2 \frac{r^2}{2} \, ||| \quad \frac{4}{3}$$

Therefore,

$$EX^2 = \frac{13}{4} + \frac{4}{6} = \frac{47}{12}$$

Hence,

$$V(X) = \frac{47}{12} - \frac{49}{16} = \frac{41}{48}$$

Examples on pmf and pdf

Before considering the examples, let us consider some theorems which will be used in subsequent chapters.

Theorem 2 *Let X be a rv with the pdf* $f(x), x \in R$.
Let $Y = g(x)$ be one-to-one function and differentiable at all x. The pdf of y is

$$f(y) = f_X(g^{-1}(y)) \left| \frac{dx}{dy} \right|,$$

where $x = g^{-1}(y)$.

Theorem 3 *Let X be an rv with pdf* $f(x); x \in R$.
Suppose $Y = g(x)$ is a many-to-one function. Let A be the set of values of X. Further, let $A_1, A_2 \ldots$ be the disjoint subsets of A such that $\bigcup_{i=1} A_i = A$. Also the transformation $Y = g(x)$ is one-to-one for every $X \in A_i, i = 1, 2 \ldots$
Then the pdf of Y is given as

$$f_Y(y) = \sum_{i=1} \left\{ f_X(g^{-1}(y)) \left| \frac{dx}{dy} \right| \right\}_{A_i}$$

Theorem 4 *Suppose X is a discrete rv with pmf* $f(X = x)$. *Let* $Y = g(x)$.
(a) If $Y = g(x)$ is a one-to-one function then

$$P[Y = y] = P[X = g^{-1}(y)]$$

(b) If $Y = g(x)$ is a many-to-one function then

$$P[Y = y] = \sum_x P[X = x, g(x) = y]$$

Theorem 5 *Let X and Y be the two rvs then $EX = EEX|Y$*

Proof Assume X and Y are continuous rvs.
Consider

$$E(X|Y) = \int_X xf(x|y)dx$$

$$= \int_X x\frac{f(x,y)}{g(y)}dx$$

$$E(EX|Y) = \int_Y \left[\int_X x\frac{f(x,y)}{g(y)}dx\right]g(y)dy$$

$$= \int_Y \left(\int_X xf(x,y)dx\right)dy$$

$$= \int_X x\left(\int_Y f(x,y)dy\right)dx$$

$$= \int_X xf(x)dx = EX$$

We can prove similarly, when X and Y are discrete rvs.

Theorem 6 *Let X and Y be two rvs then*

$$V(X) = E[V(X|Y)] + V[E(X|Y)]$$

Proof Consider,

$$E[V(X|Y)] + V[E(X|Y)]$$

$$E[(EX^2|Y) - (EX|Y)^2] + E[E(X|Y)]^2 - [E(EX|Y)]^2$$

$$E[(EX^2|Y)] - E[E(X|Y)]^2 + E[E(X|Y)]^2 - [E(EX|Y)]^2 \tag{3}$$

Now, by using Theorem 5,

$$E[(EX^2|Y)] = EX^2 \text{ and } E(EX|Y) = EX,$$

Equation (3) becomes

$$EX^2 - (EX)^2 = V(X)$$

Example 8 Let X be an rv with $\cup(-\theta, \theta)$. Find the distribution of (i) $|X|$ (ii) X^2

(i) Let $Y = |X|$ By definition

$$|X| = \begin{cases} X & ; X > 0 \\ -X & ; X < 0 \end{cases}$$

Let

$$A_1 = \{x : -\theta \le x < 0\} A_2 = \{x : 0 \le x < \theta\}$$

In this case $Y = |X|$ is one-to-one function for $X \in A_i (i = 1, 2)$

$$f_Y(y) = \sum_{i=1}^{2} \left\{ f_X(x) \left| \frac{dx}{dy} \right| \right\}_{A_i}$$

$$= \sum_{i=1}^{2} \left\{ \frac{1}{2\theta} \right\}_{A_i}$$

$$= \frac{1}{2\theta} + \frac{1}{2\theta} = \frac{1}{\theta}$$

$$f(y) = \begin{cases} \frac{1}{\theta} & ; 0 < y < \theta \\ 0 & ; \text{ otherwise} \end{cases}$$

Note One can use this result in estimating θ (see Chap. 2)

(ii) Now $Y = X^2$ is many-to-one function.
Let

$$A_1 = \{x : -\theta \le x < 0\} A_2 = \{x : 0 \le x < \theta\}$$

Now $Y = X^2$ is one-to-one function for every $x \in A_i (i = 1, 2)$

$$f_Y(y) = \sum_{i=1}^{2} \left\{ \frac{1}{(2\theta)} \frac{1}{2\sqrt{y}} \right\}_{A_i}$$

$$= \frac{1}{(2\theta)} \frac{1}{2\sqrt{y}} + \frac{1}{(2\theta)} \frac{1}{2\sqrt{y}} = \frac{1}{(2\theta)\sqrt{y}}; \ 0 < y < \theta^2$$

Therefore

$$f_Y(y) = \begin{cases} \frac{1}{2\theta\sqrt{y}} & ; \ 0 < y < \theta^2 \\ 0 & ; \ \text{otherwise} \end{cases}$$

Example 9 Let X be a discrete rv with the following pmf

$$P(X = x) = \begin{cases} k & ; \ x = 0, \pm j, j = 1, 2, \ldots, n \\ 0 & ; \ \text{otherwise} \end{cases}$$

Find the pmf of $|X|$ and X^2.

We can write as

$$P(X = x) = \begin{cases} \frac{1}{2n+1} & ; \ x = -n, -n+1, \ldots, 0, 1, 2, \ldots, n \\ 0 & ; \ \text{otherwise} \end{cases}$$

Let $Y = |X| \Rightarrow Y$ be one-to-one at $x = 0$ and many-to-one at $x \neq 0$.

$$P[Y = y] = P[|X| = y]$$

$$= P[X = y] + P[X = -y]$$

$$= \frac{2}{2n+1}$$

Hence,

$$P(Y = y) = \begin{cases} \frac{1}{2n+1} & ; \ y = 0 \\ \frac{2}{2n+1} & ; \ y = 1, 2, \ldots, n \\ 0 & ; \ \text{otherwise} \end{cases}$$

Let $Y = X^2$

$$P[Y = y] = P[X^2 = y]$$

$$= P[X = \sqrt{y}] + P[X = -\sqrt{y}]$$

$$= \frac{2}{2n+1}; \ y = 1, 4, 9, \ldots, n^2$$

$$P[Y = 0] = P[X^2 = 0] = P[X = 0]$$

$$\frac{1}{2n+1}$$

Therefore,

$$P(Y = y) = \begin{cases} \frac{1}{2n+1} & ; y = 0 \\ \frac{2}{2n+1} & ; y = 1, 4, 9, \ldots, n^2 \\ 0 & ; \text{otherwise} \end{cases}$$

Example 10 Let X be an rv with the following pdf

$$f(x) = \begin{cases} \frac{2x}{\pi^2} & ; 0 < x < \pi \\ 0 & ; \text{otherwise} \end{cases}$$

Find the cdf of (i) $\sin x$ (ii) $\cos x$

(i) Let $Y = \sin X$

$$G(y) = P[Y \le y]$$

$$= P[\sin X \le y]$$

$$= P[(0 \le X \le \sin^{-1}(y) \cup (\pi - \sin^{-1}(y) \le X \le \pi)]$$

$Y = \sin x$ is a many-to-one function

$$X = \begin{cases} \sin^{-1}(y) & ; 0 < x < \frac{\pi}{2} \\ \pi - \sin^{-1}(y) & ; \frac{\pi}{2} < x < \pi \end{cases}$$

Let

$$A_1 = \{x : 0 < x < \frac{\pi}{2}\} A_2 = \{x : \frac{\pi}{2} < x < \pi\}$$

$$f(y) = \left\{ f(x)|\frac{dx}{dy}| \right\}_{A_1} + \left\{ f(x)|\frac{dx}{dy}| \right\}_{A_2}$$

$$= \frac{2\sin^{-1}y}{\pi^2} \frac{1}{\sqrt{1-y^2}} + \frac{2(\pi - \sin^{-1}y)}{\pi^2} \frac{1}{\sqrt{1-y^2}}$$

$$= \frac{2}{\pi} \frac{1}{\sqrt{1-y^2}} ; 0 < y < 1$$

Hence,

$$f(y) = \begin{cases} \frac{2}{\pi}\frac{1}{\sqrt{1-y^2}} & ; \ 0 < y < 1 \\ 0 & ; \ \text{otherwise} \end{cases}$$

(ii) $Y = \cos x$

$$F(x) = \begin{cases} \int\limits_{0}^{x} \frac{2x}{\pi^2} = \frac{x^2}{\pi^2} & ; \ 0 < x < \pi \\ 1 & ; \ x > \pi \end{cases}$$

$$H(y) = P[Y \le y]$$

$$= P[\cos X \le y]$$

Principal value of $\cos^{-1}y$ are
$0 < \cos^{-1}y < \pi$ when $-1 < y < 1$

$$H(y) = P(\cos^{-1}y < X < \pi) = F(\pi) - F(\cos^{-1}y)$$

$$= 1 - \frac{(\cos^{-1}y)^2}{\pi^2}, -1 < y < 1;$$

See Fig. 6.
    The pdf is given by

Fig. 6 Distribution function
of cos(x)

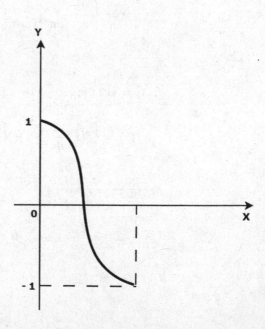

$$h(y) = \begin{cases} \dfrac{2\cos^{-1}y}{\pi^2} \cdot \dfrac{1}{\sqrt{1-y^2}} & ; -1 < y < 1 \\ 0 & ; \text{otherwise} \end{cases}$$

Example 11 Let X be an rv with the following pdf

$$f(x) = \begin{cases} \theta e^{-\theta x} & ; x > 0, \theta > 0 \\ 0 & ; \text{otherwise} \end{cases}$$

Find the pdf and df of (i) $\sin x$ (ii) $\cos x$

$$F(x) = \begin{cases} 0 & ; x < 0 \\ 1 - e^{-\theta x} & ; x \geq 0 \end{cases}$$

Consider

$$G(y) = P[Y \leq y] = P[\sin X \leq y]$$

For $0 < y < 1$; (see Fig. 7)

$$
\begin{aligned}
G(y) &= P[X \leq \sin^{-1}y] + P[\pi - \sin^{-1}y \leq X \leq 2\pi + \sin^{-1}y] \\
&+ P[3\pi - \sin^{-1}y \leq X \leq 4\pi + \sin^{-1}y] + P[5\pi - \sin^{-1}y \leq X \leq 6\pi + \sin^{-1}y] + \cdots \\
&= 1 - e^{-\theta \sin^{-1}y} + \sum_{n=1}^{\infty} P[(2n-1)\pi \sin^{-1}y \leq X \leq 2n\pi + \sin^{-1}y] \\
&= 1 - e^{-\theta \sin^{-1}y} + \sum_{n=1}^{\infty} \left[e^{-\theta[(2n-1)\pi - \sin^{-1}y]} - e^{-\theta[2n\pi + \sin^{-1}y]} \right] \\
&= 1 - e^{-\theta \sin^{-1}y} + \exp[\pi\theta + \theta\sin^{-1}y] \sum_{n=1}^{\infty} \exp[-(2\theta\pi n)] \\
&- \exp[-\theta\sin^{-1}y] \sum_{n=1}^{\infty} \exp[-2\theta\pi n] \\
&= 1 - e^{-\theta\sin^{-1}y} + (e^{\pi\theta + \theta\sin^{-1}y} - e^{-\theta\sin^{-1}y}) \left(\frac{e^{-2\pi\theta}}{1 - e^{-2\pi\theta}} \right) \\
&= \frac{1 - e^{-2\pi\theta} - e^{-\theta\sin^{-1}y} + e^{-2\pi\theta - \theta\sin^{-1}y} + e^{-\pi\theta + \theta\sin^{-1}y} - e^{-2\pi\theta - \theta\sin^{-1}y}}{1 - e^{-2\pi\theta}} \\
&= 1 + \frac{e^{-\pi\theta + \theta\sin^{-1}y} - e^{-\theta\sin^{-1}y}}{1 - e^{-2\pi\theta}}
\end{aligned}
$$

$$G(y) = \begin{cases} 1 + \dfrac{e^{-\pi\theta + \theta\sin^{-1}y} - e^{-\theta\sin^{-1}y}}{1 - e^{-2\pi\theta}} & ; 0 < y < 1 \\ 0 & ; \text{otherwise} \end{cases} \tag{4}$$

Note that for $-1 < y < 0$, the principal value of $\sin^{-1}y$ will be negative. For $-1 < y < 0$

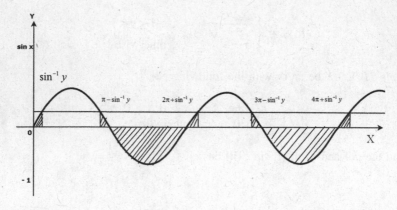

Fig. 7 Graph of $\sin(x)$

$$P[\sin X \le y] = P[\pi - \sin^{-1}y < X < 2\pi + \sin^{-1}y] + P[3\pi - \sin^{-1}y < X < 4\pi + \sin^{-1}y] + \cdots$$

$$= \sum_{n=1}^{\infty} P[(2n-1)\pi - \sin^{-1}y < X < 2n\pi + \sin^{-1}y]$$

$$= \sum_{n=1}^{\infty} \left\{ e^{-\theta[(2n-1)\pi - \sin^{-1}y]} - e^{-\theta[2n\pi + \sin^{-1}y]} \right\}$$

$$= \left(e^{\pi\theta + \theta\sin^{-1}y} - e^{-\theta\sin^{-1}y} \right) \left(\frac{e^{-2\pi\theta}}{1 - e^{-2\pi\theta}} \right)$$

$$= \frac{\left(e^{-\pi\theta + \theta\sin^{-1}y} - e^{-2\pi\theta - \theta\sin^{-1}y} \right)}{1 - e^{-2\pi\theta}}$$

$$(5)$$

From (4) and (5)

$$G(y) = \begin{cases} \dfrac{e^{-\pi\theta + \theta\sin^{-1}y} - e^{-2\pi\theta - \theta\sin^{-1}y}}{1 - e^{-2\pi\theta}} & ; \; -1 < y < 0 \\ 1 + \dfrac{e^{-\pi\theta + \theta\sin^{-1}y} - e^{-\theta\sin^{-1}y}}{1 - e^{-2\pi\theta}} & ; \; 0 < y < 1 \\ 0 & ; \; \text{otherwise} \end{cases}$$

The pdf of Y is as follows:

$$g(y) = \begin{cases} \dfrac{\theta e^{-\theta\pi}[e^{\theta\sin^{-1}y} + e^{-\pi\theta - \theta\sin^{-1}y}]}{(1 - e^{-2\pi\theta})\sqrt{1 - y^2}} & ; \; -1 < y < 0 \\ \dfrac{\theta[e^{-\pi\theta + \theta\sin^{-1}y} + e^{-\theta\sin^{-1}y}]}{(1 - e^{-2\pi\theta})\sqrt{1 - y^2}} & ; \; 0 < y < 1 \\ 0 & ; \; \text{otherwise} \end{cases}$$

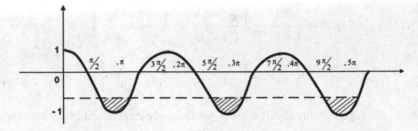

Fig. 8 Graph of cos(x)

(ii)
$$G(y) = P[Y \leq y]$$

$$= P[\cos X \leq y] = P[X \leq \cos^{-1} y]$$

For $-1 < y < 1$ for $0 < x < \infty$ Principal value of $\cos^{-1} y$ are $0 < \cos^{-1} y < \pi$, $-1 < y < 1$; (see Fig. 8)

$$G(y) = P[\cos^{-1} y < X < 2\pi - \cos^{-1} y, 2\pi + \cos^{-1} y < X < 4\pi + \cos^{-1} y, \ldots]$$

$$= \sum_{n=1}^{\infty} P[(2n-2)\pi + \cos^{-1} y < X < 2n\pi - \cos^{-1} y]$$

$$= \sum_{n=1}^{\infty} \left[1 - e^{-\theta[2n\pi - \cos^{-1} y]} - 1 + e^{-\theta[(2n-2)\pi + \cos^{-1} y]} \right]$$

$$= e^{\theta \cos^{-1} y} \sum_{n=1}^{\infty} e^{-2n\theta\pi} + e^{-\theta \cos^{-1} y} \sum_{n=1}^{\infty} e^{-\theta\pi(2n-2)}$$

where

$$\sum_{n=1}^{\infty} e^{-2n\theta\pi} = \frac{e^{-2\pi\theta}}{(1 - e^{-2\pi\theta})}$$

$$G(y) = \frac{e^{-\theta \cos^{-1} y} - e^{(\theta \cos^{-1} y - 2\pi\theta)}}{1 - e^{-2\pi\theta}}; \quad -1 < y < 0$$

Note that the same is true for $0 < y < 1$, therefore

$$G(y) = \frac{e^{-\theta \cos^{-1} y} - e^{(\theta \cos^{-1} y - 2\pi\theta)}}{1 - e^{-2\pi\theta}}; \quad -1 < y < 1$$

$$g(y) = \frac{1}{1 - e^{-2\pi\theta}} \left[\frac{\theta e^{-\theta\cos^{-1}y}}{\sqrt{1 - y^2}} + \frac{\theta e^{-2\pi\theta + \theta\cos^{-1}y}}{\sqrt{1 - y^2}} \right]$$

$$= \frac{\theta}{1 - e^{-2\pi\theta}} \left[\frac{e^{-\theta\cos^{-1}y}}{\sqrt{1 - y^2}} + \frac{e^{-2\pi\theta + \theta\cos^{-1}y}}{\sqrt{1 - y^2}} \right]$$

Example 12 A probability distribution is not uniquely determined by its moments.

Let X be an rv with lognormal distribution

$$f(x) = \begin{cases} \frac{1}{x\sqrt{2\pi}} \exp[-\frac{1}{2}(\log x)^2] & ; \; x > 0 \\ 0 & ; \; \text{otherwise} \end{cases}$$

Consider another random variable Y as

$$g(y) = \begin{cases} \{1 + b\sin(2\pi\log y)\}f(y) & ; \; y > 0, -1 < b < 1 \\ 0 & ; \; \text{otherwise} \end{cases}$$

$$EY^r = \int_0^\infty y^r f(y)dy + b \int_0^\infty y^r \sin(2\pi\log y)f(y)dy$$

$$= EX^r + \frac{b}{\sqrt{2\pi}} \int_0^\infty y^r \sin(2\pi\log y)\frac{1}{y}e^{-\frac{1}{2}(\log)^2}dy$$

Let $\log y = z \Rightarrow y = e^z$

$$= EX^r + \frac{b}{\sqrt{2\pi}} \int_{-\infty}^\infty e^{rz - \frac{z^2}{2}}\sin(2\pi z)dz$$

$$= EX^r + \frac{b}{\sqrt{2\pi}}e^{\frac{r^2}{2}} \int_{-\infty}^\infty e^{-\frac{(z-r)^2}{2}}\sin(2\pi z)dz$$

Since $z - r = t \Rightarrow \sin(2\pi z) = \sin(2\pi r + 2\pi t) = \sin 2\pi t$, r being a positive integer

$$= EX^r + \frac{be^{\frac{r^2}{2}}}{\sqrt{2\pi}} \int_{-\infty}^\infty e^{-\frac{t^2}{2}}\sin(2\pi t)dt$$

The integral is an odd function of t. Therefore the value of the integral is zero.

$$EX^r = EY^r$$

Hence, we have two different distributions and their moments are same. We can conclude that moments cannot determine the distribution uniquely.

Note If the moments of the specified order exist, then all the lower order moments automatically exist. However, the converse is not true. See the following example.

Example 13 Consider the following pdf

$$f(x) = \begin{cases} \frac{2}{x^3} & ; x \geq 1 \\ 0 & ; \text{otherwise} \end{cases}$$

$EX = 2$ and $EX^2 = \infty$

Example 14 Let

$$f(x) = \frac{(r+1)\theta^{r+1}}{(x+\theta)^{r+2}}; \ x \geq 0, \theta > 0$$

$$EX^r = (r+1)\theta^{r+1}\beta(r+1,1)$$

$$EX^{r+1} = (r+1)\theta^{r+1} \int_0^\infty \frac{x^{r+1}}{(x+\theta)^{r+2}} dx \to \infty$$

In this example moments up to rth order exist and higher order moments do not exist.

Example 15 A continuous distribution need not be symmetric even though all its central odd moments vanish.

Let

$$f(x) = \frac{1}{48}\exp(-|x|^{\frac{1}{4}})[1 - k\sin|x|^{\frac{1}{4}}]; \ -\infty < x < \infty$$

where

$$k = \begin{cases} -1 & ; x < 0 \\ 1 & ; x > 0 \end{cases}$$

In this case $EX^{2r+1} = 0$

Hence $f(x)$ is asymmetric, but all its odd moments are zero, see Churchill (1946).

Example 16 If X_1 and X_2 are independent then

$$\phi_{X_1+X_2}(t) = \phi_{X_1}(t) \times \phi_{X_2}(t),$$

where ϕ is a characteristic function. But converse is not true.

(i)

$$f(x) = \frac{1}{\pi}\frac{1}{1+x^2}; \quad -\infty < x < \infty$$

$$\phi_{X_1}(t) = e^{-|t|}$$

Let $X_1 = X_2$ almost surely $\Rightarrow P(X_1 = X_2) = 1$

$$\phi_{X_2(t)} = e^{-|t|}$$

$$\phi_{X_1+X_2}(t) = e^{-2|t|} = \phi_{X_1}(t) \times \phi_{X_2}(t)$$

But X_1 and X_2 are not independent.

(ii) Consider the joint density of (X_1, X_2)

$$f(x_1, x_2) = \begin{cases} \frac{1}{4}\{1 + x_1x_2(x_1^2 - x_2^2)\} & ; \ |x_1| < 1 \text{ and } |x_2| < 1 \\ 0 & ; \ \text{otherwise} \end{cases}$$

Marginal pdf of $X_i(i = 1, 2)$

$$g(x_1) = \begin{cases} \frac{1}{2} & ; \ |x_1| \leq 1 \\ 0 & ; \ \text{otherwise} \end{cases}$$

$$g(x_2) = \begin{cases} \frac{1}{2} & ; \ |x_2| \leq 1 \\ 0 & ; \ \text{otherwise} \end{cases}$$

$$\phi_{X_1}(t) = \int_{-1}^{1} e^{-itx_1} \frac{dx_1}{2} = \frac{e^{it} - e^{-it}}{2it} = \frac{\sin t}{t}$$

$$\phi_{X_2}(t) = \frac{\sin t}{t}$$

$$\phi_{X_1+X_2}(t) = \frac{(\sin t)^2}{t^2}$$

Next, $Z = X_1 + X_2$ then

$$h(z) = \int f(u, z-u)\,du$$

$$= \frac{1}{4}\int [1 + u(z-u)\{u^2 - (z-u)^2\}]\,du$$

$$= \frac{1}{4}\int [1 + 3z^2u^2 - 2zu^3 - z^3u]\,du$$

Since $X_1 = u$ and $X_2 = z - u$

The limits of integration for u in terms of z is given by $-1 \leq u \leq z+1$; $u \leq 0$ and $z - 1 \leq u \leq 1$; $u > 0$

$$h(z) = \frac{1}{4}\int_{-1}^{z+1} [1 + 3z^2u^2 - 2zu^3 - z^3u]\,du = \frac{2+z}{4}; \quad -2 \leq z \leq 0$$

$$= \frac{1}{4}\int_{z-1}^{1} [1 + 3z^2u^2 - 2zu^3 - z^3u]\,du = \frac{2-z}{4}; \quad 0 < z \leq 2$$

$$\phi_{X_1+X_2}(t) = \int_{-2}^{2} e^{itz}h(z)\,dz$$

$$= \int_{-2}^{0} \frac{2+z}{4}e^{itz}\,dz + \int_{0}^{2} \frac{2-z}{4}e^{itz}\,dz$$

$$= \int_{0}^{2} \frac{2-z}{4}e^{-itz}\,dz + \int_{0}^{2} \frac{2-z}{4}e^{itz}\,dz$$

$$= \int_{0}^{2} (e^{-itz} + e^{itz})\frac{2-z}{4}\,dz$$

$$= \frac{1}{2}\int_{0}^{2} (2-z)\cos(tz)\,dz$$

$$= \frac{2 - 2\cos(2t)}{4t^2} = \frac{1 - \cos(2t)}{2t^2} = \left(\frac{\sin t}{t}\right)^2$$

$$= \phi_{X_1}(t) \times \phi_{X_2}(t)$$

But $f(x_1, x_2) \neq g(x_1) \times g(x_2)$

Example 17 If X_1, X_2, X_3, X_4 are independent $N(0,1)$ random variables, show that

(a) $Z = |X_1 X_2 + X_3 X_4|$ had exponential pdf e^{-x} for $x > 0$.

(b) $Z_1 = \frac{X_1}{X_2}$ has Cauchy distribution.

$$M_{X_1 X_2}(t) = \int_{-\infty}^{\infty} \int_{-\infty}^{\infty} \frac{\exp(tx_1 x_2)}{2\pi} \exp\left[-\frac{1}{2}(x_1^2 + x_2^2)\right] dx_1 dx_2$$

$$= \frac{1}{\sqrt{2\pi}} \int_{-\infty}^{\infty} \exp - \left(\frac{x_1^2}{2}\right) \left(\frac{1}{\sqrt{2\pi}} \int_{-\infty}^{\infty} \exp\left[tx_1 x_2 - \frac{x_2^2}{2}\right] dx_2\right) dx_1$$

Now,

$$\exp\left[-\frac{1}{2}(x_2^2 - 2tx_1 x_2 + t^2 x_1^2 - t^2 x_1^2)\right] = \exp\left[-\frac{1}{2}(x_2 - tx_1)^2 + \frac{t^2 x_1^2}{2}\right]$$

Hence

$$\frac{1}{\sqrt{2\pi}} \int_{-\infty}^{\infty} \exp\left[tx_1 x_2 - \frac{x_2^2}{2}\right] dx_2 = e^{\frac{t^2 x_1^2}{2}} \int_{-\infty}^{\infty} \frac{1}{\sqrt{2\pi}} e^{-\frac{1}{2}(x_2 - tx_1)^2} dx_2 = e^{\frac{t^2 x_1^2}{2}}$$

$$M_{X_1 X_2}(t) = \int_{-\infty}^{\infty} \frac{1}{\sqrt{2\pi}} \exp\left[-\frac{x_1^2}{2} + \frac{t^2 x_1^2}{2}\right] dx_2 = \int_{-\infty}^{\infty} \frac{1}{\sqrt{2\pi}} \exp\left[-\frac{x_1^2}{2}(1 - t^2)\right] dx_1$$

$$= \frac{1}{\sqrt{1 - t^2}}$$

Similarly,

$$M_{X_3 X_4}(t) = \frac{1}{\sqrt{1 - t^2}}$$

Hence

$$M_{X_1 X_2 + X_3 X_4}(t) = \frac{1}{1 - t^2}$$

This is the mgf of Z with Laplace distribution.

$$f(x) = \frac{1}{2}\exp(-|x|); \quad -\infty \le x \le \infty$$

To verify this

$$M_X(t) = \frac{1}{2}\int\limits_{-\infty}^{0} e^x e^{tx} dx + \frac{1}{2}\int\limits_{0}^{\infty} e^{-x} e^{tx} dx = \frac{1}{1-t^2}$$

$$f_{X_1 X_2 + X_3 X_4}(x) = e^{-x}; \quad x > 0$$

(b) Let $Z_1 = \frac{X_1}{X_2}$, $Z_2 = X_2 \Rightarrow X_1 = Z_1 Z_2$

$$\frac{\partial(X_1, X_2)}{\partial(Z_1, Z_2)} = Z_2$$

$$f(Z_1, Z_2) = f(X_1, X_2)Z_2 = \frac{1}{2\pi}\exp\left[-\frac{1}{2}\{z_1^2 z_2^2 + z_2^2\}\right]z_2 = \frac{z_2}{2\pi}\exp\left[-\frac{z_2^2}{2}(1+z_1^2)\right]$$

$$f(z_1) = \int\limits_{-\infty}^{\infty} \frac{z_2}{2\pi}\exp\left[-\frac{z_2^2}{2}(1+z_1^2)\right]dz_2$$

$$= 2\int\limits_{0}^{\infty} \frac{z_2}{2\pi}\exp\left[-\frac{z_2^2}{2}(1+z_1^2)\right]dz_2$$

$$= \int\limits_{0}^{\infty} \frac{1}{\pi}\exp[-w(1+z_1^2)]dw = \frac{1}{\pi}\frac{1}{(1+z_1^2)}$$

Hence Z_1 has a Cauchy distribution.

Example 18 If the rv $X \sim B(n,p)$ and the rv Y has negative binomial distribution with parameters r and p, prove that

$$P[X \le r-1] = P[Y > n-r]$$

Now,

$$P[Y > n-r] = \sum_{y=n-r+1}^{\infty} \binom{y+r-1}{r-1} p^r q^y$$

Let $Z = y - (n - r + 1)$

$$= p^r q^{n-r+1} \sum_{z=0}^{\infty} \binom{z+n}{r-1} q^z$$

Now

$$\binom{z+n}{r-1} = \sum_{k=0}^{r-1} \binom{n}{k}\binom{z}{r-1-k}$$

$$P[Y > n - r] = p^r q^{n-r+1} \sum_{z=0}^{\infty} \left[\sum_{k=0}^{r-1} \binom{n}{k}\binom{z}{r-1-k} \right] q^z$$

Now, $\binom{m}{r} = 0$ if $m < r$

$$= p^r q^{n-r+1} \sum_{k=0}^{r-1} \binom{n}{k} \sum_{z=r-1-k}^{\infty} \binom{z}{r-1-k} q^z$$

Let $t = z - (r - 1 - k)$

$$= p^r q^{n-r+1} \sum_{k=0}^{r-1} \left[\binom{n}{k} \sum_{t=0}^{\infty} \binom{t+r-1-k}{r-1-k} q^{t+r-1-k} \right]$$

$$= p^r q^n \sum_{k=0}^{r-1} \left[\binom{n}{k} \sum_{t=0}^{\infty} \binom{t+r-1-k}{r-1-k} q^{t-k} \right]$$

$$= p^r q^n \sum_{k=0}^{r-1} \left[\binom{n}{k} q^{-k} \sum_{t=0}^{\infty} \binom{t+r-1-k}{t} q^t \right]$$

$$= p^r q^n \sum_{k=0}^{r-1} \left[\binom{n}{k} q^{-k} (1-q)^{-(r-k)} \right]$$

$$= \sum_{k=0}^{r-1} \left[\binom{n}{k} p^k q^{n-k} \right] = P[X \leq r - 1]$$

Example 19 Let X be an rv with $B(n, p)$. Prove that

$$F_{n+1}(y) = pF_n(y - 1) + qF_n(y),$$

where,

$$F_n(y) = P[Y \le y] \quad \sum_{x=0}^{y} \binom{n}{x} p^x q^{n-x}$$

Consider $F_{n+1}(y) = pF_n(y-1) + qF_n(y)$

$$= p\sum_{x=0}^{y-1} \binom{n}{x} p^x q^{n-x} + q\sum_{x=0}^{y} \binom{n}{x} p^x q^{n-x}$$

$$= \sum_{x=0}^{y-1} \binom{n}{x} p^{x+1} q^{n-x} + \sum_{x=0}^{y} \binom{n}{x} p^x q^{n-x+1}$$

$$= [pq^n + \binom{n}{1} p^2 q^{n-1} + \binom{n}{2} p^3 q^{n-2} + \cdots + \binom{n}{y-1} p^y q^{n-y+1}]$$

$$+ [q^{n+1} + \binom{n}{1} pq^n + \binom{n}{2} p^2 q^{n-1} + \cdots + \binom{n}{y} p^y q^{n-y+1}]$$

$$= q^{n+1} + \left[\binom{n}{0} + \binom{n}{1}\right] pq^n + \left[\binom{n}{1} + \binom{n}{2}\right] p^2 q^{n-1} + \left[\binom{n}{2} + \binom{n}{3}\right] p^3 q^{n-2}$$

$$+ \cdots + \left[\binom{n}{y-1} + \binom{n}{y}\right] p^y q^{n-y+1}$$

$$= q^{n+1} + \left[\binom{n+1}{1}\right] pq^n + \left[\binom{n+1}{2}\right] p^2 q^{n-1} + \left[\binom{n+1}{3}\right] p^3 q^{n-2} + \cdots + \left[\binom{n+1}{y}\right] p^y q^{n-y+1}$$

$$P_{n+1}[X \le y] = \sum_{x=0}^{y} \binom{n+1}{x} p^x q^{n-x+1} = F_{n+1}(y)$$

Example 20 Let X be an rv with $B(n,p)$. Prove that

$$P[X \le k] = (n-k) \binom{n}{k} \int_{0}^{q} t^{n-k-1}(1-t)^k dt$$

Now,

$$P[X \le k] = \sum_{r=0}^{k} \binom{n}{r} p^r q^{n-r}$$

$$\frac{dP[X \le k]}{dq} = \sum_{r=0}^{k} \binom{n}{r} [rp^{r-1}(-1)q^{n-r} + p^r(n-r)q^{n-r-1}]$$

$$= \sum_{r=0}^{k} -\frac{n(n-1)!}{(r-1)!(n-r)!} p^{r-1} q^{n-r} + \frac{n(n-1)!}{r!(n-r-1)!} p^r q^{n-r-1}$$

$$= \sum_{r=0}^{k} n\left[\binom{n-1}{r}p^r q^{n-r-1} - \binom{n-1}{r-1}p^{r-1}q^{n-r}\right]$$

Let $A_r = \binom{n-1}{r}p^r q^{n-r-1}$

$$= \sum_{r=0}^{k} n[A_r - A_{r-1}] = nA_k \tag{6}$$

$$\frac{dP}{dq} = n\binom{n-1}{k}p^k q^{n-k-1}$$

On integrating both sides

$$P[X \leq k] = n\binom{n-1}{k}\int_0^q (1-u)^k u^{n-k-1}du$$

$$= (n-k)\binom{n}{k}\int_0^q (1-u)^k u^{n-k-1}du$$

Example 21 If the rv X has $B(n,p)$ and Y has Beta distribution with parameters k and $(n-k-1)$ then prove that

$$P[Y \leq p] = P[X \geq k]$$

Now, $P[X \geq k] = \sum_{r=k}^{n}\binom{n}{r}p^r q^{n-r}$

$$\frac{dP[X \geq k]}{dp} = nT_{k-1}, \text{ where } T_r = \binom{n-1}{r}p^r q^{n-r-1}$$

$$= n\binom{n-1}{k-1}p^{k-1}q^{n-k}$$

On integrating both sides

$$P[X \geq k] = n\binom{n-1}{k-1}\int_0^p u^{k-1}(1-u)^{n-k}du$$

$$= \frac{1}{\beta(k, n-k+1)}\int_0^p u^{k-1}(1-u)^{n-k}du$$

Example 22 Let X be a Poisson Distribution with parameter λ, prove that

$$P[X \leq x] = \int_{\lambda}^{\infty} \frac{e^{-t}t^x}{\Gamma(x+1)} dt$$

Let

$$I_x = \int_{\lambda}^{\infty} \frac{1}{x!} e^{-t} t^x dt$$

$$= \frac{1}{x!} \left([-e^{-t}t^x]_{\lambda}^{\infty} + \int_{\lambda}^{\infty} xt^{x-1}e^{-t} dt \right)$$

$$= \frac{e^{-\lambda}\lambda^x}{x!} + \frac{1}{(x-1)!} \int_{\lambda}^{\infty} t^{x-1}e^{-t} dt$$

$$= \frac{\lambda^x e^{-\lambda}}{x!} + I_{x-1}$$

$$= \frac{e^{-\lambda}\lambda^x}{x!} + \frac{e^{-\lambda}\lambda^{x-1}}{(x-1)!} + I_{x-2}$$

$$= \frac{e^{-\lambda}\lambda^x}{x!} + \frac{e^{-\lambda}\lambda^{x-1}}{(x-1)!} + \frac{e^{-\lambda}\lambda^{x-2}}{(x-2)!} + I_{x-3}$$

$$= \frac{e^{-\lambda}\lambda^x}{x!} + \frac{e^{-\lambda}\lambda^{x-1}}{(x-1)!} + \cdots + \frac{e^{-\lambda}\lambda}{1!} + e^{-\lambda}$$

$$= \sum_{i=0}^{x} \frac{e^{-\lambda}\lambda^i}{i!} = P[X \leq x]$$

Example 23 Let X_1, X_2 be the iid rvs with $N(0, 1)$.
Find the pdf of $Z_1 = \sqrt{X_1^2 + X_2^2}$ and $Z_2 = \tan^{-1}(\frac{X_1}{X_2})$.

$$Z_1^2 = X_1^2 + X_2^2, \quad \tan Z_2 = \frac{X_1}{X_2} \Rightarrow Z_1^2 = X_2^2(\tan Z_2)^2 + X_2^2$$

$$X_2^2 = \frac{Z_1^2}{1 + \tan^2 Z_2} = \frac{Z_1^2}{\sec^2 Z_2}$$

$$X_2 = \pm \frac{Z_1}{\sec Z_2} = \pm Z_1 \cos Z_2, \quad X_1 = \pm Z_1 \sin Z_2$$

Therefore $X_1 = Z_1 \sin Z_2$ or $-Z_1 \sin Z_2$
$X_2 = Z_1 \cos Z_2$ or $-Z_1 \cos Z_2$

$$\frac{\partial(X_1, X_2)}{\partial(Z_1, Z_2)} = \begin{pmatrix} \pm\sin z_2 & \pm z_1 \cos z_2 \\ \pm\cos z_2 & \mp z_1 \sin z_2 \end{pmatrix} = z_1 \sin^2 z_2 + z_1 \cos^2 z_2 = z_1$$

$$|J| = Z_1 + Z_1 = 2Z_1$$

$$f(x_1, x_2) = \frac{1}{2\pi} \exp\left[-\left(\frac{x_1^2 + x_2^2}{2}\right)\right]$$

$$f(z_1, z_2) = \frac{1}{2\pi} e^{-\frac{z_1^2}{2}} (2z_1)$$

$$= \frac{z_1}{\pi} \exp\left[-\frac{z_1^2}{2}\right]$$

$X_1 \in (-\infty, \infty)$ and $X_2 \in (-\infty, \infty)$
then $Z_1 \in (0, \infty)$ and $Z_2 \in (-\frac{\pi}{2}, \frac{\pi}{2})$

$$h(z_1) = \int_{-\frac{\pi}{2}}^{\frac{\pi}{2}} \frac{z_1}{\pi} \exp\left[-\frac{z_1^2}{2}\right] dz_2$$

$$= \frac{z_1}{\pi} \exp\left[-\frac{z_1^2}{2}\right] \pi = z_1 \exp\left[-\frac{z_1^2}{2}\right]; \quad 0 < z_1 < \infty$$

$$g(z_2) = \begin{cases} \frac{1}{\pi} & ; \quad -\frac{\pi}{2} < z_2 < \frac{\pi}{2} \\ 0; & ; \quad \text{otherwise} \end{cases}$$

Example 24 Let $X_1, X_2, \ldots, X_n$ be iid rvs from $B(1, p)$. Find the distribution of $S^2 = \sum (X_i - \bar{X})^2$.

$$S^2 = \sum (X_i - \bar{X})^2 = \sum X_i^2 - n \frac{\left(\sum X_i\right)^2}{n^2} = \sum X_i^2 - \frac{\left(\sum X_i\right)^2}{n}$$

Since $X_i = 0$ or 1 for $i = 1, 2, \ldots, n$, $\sum X_i^2 \sim B(n, p)$ and $\sum X_i \sim B(n, p)$
Let $Y = \sum X_i^2 = \sum X_i$

$$S^2 = y - \frac{y^2}{n} = y\left(1 - \frac{y}{n}\right) = y\left(\frac{n-y}{n}\right)$$

$$P[S^2 = 0] = P(y = 0) + P(y = n) = q^n + p^n$$

$$P\left(S^2 = \frac{n-1}{n}\right) = P(y = 1) + P(y = n - 1) = npq^{n-1} + np^{n-1}q$$

$$= npq(q^{n-2} + p^{n-2})$$

$$P\left(S^2 = \frac{2(n-2)}{n}\right) = P(y = 2) + P(y = n - 2) = \binom{n}{2}\left[p^2 q^{n-2} + p^{n-2} q^2\right]$$

$$= \binom{n}{2}p^2 q^2 [q^{n-4} + p^{n-4}]$$

$$P\left(S^2 = \frac{3(n-3)}{n}\right) = P(y = 3) + P(y = n - 3) = \binom{n}{3}\left[p^3 q^{n-3} + p^{n-3} q^3\right]$$

$$= \binom{n}{3}p^3 q^3 [q^{n-6} + p^{n-6}]$$

In general

$$P\left(S^2 = \frac{i(n-i)}{n}\right) = P(y = i) + P(y = n - i) = \binom{n}{i}\left[p^i q^{n-i} + p^{n-i} q^i\right]$$

$$= \binom{n}{i}p^i q^i [q^{n-2i} + p^{n-2i}]; \quad i = 0, 1, 2, \ldots, n$$

Example 25 Let $X_1, X_2, \ldots, X_n$ be independent rvs with exponential distribution having mean one. Prove that the following rvs have the same distribution.

$$Z = \max(X_1, X_2, \ldots, X_n)$$

$$W = X_1 + \frac{X_2}{2} + \frac{X_3}{3} + \cdots + \frac{X_n}{n}$$

$$P[Z \leq z] = (F(z))^n = (1 - e^{-z})^n$$

$$f(z) = n(1 - e^{-z})^{n-1} e^{-z}$$

$$M_z(t) = \int_0^\infty e^{tz} n(1 - e^{-z})^{n-1} e^{-z} dz$$

Let $u = 1 - e^{-z} \Rightarrow du = e^{-z} dz$ and $e^{tz} = (1 - u)^{-t}$

$$z = 0 \Rightarrow u = 0, \ z = \infty \Rightarrow u = 1$$

$$= \int_0^1 n u^{n-1}(1-u)^{-t} du = n\beta(n, 1-t)$$

$$= \frac{n\Gamma(n)\Gamma(1-t)}{\Gamma(n+1-t)} = \frac{n!(-t)!}{(n-t)!}$$

$$= \frac{n!(-t)!}{(n-t)(n-t-1)\dots(2-t)(1-t)(-t)!}$$

$$M_z(t) = \frac{n!}{(n-t)(n-t-1)\cdots(2-t)(1-t)}$$

$$M_w(t) = M_{X_1}(t) \times M_{X_{\frac{2}{2}}}(t) \times \dots \times M_{X_{\frac{n}{n}}}(t)$$

$$= \frac{1}{1-t} \times \frac{2}{2-t} \times \cdots \times \frac{n}{n-t}$$

$$= \frac{n!}{(n-t)(n-t-1)\dots(2-t)(1-t)}$$

Hence, Z and W have the same distribution.

Example 26 Let X_1 and X_2 be independent rvs with $\cup(0, \theta_i), i = 1, 2$ respectively. Let $Z_1 = \min(X_1, X_2)$ and

$$Z_2 = \begin{cases} 0 & ; Z_1 = X_1 \\ 1 & ; Z_1 = X_2 \end{cases}$$

Show that Z_1 and Z_2 are independent rvs.

(i) $\theta_1 < \theta_2$

$$P[Z_1 \le z_1, Z_2 = 0] = P[\min(X_1, X_2) \le z_1, Z_2 = 0]$$

$$= P[(X_1 \le z_1, X_1 < X_2]$$

$$= \int_0^{z_1} \int_{x_1}^{\theta_2} \frac{1}{\theta_1 \theta_2} dx_1 dx_2$$

$$= \frac{1}{\theta_1 \theta_2}\left(z_1 \theta_2 - \frac{z_1^2}{2}\right); \quad 0 < z_1 < \theta_1 \tag{7}$$

$$= \int_0^{z_1} \frac{1}{\theta_1}\left(\frac{\theta_2 - x_1}{\theta_2}\right) dx_1$$

$$= \frac{z_1}{\theta_1 \theta_2}\left(\theta_2 - \frac{z_1}{2}\right)$$

$$P[Z_2 = 0] = \int\limits_{0}^{\theta_1}\int\limits_{x_1}^{\theta_2} \frac{1}{\theta_1\theta_2} dx_1 dx_2$$

$$= \frac{1}{\theta_1\theta_2}\left[\int\limits_{0}^{\theta_1} (\theta_2 - x_1)dx_1\right] = 1 - \frac{\theta_1}{2\theta_2} \qquad (8)$$

$$P[Z_1 \le z_1, Z_2 = 1] = P[\min(X_1, X_2) \le z_1, Z_2 = 1]$$

$$= P[X_2 \le z_1, X_2 < X_1]$$

$$= \int\limits_{0}^{z_1}\int\limits_{x_2}^{\theta_1} \frac{dx_1 dx_2}{\theta_1\theta_2}$$

$$= \int\limits_{0}^{z_1} \frac{(\theta_1 - x_1)}{\theta_1\theta_2} dx_1$$

$$= \frac{1}{\theta_1\theta_2}\left(z_1\theta_1 - \frac{z_1^2}{2}\right)$$

$$= \frac{z_1}{\theta_1\theta_2}\left(\theta_1 - \frac{z_1}{2}\right) \qquad (9)$$

$$P[Z_2 = 1] = P[X_2 \le X_1]$$

$$= \int\limits_{0}^{\theta_1}\int\limits_{x_2}^{\theta_1} \frac{1}{\theta_1\theta_2} dx_1 dx_2$$

$$= \frac{1}{\theta_1\theta_2}\left[\int\limits_{0}^{\theta_1} (\theta_1 - x_2)dx_2\right]$$

$$= \frac{1}{\theta_1\theta_2}\left[\theta_1 x_2 - \frac{x_2^2}{2}\right]_{0}^{\theta_1}$$

$$= \frac{1}{\theta_1\theta_2}[\theta_1^2 - \frac{\theta_1^2}{2}]$$

$$= \frac{\theta_1}{2\theta_2}; \ \theta_1 < \theta_2 \qquad (10)$$

From (7) and (9)

$$P[Z_1 \leq z_1] = \frac{z_1}{\theta_1 \theta_2} [\theta_1 + \theta_2 - z_1] = \frac{\theta_1 + \theta_2}{\theta_1 \theta_2} z_1 - \frac{z_1^2}{\theta_1 \theta_2}$$

$$f(z_1) = \frac{\theta_1 + \theta_2}{\theta_1 \theta_2} - \frac{2z_1}{\theta_1 \theta_2}; \ 0 < z_1 < \theta_1 \tag{11}$$

$$P[Z_2 = 0] = 1 - \frac{\theta_1}{2\theta_2} \tag{12}$$

$$P[Z_2 = 1] = \frac{\theta_1}{2\theta_2} \tag{13}$$

(ii) $\theta_1 > \theta_2$
Similarly, as before

$$f(z_1) = \frac{\theta_1 + \theta_2}{\theta_1 \theta_2} - \frac{2z_1}{\theta_1 \theta_2}; \ 0 < z_1 < \theta_2 \tag{14}$$

$$P[Z_2 = 0] = 1 - \frac{\theta_2}{2\theta_1} \tag{15}$$

$$P[Z_2 = 1] = \frac{\theta_2}{2\theta_1} \tag{16}$$

(iii) $\theta_1 = \theta_2 = \theta$
Similarly, as before

$$f(z_1) = \frac{2}{\theta} - \frac{2z_1}{\theta^2}; \ 0 < z_1 < \theta \tag{17}$$

$$P[Z_2 = 0] = \frac{1}{2} \tag{18}$$

$$P[Z_2 = 1] = \frac{1}{2} \tag{19}$$

From (7), (9), (11), (13), (14), (15), (17), (18) and (19), Z_1 and Z_2 are independent rvs.

Order Statistics

Many functions of random variable of interest, in practice, depend on the relative magnitude of the observed variables. For instance, we may be interested in the fastest time in an automobile race or the heaviest mouse among those found on a certain diet. Thus we often order observed random variables according to their magnitudes. The resulting ordered variables are called order statistics.

Definition 2 If the random variables $X_1, X_2, \cdots, X_n$ are arranged in ascending order of magnitude and then written as $X_{(1)} \leq X_{(2)} \leq \ldots \leq X_{(r)} \leq X_{(r+1)} \leq \cdots \leq X_{(n)}$, $X_{(r)}$ is called the rth order statistic ($r = 1, 2, \ldots, n$) and $(X_{(1)}, X_{(2)}, \ldots, X_{(n)})$ is called the set of order statistics.

Let $X_1, X_2, \cdots, X_n$ be n independent random variables. Let $F_r(x)$ be the distribution function (df) of the rth order statistic $X_{(r)}$.

Hence,

$$F_r(x) = P[X_{(r)} \leq x]$$

$$= P[\text{Atleast } r \text{ of } X_i \text{ are less than or equal to } x]$$

$$= \sum_{i=r}^{n} \binom{n}{i} F^i(x)[1 - F(x)]^{n-i}, \tag{20}$$

where $F(x) = $ df of $X = P[X \leq x]$.

Theorem 7 *The distribution of rth order statistic is*

$$f_{X_{(r)}}(x) = \frac{n!}{(r-1)!(n-r)!} F^{r-1}(x)[1 - F(x)]^{n-r} f(x); \; x \in R \tag{21}$$

Proof Differentiating (20) with respect to x, let $\bar{F}(x) = 1 - F(x)$

$$f_{X_{(r)}}(x) = \sum_{i=r}^{n} \binom{n}{i} i F^{i-1}(x) \bar{F}^{n-i}(x) f(x) - \sum_{i=r}^{n} \binom{n}{i} F^i(x)(n-i) \bar{F}^{n-i-1}(x) f(x)$$

$$= n \sum_{i=r}^{n} \binom{n-1}{i-1} F^{i-1}(x) \bar{F}^{n-i}(x) f(x) - n \sum_{i=r}^{n} \binom{n-1}{i} F^i(x) \bar{F}^{n-i-1}(x) f(x)$$

$$\text{Let } A_i = \binom{n-1}{i} F^i(x) \bar{F}^{n-i-1}(x) f(x).$$

$$\text{Therefore, } f_{X_{(r)}}(x) = n \sum_{i=r}^{n} (A_{i-1} - A_i) = n(A_{r-1} - A_n)$$

$$\text{But } A_n = 0$$

$$f_{X_{(r)}}(x) = n A_{r-1}$$

$$= n \binom{n-1}{r-1} F^{r-1}(x) \bar{F}^{n-r}(x) f(x)$$

Corollary 1 *The distribution of* $X_{(n)} = \max_i X_i$ *is*

$$f_{X_{(n)}}(x) = nF^{n-1}(x)f(x); \; x \in R \tag{22}$$

Proof

$$\begin{aligned}
f_{X_{(n)}}(x) &= P[X_{(n)} \leq x] \\
&= P[\text{All } X_i's \leq x] \\
&= \prod_{i=1}^{n} P[X_i \leq x] = F^n(x) \quad (X_i's \text{ are } iid)
\end{aligned}$$

Hence

$$f_{X_{(n)}}(x) = nF^{n-1}(x)f(x); \; x \in R \tag{23}$$

Corollary 2 *The distribution of* $X_{(1)} = \min_i X_i$ *is,*

$$f_{X_{(1)}}(x) = n[1 - F(x)]^{n-1}f(x); \; x \in R \tag{24}$$

Proof

$$\begin{aligned}
F_{X_{(1)}}(x) &= P[X_{(1)} \leq x] \\
&= 1 - P[X_{(1)} > x] \\
&= 1 - P[\text{All } X_i's > x] \\
&= 1 - \prod_{i=1}^{n} P[X_i > x] \\
&= 1 - [1 - F(x)]^n \quad (X_i's \text{ are } iid)
\end{aligned}$$

Hence

$$f_{X_{(1)}}(x) = n[1 - F(x)]^{n-1}f(x); \; x \in R$$

Theorem 8 *The joint pdf of* $(X_{(1)}, X_{(2)}, \ldots, X_{(n)})$ *is given by*

$$g(X_{(1)}, X_{(2)}, \ldots X_{(n)}) = \begin{cases} n! \prod_i^n f(x_{(i)}) & ; \; X_{(1)} < X_{(2)} < \cdots < X_{(n)} \\ 0 & ; \; \text{otherwise} \end{cases} \tag{25}$$

Corollary 3 *The marginal pdf of* $X_{(r)}$ *from (25) is given by (21).*

Proof

$$g_{X_{(r)}}(x_r) = n!f(x_r) \int\limits_{-\infty}^{x_r} \int\limits_{-\infty}^{x_{r-1}} \cdots \int\limits_{-\infty}^{x_2} \int\limits_{x_r}^{\infty} \int\limits_{x_{r+1}}^{\infty} \cdots \int\limits_{x_{n-1}}^{\infty}$$

$$\times \prod_{i \neq r}^{n} f(x_{(i)}) dx_1 dx_2 \ldots dx_{(r-1)} dx_{(r+1)} \ldots dx_n \tag{26}$$

Consider

$$\int\limits_{x_{n-1}}^{\infty} f(x_n) dx_n = 1 - F(x_{n-1})$$

$$\int\limits_{x_{n-2}}^{\infty} f(x_{n-1})[1 - F(x_{n-1})] dx_{n-1} = \frac{1}{2}[1 - F(x_{n-2})]^2$$

Hence

$$\int\limits_{x_r}^{\infty} \int\limits_{x_{r+1}}^{\infty} \cdots \int\limits_{x_{n-1}}^{\infty} f(x_n)f(x_{n-1}) \cdots f(x_{r+1}) dx_n dx_{n-1} \cdots dx_{r+1}$$

$$= \frac{1}{(n-r)!}[1 - F(x_r)]^{n-r} \tag{27}$$

Next,

$$\int\limits_{-\infty}^{x_2} f(x_1) dx_1 = F(x_2)$$

$$\int\limits_{-\infty}^{x_3} F(x_2)f(x_2) dx_2 = \frac{F^2(x_3)}{2}$$

$$\int\limits_{-\infty}^{x_r} f(x_{r-1})F^{r-2}(x_{r-1}) dx_{r-1} = \frac{F^{r-1}(x_{r-1})}{(r-1)!} \tag{28}$$

Hence, from (25), (26), (27), and (28), we get the distribution of X_r as given in (21).

Theorem 9 *The joint distribution of X_r and X_s, $(r < s)$, is given as*

$$g_{rs}x_r, x_s) = \begin{cases} \frac{n!}{(r-1)!(s-r-1)!(n-s)!} F^{r-1}(x_r)[F(x_s) - F(x_r)]^{s-r-1}[1 - F(x_s)]^{n-s} f(x_r) f(x_s); & x_r < x_s \\ 0; & \text{otherwise} \end{cases}$$

(29)

Proof Note that

$$-\infty < x_1 < x_2 \cdots x_{r-1} < x_r < x_{r+1}, \cdots < x_{s-2} < x_{s-1} < x_s < x_{s+1} \ldots < x_n < \infty$$

Consider three parts
(a) Omitting x_r,

$$-\infty < x_1 < x_2, -\infty < x_2 < x_3, -\infty < x_3 < x_4, \cdots, -\infty < x_{r-2} < x_{r-1},$$

$$- < \infty < x_{r-1} < x_r$$

(30)

(b) Omitting x_s,

$$x_r < x_{r+1} < x_s, x_{r+1} < x_{r+2} < x_s, x_{r+2} < x_{r+3} < x_s, \ldots, x_{s-3} < x_{s-2} < x_s,$$

$$x_{s-2} < x_{s-1} < x_s$$

(31)

(c)

$$x_s < x_{s+1} < \infty, x_{s+1} < x_{s+2} < \infty, x_{s+2} < x_{s+3} < \infty, \ldots, x_{n-2} < x_{n-1} < \infty,$$

$$x_{n-1} < x_n < \infty$$

(32)

Consider (a) omitting x_r

$$\int_{-\infty}^{x_2} f(x_1) dx_1 = F(x_2), \quad \int_{-\infty}^{x_3} F(x_2) f(x_2) dx_2 = \frac{F^2(x_3)}{2}$$

$$\int_{-\infty}^{x_4} \frac{F^2(x_3)}{2} f(x_3) dx_3 = \frac{F^3(x_4)}{3!}$$

In general,

$$\int_{-\infty}^{x_r} \frac{F^{r-2}(x_{r-1})}{(r-2)!} f(x_{r-1})dx_{r-1} = \frac{F^{r-1}(x_r)}{(r-1)!} \tag{33}$$

Consider (b) omitting x_s

$$\int_{x_{s-2}}^{x_s} f(x_{s-1})dx_{s-1} = F(x_s) - F(x_{s-2})$$

$$\int_{x_{s-3}}^{x_s} [F(x_s) - F(x_{s-2})]f(x_{s-2})dx_{s-2} = \frac{1}{2}[F(x_s) - F(x_{s-3})]^2,$$

$$\int_{x_{s-4}}^{x_s} \frac{[F(x_s) - F(x_{s-3})]^2}{2} f(x_{s-3})dx_{s-3} = \frac{[F(x_s) - F(x_{s-3})]^3}{3!},$$

$$\int_{x_{r+1}}^{x_s} \frac{[F(x_s) - F(x_{r+2})]^{s-r-3}}{(s-r-3)!} f(x_{r+2})dx_{r+2} = \frac{[F(x_s) - F(x_{r+1})]^{s-r-2}}{(s-r-2)!},$$

$$\int_{x_r}^{x_s} \frac{[F(x_s) - F(x_{r+1})]^{s-r-2}}{(s-r-2)!} f(x_{r+1})dx_{r+1} = \frac{[F(x_s) - F(x_r)]^{s-r-1}}{(s-r-1)!},$$

Consider (c)

$$\int_{x_{n-1}}^{\infty} f(x_n)dx_n = 1 - F(x_{n-1})$$

$$\int_{x_{n-2}}^{\infty} [1 - F(x_{n-1})f(x_{n-1})dx_{n-1} = \frac{[1 - F(x_{n-2})]^2}{2}$$

$$\int_{x_{n-3}}^{\infty} \frac{[1 - F(x_{n-2})]^2}{2} f(x_{n-2})dx_{n-2} = \frac{[1 - F(x_{n-2})]^3}{3!}$$

$$\int\limits_{x_{s+1}}^{\infty} \frac{[1 - F(x_{s+2})]^{n-s-2}}{(n-s-2)!} f(x_{s+2})dx_{s+2} = \frac{[1 - F(x_{s+1})]^{n-s-1}}{(n-s-1)!}$$

$$\int\limits_{x_s}^{\infty} \frac{[1 - F(x_{s+1})]^{n-s-1}}{(n-s-1)!} f(x_{s+1})dx_{s+1} = \frac{[1 - F(x_s)]^{n-s}}{(n-s)!} \tag{34}$$

Hence from (a)

$$\int\limits_{-\infty}^{x_2} \int\limits_{-\infty}^{x_3} \int\limits_{-\infty}^{x_4} \cdots \int\limits_{-\infty}^{x_r} f(x_1)f(x_2)\cdots f(x_{r-1})dx_1 dx_2 \cdots dx_{r-1} = \frac{F^{r-1}(x_r)}{(r-1)!} \tag{35}$$

From (b)

$$\int\limits_{x_{s-2}}^{x_s} \int\limits_{x_{s-3}}^{x_s} \cdots \int\limits_{x_{r+1}}^{x_s} \int\limits_{x_r}^{x_s} f(x_{s-1})f(x_{s-2})\cdots f(x_{r+2})f(x_{r+1})dx_{s-1}dx_{s-2}\cdots dx_{r+2}dx_{r+1}$$

$$= \frac{[F(x_s) - F(x_r)]^{s-r-1}}{(s-r-1)!} \tag{36}$$

From (c)

$$\int\limits_{x_{n-1}}^{\infty} \int\limits_{x_{n-2}}^{\infty} \cdots \int\limits_{x_{s+1}}^{\infty} \int\limits_{x_s}^{\infty} f(x_n)f(x_{n-1})f(x_{n-2})\cdots f(x_{s+1})dx_n dx_{n-1}\cdots dx_{s+1}$$

$$= \frac{[1 - F(x_s)]^{n-s}}{(n-s)!} \tag{37}$$

Therefore

$$= f(x_r)f(x_s) \int\limits_{-\infty}^{x_2} \int\limits_{-\infty}^{x_3} \int\limits_{-\infty}^{x_r} \int\limits_{x_r}^{x_s} \int\limits_{x_{r+1}}^{x_s} \int\limits_{x_{r+2}}^{x_s} \int\limits_{x_s}^{\infty} \int\limits_{x_{s+1}}^{\infty} \cdots \int\limits_{x_{n-2}}^{\infty} \int\limits_{x_{n-1}}^{\infty}$$

$$\times n! \prod_{i \neq r,s}^{n} f(x_i)dx_1 dx_2 \cdots dx_{r-1}dx_{r+1}dx_{r+2}\cdots dx_{s-1}dx_{s+1}\cdots dx_n \tag{38}$$

By using (35), (36), and (37), the joint distribution of Y_r and X_s $(r < s)$, is given as

$$c_{rs}f(x_r)F^{r-1}(x_r)[F(x_s) - F(x_r)]^{s-r-1}[1 - F(x_s)]^{n-s}f(x_s), \tag{39}$$

where $c_{rs} = \frac{n!}{(r-1)!(s-r-1)!(n-s)!}$

Distribution of Range

Let $W_{rs} = X_s - X_r$

$$f(w_{rs}) = c_{rs} \int\limits_{-\infty}^{\infty} f(x)F^{r-1}(x)[F(x+w_{rs}) - F(x)]^{s-r-1}[1 - F(x+w_{rs})]^{n-s}f(x+w_{rs})dx,$$

Put $r = 1$, $s = n$, then call $w_{rs} = w$,

$$f(w) = n(n-1) \int\limits_{-\infty}^{\infty} f(x)[F(x+w) - F(x)]^{n-2}f(x+w)dx \tag{40}$$

The cdf of w is more simpler

$$F(w) = n \int\limits_{-\infty}^{\infty} f(x) \int\limits_{0}^{w} (n-1)[F(x+w) - F(x)]^{n-2}dwdx$$

$$= n \int\limits_{-\infty}^{\infty} f(x)[F(x+w) - F(x)]^{n-1}dx$$

Example 27 Let $X_1, X_2, \ldots, X_n$ are $\cup(0,1)$

(i) Distribution of rth order statistic.

$$f_r(x) = \frac{x^{r-1}(1-x)^{n-r}}{\beta(r, n-r+1)}; \ 0 < x < 1$$

(ii) Joint distribution of rth and sth order statistics.

$$f_{rs}(x,y) = \begin{cases} c_{rs}x^{r-1}(y-x)^{s-r-1}(1-y)^{n-s} & ; \ 0 \leq x \leq y \leq 1 \\ 0 & ; \ \text{otherwise} \end{cases}$$

(iii) Distribution of W_{rs}.

$$f_{W_{rs}}(w) = \frac{w^{s-r-1}(1-w)^{n-s+r}}{\beta(s-r, n-s+r+1)}; \ 0 \leq w < 1$$

(iv) Distribution of range ($s = n$, $r = 1$ in (iii)).

$$f(w) = \frac{w^{n-2}(1-w)}{\beta(n-1, 2)}; \ 0 < w < 1$$

Chapter 1
Sufficiency and Completeness

1.1 Introduction

Suppose that a random variable (rv) X is known to have a Gamma distribution $G(p, \sigma)$ but we do not know one of the parameters, say, σ. Suppose further that a sample $X_1, X_2, \ldots, X_n$ is taken on X. The problem of point estimation is to pick a (one dimensional) Statistic $T(X_1, X_2, \ldots, X_n)$ that best estimates the parameter σ. The numerical value of T when the realization is $x_1, x_2, \ldots, x_n$, is known as estimate of σ. From the Wikipedia, we quote the difference between the estimator and the estimate. In statistics, an estimator is a rule for calculating an estimate of a given quantity based on observed data. Hence rule is an estimator and result is the estimate. Here, we will use the same word "estimate" for both the function of T and its numerical value. If both p and σ are unknown, we find a joint Statistic $T = (W_1, W_2)$ as an estimate of (p, σ).

Let X be a random variable with a distribution function (df) F which depends on a set of parameters. Suppose further that the functional form of F is known except perhaps for a finite number of these parameters. Let θ be the vector of (unknown) parameters associated with F.

Definition 1.1.1 The set of all admissible values of the parameters of a distribution function F is called the parameter space.

Let $F(x, \theta) = $ df of X if θ is the vector of parameter associated with the df of X. Denote the parameter set as Θ. Then $\theta \in \Theta$. Hence the set $\{F(x, \theta) : \theta \in \Theta, x \in \Re\}$ is called the family of df of X.

Example 1.1.1 Let the rv X have Poisson distribution with parameter λ denoted as $P(\lambda)$ where λ is unknown. Then $\Theta = \{\lambda : \lambda > 0\}$ and $\{P(\lambda) : \lambda > 0\}$ is the family of probability mass function's (pmf) of X.

Example 1.1.2 Let the rv X have a binomial distribution with parameters n and p. Then it is denoted as $X \sim B(n, p)$. Note that only p is unknown.

© Springer Science+Business Media Singapore 2016
U.J. Dixit, *Examples in Parametric Inference with R*,
DOI 10.1007/978-981-10-0889-4_1

Hence, $\Theta = \{p : 0 < p < 1\}$ and $\{B(n, p), 0 < p < 1\}$ is the family of pmf's of X.

Example 1.1.3 Let the rv X have a normal distribution with mean μ and variance σ^2. It is denoted as $X \sim N(\mu, \sigma^2)$. Assume that μ and σ are unknown. Then $\Theta = \{(\mu, \sigma^2) : -\infty < \mu < \infty, \sigma > 0\}$.
If $\mu = \mu_0$ and σ is unknown,

$$\Theta = \{(\mu_0, \sigma^2)\} : \sigma > 0\}$$

If $\sigma = \sigma_0$ and μ is unknown

$$\Theta = \{(\mu, \sigma_0^2) : -\infty < \mu < \infty\}$$

Definition 1.1.2 Let $X_1, X_2, \ldots, X_n$ be iid rvs from $F(x, \theta)$, where $\theta = (\theta_1, \theta_2, \ldots, \theta_n)$ is the vector of unknown parameters. Also $\theta \in \Theta \subseteq \Re_n$, where $\Re_n$ is the set of real numbers. A Statistic $T(X_1, X_2, \ldots X_n)$ is said to be a point estimate of θ if T maps $\Re_n$ into Θ, i.e.,

$$T : \Re_n \to \Theta.$$

Example 1.1.4 Let $X_1, X_2, \ldots X_n$ be a sample from $N(\theta, 1)$, where θ is an unknown parameter. Then we get several estimators for θ. Let T be

(1) $T = \bar{X}$, where $\bar{X} = n^{-1} \sum_{i=1}^{n} X_i$ an estimator for θ.

(2) $T = \dfrac{1}{n(n+1)} \sum_{i=1}^{n} i X_i$ is an estimate of θ.

(3) Any X_i is an estimate of θ.

It implies $T = X_i$ $(i = 1, 2, \ldots, n)$.

Example 1.1.5 Let $X_1, X_2, \ldots X_n$ be a sample from $B(1, p)$, where p is unknown. Here also we get several estimators

1. $T = \dfrac{\bar{X}}{n}$

2. $T = \dfrac{1}{3}$

3. $T = \dfrac{X_1}{n}$

4. $T = \dfrac{X_1 + X_2 + X_3}{3n}$

are all estimates of p.

From these examples, it is clear that we need some criterion to choose among all possible estimates.

1.2 Sufficient Statistics

Why do we require sufficient statistics?

The answer is in the meaning of sufficiency. One of the important objectives in the primary stage of statistical analysis is to process the observed data and transform it to a form most suitable for decision-making. The primary data processing generally reduces the dimensionality of the original sets of variables. It is desired, that no information relevant to the decision process will be lost in this primary data reduction. As shown later in the chapter, not all families of distribution function allow such a reduction without losing information. On the other hand, there are families of df for which all set of sample values can give a real-valued statistics. The theory of sufficient statistics enables us to characterize families of df and provides corresponding transformation, which yields sufficient statistics.

A sufficient statistic, for a parameter θ is a statistic that, in a certain sense, captures all the information about θ contained in the sample. Any additional information in the sample, besides the value of sufficient statistics, does not contain any more information about θ. These considerations lead to the data reduction technique known as the sufficient principle.

We start with a heuristic definition of a sufficient statistic. We say T is a sufficient statistic if the statistician who knows the value of T can do just a good job of estimating the unknown parameter θ as the statistician who knows the entire random sample.

Definition 1.2.1 Let $X = (X_1, X_2, \ldots, X_n)$ be a sample from $\{F(x, \theta) : \theta \in \Theta\}$. A statistic $T = T(X)$ is sufficient for θ if and only if the conditional distribution of the sample X, given $T(X) = t$ does not depend on θ.

To motivate the mathematical definition, we consider the following experiment. Let $T(X)$ be a sufficient statistics. There are two statisticians; we will call them A and B. Statistician A knows the entire random sample $X_1, X_2, \ldots, X_n$, but Statistician B only knows the value of T, call it t. Since conditional distribution of $X_1, X_2, \ldots, X_n$ given T does not depend on θ, statistician B knows this conditional distribution. So he can use computer to generate a random sample $X'_1, X'_2, \ldots, X'_n$, which has this conditional distribution. But then his random sample has the same distribution as a random sample drawn from the population (with its unknown value of θ). So statistician B can use his random sample $X'_1, X'_2, \ldots, X'_n$ to compute whatever statistician A computes using his random sample $X_1, X_2, \ldots, X_n$, and he will (on average) do as well as statistician A.

Definition 1.2.2 Let $X = (X_1, X_2, \ldots, X_n)$ be a sample from $\{F(x, \theta) : \theta \in \Theta\}$. A statistic $T = T(X)$ is sufficient for θ if and only if the conditional distribution of $X_i (i = 1, 2, \ldots, n)$, given the value $T(X) = t$ does not depend on θ.

Example 1.2.1 (normal population, unknown mean and known variance) Let the rvs $X_i, i = 1, 2, \ldots, n$ have normal distribution with mean μ and variance σ^2. Now σ^2 is known, say σ_0^2.

Now we say that $T = \sum X_i$ is sufficient for μ.

$$f(x_1, x_2, \ldots x_n, \mu) = \frac{1}{\sigma_0\sqrt{2\pi}} \exp\left[-\frac{1}{2\sigma_0^2}\sum(x_i - \mu)^2\right]. \qquad (1.2.1)$$

Distribution of T is $N(n\mu, n\sigma_0^2)$,

$$f(t) = \frac{1}{\sigma_0\sqrt{2\pi n}} \exp\left[-\frac{1}{2n\sigma_0^2}(t - n\mu)^2\right] \qquad (1.2.2)$$

Using Definition 1.2.1,

$$f(x_1, x_2, \ldots x_n | T = t) = \frac{f(x_1, x_2, \ldots x_n, T = t)}{f(t)} \qquad (1.2.3)$$

Note that $\{X = x\}$ is a subset of $\{T(X) = t\}$

$$= \frac{f(x_1, x_2, \ldots x_n)}{f(t)} \qquad (1.2.4)$$

$$= \frac{(\frac{1}{\sigma_0\sqrt{2\pi}})^n \exp\left[-\frac{1}{2\sigma_0^2}\sum_{i=1}^n (x_i - \mu)^2\right]}{\frac{1}{\sigma_0\sqrt{2\pi n}} \exp\left[-\frac{1}{2n\sigma_0^2}(t - n\mu)^2\right]} \qquad (1.2.5)$$

Consider

$$= \exp\left[-\frac{1}{2\sigma_0^2}\left(\sum_{i=1}^n x_i^2 - 2\mu\sum_{i=1}^n x_i + n\mu^2\right)\right]$$

$$= \exp\left[-\frac{1}{2\sigma_0^2}\sum_{i=1}^n x_i^2 + \frac{2\mu t}{2\sigma_0^2} - \frac{n\mu^2}{2\sigma_0^2}\right]$$

Next

$$\exp\left[-\frac{1}{2n\sigma_0^2}(t - n\mu)^2\right] = \exp\left[-\frac{t^2}{2n\sigma_0^2} + \frac{2n\mu t}{2n\sigma_0^2} - \frac{n^2\mu^2}{2n\sigma_0^2}\right]$$

Then

$$f(x \mid T = t) = c\exp\left[-\frac{1}{2\sigma_0^2}\left(\sum_{i=1}^n x_i^2 - \frac{t^2}{n}\right) + \frac{\mu t}{\sigma_0^2} - \frac{\mu t}{\sigma_0^2} - \frac{n\mu^2}{2\sigma_0^2} + \frac{n\mu^2}{2\sigma_0^2}\right],$$

where c is a constant, $x = (x_1, x_2, \ldots, x_n)$

$$f(x \mid T = t) = \cdots \exp\left[-\frac{1}{2\sigma_0^2}\left(\sum_{i=1}^{n} v_i^2 \quad \frac{t^2}{n^2}\right)\right] \tag{1.2.6}$$

We can see that (1.2.6) does not involve μ.

Therefore, $f(x \mid T = t)$ does not depend on μ. Using Definition 1.2.1, we can claim that T is sufficient for μ.

Example 1.2.2 Let the random variables X_1 and X_2 have Poisson distribution with parameter λ. Find sufficient statistic for λ.

$$f(X_1 = x_1, X_2 = x_2, \mid \lambda) = \frac{e^{-\lambda}\lambda^{x_1}}{x_1!}\frac{e^{-\lambda}\lambda^{x_2}}{x_2!} \tag{1.2.7}$$

Let $T = X_1 + X_2$ and T has $P(2\lambda)$.

$$f(X_1 = x_1, X_2 = x_2 \mid T = t) = \frac{e^{-\lambda}\lambda^{x_1}}{x_1!}\frac{e^{-\lambda}\lambda^{x_2}}{x_2!}\frac{t!}{e^{-2\lambda}(2\lambda)^t}$$

$$= \frac{t!}{x_1!x_2!}\frac{1}{2^t}; \quad x_i \le t \ (i = 1, 2) \tag{1.2.8}$$

Therefore, $f(x \mid T = t)$ does not depend on λ.

Using Definition 1.2.1, we can claim that T is sufficient for λ.

Example 1.2.3 Consider the Example 1.2.1 with $\sigma_0 = 1$. In this example, we can say that $(X \mid T = t)$ is multivariate normal with mean vector $\theta 1_n$ and covariance matrix I_n where $1_n = (1, 1, \ldots, 1)$ and I_n is an identity matrix of order n.

Consider the transformation $Y = HX$, where H is an orthogonal matrix such that $Y_n = \sqrt{n}\bar{x}$,

$$Y_k = \frac{kx_{k+1} - (x_1 + x_2 + \cdots + x_k)}{\sqrt{k(k+1)}}, k = 1, 2, \ldots, n-1 \tag{1.2.9}$$

$$H = \begin{pmatrix} -\frac{1}{\sqrt{2}} & \frac{1}{\sqrt{2}} & 0 & 0 .. 0 & 0 \\ -\frac{1}{\sqrt{6}} & -\frac{1}{\sqrt{6}} & \frac{2}{\sqrt{6}} & 0 .. 0 & 0 \\ \cdot & \cdot & \cdot & \cdots & \cdot \\ \cdot & \cdot & \cdot & \cdots & \cdot \\ -\frac{1}{\sqrt{n(n-1)}} & -\frac{1}{\sqrt{n(n-1)}} & \cdot & \cdots \frac{n-1}{\sqrt{n(n-1)}} & 0 \\ \frac{1}{\sqrt{n}} & \frac{1}{\sqrt{n}} & \cdot & \cdots \frac{1}{\sqrt{n}} & \frac{1}{\sqrt{n}} \end{pmatrix}, \tag{1.2.10}$$

The transformation $H : X \to Y$ is known in a particular form as the Helmert transformation.

$$E(Y) = HE(X) = H\theta 1_n \qquad (1.2.11)$$

$$V(Y) = H'HV(X) = I_n \qquad (1.2.12)$$

$$H1_n = \begin{pmatrix} 0 \\ 0 \\ . \\ . \\ \sqrt{n} \end{pmatrix},$$

Hence $Y_1, Y_2, \ldots Y_{n-1}$ are iid rvs having $N(0, 1)$ distribution independent of Y_n.

Hence, the conditional distribution of $Y_1, Y_2, \ldots, Y_{n-1}$ given $Y_n = \sqrt{n}\bar{x}$ is $N(0, I_{n-1})$. Using Definition 1.2.1, we can claim that Y_n is sufficient for μ.

Example 1.2.4 Consider the Example 1.2.2 with $T = X_1 + 2X_2$.

$$P(X_1 = 1, X_2 = 1|T = 3) = \frac{P(X_1 = 1, X_2 = 1)}{P(X_1 = 1, X_2 = 1) + P(X_1 = 3, X_2 = 0)}$$

$$= \frac{(e^{-\lambda}\lambda)(e^{-\lambda}\lambda)}{e^{-2\lambda}\lambda^2 + \frac{e^{-\lambda}(\lambda)^3}{3!}e^{-\lambda}}$$

$$= \frac{e^{-2\lambda}\lambda^2}{e^{-2\lambda}[\lambda^2 + \frac{\lambda^3}{6}]} = \frac{1}{1 + \frac{\lambda}{6}} \qquad (1.2.13)$$

Equation (1.2.13) depends on λ. Hence we cannot say that T is sufficient for λ.

Note that the distribution of $X_1 + 2X_2$ is not $P(3\lambda)$.

Example 1.2.5 Let $X_1, X_2, \ldots X_m$ be a random sample from $B(n, p)$. Then the distribution of $T = \sum_{i=1}^{m} X_i$ has $B(nm, p)$. Hence we will find the distribution of

1. $f(x_1|T)$
2. $f(x_1, x_2|T)$
3. $f(x_1, x_2, \ldots, x_r|T)(r < m)$

$$\text{(i)} f(X_1 = x_1|T = t) = \frac{f(X_1 = x_1, T = t - x_1)}{P(T = t)}$$

$$f(X_1 = x_1|T = t) = \frac{\binom{n}{x_1} p^{x_1} q^{n-x_1} \binom{n(m-1)}{t-x_1} p^{t-x_1} q^{n(m-1)-t+x_1}}{\binom{nm}{t} p^t q^{nm-t}} \qquad (1.2.14)$$

$$= \frac{\binom{n}{x_1}\binom{n(m-1)}{t-x_1}}{\binom{nm}{t}} \quad ; \ x_1 = 0, 1, \ldots, t \qquad (1.2.15)$$

(ii) As in (i), we can write $f(X_1, X_2 | T = t)$

$$f(x_1, x_2 | t) = \frac{\binom{n}{x_1} \binom{n}{x_2} \binom{n(m-2)}{t-x_1-x_2}}{\binom{nm}{t}} \quad ; \; x_1 + x_2 \leq t \qquad (1.2.16)$$

$$\text{(iii)} \; f(x_1, x_2, \ldots, x_r | T = t) = \frac{\prod_{i=1}^{r} \binom{n}{x_i} \binom{n(m-r)}{t-\sum_{i=1}^{r} x_i}}{\binom{nm}{t}} \quad ; \sum_{i=1}^{r} x_i \leq t \quad (1.2.17)$$

We can see that (1.2.15), (1.2.16) and (1.2.17) are free from the parameter p. Hence by Definition 1.2.2 we can conclude that, T is sufficient for p.

Example 1.2.6 Let $X_1, X_2, \ldots, X_n$ be iid rvs from an exponential distribution with mean θ. Then the distribution of $T = \sum_{i=1}^{n} x_i$ is gamma with mean $n\theta$, where n is a shape parameter. The distribution of T is gamma and denoted as $G(n, \frac{1}{\theta})$. Now we will find the distribution of

1. $f(X_1 | T)$
2. $f(X_1, X_2 | T)$
3. $f(X_1, X_2, \ldots X_n | T)$.

(i) Let

$$f(x_i) = \frac{1}{\theta} e^{-\frac{x_i}{\theta}} \quad ; x_i > 0, \; \theta > 0, \; i = 1, 2, \ldots, n$$

$$f(t) = \frac{e^{\frac{-t}{\theta}} t^{n-1}}{\theta^n \Gamma(n)} \quad ; \; t > 0, \; \theta > 0$$

$$f(x_1 | t) = \frac{f(x_1) f(t_2)}{f(t)} \quad ; \text{ where } T_2 = \sum_{i=2}^{n} x_i$$

$$= \frac{(n-1)(t-x_1)^{n-2}}{t^{n-1}} \quad ; 0 < x_1 < t \qquad (1.2.18)$$

(ii) $f(x_1, x_2 | t) = \dfrac{(n-1)(n-2)(t-x_1-x_2)^{n-3}}{t^{n-1}} \quad ; \; 0 < x_1 + x_2 < t \; (1.2.19)$

(iii) $f(x_1, x_2, \ldots x_n | t) = \dfrac{\Gamma(n)}{t^{n-1}} \quad ; \; \sum_{i=1}^{n} x_i < t \qquad (1.2.20)$

Using Definitions 1.2.1 and 1.2.2, we can conclude that T is sufficient.

Example 1.2.7 Let $X_1, X_2, \ldots, X_n$ be iid random sample from $U(0, \theta)$. Then we will find the distribution of

1. $f(x_i|T)$
2. $f(x_i, x_j|T = t)$ $i \neq j$
3. $f(x_1, x_2, \ldots x_n|T = t)$, where $T = \text{Max}(X_1, X_2, \ldots X_n)$

$$(i)\ f(x_i|t) = \frac{1}{t}\ \ ;\ 0 < x_i < t,\ i = 1, 2, \ldots, n \tag{1.2.21}$$

$$(ii)\ f(x_i, x_j|t) = \frac{1}{t^2}\ \ ;\ 0 < x_i < t,\ 0 < x_j < t,\ i \neq j \tag{1.2.22}$$

$$(iii)\ f(x_1, x_2, \ldots x_n|t) = \frac{1}{nt^{n-1}}\ \ ;\ 0 < x_i < t,\ i = 1, 2, \ldots, n \tag{1.2.23}$$

In all the above examples, T is sufficient for θ.

Lemma 1.2.1 *Let $X_1, X_2, \ldots X_n$ be iid discrete rvs with df $F(x)$. Suppose $X_{(m)} = Max(X_1, X_2, \ldots X_n)$*

$$h(t) = P[X_{(m)} = t] = H(t) - H(t - 1). \tag{1.2.24}$$

where $H(t) = [F(t)]^m$

Proof

$$H(t) = P[X_{(m)} \leq t]$$
$$= P[\text{All } X_i's \leq t]$$
$$= \prod_{i=1}^{m} P[X_i's \leq t] = [F(t)]^m$$
$$h(t) = P[X_{(m)} = t] = [F(t)]^m - [F(t - 1)]^m$$
$$= [H(t)] - [H(t - 1)]$$

Lemma 1.2.2

$$P[X_i = r|X_{(m)}] = \begin{cases} P(X_i = r)\frac{h_1(t)}{h(t)} & r = 0, 1, 2, \ldots t - 1, \\ P(X_i = r)\frac{h_2(t)}{h(t)} & r = t, \end{cases} \tag{1.2.25}$$

where

$$h_1(t) = F(t)^{m-1} - F(t - 1)^{m-1}$$

$$h_2(t) = F(t)^{m-1}$$

Proof Let

$$T_{max} = Max(X_1, X_2, \ldots, X_{i-1}, X_{i+1}, \ldots X_m)$$

$$P[X_i = r/X_{(m)}] = \begin{cases} \frac{P(X_i = r, T_{max} = t)}{P(X_{(m)} = t)} & r = 0, 1, 2, \ldots t - 1, \\ \frac{P(X_i = r, X_j \le t)}{P(X_{(m)} = t)} & r = t, i \ne j = 1, 2, \ldots, n \end{cases} \tag{1.2.26}$$

$$= \begin{cases} P(X_i = r)\frac{h_1(t)}{h(t)} & r = 0, 1, 2, \ldots t - 1 \\ P(X_i = r)\frac{h_2(t)}{h(t)} & r = t. \end{cases} \tag{1.2.27}$$

One can see further details from Dixit and Kelkar (2011).

Note 1: Let $X_1, X_2, \ldots, X_m$ be iid Binomial or Poisson rvs with df $F(x)$ and $G(x)$ respectively. Then $P[X_i = r | X_{(m)}]$ depends on parameters of Binomial or Poisson distribution. Hence $X_{(m)}$ is not sufficient for the parameters of Binomial or Poisson distribution.

Note 2: In Example 1.2.7, we have shown that $X_{(n)} = Max(X_1, X_2, \ldots, X_n)$ is sufficient for θ. Moreover, we can show that order statistics are sufficient statistics but we should not expect more with so little information about the density f. However, even if we do specify more about the density we still may not be able to get much of sufficiency reduction. For example, suppose that f is a Cauchy pdf

$$f(x \mid \theta) = \frac{1}{\pi[1 + (x - \theta)^2]} \; ; x \in \Re, \theta \in \Re.$$

the logistic pdf

$$f(x \mid \theta) = \frac{\exp[-(x - \theta]}{[1 + \exp(x - \theta)]^2} \; ; x \in \Re, \theta \in \Re.$$

For details see Lehman and Casella (1998) or Casella and Berger (2002).

Note 3: Definitions 1.2.1 and 1.2.2 are not very useful because we have to guess statistic T and check to see whether T is sufficient. Moreover, the procedure for checking that T is sufficient is quite time consuming. Hence, using the next theorem due to Halmos and Savage (1949), we can find a sufficient statistics by simple inspection of the pdf or pmf of the sample.

Theorem 1.2.1 (Factorization Theorem) *Let $X_1, X_2, \ldots, X_n$ be a random sample with joint density $f(x_1, x_2, \ldots, x_n | \theta)$. A statistics $T(X)$ is sufficient if and only if the joint density can be factored as follows:*

$$f(x_1, x_2, \ldots, x_n \mid \theta) = h(x_1, x_2, \ldots, x_n)g[T(X) \mid \theta] \tag{1.2.28}$$

We give proof only for the discrete distributions.

Proof Assume $T(X)$ is sufficient. Using definition, we can use

$$h(x_1, x_2, \ldots, x_n) = f(x_1, x_2, \ldots, x_n \mid T = t) \tag{1.2.29}$$

because $h(x)$ does not depend on θ.

$$
\begin{aligned}
f(x_1, x_2, \ldots, x_n \mid \theta) &= f(x_1, x_2, \ldots, x_n \text{ and } T = t) \\
&= f(x_1, x_2, \ldots, x_n \mid T = t)g(t \mid \theta) \\
&= h(x_1, x_2, \ldots, x_n)g(t \mid \theta)
\end{aligned}
$$

Now assume that factorization (1.2.28) holds.

$$
\begin{aligned}
f[T = t_0 \mid \theta] &= \sum_{T=t_0} f(X = x \mid \theta) \\
&= \sum_{T=t_0} g(t|\theta)h(x_1, x_2, \ldots, x_n) \\
&= g(t_0|\theta) \sum_{T=t_0} h(x_1, x_2, \ldots, x_n)
\end{aligned}
$$

Suppose that $f[T = t_0|\theta] > 0$ for some $\theta > 0$,

$$
\begin{aligned}
f(X = x|T = t_0) &= \frac{f[X = x, T = t_0]}{f(T = t_0)} = \frac{f(X = x)}{f(T = t_0)} \\
&= \frac{g(t_0|\theta)h(x_1, x_2, \ldots, x_n)}{g(t_0|\theta) \sum_{T=t_0} h(x_1, x_2, \ldots, x_n)} \\
&= \frac{h(x_1, x_2, \ldots, x_n)}{\sum_{T=t_0} h(x_1, x_2, \ldots, x_n)} \\
&= \frac{h(x)}{\sum_{T=t_0} h(x)}
\end{aligned}
$$

Since the ratio does not depend on θ. By Definition 1.2.1, $T(X)$ is sufficient for θ.

Example 1.2.8 For normal model described in Example 1.2.1 consider

$$\sum(x_i - \mu)^2 = \sum(x_i - \bar{x} + \bar{x} - \mu)^2$$

$$= \sum(x_i - \bar{x})^2 + n(\bar{x} - \mu)^2$$

$$f(x|\mu) = (2\pi\sigma_0^2)^{\frac{-n}{2}} \exp\left[\frac{-1}{2\sigma_0^2}\left\{\sum(x_i - \bar{x})^2 + n(\bar{x} - \mu)^2\right\}\right]$$

$$f(x|\mu) = (2\pi\sigma_0^2)^{\frac{-n}{2}} \exp\left[\frac{-\sum(x_i - \bar{x})^2}{2\sigma_0^2}\right] \exp\left[-\frac{n(\bar{x} - \mu)^2}{2\sigma_0^2}\right] \quad (1.2.30)$$

Using factorization theorem,

$$h(x) = (2\pi\sigma_0^2)^{\frac{-n}{2}} \exp\left[\frac{-\sum(x_i - \bar{x})^2}{2\sigma_0^2}\right] \quad (\sigma_0 \text{ is given})$$

which does not depend on the unknown parameter μ. The factor contained in (1.2.30) is a function of T and μ. In this case $T = \bar{x}$. Therefore,

$$g(t|\mu) = \exp\left[\frac{-n(\bar{x} - \mu)^2}{2\sigma_0^2}\right].$$

Thus, By factorization theorem, $T(X) = \bar{X}$ is sufficient for μ.

Example 1.2.9 Consider a Binomial distribution described in Example 1.2.5

$$f(X|p) = \prod_{i=1}^{m} \binom{n}{x_i} p^{\sum_{i=1}^{m} x_i} q^{mn - \sum_{i=1}^{m} x_i} \quad ; \quad q = 1 - p$$

$$= p^t q^{mn-t} \prod_{i=1}^{m} \binom{n}{x_i}$$

$$= g(t|p)h(x)$$

where $T = \sum_{i=1}^{m} x_i$.

Using factorization theorem, we can say that T is sufficient for p.

Example 1.2.10 Let $X_1, X_2, \ldots, X_n$ be independent identically distributed as discrete uniform random variables.

$$f[X_i = x_i] = \begin{cases} \frac{1}{N} & ; \ x_i = 1, 2, \ldots N, \ i = 1, 2, \ldots, n \\ 0 & ; \text{otherwise} \end{cases}$$

$$= \frac{1}{N^n} \quad ; \quad x_i = 1, 2, \ldots, N$$

$$= \frac{1}{N^n} \prod_{i=1}^{n} I_A(x_i) I_B(x_i),$$

Let $X_{(1)} = \min X_i$ and $X_{(n)} = \max X_i$ where $A = \{1 \leq X_{(1)} \leq N\}$ and $B = \{X_{(n)} \leq N \leq \infty\}$ where

$$I_A(x_i) = \begin{cases} 1 \; ; \; x_i \in A \\ 0 \; ; \; \text{otherwise} \end{cases}$$

$$I_B(x_i) = \begin{cases} 1 \; ; \; x_i \in B \\ 0 \; ; \; \text{otherwise} \end{cases}$$

$$f(X|N) = \prod_{i=1}^{n} I_A(x_i) N^{-n} \prod_{i=1}^{n} I_B(x_i),$$

$$= h(x) g(t|N)$$

where $T = X_{(n)} = \max X_i$. Hence $X_{(n)}$ is sufficient for N.

Example 1.2.11 Let $X_1, X_2, \ldots, X_n$ be a sample from the following pdf

$$f(x|\theta) = \begin{cases} \frac{1}{\theta} \; ; \; -\frac{\theta}{2} < x < \frac{\theta}{2} \\ 0 \; ; \; \text{otherwise} \end{cases}$$

The joint pdf of $X_1, X_2, \ldots, X_n$ is given by

$$f(x|\theta) = \theta^{-n} I_A(x)$$

where

$$A = \left\{ (x) : -\frac{\theta}{2} \leq X_{(1)} \leq X_{(n)} \leq \frac{\theta}{2} \right\}$$

$$= h(x) g(t|\theta)$$

where $h(x) = 1$ and $g(t|\theta) = \theta^{-n} T_A(x)$.

By factorization theorem $(X_{(1)}, X_{(n)})$ are sufficient for θ.

Example 1.2.12 Consider the Example 1.2.1, where σ is unknown.

$$f(X|\mu, \sigma) = (2\pi\sigma^2)^{-\frac{n}{2}} \exp\left[-\frac{1}{2\sigma^2} \sum (x_i - \mu)^2 \right]$$

$$= (2\pi\sigma^2)^{-\frac{n}{2}} \exp\left[-\frac{1}{2\sigma^2} \left(\sum x_i^2 - 2\mu \sum x_i + n\mu^2 \right) \right]$$

$$= (2\pi\sigma^2)^{-\frac{n}{2}} \exp\left[-\frac{\sum x_i^2}{2\sigma^2} + \frac{\mu \sum x_i}{\sigma^2} - \frac{n\mu^2}{2\sigma^2} \right]$$

In this case $h(x) = 1$ and $T(x) = \left(\sum x_i, \sum x_i^2 \right)$ is jointly sufficient for (μ, σ^2).

Note 1: An equivalent sufficient statistic that is frequently used is $T(x) = (\bar{X}, S^2)$ where $S^2 = \sum (X_i - \bar{X})^2$.

Note 2: $\bar{X}$ is not sufficient for μ if σ^2 is unknown and S^2 is not sufficient for σ^2 if μ is unknown.

Note 3: If, however, σ^2 is known, $\bar{X}$ is sufficient for μ. If $\mu = \mu_0$ is known then $\sum(x_i - \mu_0)^2$ is sufficient for σ^2.

Note 4: Theorem 1.2.1 holds if θ is a vector of parameters and T is a random vector, and we say that T is jointly sufficient for θ. Further even if θ is a scalar, T, may be a vector, still sufficiency holds (see Example 1.2.11).

Remark: If T is sufficient for θ, any one to one function of T is also sufficient. In Example 1.2.1, $\bar{X}$ is sufficient for μ but $\bar{X}^2$ is not sufficient for μ^2. Further $\bar{X}$ is sufficient for μ^2 or any other function of μ. Theorem 1.2.1 can not be used to show that a given statistic T is not sufficient. Mostly, we can use Definition 1.2.1 or 1.2.2.

Stigler (1972) had given some theorems to show that the given statistics is not sufficient.

1.3 Minimal Sufficiency

We will start with an example so that our idea on Minimal sufficiency will be clear.

Example 1.3.1 (Zacks 1971) Consider the case of n independent Bernoulli experiments. Let

$$X_i = \begin{cases} 1 & \text{if ith experiment is successful} \\ 0 & \text{; otherwise} \end{cases}$$

The sample space Ω is a discrete one and contains the 2^n points $(X_1, X_2, \ldots, X_n)$. The joint density of $(X_1, X_2, \ldots, X_n)$ is

$$f(X_1, X_2, \ldots, X_n; p) = \begin{cases} p^{t_n}(1-p)^{n-t_n} & X_i = 0, 1, t_n = \sum_{i=1}^{n} X_i, \ 0 < p < 1 \\ 0 & \text{; otherwise} \end{cases}$$

(i) According to factorization theorem, the sample point $S_1 = (X_1, X_2, \ldots, X_n)$ is a sufficient statistic for p.

(ii) According to same theorem $S_2 = (X_1 + X_2, X_3, \ldots, X_n)$, $S_3 = (X_1 + X_2 + X_3, X_4, \ldots, X_n)$, $\ldots S_{n-1} = (X_1 + X_2 + \cdots + X_{n-1}, X_n)$ and $S_n = \sum_{i=1}^{n} X_i$ are all sufficient for p.

The Statistic S_1 partitions Ω into 2^n disjoint exhaustive sets, each of which contains one of the original points $(X_1, X_2, \ldots X_n)$.

The Statistic S_2 partitions Ω into $3 \times 2^{n-2}$ sets; The number 3 we can get from $X_1 + X_2 = 0$, $X_1 + X_2 = 1$, $X_1 + X_2 = 2$, i.e., we get three points of the form $\{(0, 0, x_3, x_4, \ldots, x_n)\}$, $\{(0, 1, x_3, x_4, \ldots x_n), (1, 0, x_3, x_4, \ldots x_n)\}$ and $\{(1, 1, x_3, \ldots x_n)\}$.

We remark that S_2 is a function of S_1 say $T(S_1)$ which adds the first components of S_1 and leaves the other components unchanged. We further notice that each set of the partitions induced by S_1 is included in one of the sets of the partitions induced by S_2 but not vice versa.

In a similar manner $S_3 = T(S_2)$ partitions Ω into $4 \times 2^{n-3}$ sets. The number 4 we can get from $X_1 + X_2 + X_3 = 0$, $X_1 + X_2 + X_3 = 1$, $X_1 + X_2 + X_3 = 2$, $X_1 + X_2 + X_3 = 3$.

The partitions associated with S_i is $(i+1)2^{n-i}$ and S_j is $(j+1)2^{n-j}$ for $(i < j)$. For every $n \geq 2$, we have $(j+1)2^{n-j} < (i+1)2^{n-i}$. Thus we have the following:

Statistics	Number of sets
S_1	2^n
S_2	$3 \times 2^{n-2}$
S_3	$4 \times 2^{n-3}$
.	.
.	.
S_i	$(i+1) \times 2^{n-i}$
S_j	$(j+1) \times 2^{n-j}$
.	.
S_n	$n+1$

We can say that S_n contains minimum number of sets. S_n is a function of S_{n-1} and S_{n-1} is a function of S_{n-2}, etc. Therefore, the statistics S_n is considered minimal sufficient statistic in the sense that the number of sets it forms is minimal and it is a function of any other sufficient statistics.

Definition 1.3.1 $T(X)$ is a minimal sufficient statistic for $\theta \in \Theta$ if

1. $T(X)$ is a sufficient statistic
2. $T(X)$ is a function of any other sufficient statistic.

Sampson and Spencer (1976) had extensively discussed the technique that in a particular statistical model a given statistic is not sufficient or that a given sufficient statistic is not minimal.

Lemma 1.3.1 *Let $K(\lambda)$ and $Q(\lambda)$ be the functions (possibly vector valued) defined on the same domain D. A necessary and sufficient condition for Q to be a function of K, i.e., for any $\lambda_1, \lambda_2 \in D$ satisfying $K(\lambda_1) = K(\lambda_2)$, it follows that $Q(\lambda_1) = Q(\lambda_2)$.*

Definition 1.3.2 The points $x, y \in \Omega$ are said to be likelihood equivalent if there exists $h(x, y) > 0$ such that for all $\theta \in \Theta$, $f(x|\theta) = h(x, y)f(y|\theta)$.

Lemma 1.3.2 *In order that $x, y \in \Omega$ be likelihood equivalent, it is necessary and sufficient that for all $\theta_1, \theta_2 \in \Theta$, $f(x|\theta_1)f(y|\theta_2) = f(x|\theta_2)f(y|\theta_1)$*

Lemma 1.3.3 *Let $T(X)$ be any sufficient statistic for θ. If $x, y \in \Omega$ are any points such that $T(x) = T(y)$, then x and y are likelihood equivalent.*

Theorem 1.3.1 *Suppose $T(X)$ is sufficient statistic for θ. $T(X)$ is minimal sufficient if for any x, y that are likelihood equivalent, it follows that $T(x) = T(y)$*

Proof Let $S(X)$ be a sufficient statistic for θ. Suppose that $x, y \in \Omega$ satisfying the condition $S(x) = S(y)$. Now by Lemma 1.3.3, it follows that X and Y are likelihood equivalent. Given that $T(x) = T(y)$, using Lemma 1.3.1, we get $T(X)$ is a function of $S(X)$. By Definition 1.3.1, we say that $T(X)$ is a minimal sufficient.

Theorem 1.3.2 *Let $T(X)$ be any statistic, if there exists some θ_1, $\theta_2 \in \Theta$ and $x, y \in \Omega$, such that*

1. $T(x) = T(y)$
2. $f(x|\theta_1)f(y|\theta_2) \neq f(x|\theta_2)f(y|\theta_1)$

Then $T(X)$ is not a sufficient statistic.

Proof Let $T(X)$ be a sufficient for θ.
For any $x, y \in \Omega$, we have from 1 $T(x) = T(y)$. From Lemma 1.3.3, x and y are likelihood equivalent. From Lemma 1.3.2,

$$f(x|\theta_1)f(y|\theta_2) = f(x|\theta_2)f(y|\theta_1),$$

which is a contradiction to the condition 2. Therefore, $T(X)$ is not a sufficient statistic.

Theorem 1.3.3 *Let $S(X)$ be a statistic. Suppose $T(X)$ is a minimal sufficient statistics. If there exists $x, y \in \Omega$ such that $S(x) = S(y)$ and $T(x) \neq T(y)$ then $S(X)$ is not a sufficient statistic.*

Proof Let, if possible, $S(X)$ is a sufficient statistic. For any $x, y \in \Omega$, we have $S(x) = S(y)$. Now $T(X)$ is a minimal sufficient statistic, we have, by Definition 1.3.1, $T(x) = T(y)$, which is a contradiction.
  Therefore, $S(X)$ is not a sufficient statistic.

Theorem 1.3.4 *Let $T(X)$ and $S(X)$ be sufficient statistics. If there exists $x, y \in \Omega$, such that*

1. $T(x) \neq T(y)$
2. $S(x) = S(y)$

then $S(X)$ is not a minimal sufficient statistic.

Proof Let, if possible, $S(X)$ is a minimal sufficient statistics. For $x, y \in \Omega$, we have $S(x) = S(y)$. Since $S(X)$ is a minimal sufficient statistic, therefore by Definition 1.3.1, it is a function of $T(X)$, $T(X)$ is a sufficient statistic.
  By Lemma 1.3.1 $T(x) = T(y)$, which is a contradiction to 1. Therefore, $S(X)$ is not a minimal sufficient statistic.

Example 1.3.2 Consider the normal model described in (1.2.1), where μ and σ^2 are both unknown.

Suppose X is likelihood equivalent to Y, i.e., there exists $h(x, y) > 0$, such that $\forall \theta$,

$$\frac{f(x|\mu, \sigma^2)}{f(y|\mu, \sigma^2)} = h(x, y)$$

$$\frac{f(x|\mu, \sigma^2)}{f(y|\mu, \sigma^2)} = \frac{(2\pi\sigma^2)^{\frac{-n}{2}} \exp\left[-\frac{1}{2\sigma^2}\{n(\bar{x} - \mu)^2 + (n-1)S_x^2\}\right]}{(2\pi\sigma^2)^{\frac{-n}{2}} \exp\left[-\frac{1}{2\sigma^2}\{n(\bar{y} - \mu)^2 + (n-1)S_y^2\}\right]}$$

This ratio will be constant if and only if $\bar{x} = \bar{y}$ and $S_x^2 = S_y^2$.
Hence, by Theorem 1.3.1, $(\bar{x}, S_x^2)$ is minimal sufficient for μ and σ^2.

Example 1.3.3 Suppose $(X_1, X_2, \ldots, X_n)$ are iid uniform rvs on the interval $(\theta, \theta + 1)$, $-\infty < \theta < \infty$.
Then the joint pdf of X is

$$f(x|\theta) = \begin{cases} 1 \; ; \; \theta < x_i < \theta + 1, \; i = 1, 2, \ldots, n \\ 0 \; ; \; \text{otherwise} \end{cases}$$

This is same as

$$f(x|\theta) = \begin{cases} 1 \; ; \; x_{(n)} - 1 < x_i < x_{(1)} \\ 0 \; ; \; \text{otherwise} \end{cases}$$

$$\frac{f(x|\theta)}{f(x|\theta)} = \frac{I(\theta - x_{(n)} + 1)I(x_{(1)} - \theta)}{I(\theta - y_{(n)} + 1)I(y_{(1)} - \theta)}$$

This ratio will be constant if and only if $x_{(n)} = y_{(n)}$ and $x_{(1)} = y_{(1)}$.
Hence, by Theorem 1.3.1, $T(X) = (x_{(1)}, x_{(n)})$ is a minimal sufficient statistic.

Example 1.3.4 Let X_1, X_2, X_3, X_4 are iid rvs having the following pmf:

$$f(X_1 = x_1, X_2 = x_2, X_3 = x_3) = \frac{n! \; p_1^{x_1} p_2^{x_2} p_3^{x_3} p_4^{(n-x_1-x_2-x_3)}}{x_1! x_2! x_3! (n - x_1 - x_2 - x_3)!}$$

where $p_4 = 1 - p_1 - p_2 - p_3$, $\sum_{i=1}^{3} X_i \leqslant n$

Consider

$$\frac{f(X_1 = x_1, X_2 = x_2, X_3 = x_3)}{f(Y_1 = y_1, Y_2 = y_2, Y_3 = y_3)} = \frac{\dfrac{n! \; p_1^{x_1} p_2^{x_2} p_3^{x_3} p_4^{(n-x_1-x_2-x_3)}}{x_1! x_2! x_3! (n - x_1 - x_2 - x_3)!}}{\dfrac{n! \; p_1^{n_1} p_2^{n_2} p_3^{n_3} p_4^{(n-y_1-y_2-y_3)}}{y_1! y_2! y_3! (n - y_1 - y_2 - y_3)!}}$$

This ratio is independent of $p_i (i = 1, 2, 3, 4)$ if $X_1 = Y_1$, $X_2 = Y_2$ and $X_3 = Y_3$.
By Theorem 1.3.1 (X_1, X_2, X_3) is minimal sufficient.

Example 1.3.5 Consider the normal model described in Example 1.2.1, where μ and σ^2 are both unknown and $\sigma^2 = \mu$.

As before,

$$\frac{f(x \mid \mu, \mu)}{f(y \mid \mu, \mu)} = \frac{(2\pi\mu)^{\frac{-n}{2}} \exp\left[-\frac{\sum x_i^2}{2\mu^2} + \sum x_i - n\mu\right]}{(2\pi\mu)^{\frac{-n}{2}} \exp\left[-\frac{\sum y_i^2}{2\mu^2} + \sum y_i - n\mu\right]}$$

This ratio will be independent of μ if and only if $\sum x_i^2 = \sum y_i^2$. Hence, By Theorem 1.3.1, $\sum x_i^2$ is minimal sufficient statistic for μ.

Remark 1: In a normal model described in Example 1.2.1, where $\sigma^2 = \mu^2$, $(\sum x_i, \sum x_i^2)$ is minimal sufficient in $N(\mu, \mu^2)$.

Example 1.3.6 Let $X_1, X_2, \ldots X_n$ be iid rvs with the following pmf having the density function:

$$f(X = x) = (1 - \theta)\theta^{(x-\mu)} \quad ; x = \mu, \mu + 1 \ldots, 0 < \theta < 1 \qquad (1.3.1)$$

To find the minimal sufficient statistic if μ and θ are unknown.

$$\frac{P(x_{(1)}, x_{(2)}, \ldots, x_{(n)})}{P(y_{(1)}, y_{(2)}, \ldots, y_{(n)})} = \frac{\theta^{[\sum_{i=1}^{n} x_{(i)} - nx_{(1)} + nx_{(1)} - n\mu]}}{\theta^{[\sum_{i=1}^{n} y_{(i)} - ny_{(1)} + ny_{(1)} - n\mu]}}$$

$$= \frac{\theta^{[\sum_{i=1}^{n}(x_{(i)} - x_{(1)}) + n(x_{(1)} - \mu)]}}{\theta^{[\sum_{i=1}^{n}(y_{(i)} - y_{(1)}) + n(y_{(1)} - \mu)]}}$$

The ratio is independent of θ and μ if
$\sum_{i=1}^{n}(x_{(i)} - x_{(1)}) = \sum_{i=1}^{n}(y_{(i)} - y_{(1)})$ and $x_{(1)} = y_{(1)}$
Hence $\sum_{i=2}^{n}(x_{(i)} - X_{(1)})$ and $x_{(1)}$ are jointly sufficient for (θ, μ).

Example 1.3.7 Let $Y_i (i = 1, 2, \ldots, n)$ are iid with $N(\alpha + \beta X_i, \sigma^2)$, where $-\infty < \alpha, \beta < \infty, \sigma > 0$ are unknown parameters and $(X_1, X_2, \ldots, X_n)$ are known.

$$\frac{f(y|\alpha, \beta, \sigma^2)}{f(z|\alpha, \beta, \sigma^2)} = \frac{\exp\left[-\frac{1}{2\sigma^2}\left\{\sum y_i^2 - 2\alpha \sum y_i - 2\beta \sum x_i y_i + n\alpha^2 + 2\alpha\beta \sum x_i + \beta^2 \sum x_i^2\right\}\right]}{\exp\left[-\frac{1}{2\sigma^2}\left\{\sum z_i^2 - 2\alpha \sum z_i - 2\beta \sum x_i z_i + n\alpha^2 + 2\alpha\beta \sum x_i + \beta^2 \sum x_i^2\right\}\right]}$$

This ratio is independent if and only if $\sum y_i^2 = \sum z_i^2$, $\sum y_i = \sum z_i$ and $\sum x_i y_i = \sum x_i z_i$.

Therefore, $(\sum x_i, \sum y_i^2, \sum x_i y_i)$ is minimal sufficient.

Example 1.3.8 (Sampson and Spencer (1976)): Let $(X_1, X_2, \ldots, X_n)$ be iid rvs according to $f(x|\theta)$, where

$$f(x|\theta) = \begin{cases} \frac{x}{\theta^2} & ; \ 0 < x \leq \theta \\ \frac{2\theta - x}{\theta^2} & ; \ \theta \leq x \leq 2\theta \\ 0 & ; \ \text{otherwise} \end{cases}$$

Here we will show that for $n \geq 5$, median X_i and max X_i are not jointly sufficient.

Let $\theta_1 = \frac{9}{8}$ and $\theta_2 = 1$

Consider

$$X = \left(\frac{1}{4}, \frac{1}{4}, \frac{5}{4}, \frac{3}{2}, 1, 1, \ldots, 1\right)$$

$$Y = \left(\frac{1}{2}, \frac{3}{2}, 1, 1, \ldots, 1\right)$$

max $X = \frac{3}{2}$, Median $X = 1$, max $Y = \frac{3}{2}$ and Median $Y = 1$

$$f\left(x \mid \frac{9}{8}\right) = \begin{cases} \frac{64}{81}x & ; \ 0 < x < \frac{9}{8} \\ \frac{16}{81}(9 - 4x) & ; \ \frac{9}{8} \leq x < \frac{9}{4} \\ 0 & ; \ \text{otherwise} \end{cases}$$

$$f(x \mid 1) = \begin{cases} x & ; \ 0 \leq x \leq 1 \\ 2 - x & ; \ 1 \leq x \leq 2 \\ 0 & ; \ \text{otherwise} \end{cases}$$

$$f\left(x \mid \frac{9}{8}\right) = \left(\frac{64}{81} \times \frac{1}{4}\right)\left(\frac{64}{81} \times \frac{1}{4}\right)\left(\frac{16}{81} \times 4\right)\left(\frac{16}{81} \times 3\right)\left(\frac{64}{81}\right)^{n-4}$$

$$= \left(\frac{(64)^{n-1}}{81^n} \times 3\right)$$

$$f\left(y \mid \frac{9}{8}\right) = \left(\frac{(64)^{n-1}}{81^n} \times 24\right)$$

$$f(x \mid 1) = \left(\frac{3}{128}\right)$$

$$f(y \mid 1) = \left(\frac{1}{4}\right)$$

Therefore,

$$f\left(x \mid \frac{9}{8}\right) f(y \mid 1) = \left(\frac{(64)^{n-1}}{81^n} \times \frac{3}{4}\right)$$

and

$$f(x \mid 1) f\left(y \mid \frac{9}{8}\right) = \left(\frac{(9)(64)^{n-1}}{81^n}\right)$$

$$f\left(x\mid\frac{9}{x}\right)f(y\mid1)\neq f(r\mid1)f\left(y\mid\frac{9}{n}\right)$$

By Theorem 1.3.4, Median X_i and Max X_i are not jointly sufficient.

Example 1.3.9 Let $(X_1, X_2, \ldots, X_n)$ be iid rvs with left truncated exponential distribution.

$$f(x_i\mid\alpha, \beta) = \begin{cases} \frac{1}{\beta}\exp[\frac{-(x-\alpha)}{\beta}] \; ; & x_i \geq \alpha \\ 0 & ; \text{ otherwise} \end{cases}$$

This model is used in systems reliability theory. Suppose we have n independent and identical systems. Let the random variable X be represents the failure time of a system. The data consist of order statistics $X_{(1)} \leq \cdots \leq X_{(n)}$, where the ith order statistic $X_{(i)}$ represent the failure time of a system which is failed at ith time point, $(i = 1, 2, \ldots, n)$.

$$f(x_{(1)}, x_{(2)}, \ldots, x_{(n)}) = \frac{n!}{\beta^n}\exp\left[-\frac{\sum_{i=1}^{n}(x_{(i)}-\alpha)}{\beta}\right]$$

$$= \frac{n!}{\beta^n}\exp\left[-\frac{1}{\beta}\sum_{i=1}^{n}(x_{(i)}-x_{(1)}+x_{(1)}-\alpha)\right]$$

$$= \frac{n!}{\beta^n}\exp\left[-\frac{1}{\beta}\sum_{i=1}^{n}(x_{(i)}-x_{(1)})+\frac{n(x_{(1)}-\alpha)}{\beta}\right]$$

Next,

$$\frac{f(y_{(1)}, y_{(2)}, \ldots, y_{(n)}\mid\alpha, \beta)}{f(x_{(1)}, x_{(2)}, \ldots, x_{(n)}\mid\alpha, \beta)} = \frac{\exp\left[-\frac{1}{\beta}\sum_{i=1}^{n}(y_{(i)}-y_{(1)})+\frac{n(y_{(1)}-\alpha)}{\beta}\right]}{\exp\left[-\frac{1}{\beta}\sum_{i=1}^{n}(x_{(i)}-x_{(1)})+\frac{n(x_{(1)}-\alpha)}{\beta}\right]}$$

This ratio will be independent of α and β if and only if $\sum_{i=1}^{n}(y_{(i)} - y_{(1)}) = \sum_{i=1}^{n}(x_{(i)}-x_{(1)})$ and $y_{(1)} = x_{(1)}$. This implies that $[x_{(1)}, \sum_{i=1}^{n}(x_{(i)}-X_{(1)})]$ is minimal sufficient for (α, β).

Remark A minimal sufficient statistics is not unique. Any one-to-one function of a minimal sufficient statistic is also a minimal sufficient statistic. For example, in $\cup(\theta, \theta+1)$, $(X_{(1)}, X_{(n)})$, $(X_{(n)} - X_{(1)}, X_{(n)} + X_{(1)})$, $\left(\frac{X_{(n)} - X_{(1)}}{2}, \frac{X_{(n)} + X_{(1)}}{2}\right)$, etc., are all minimal sufficient statistics.

1.4 Ancillary Statistics

Fisher had introduced the term ancillary statistics. Many statisticians had given the definition of ancillary statistics, this include Basu (1959), Cox and Hinkley (1996) etc.

Definition 1.4.1 A statistic $T(X)$ whose distribution does not depend on the parameter θ, is called an ancillary statistic.

Here, we will consider some examples of ancillary statistics.

Example 1.4.1 Let the rvs X_1 and X_2 are distributed as $N(\theta, 1)$. Then the distribution of $W = X_2 - X_1$ is $N(0, 2)$, which does not depend on θ. Hence W is a ancillary statistics.

Example 1.4.2 Let $X_1 X_2, \ldots, X_n$ are iid rvs with $\cup \left(\theta - \frac{1}{2}, \theta + \frac{1}{2}\right)$. Then

$$f(x|\theta) = \prod_{i=1}^{n} I_{[\theta - \frac{1}{2}, \theta + \frac{1}{2}]} x_i = I_{(\theta - \frac{1}{2}, \infty)} x_{(1)} I_{(-\infty, \theta + \frac{1}{2})} x_{(n)} \qquad (1.4.1)$$

By Theorem 1.3.1, $T = (T_1, T_2) = \left(x_{(n)}, x_{(n)} - x_{(1)}\right)$ is minimal sufficient.
$x_{(1)} = T_1 - T_2$
$\theta - \frac{1}{2} < T_1 - T_2 < T_1 < \theta + \frac{1}{2}$
$\theta - \frac{1}{2} < T_1 - T_2$ and $T_1 < \theta + \frac{1}{2}$
$\theta - \frac{1}{2} + T_2 < T_1$ and $T_1 < \theta + \frac{1}{2}$
$\Rightarrow \theta - \frac{1}{2} + T_2 < T_1 < \theta + \frac{1}{2}$

Now, the joint pdf of T_1 and T_2 is

$$f(t_1, t_2|\theta) = n(n-1)t_2^{n-2} I_{[0,1]}(t_2) I_{[\theta - \frac{1}{2} + t_2, \theta + \frac{1}{2}]} t_1$$

The marginal density of T_2 is

$$f(t_2|\theta) = n(n-1)t_2^{n-2}(1 - t_2); \quad 0 < t_2 < 1 \qquad (1.4.2)$$

This does not depend on θ.

Hence T_2 is ancillary. Further, we can say that the statistic T_2 does not give any information about θ.

But interestingly, conditional density of T_1 given T_2 is

$$f(t_1|t_2, \theta) = \frac{1}{1 - t_2} I_{[\theta - \frac{1}{2} + t_2, \theta + \frac{1}{2}]} t_1, \qquad (1.4.3)$$

which depends on θ.

Example 1.4.3 Suppose X_1 and X_2 are iid observations from the pdf

$$f(x|\alpha) = \alpha x^{\alpha-1} \exp[-x^\alpha], x > 0, \alpha > 0$$

We will show that $\dfrac{\log X_1}{\log X_2}$ is an ancillary statistic.

Let $W_1 = \log X_1$ and $W_2 = \log X_2$. Since X_1 and X_2 are iid rvs then $W_1 + W_2$ are also iid rvs. Suppose $W = W_1 + W_2$
Hence,

$$f(w|\alpha) = \alpha \exp[\alpha w - e^{\alpha w}], \quad -\infty < w < \infty.$$

Let $Z = \alpha W$, then

$$f(z) = \exp[z - e^z], \ z > 0$$

Hence Z_1 and Z_2 are iid rvs with $f(z|\alpha = 1)$.
Then

$$\frac{\log X_1}{\log X_2} = \frac{W_1}{W_2} = \frac{z_1/\alpha}{z_2/\alpha} = \frac{z_1}{z_2}$$

Since the distribution of z is independent of α, then the distribution $\frac{z_1}{z_2}$ does not dependent on α.
Thus $\frac{\log X_1}{\log X_2}$ is an ancillary statistic.

Example 1.4.4 Consider the normal model in Example 1.2.1, where $\sigma^2 = 1$.
Define $T(X) = (n-1)S^2 = \sum_{i=1}^{n}(X_i - \bar{X})^2$
The distribution of $T(X)$ is χ^2_{n-1}. Thus $T(X)$ is independent of μ. Hence $T(X)$ is an ancillary statistic.

Note: In Example 1.2.1, $\mu = 0$, the distribution of $\bar{X}$ is $N\left(0, \frac{\sigma^2}{n}\right)$.

Hence $\bar{X}$ is not ancillary.

1.5 Completeness

Definition 1.5.1 Let $\{f_\theta(x) : \theta \in \Theta\}$ be a family of pdf's (or pmf's). We say that the family is complete if

$$Eg(x) = 0 \ \ \forall \ \theta \in \Theta \tag{1.5.1}$$

then $P[g(x) = 0] = 1 \quad \forall \ \theta \in \Theta$

Definition 1.5.2 A Statistics $T(X)$ is said to be complete if the family of distribution of T is complete.

Example 1.5.1 Let $X_i (i = 1, 2, \ldots, m)$ be iid rvs as $B(n, p)$. Then the distribution of $T = \sum_{i=1}^{m} X_i$ is $B(nm, p)$.

Now, we will show that T is complete. Let $g(t)$ be a function of t.

$$Eg(t) = \sum_{t=0}^{mn} g(t) \binom{mn}{t} p^t (1 - p)^{mn-t}$$

Since $Eg(t) = 0$
Therefore,

$$\sum_{t=0}^{mn} g(t) \binom{mn}{t} \left(\frac{p}{1 - p}\right)^t = 0$$

Let $r = \frac{p}{1-p}$, then

$$\sum_{t=0}^{mn} g(t) \binom{mn}{t} r^t = 0$$

Hence, RHS of the above expression is a polynomial of degree mn in r.
Since LHS is zero then each term is zero. Therefore, $g(t)\binom{mn}{t}r^t = 0$.
But $\binom{mn}{t}r^t \neq 0$. Therefore, $g(t) = 0$ with probability 1.

Example 1.5.2 Consider the rv X, where each $X_i (i = 1, 2, \ldots, n)$ has $N(\mu, 1)$. Hence $\bar{X}$ has $N(\mu, \frac{1}{n})$. Now we will show that $\bar{X}$ is complete.

Let $Y = \bar{X}$. Consider the function $g(y)$. Since $Eg(y) = 0$. This implies

$$\int_{-\infty}^{\infty} g(y) (\frac{n}{2\pi})^{\frac{1}{2}} \exp\left[-\frac{n}{2}(y - \mu)^2\right] dy = 0$$

By removing nonzero constants and $n\mu = S$

$$\int_{-\infty}^{\infty} h(y) e^{-\frac{ny^2}{2}} e^{Sy} dy = 0$$

This equation states that the Laplace transform of the function $h(y)e^{-\frac{ny^2}{2}}$ is zero identically.

But the function zero also has the transform which is zero identically, Hence by uniqueness property of Laplace transform it follows that $h(y)e^{\frac{n y^2}{2}} = 0$. This implies $h(y) = 0$ with probability 1 for all μ.

Example 1.5.3 Let X be the rv from the pmf

$$f(x|\theta) = \left(\frac{\theta}{2}\right)^{|x|} (1-\theta)^{1-|x|}; \quad x = -1, 0, 1, \quad 0 \le \theta \le 1$$

$$EX = -\frac{\theta}{2} + \frac{\theta}{2}$$

But $X = 0$ is not with probability 1. Hence X is not complete. But $|X|$ is complete and its distribution is Bernoulli.

In Example 1.5.1, we have shown that Binomial family is complete. Bernoulli is a particular case of Binomial. Hence $|X|$ is a complete statistics.

Example 1.5.4 Let $X_1, X_2, \ldots, X_n$ be iid $N(\theta, \alpha\theta^2)$, where $\alpha > 0$, a known constant and $\theta > 0$. We can easily show that $(\bar{X}, S^2)$ is minimal sufficient statistic Using the Theorem 1.3.1 for fixed $\alpha > 0$.

In this example $(n-1)S^2 = \sum(X_i - \bar{X})^2$. Therefore, $\frac{(n-1)S^2}{\sigma^2}$ has χ^2 with $(n-1)$ df and $\sigma^2 = \alpha\theta^2$. Hence $E(S^2) = \alpha\theta^2$, $E[\bar{X}^2] = \frac{(n+\alpha)\theta^2}{n}$.

$\Rightarrow E[\frac{S^2}{\alpha}] = \theta^2$ and $E[\frac{n\bar{X}^2}{(n+\alpha)}] = \theta^2$

Consider $T(X) = \frac{n\bar{X}^2}{\alpha+n} - \frac{S^2}{\alpha}$ then $E[T(X)] = 0 \,\forall\, \theta$. But $T(X)$ is not identically zero. Hence $(\bar{X}, S^2)$ is not complete.

Example 1.5.5 Stigler (1972) had given an interesting example of an incomplete family. If we remove one probability distribution (say P_n) then the family $\wp - P_n$ is not a complete, where $\wp$ is a family of discrete uniform probability distributions.

Consider a function

$$g(x) = \begin{cases} 0 & ; x = 1, 2, \ldots, n-1, n+2, n+3, \ldots \\ a & ; x = n \\ -a & ; x = n+1 \end{cases}$$

where a is nonzero constant.
Consider the following case:
(i) $N < n$

$$E[g(X)] = \sum_{x=1}^{N} g(x)\frac{1}{N} = 0$$

because $g(x) = 0$ for $x = 1, 2, \ldots, n-1$

(ii) $N > n$

$$E[g(X)] = \sum_{x=1}^{N} \frac{g(x)}{N}$$

$$= \frac{1}{N} [g(1) + g(2) + \cdots + g(n-1) + g(n) + g(n+1) + g(n+2) + \cdots]$$

$$= \frac{1}{N} [0 + a + (-a) + 0] = 0$$

(iii) $N = n$

$$E[g(X)] = \sum_{x=1}^{n} \frac{g(x)}{N}$$

$$= \frac{1}{N} [g(1) + g(2) + \cdots + g(n)]$$

$$= \frac{1}{N} [0 + a] = \frac{a}{N}$$

Therefore,

$$E[g(X)] = \begin{cases} 0 & ; \ \forall \ N \neq n \\ \frac{a}{N} & ; \ \forall \ N = n \end{cases}$$

Therefore, we conclude that the family $\wp - P_n$ is not complete.

Example 1.5.6 Consider the family of distributions:

$$P = P[X = x|\lambda] = \frac{\lambda^x e^{-\lambda}}{x!}; x = 0, 1, 2, \ldots \ \text{and} \ \lambda = 1 \ \text{or} \ 2$$

If $\lambda = 1$

$$E[g(X)] = e^{-1}g(0) + e^{-1} \sum_{x=1}^{\infty} \frac{g(x)}{x!}$$

Let $g(0) = 0$ and $\sum_{x=1}^{\infty} \frac{g(x)}{x!} = 0$ then $E[g(x)] = 0$ but $g(x) \neq 0$ because $g(1) = 1$, $g(2) = -2$, and $g(0) = 0$ if $x > 2$

If $\lambda = 2$

$$E[g(X)] = e^{-2}g(0) + e^{-2}\sum_{x=1}^{\infty}\frac{g(x)}{x!}$$

By the above argument $E[g(X)] = 0$ and $g(x) \neq 0$ identically.

Hence, the Poisson family with λ restricted is not complete.

Example 1.5.7 Let $X_1, X_2, \ldots, X_n$ be iid rvs having the following uniform distributions

$$(i) f_1(x|\theta) = \begin{cases} \frac{1}{\theta} \;;\; 0 < x < \theta \\ 0 \;;\; \text{otherwise} \end{cases}$$

$$(ii) f_2(x|\theta) = \begin{cases} 1 \;;\; \theta < x < \theta + 1 \\ 0 \;;\; \text{otherwise} \end{cases}$$

$$(iii) f_3(x|\theta) = \begin{cases} \frac{1}{\theta} \;;\; \theta < x < 2\theta \\ 0 \;;\; \text{otherwise} \end{cases}$$

$$(iv) f_4(x|\theta) = \begin{cases} \frac{1}{2\theta} \;;\; \theta < x < 3\theta \\ 0 \;;\; \text{otherwise} \end{cases}$$

$$(v) f_5(x|\theta) = \begin{cases} 1 \;;\; \theta - \frac{1}{2} < x < \theta + \frac{1}{2} \\ 0 \;;\; \text{otherwise} \end{cases}$$

(i) In this example, we have to show that $X_{(n)}$ is minimal sufficient. Let $T = X_{(n)}$. Then the pdf of T is

$$h(t|\theta) = \begin{cases} \frac{nt^{n-1}}{\theta^n} \;;\; 0 < t < \theta \\ 0 \;\;\;\;\;\; ;\; \text{otherwise} \end{cases}$$

Let $g(T)$ be a function such that $E[g(T)] = 0 \;\; \forall\, \theta$. Since $E[g(T)]$ is constant as a function of θ, its derivative is zero.

$$E[g(T)] = \int_0^{\theta} g(t)\frac{nt^{n-1}}{\theta^n}dt = 0$$

$$\Rightarrow \int_0^{\theta} g(t)t^{n-1}dt = 0 \tag{1.5.2}$$

$$\Rightarrow \frac{d}{d\theta} \int\limits_{0}^{\theta} g(t)t^{n-1}dt = 0$$

Using the following rule, if range is not independent of θ and f is zero at the extremes of the range, i.e., $f(a, \theta) = 0 = f(b, \theta)$ then

$$\frac{\partial}{\partial\theta} \int\limits_{a}^{b} f dx = \int\limits_{a}^{b} \frac{\partial f}{\partial\theta} dx - f(a, \theta)\frac{\partial a}{\partial\theta} + f(b, \theta)\frac{\partial b}{\partial\theta}$$

From (1.5.2),

$$= g(\theta)\theta^{n-1} = 0$$

Therefore, $g(\theta) = 0 \ \forall \ \theta > 0$.

Hence T is complete.

(ii) Using Theorem 1.3.1, we can show that $(X_{(1)}, X_{(n)})$ is minimal sufficient. The joint distribution of $(X_{(1)}, X_{(n)})$ is given by

$$f(x, y|\theta) = \begin{cases} n(n-1)(y-x)^{n-2} \ ; \ \theta < x < y < \theta+1 \\ 0 \qquad\qquad\qquad ; \ \text{otherwise} \end{cases}$$

Let $T_1 = X_{(n)} - X_{(1)}$ and $T_2 = \frac{X_{(1)}+X_{(n)}}{2}$

$\Rightarrow X_{(1)} = \frac{2T_2 - T_1}{2}$ and $X_{(n)} = \frac{2T_2 + T_1}{2}$ and $|J| = 1 \ \ \theta < X_{(1)} < X_{(n)} < \theta+1$

$\theta < \frac{2T_2 - T_1}{2}$ and $\frac{2T_2 + T_1}{2} < \theta+1$

$2\theta < 2T_2 - T_1$ and $2T_2 + T_1 < 2\theta+2$

$2\theta + T_1 < 2T_2$ and $2T_2 < 2\theta + 2 - T_1$

$\theta + \frac{T_1}{2} < T_2$ and $T_2 < \theta + 1 - \frac{T_1}{2}$

$\Rightarrow \theta + \frac{T_1}{2} < T_2 < \theta + 1 - \frac{T_1}{2}$

Trivially, $0 < T_1 < 1$

$$h(t_1, t_2|\theta) = \begin{cases} n(n-1)t_1^{n-2} \ ; \ \ 0 < t_1 < 1, \ \ \theta + \frac{t_1}{2} < t_2 < \theta + 1 - \frac{t_1}{2} \\ 0 \qquad\qquad\quad ; \ \text{otherwise} \end{cases}$$

$$h(t_1|\theta) = n(n-1)t_1^{n-2}(1 - t_1); \quad 0 < t_1 < 1$$

$$E(T_1) = \frac{n-1}{n+1},$$

Consider the $T(X, Y) = y - x - \frac{n-1}{n+1}$

Therefore, $E[T(X, Y)] = 0$ but $T(X, Y)$ is not identically zero.

We can conclude that R is ancillary but $(X_{(1)}, X_{(n)})$ is not complete.

The reader should show that if a function of a sufficient statistics is ancillary, then the sufficient statistics is not complete.

(iii) Using Theorem 1.3.1, we can show that $(X_{(1)}, X_{(n)})$ is minimal sufficient statistics.

Let $W_i = \frac{X_i}{\theta}$, $i = 1, 2, \ldots, n$.

Then the joint distribution of $W_1, W_2, \ldots, W_n$ is given as

$$f(w_1, w_2, \ldots, w_n) = \begin{cases} 1 \; ; \; 1 < w_i < 2, \; i = 1, 2, \ldots, n \\ 0 \; ; \; \text{otherwise} \end{cases}$$

Let $W_{(1)} = \dfrac{X_{(1)}}{\theta}$ and $W_{(n)} = \dfrac{X_{(n)}}{\theta}$

So $\dfrac{X_{(1)}}{X_{(n)}} = \dfrac{W_{(1)}}{W_{(n)}}$, which is free from θ.

Hence, $\dfrac{X_{(1)}}{X_{(n)}}$ is ancillary and a function of sufficient statistic. Therefore, $(X_{(1)}, X_{(n)})$ is not complete.

In (iv) and (v) we can easily show using the same argument by (ii) and (iii). $(X_{(1)}, X_{(n)})$ is not complete. Similarly, we can find ancillary statistic for (iv) and (v).

Example 1.5.8 Let $P[X = -1] = \theta$, $P[X = x] = (1 - \theta)^2 \theta^x$, $x = 0, 1, \ldots$
Using Theorem 1.3.1

$$\frac{P[X = x]}{P[X = y]} = \frac{(1 - \theta)^2 \theta^x}{(1 - \theta)^2 \theta^y}$$

This is independent of θ if $X = Y$.

Hence X is minimal sufficient.

But X is not complete, because

$$\begin{aligned} E(X) &= -\theta + \sum_{x=0}^{\infty} x(1 - \theta)^2 \theta^x \\ &= -\theta + (1 - \theta)^2 [\theta + 2\theta^2 + 3\theta^3 + \cdots] \\ &= -\theta + \theta(1 - \theta)^2 [1 + 2\theta + 3\theta^2 + \cdots] \\ &= -\theta + \frac{\theta(1 - \theta)^2}{(1 - \theta)^2} = 0 \end{aligned}$$

But $X \neq 0$ identically.

Example 1.5.9 Dahiya and Kleyle (1975) have studied the estimation of parameters of this mixed failure time distribution (MFTD)

$$F(x) = \begin{cases} 1 - pe^{-\frac{x}{\theta}} & ; \ x \geq 0, \ \theta > 0, \ 0 < p \leq 1 \\ 0 & ; \ x < 0 \end{cases} \tag{1.5.3}$$

from Type I censored data.

It may be noted that the parameter θ cannot be estimated when all the observations in the sample come from degenerate distributions. In such a situation, they have modified the form of the estimator and studied the properties of modified estimator. The corresponding pdf of (1.5.3) can be written as

$$f(x|p, \theta) = \begin{cases} (1-p)^{I(x)}[\frac{p}{\theta}e^{-\frac{x}{\theta}}]^{1-I(x)} & ; \ x > 0, \ \theta > 0, \ 0 < p \leq 1 \\ 0 & ; \ \text{otherwise} \end{cases} \tag{1.5.4}$$

where

$$I(x) = \begin{cases} 1 & ; \ x = 0 \\ 0 & ; \ x > 0 \end{cases}$$

The joint distribution of $(X_1, X_2, \ldots, X_n)$ is given as

$$f(x_1, x_2, \ldots, x_n|p, \theta) = \begin{cases} (1-p)^{\sum_{i=1}^{n} I(x_i)} \prod_{i=1}^{n}[\frac{p}{\theta}e^{-\frac{x_i}{\theta}}]^{1-I(x_i)} & ; \ x_i > 0, \ \theta > 0, \ 0 < p \leq 1 \\ 0 & ; \ \text{otherwise} \end{cases}$$

$$\tag{1.5.5}$$

Let $r = n - \sum_{i=1}^{n} I(x_i)$, r denotes the number of positive observations in the sample. Then (1.5.5) can be written as

$$f(x_1, x_2, \ldots, x_n|p, \theta) = \begin{cases} (1-p)^{n-r}(\frac{p}{\theta})^r e^{-\frac{\sum x_i}{\theta}} & ; \ x_i \geq 0 \\ 0 & ; \ \text{otherwise} \end{cases} \tag{1.5.6}$$

One can easily see that by Theorem 1.2.1, (r, z_r) is sufficient for (p, θ), where $z_r = \sum_{i=1}^{n} x_i$. For details, see Dixit (1993).
Next lemma is given by Dixit (1993).

Lemma 1.5.1 *Let (X_1, X_2) be a rv having a joint distribution with parameters (θ_1, θ_2). Further, (X_1, X_2) may be vectors. Suppose the marginal distribution of X_1 is discrete which depends on θ_1 only and belongs to a complete family of distributions. Further suppose the conditional distribution of X_2 given X_1 depends only on θ_2 and belongs to a complete family of distributions, then the family of joint distribution (X_1, X_2) is complete.*

Proof Let S_1 be the support of X_1. Since X_1 is discrete, S_1 is at most countable. Let h be a function such that

$$E[h(X_1, X_2)] = 0 \quad \forall \ (\theta_1, \theta_2) \in \Theta$$

That is,

$$\int \int h(x_1, x_2) dF(x_1, x_2) = 0 \quad \forall \ (\theta_1, \theta_2) \in \Theta$$

$$\sum_{x_1 \in S_1} P[X_1 = x_1] \int h(x_1, x_2) dF(x_2|x_1) = 0 \quad \forall \ (\theta_1, \theta_2) \in \Theta$$

We are given that marginal distribution of X_1 belongs to a complete family.

$$\sum_{x_1 \in S_1} k(x_1) P[X_1 = x_1] = 0$$

where

$$k(x_1) = \int h(x_1, x_2) dF(x_2|x_1)$$

It may be noted that S_1^c is a P_{θ_1}-null set for all θ_1 and θ_2.

Now, for each fixed $x_1 \in S_1$, $k(x_1) = 0$,
it implies that $\int h(x_1, x_2) dF(x_2|x_1) = 0$.
Again given that conditional distribution of X_2 given $X_1 = x_1$ is complete for each $x_1 \in S_1$, we get that

$$h(x_1, x_2) = 0 \quad \forall \ x_2 \in N_{x_1}^c,$$

where $P_{\theta_2}(N_{x_1}) = 0 \ \forall \ x_1 \in S_1$.
Now since S_1 is countable, $N = \cup N_{x_1}$ is a P_{θ_2}-null set and we get that

$$h(x_1, x_2) = 0 \quad \forall \ x_2 \in N^c \text{ and } \forall \ x_1 \in S_1$$

Hence,

$$P_{\theta_1, \theta_2}[h(x_1, x_2) = 0] = 1 \quad \forall \ \theta_1, \theta_2.$$

Now, we consider the completeness about the family in (1.5.4).

The marginal distribution of r is binomial with (n, p), which is a complete family of distribution and the conditional distribution of z giver r

$$f(z|r) = \begin{cases} \frac{e^{-\frac{z}{\theta}} z^{r-1}}{\Gamma(r)\theta^r} & ; \ z > 0, \ r > 0, \theta > 0 \\ 0 & ; \ \text{otherwise} \end{cases} \tag{1.5.7}$$

Which depends only on θ and is a complete family of distribution. Hence, from Lemma 1.5.1, (r, z) is complete for (P, θ).

Theorem 1.5.1 (Basu's Theorem) *In a complete family every ancillary statistic is independent of the minimal sufficient statistic.*

Proof Let T be a complete sufficient statistic.
Let S be an ancillary statistic.
By definition, for any event A, $P\{S \in A\}$ does not depend on θ. Further, $P\{S \in A|T = t\}$ also does not depend on θ because T is sufficient.

Consider a function

$$g(T) = P[S \in A] - P[S \in A|T = t]$$

$$E[g(T)] = P[S \in A] - EP[S \in A|T = t] = 0 \ \forall \ \theta$$

By the assumption of completeness $P[g(T) = 0] = 1$

$$\Rightarrow P[S \in A] = P[S \in A|T = t]$$

This implies S and T are independent. Thus

$$P[S \in A, T \in B] = P[S \in A]P[T \in B]$$

Hence, for any sets A and B, S and T are independent.

Definition 1.5.3 A family of distributions $\{F(t|\theta) : \theta \in \Theta\}$ is boundedly complete if

$$E[g(T)] = \int g(t)f(t)dt = 0 \ \forall \ \theta$$

and real statistics $g(t)$ satisfying $|g(t)| < M$, then $g(t) = 0$.

Theorem 1.5.2 *If a family of distributions is complete then it is boundedly complete.*

Remark The converse of the theorem is not true.

Example 1.5.10 Let T be a random variable with the following probability distribution:

$$P[T = 0] = q \quad \text{and} \quad P[T = i + 1] = p^2q^i, \ i = 0, 1, 2, \ldots, \ 0 < p < 1, \ q = 1 - p$$

Let $E[g(T)] = 0$ then

$$g(0)q + g(1)p^2 + g(2)p^2q + g(3)p^2q^2 + \cdots = 0$$

$$g(1) + g(2)q + g(3)q^2 + \cdots = -g(0)qp^{-2}$$
$$= -g(0)q(1 - q)^{-2}$$
$$= -g(0)[q + 2q^2 + 3q^3 + \cdots]$$

This implies that $g(1) = 0$, $g(2) = -g(0)$, $g(3) = -2g(0)$, etc. Hence,

$$g(i) = -(i-1)g(0)$$

If $g(0) = 0$ then $g(t) = 0$ at all nonnegative integers. Otherwise, the function $g(t)$ is unbounded.

Therefore, there are nondegenerate unbiased estimates of zero but they are none that are bounded. Hence, we conclude that the family of distributions is boundedly complete but not complete.

1.6 Exponential Class Representation: Exponential Family

Let X be a vector valued random variable with pdf/pmf $\{f(x|\theta), \theta \in \Theta\}$ and θ is a vector of parameters. We say that X belongs to the exponential family

1. Θ contains an open rectangle.
2. $x : f(x \mid \theta) > 0$ is independent of θ.
3. $\log f(x, \theta) = \sum_{i=1}^{k} u_i(\theta)T_i(x) + v(\theta) + w(x)$.
4. The partial derivatives $\frac{\partial u_i}{\partial \theta_j} (i = 1, 2, \ldots, n, j = 1, 2, \ldots, k)$ are continuous and the jacobian

$$|J| = \left| \frac{\partial(u_1, u_2, \ldots, u_n)}{\partial(\theta_1, \theta_2, \ldots, \theta_k)} \right| \neq 0$$

5. $\{T_1(x), T_2(x), \ldots, T_n(x), 1\}$ are linearly independent.

Example 1.6.1 Let $X_1, X_2, \ldots, X_n$ be independent random variables each having distribution $G(p, \alpha)$.

Let

$$f(x|p, \alpha) = \frac{x^{p-1}e^{-\frac{x}{\sigma}}}{\sigma^p \Gamma(p)}; x > 0, p > 0, \sigma > 0 \tag{1.6.1}$$

The joint pdf of $X_1, X_2, \ldots, X_n$ is $f(x_1, x_2, \ldots, x_n \mid p, \sigma) = \dfrac{\prod_{i=1}^{n} x_i^{p-1} \exp[-\frac{\sum_{i=1}^{n} X_i}{\sigma}]}{\sigma^{np}(\Gamma p)^n}$

$$\log f(x \mid p, \sigma) = -n \log \Gamma(p) - np \log \sigma + (p-1) \sum_{i=1}^{n} \log x_i - \frac{1}{\sigma} \sum_{i=1}^{n} x_i \tag{1.6.2}$$

1. $\Theta = \{(\sigma, p) | \sigma, p > 0\}$ contains an open rectangle.
2. $\{(x_1, x_2, \ldots, x_n) | x_1 > 0, x_2 > 0, \ldots, x_n > 0\}$ is independent of parameters σ and p.
3. $T_1(x) = \sum_{i=1}^{n} \log x_i$, $T_2(x) = \sum_{i=1}^{n} x_i$, $u_1(p) = p - 1$ and $u_2(p) = -\frac{1}{\sigma}$.

$$J = \begin{pmatrix} \frac{\partial u_1}{\partial p} & \frac{\partial u_1}{\partial \sigma} \\ \frac{\partial u_2}{\partial p} & \frac{\partial u_2}{\partial \sigma} \end{pmatrix}$$

$$= \begin{pmatrix} 1 & 0 \\ 0 & \frac{1}{\sigma^2} \end{pmatrix} = \frac{1}{\sigma^2}$$

and $|J| > 0$

4. Linear Independence: $\{T_1, T_2, 1\}$

$$a \sum_{i=1}^{n} \log x_i + b \sum_{i=1}^{n} x_i + c = 0 \qquad \text{(1.6.3)}$$

Consider $X = (e^{\alpha}, 1, 1, 1 \ldots, 1)$ for $\alpha \neq 0$

$$a\alpha + b(e^{\alpha} + n - 1) + c = 0$$

$$(c + nb) + (a + b)\alpha + b \left[\sum_{r=2}^{\infty} \frac{\alpha^r}{r!} \right] = 0$$

This implies that $c + nb = 0$, $a + b = 0$ and $c = 0$
Then $a = 0$, $b = 0$ and $c = 0$.
Thus, gamma distribution defined in (1.6.1) belongs to exponential family.
If a probability distribution belongs to an exponential family then one can get complete sufficient statistics. Here, we will give a proof which is within the scope of this book. General proof is given by Lehman and Casella (1998).

Theorem 1.6.1 *Let* $\{f(x|\theta), \theta \in \Theta\}$ *be a k-parameter exponential family given by*

$$f(x|\theta) = \exp[\sum_{i=1}^{k} u_i(\theta) T_i(x) + v(\theta) + w(x)] \qquad \text{(1.6.4)}$$

Then $T = (T_1(X), T_2(X), \ldots, T_k(X))$ *is a complete sufficient statistic, where* $u(\theta) = (u_1, u_2, \ldots, u_k)$

Proof This is given by Rohatagi and Saleh (2001) for $k = 1$.
Let X be a discrete random variable. By factorization theorem, we can show that T is sufficient.

Let $k = 1$ and $u(\theta) = \theta$ in (1.6.4).

$$Eg(T) = \sum_{t}' g(t) P[T(X) = t]$$

$$= \sum_{t} g(t) \exp[\theta t + v(\theta) + w^*(t)] = 0 \;\; \forall \; \theta > 0 \qquad (1.6.5)$$

where $P(T(X) = t) = \exp[\theta t + v(\theta) + w^*(t)]$
Now, we have to show $g(t) = 0 \;\; \forall \; \theta$
Let

$$x^+ = \begin{cases} x \; ; \; x \geq 0 \\ 0 \; ; \; x < 0 \end{cases}$$

and

$$x^- = \begin{cases} -x \; ; \; x < 0 \\ 0 \;\;\; ; \; x \geq 0 \end{cases}$$

Then $g(t) = g^+(t) - g^-(t)$.
Further, g^+ and g^- are both nonnegative functions.
From (1.6.5),

$$E[g(T)] = 0 \Rightarrow E\left[g^+(T)\right] = E\left[g^-(T)\right] \qquad (1.6.6)$$

Therefore,

$$\sum_{t} g^+(t) \exp[\theta t + v(\theta) + w^*(t)] = \sum_{t} g^-(t) \exp[\theta t + v(\theta) + w^*(t)] \;\; \forall \; \theta$$

$$(1.6.7)$$

For fixed $\theta = \theta_0 \in (\alpha, \beta)$

$$P^+(t) = \frac{g^+(t) \exp[\theta_0 t + w^*(t)]}{\sum_{t} g^+(t) \exp[\theta_0 t + w^*(t)]} \qquad (1.6.8)$$

$$P^-(t) = \frac{g^-(t) \exp[\theta_0 t + w^*(t)]}{\sum_{t} g^-(t) \exp[\theta_0 t + w^*(t)]} \qquad (1.6.9)$$

From (1.6.7), (1.6.8) and (1.6.9),

$$\sum_{t} P^+(t) \left[\sum_{t} g^+(t) \exp(\theta_0 t + w^*(t))\right] \exp(\theta - \theta_0)t$$

$$= \sum_{t} P^-(t) \left[\sum_{t} g^-(t) \exp(\theta_0 t + w^*(t))\right] \exp(\theta - \theta_0)t$$

Now $\alpha < \theta < \beta \Rightarrow \alpha - \theta_0 < \theta - \theta_0 < \beta - \theta_0$, we get,

$$\sum_t e^{\delta t} P^+(t) = \sum_t e^{\delta t} P^-(t) \ \forall \ \delta \in (\alpha - \theta_0, \beta - \theta_0) \qquad (1.6.10)$$

By uniqueness of MGF if it exists, (1.6.10) implies that, $P^+(t) = P^-(t) \ \forall \ t$
Then $g^+(t) = g^-(t) \Rightarrow g(t) = 0 \ \forall \ \theta$

Remark 1 From Example 1.6.1, we can see that $\left(\sum \log X_i, \sum X_i\right)$ is complete sufficient statistic for (p, σ). But one should not conclude that $\sum \log X_i$ and $\sum X_i$ are individually complete sufficient for p and σ respectively.

Example 1.6.2 Let $X_1, X_2, \ldots, X_n$ are iid with $N(\mu, \sigma^2)$. The joint pdf of $X_1, X_2, \ldots, X_n$ is

$$f(x|\mu, \sigma^2) = \left(\frac{1}{\sigma\sqrt{2\pi}}\right)^n \exp\left[-\frac{1}{2\sigma^2} \sum_{i=1}^{n}(x_i - \mu)^2\right]$$

$$= \left(\frac{1}{\sigma\sqrt{2\pi}}\right)^n \exp\left[-\frac{1}{2\sigma^2}\left(\sum x_i^2 - 2\mu \sum x_i + n\mu^2\right)\right]$$

$$= \left(\frac{1}{\sigma\sqrt{2\pi}}\right)^n \exp\left[-\frac{\sum x_i^2}{2\sigma^2} + \frac{\mu \sum x_i}{\sigma^2} - \frac{n\mu^2}{2\sigma^2}\right]$$

This belongs to exponential family.
Thus, from Theorem 1.6.1, we can conclude that (T_1, T_2) is complete sufficient statistic for (μ, σ^2) where $T_1 = \sum X_i$ and $T_2 = \sum X_i^2$.

Remark 2: If $\sigma^2 = \mu$ then in Example 1.6.2, $\sum X_i^2$ is complete sufficient statistic for μ.

1.7   Exercise 1

1. Let $X_1, X_2, \ldots, X_n$ be the independent rvs having the following uniform distribution

(i)

$$f(x|\theta) = \begin{cases} \frac{1}{2\theta} \ ; -\theta < x < \theta \\ 0 \ ; \text{ otherwise} \end{cases}$$

Show that $Y_{(n)} = \max_i |X_i|$ is sufficient statistic.

Further, find the distribution of Y_i given $Y_{(n)}$.

(ii)

$$f(x|\theta) = \begin{cases} \frac{1}{\theta} & ; 0 < x < \theta \\ 0 & ; \text{ otherwise} \end{cases}$$

Show that $X_{(n)}$ is sufficient statistic.

(iii)

$$f(x|\theta_1, \theta_2) = \begin{cases} \frac{1}{\theta_2 - \theta_1} & ; \theta_1 < x < \theta_2 \\ 0 & ; \text{ otherwise} \end{cases}$$

Show that $(X_{(1)}, X_{(n)})$ is jointly sufficient statistic.

(iv)

$$f(x|\theta) = \begin{cases} 1 & ; \theta < x < \theta + 1 \\ 0 & ; \text{ otherwise} \end{cases}$$

Show that $(X_{(1)}, X_{(n)})$ is jointly sufficient statistic.

(v)

$$f(x_i|\theta) = \begin{cases} \frac{1}{2i\theta} & ; -i(\theta - 1) < x_i < i(\theta + 1) \\ 0 & ; \text{ otherwise} \end{cases}$$

Prove that $(\frac{\text{Min}X_i}{i}, \frac{\text{Max}X_i}{i})$ is jointly sufficient statistic.

2. Let $(X_1, Y_1), (X_2, Y_2), \ldots, (X_n, Y_n)$ be iid rvs with uniform bivariate rvs as follows

$$f(x, y|\alpha, \beta, a, b) = \begin{cases} \frac{1}{(\beta - \alpha)(b - a)} & ; \alpha < x < \beta, a < y < b \\ 0 & ; \text{ otherwise} \end{cases}$$

Prove that $(X_{(1)}, X_{(n)}, Y_{(1)}, Y_{(n)})$ is jointly sufficient for (α, β, a, b).

3. The random variable $X_1, X_2, \ldots, X_n$ be iid rvs with a common Laplace distribution with density

$$f(x|\theta) = \begin{cases} \frac{1}{2\theta} \exp[-\frac{|x|}{\theta}] & ; -\infty < x < \infty, \theta > 0 \\ 0 & ; \text{ otherwise} \end{cases}$$

Prove that $T = \sum_{i=1}^n |X_i|$ is complete sufficient statistic.

4. Let $X_1, X_2, \ldots, X_n$ be iid rvs having the following distribution

(a) $\quad P(X = x|\lambda) = \dfrac{e^{-\lambda}\lambda^x}{x!}; x = 0, 1, 2, \ldots, \lambda > 0$

(b) $\quad P(X = x|p) = \dbinom{n}{x} p^x q^{n-x}; x = 0, 1, 2, \ldots, n, 0 < p < 1$ and n is known

Find (i) $P[X_1 = x|T = t]$ (ii) $P[X_1 = x_1, X_2 = x_2|T = t]$, where $T = \sum X_i$

5. Let $\{Y_{ij}\}, i = 1, 2, \ldots, p, j = 1, 2, \ldots, q$ be independent rvs and

$$Y_{ij} = \mu + \alpha_i + \epsilon_{ij}; i = 1, 2, \ldots, p, j = 1, 2, \ldots, q,$$

where $\alpha_1, \alpha_2, \ldots, \alpha_p$ are iid rvs with $N(0, \sigma_1^2)$ and $\epsilon_{ij} \sim N(0, \sigma^2)$. The parameters μ, σ_1 and σ are unknown. Show that (T, T_e, T_α) is jointly sufficient, where

$$T = \frac{1}{pq} \sum_{i=1}^{p} \sum_{j=1}^{q} Y_{ij}, T_e = \frac{1}{pq} \sum_{i=1}^{p} \sum_{j=1}^{q} (Y_{ij} - \bar{Y}_{i.})^2, \bar{Y}_{i.} = \frac{1}{q} \sum_{j=1}^{q} Y_{ij}, T_\alpha = q \sum_{i=1}^{p} (\bar{Y}_{i.} - T)^2$$

Remark: In linear model, it is called as one-way analysis.

6. Let $X_1, X_2, \ldots, X_n$ be iid rvs having the following pdf as

$$f(x|\theta) = \begin{cases} \frac{2}{\theta^2}(\theta - x) \; ; \; 0 \leq x \leq \theta \\ 0 \qquad\quad ; \; \text{otherwise} \end{cases}$$

Find the minimal sufficient statistic.

7. Let $X_1, X_2, \ldots, X_n$ be iid rvs with the following uniform distribution

$$f(x|N_1, N_2) = \begin{cases} \frac{1}{N_2 - N_1} \; ; \; x = N_1 + 1, \ldots, N_2 \\ 0 \qquad\; ; \; \text{otherwise} \end{cases}$$

Find the sufficient statistics for (N_1, N_2).

8. From the problem 1(iii) find the distribution of $f(X_1, X_2, X_3|T)$, where $T = (X_{(1)}, X_{(n)})$.

9. From the problem 8, find the distribution of $(X_1, X_2, \ldots, X_n|T)$, where $T = (X_{(1)}, X_{(n)})$. Further find $f(X_1|T)$ if it exists.

10. Find the sufficient statistics from the following distribution based on a sample of size n.

$$f(x|\theta) = a(\theta)2^{-\frac{x}{\theta}}; x = \theta, \theta + 1, \ldots, \theta > 0,$$

where $a(\theta)$ is constant.

Further find the distribution of $(X_1, X_2, \ldots, X_n|T)$, and $(X_i|T)$, where $i = 1, 2, \ldots, n$ and T is a sufficient statistics.

11. Let $X_1, X_2, \ldots, X_n$ be iid rvs from $\cup(\theta - \frac{1}{2}, \theta + \frac{1}{2})$, $\theta \in \Theta$, $T = (X_{(1)}, X_{(n)})$ is sufficient for θ. By using ancillary statistic, show that it is not complete.

12. Let $X_1, X_2, \ldots, X_n$ be a random sample from the inverse Gaussian distribution with the following pdf:

$$f(x|\mu, \theta) = \left(\frac{\theta}{2\pi x^3}\right)^{\frac{1}{2}} \exp\left[-\frac{\theta(x - \mu)^2}{2x\mu^2}\right]; x > 0, \mu > 0, \lambda > 0,$$

Prove that the statistics $\bar{X}$ and $T = \frac{n}{\sum_{i=1}^{n}(\frac{1}{Y_i} - \bar{X})}$ are sufficient and complete.

13. Let $X_1, X_2, \ldots, X_n$ be a random sample from the pdf

$$f(x|\mu) = \exp[-(x - \mu)]; \; x > \mu, \; \mu \in \Re$$

(i) Prove that $X_{(1)}$ is a complete sufficient statistic.

(ii) Using Basu's Theorem prove that $X_{(1)}$ and $S^2 = \left(\sum X_i - \bar{X}\right)^2$ are independent.
(Hint: Let $Z_i = X_i - \mu$, then $S^2 = \sum(Z_i - \bar{Z})^2$.

14. Let $X_1, X_2, \ldots, X_n$ be a random sample from the following Pareto distribution

$$f(x|\alpha, \theta) = \frac{\alpha\theta^\alpha}{x^{\alpha+1}}; \; x > \theta, \; \theta > 0, \; \alpha > 0$$

For known θ, prove that $\prod_{i=1}^{n} X_i$ is complete sufficient statistic for α.

15. Let $X_1, X_2, \ldots, X_n$ be iid rvs with $N(\mu, 1)$. Prove that $\bar{X}^2$ is not sufficient for μ^2.

16. Let $X_1, X_2, \ldots, X_n$ be iid rvs with (i) $\cup(\theta, 2\theta)$ (ii) $\cup(\theta, 3\theta)$
Show that $(X_{(1)}, X_{(n)})$ is not complete.

17. Let $X_1, X_2, \ldots, X_n$ be a random sample from the pdf $f(x|\theta)$

$$f(x|\theta) = \begin{cases} \theta x^{\theta-1} ; & 0 < x < 1 \\ 0 & ; \text{ otherwise} \end{cases}$$

Find sufficient and complete statistic for θ.

18. (Zacks 1971) Let X_1 and X_2 be random variables having the density function

$$f(x|\sigma) = \begin{cases} (1 + \sigma\sqrt{2\pi})^{-1} \exp[-\frac{x^2}{2\sigma^2}] & ; x < 0 \\ (1 + \sigma\sqrt{2\pi})^{-1} & ; 0 \leqslant x \leqslant 0 \\ (1 + \sigma\sqrt{2\pi})^{-1} \exp[-\frac{(x-1)^2}{2\sigma^2}] & ; 1 \leqslant x \end{cases}$$

Find the minimal sufficient statistic for σ.

References

Basu D (1959) The family of ancillary statistics. Sankhya 21:247–256

Casella G, Berger RL (2002) Statistical inference. Duxbury

Cox DR, Hinkley DV (1996) Theoretical statistics. Chapman and Hall

Dahiya RC, Kleyle RM (1975) Estimation of parameters and the mean life of a mixed failure time distribution. Commun Stat Theor Meth 29(11):2621–2642

Dixit VU (1993) Statistical inference for AR(I) process with mixed errors. Unpublished Ph.D Thesis, Department of Statistics, Shivaji University, Kollhapur

Dixit UJ, Kelkar SG (2011) Estimation of the parameters of binomial distribution in the presence of outliers. J Appl Stat Sci 19(4):489–498

Halmos PR, Savage LJ (1949) Application of the Radon-Nikodyn theorem to the theory of sufficient statistics. Ann Math Stat 20:225–241

Lehman EL, Casella G (1998) Theory of point estimation, 2nd edn. Springer, New York

Rohatagi VK, Saleh EAK (2001) An introduction to probability and statistics. Wiley

Sampson A, Spencer B (1976) Sufficiency, minimal sufficiency and the lack thereof. Am Stat 30(1):34–35

Stigler SM (1972) Completeness and unbiased estimation. Am Stat 26(2):28–29

Zacks S (1971) The theory of statistical inference. Wiley

Chapter 2
Unbiased Estimation

If the average estimate of several random samples is equal to the population parameter then the estimate is unbiased. For example, if credit card holders in a city were repetitively random sampled and questioned what their account balances were as of a specific date, the average of the results across all samples would equal the population parameter. If, however, only credit card holders in one specific business were sampled, the average of the sample estimates would be biased estimator of all account balances for the city and would not equal the population parameter.

If the mean value of an estimator in a sample equals the true value of the population mean then it is called an unbiased estimator. If the mean value of an estimator is either less than or greater than the true value of the quantity it estimates, then the estimator is called a biased estimator. For example, suppose you decide to choose the smallest or largest observation in a sample to be the estimator of the population mean. Such an estimator would be biased because the average of the values of this estimator would be always less or more than the true population mean.

2.1 Unbiased Estimates and Mean Square Error

Definition 2.1.1 A statistics $T(X)$ is called an unbiased estimator for a function of the parameter $g(\theta)$, provided that for every choice of θ,

$$ET(X) = g(\theta) \tag{2.1.1}$$

Any estimator that is not unbiased is called biased. The bias is denoted by $b(\theta)$.

$$b(\theta) = ET(X) - g(\theta) \tag{2.1.2}$$

© Springer Science+Business Media Singapore 2016
U.J. Dixit, *Examples in Parametric Inference with R*,
DOI 10.1007/978-981-10-0889-4_2

We will now define mean square error (mse)

$$
\begin{aligned}
MSE[T(X)] &= \mathrm{E}[T(X) - g(\theta)]^2 \\
&= \mathrm{E}[T(X) - \mathrm{E}T(X) + b(\theta)]^2 \\
&= \mathrm{E}[T(X) - \mathrm{E}T(X)]^2 + 2b(\theta)\mathrm{E}[T(X) - \mathrm{E}T(X)] + b^2(\theta) \\
&= \mathrm{V}[T(X)] + b^2(\theta) \\
&= \text{Variance of } [T(X)] + [\text{bias of } T(X)]^2
\end{aligned}
$$

Example 2.1.1 Let $(X_1, X_2, \ldots, X_n)$ be Bernoulli rvs with parameter θ, where θ is unknown. $\bar{X}$ is an estimator for θ. Is it unbiased ?

$$
\mathrm{E}\bar{X} = \frac{1}{n} \sum_{i=1}^{n} X_i = \frac{n\theta}{n} = \theta
$$

Thus, $\bar{X}$ is an unbiased estimator for θ.
We denote it as $\hat{\theta} = \bar{X}$.

$$
\mathrm{Var}(\bar{X}) = \frac{1}{n^2} \sum_{i=1}^{n} \mathrm{V}(X_i) = \frac{n\theta(1-\theta)}{n^2} = \frac{\theta(1-\theta)}{n}
$$

Example 2.1.2 Let $X_i (i = 1, 2, \ldots, n)$ be iid rvs from $N(\mu, \sigma^2)$, where μ and σ^2 are unknown.

Define $nS^2 = \sum_{i=1}^{n}(X_i - \bar{X})^2$ and $n\sigma^2 = \sum_{i=1}^{n}(X_i - \mu)^2$
Consider

$$
\begin{aligned}
\sum_{i=1}^{n}(X_i - \mu)^2 &= \sum_{i=1}^{n}(X_i - \bar{X} + \bar{X} - \mu)^2 \\
&= \sum_{i=1}^{n}(X_i - \bar{X})^2 + 2\sum_{i=1}^{n}(X_i - \mu)(\bar{X} - \mu) + n(\bar{X} - \mu)^2 \\
&= \sum_{i=1}^{n}(X_i - \bar{X})^2 + n(\bar{X} - \mu)^2
\end{aligned}
$$

Therefore,

$$
\sum_{i=1}^{n}(X_i - \bar{X})^2 = \sum_{i=1}^{n}(X_i - \mu)^2 - n(\bar{X} - \mu)^2
$$

$$E\left[\sum_{i=1}^{n}(X_i - \bar{X})^2\right] = E\left[\sum_{i=1}^{n}(X_i - \mu)^2\right] - nE[(\bar{X} - \mu)^2]$$

$$= n\sigma^2 - \frac{n\sigma^2}{n} = n\sigma^2 - \sigma^2$$

Hence,

$$E(S^2) = \sigma^2 - \frac{\sigma^2}{n} = \sigma^2\left(\frac{n-1}{n}\right)$$

Thus, S^2 is a biased estimator of σ^2.
Hence

$$b(\sigma^2) = \sigma^2 - \frac{\sigma^2}{n} - \sigma^2 = -\frac{\sigma^2}{n}$$

Further, $\frac{nS^2}{n-1}$ is an unbiased estimator of σ^2.

Example 2.1.3 Further, if $(n-1)S^2 = \sum_{i=1}^{n}(X_i - \bar{X})^2$, then $\frac{(n-1)S^2}{\sigma^2}$ has χ^2 with $(n-1)$ df. Here, we examine whether S is an unbiased estimator of σ.

Let $\frac{(n-1)S^2}{\sigma^2} = w$
Then

$$E(\sqrt{w}) = \int_0^\infty \frac{w^{\frac{1}{2}} e^{-\frac{w}{2}} w^{\frac{n-1}{2}-1}}{\Gamma\left(\frac{n-1}{2}\right) 2^{\frac{n-1}{2}}} dw$$

$$= \frac{\Gamma\left(\frac{n}{2}\right) 2^{\frac{n}{2}}}{\Gamma\left(\frac{n-1}{2}\right) 2^{\frac{n-1}{2}}} = \frac{\Gamma\left(\frac{n}{2}\right) 2^{\frac{1}{2}}}{\Gamma\left(\frac{n-1}{2}\right)}$$

$$E\left[\frac{(n-1)^{\frac{1}{2}}S}{\sigma}\right] = \frac{2^{\frac{1}{2}}\Gamma\left(\frac{n}{2}\right)}{\Gamma\left(\frac{n-1}{2}\right)}$$

Hence

$$E(S) = \frac{2^{\frac{1}{2}}\Gamma\left(\frac{n}{2}\right)}{\Gamma\left(\frac{n-1}{2}\right)} \frac{\sigma}{(n-1)^{\frac{1}{2}}} = \left(\frac{2}{n-1}\right)^{\frac{1}{2}} \frac{\Gamma\left(\frac{n}{2}\right)}{\Gamma\left(\frac{n-1}{2}\right)} \sigma$$

Therefore,

$$E\left(\frac{S}{\sigma}\right) = \left(\frac{2}{n-1}\right)^{\frac{1}{2}} \frac{\Gamma\left(\frac{n}{2}\right)}{\Gamma\left(\frac{n-1}{2}\right)}$$

Therefore,

$$\text{Bias}(S) = \sigma \left[\left(\frac{2}{n-1} \right)^{\frac{1}{2}} \frac{\Gamma\left(\frac{n}{2}\right)}{\Gamma\left(\frac{n-1}{2}\right)} - 1 \right]$$

Example 2.1.4 For the family (1.5.4), $\hat{p}$ is U-estimable and $\hat{\theta}$ is not U-estimable. For (p, θ), it can be easily seen that $\hat{p} = \frac{r}{n}$ and $E\hat{p} = p$. Next, we will show $\hat{\theta}$ is not U-estimable.

Suppose there exist a function $h(r, z)$ such that

$$Eh(r, z) = \theta \quad \forall \ (p, \theta) \in \Theta.$$

Since

$$EE[h(r, z)|r] = \theta$$

We get

$$\sum_{r=1}^{n} \binom{n}{r} p^r q^{n-r} \int_0^\infty h(r, z) \frac{e^{-\frac{z}{\theta}} z^{r-1} dz}{\theta^r \Gamma(r)} + q^n h(0, 0) = \theta$$

Substituting $\frac{p}{q} = \Psi$, and dividing q^n on both sides

$$\sum_{r=1}^{n} \Psi^r \binom{n}{r} \int_0^\infty h(r, z) \frac{e^{-\frac{z}{\theta}} z^{r-1} dz}{\theta^r \Gamma(r)} + h(0, 0) = \theta(1 + \Psi)^n, \quad \text{Since} \ q = (1 + \Psi)^{-1}$$

Comparing the coefficients of Ψ^r in both sides, we get, $h(0, 0) = \theta$, which is a contradiction.

Hence, there does not exist any unbiased estimator of θ. Thus θ is not U-estimable.

Example 2.1.5 Let X is $N(0, \sigma^2)$ and assume that we have one observation. What is the unbiased estimator of σ^2?

$$E(X) = 0$$

$$V(X) = EX^2 - (EX)^2 = \sigma^2$$

Therefore,

$$E(X^2) = \sigma^2$$

Hence X^2 is an unbiased estimator of σ^2.

Example 2.1.6 Sometimes an unbiased estimator may be absurd.

Let the rv X be $P(\lambda)$ and we want to estimate $\Psi(\lambda)$, where

$$\Psi(\lambda) = \exp[-k\lambda]; \quad k > 0$$

Let $T(X) = [-(k-1)]^x; k > 1$

$$E[T(X)] = \sum_{x=0}^{\infty} [-(k-1)]^x \frac{e^{-\lambda}\lambda^x}{x!}$$

$$= e^{-\lambda} \sum_{x=0}^{\infty} \frac{[-(k-1)\lambda]^x}{x!}$$

$$= e^{-\lambda} e^{[-(k-1)\lambda]}$$

$$= e^{-k\lambda}$$

$$T(x) = \begin{cases} [-(k-1)]^x > 0; & x \text{ is even and } k > 1 \\ [-(k-1)]^x < 0; & x \text{ is odd and } k > 1 \end{cases}$$

which is absurd since $\Psi(\lambda)$ is always positive.

Example 2.1.7 Unbiased estimator is not unique.

Let the rvs X_1 and X_2 are $N(\theta, 1)$. X_1, X_2, and $\alpha X_1 + (1-\alpha)X_2$ are unbiased estimators of θ, $0 \leq \alpha \leq 1$.

Example 2.1.8 Let $X_1, X_2, \ldots, X_n$ be iid rvs from Cauchy distribution with parameter θ. Find an unbiased estimator of θ.

Let

$$f(x|\theta) = \frac{1}{\pi[1 + (x-\theta)^2]}; \quad -\infty < x < \infty, -\infty < \theta < \infty$$

$$F(x|\theta) = \int_{-\infty}^{x} \frac{dy}{\pi[1 + (y-\theta)^2]}$$

$$= \frac{1}{2} + \frac{1}{\pi} \tan^{-1}(x - \theta)$$

Let $g(x_{(r)})$ be the pdf of $X_{(r)}$, where $X_{(r)}$ is the rth order statistics.

$$g(x_{(r)}) = \frac{n!}{(n-r)!(r-1)!} f(x_{(r)})[F(x_{(r)})]^{r-1}[1 - F(x_{(r)})]^{n-r}$$

$$= \frac{n!}{(n-r)!(r-1)!} \left[\frac{1}{\pi} \frac{1}{[1 + (x_{(r)} - \theta)^2]} \right] \left[\frac{1}{2} + \frac{1}{\pi} \tan^{-1}(x_{(r)} - \theta) \right]^{r-1} \left[\frac{1}{2} - \frac{1}{\pi} \tan^{-1}(x_{(r)} - \theta) \right]^{n-r}$$

$$E(X_{(r)} - \theta) = \frac{n!}{(n-r)!(r-1)!} \frac{1}{\pi} \int_{-\infty}^{\infty} \frac{x_{(r)} - \theta}{[1 + (x_{(r)} - \theta)^2]} \left[\frac{1}{2} + \frac{1}{\pi} \tan^{-1}(x_{(r)} - \theta) \right]^{r-1}$$

$$\times \left[\frac{1}{2} - \frac{1}{\pi} \tan^{-1}(x_{(r)} - \theta)^{n-r} \right] dx_{(r)}$$

Let $(x_{(r)} - \theta) = y$

$$E(X_{(r)} - \theta) = C_{rn} \frac{1}{\pi} \int_{-\infty}^{\infty} \frac{y}{1 + y^2} \left[\frac{1}{2} + \frac{1}{\pi} \tan^{-1} y \right]^{r-1} \left[\frac{1}{2} - \frac{1}{\pi} \tan^{-1} y \right]^{n-r} dy,$$

where $C_{rn} = \frac{n!}{(n-r)!(r-1)!}$
Let

$$u = \frac{1}{2} + \frac{1}{\pi} \tan^{-1} y \Rightarrow u - \frac{1}{2} = \frac{1}{\pi} \tan^{-1} y$$

$$\Rightarrow \left(u - \frac{1}{2} \right) \pi = \tan^{-1} y \Rightarrow y = \tan \left(u - \frac{1}{2} \right) \pi \Rightarrow y = -\cot \pi u$$

$$dy = \pi \left[\frac{(\cos \pi u)(\cos \pi u)}{\sin^2 \pi u} + \frac{\sin \pi u}{\sin \pi u} \right] du$$

$$= \pi[\cot^2 \pi u + 1] = \pi[y^2 + 1]du$$

$$E(X_{(r)} - \theta) = -\frac{n!}{(n-r)!(r-1)!} \int_0^1 u^{r-1}(1 - u)^{n-r} \cot \pi u du$$

$$= -C_{rn} \int_0^1 u^{r-1}(1 - u)^{n-r} \cot \pi u du$$

Replace r by $n - r + 1$

$$E(X_{(n-r+1)} - \theta) = -\frac{n!}{(n-r)!(r-1)!} \int_0^1 \cot(\pi u) u^{n-r}(1-u)^{r-1} du$$

Let $1 - u = w$

$$= -\frac{n!}{(n-r)!(r-1)!} \int_0^1 (-1) \cot[\pi(1-w)](1-w)^{n-r} w^{r-1} dw$$

$$= \frac{n!}{(n-r)!(r-1)!} \int_0^1 \cot(\pi w)(1-w)^{n-r} w^{r-1} dw$$

Now

$$\int_0^1 u^{r-1}(1-u)^{n-r} \cot \pi u du = \int_0^1 \cot(\pi w)(1-w)^{n-r} w^{r-1} dw$$

$$E[(x_{(r)} - \theta) + (x_{(n-r+1)} - \theta)] = 0$$

$$E[X_{(r)} + X_{(n-r+1)}] = 2\theta$$

$$\hat{\theta} = \frac{x_{(r)} + x_{(n-r+1)}}{2}$$

Therefore, $\frac{x_{(r)} + x_{(n-r+1)}}{2}$ is an unbiased estimator of θ.

Note: Moments of Cauchy distribution does not exist but still we get an unbiased estimator of θ.

Example 2.1.9 Let X be rv with $B(1, p)$. We examine whether p^2 is U-estimable.

Let $T(x)$ be an unbiased estimator of p^2

$$\sum_{x=0}^1 T(x) p^x (1-p)^{1-x} = p^2$$

$$T(0)(1-p) + T(1)p = p^2$$

$$p[T(1) - T(0)] + T(0) = p^2$$

Coefficient of p^2 does not exist.

Hence, an unbiased estimator of p^2 does not exist.

Empirical Distribution Function

Let $X_1, X_2, \ldots, X_n$ be a random sample from a continuous population with df F and pdf f. Then the order statistics $X_{(1)} \leq X_{(2)} \leq \cdots \leq X_{(n)}$ is a sufficient statistics. Define $\hat{F}(x) = \frac{\text{Number of } X_i's \leq x}{n}$, same thing can be written in terms of order statistics as,

$$\hat{F}(x) = \begin{cases} 0 \ ; \ X_{(1)} > x \\ \frac{k}{n} \ ; \ X_{(k)} \leq x < X_{(k+1)} \\ 1 \ ; \ x \geq X_{(n)} \end{cases}$$

$$= \frac{1}{n} \sum_{j=1}^{n} \mathbf{I}(x - X_{(j)})$$

where

$$I(y) = \begin{cases} 1; \ y \geq 0 \\ 0; \ \text{otherwise} \end{cases}$$

Example 2.1.10 Show that empirical distribution function is an unbiased estimator of $F(x)$

$$\hat{F}(x) = \frac{1}{n} \sum_{j=1}^{n} \mathbf{I}(x - X_{(j)})$$

$$\mathrm{E}\hat{F}(x) = \frac{1}{n} \sum_{j=1}^{n} P[X_{(j)} \leq x]$$

$$= \frac{1}{n} \sum_{j=1}^{n} \sum_{k=j}^{n} \binom{n}{k} [F(x)]^k [1 - F(x)]^{n-k} \text{ (see (Eq. 20 in "Prerequisite"))}$$

$$= \frac{1}{n} \sum_{j=1}^{k} \sum_{k=1}^{n} \binom{n}{k} [F(x)]^k [1 - F(x)]^{n-k}$$

$$= \frac{1}{n} \sum_{k=1}^{n} \binom{n}{i} [F(x)]^k [1 - F(x)]^{n-k} \sum_{j=1}^{k} (1)$$

$$= \frac{1}{n} \sum_{k=1}^{n} k \binom{n}{k} [F(x)]^k [[1 - F(x)]^{n-k}$$

$$= \frac{1}{n} [nF(x)] = F(x)$$

Note: One can see that $I(x - X_{(j)})$ is a Bernoulli random variable. Then $EI(x - X_{(j)}) = F(x)$, so that $E\hat{F}(x) = F(x)$. We observe that $\hat{F}(x)$ has a Binomial distribution with mean $F(x)$ and variance $\frac{F(x)[1-F(x)]}{n}$. Using central limit theorem, for iid rvs, we can show that as $n \to \infty$

$$\sqrt{n} \left[\frac{\hat{F}(x) - F(x)}{\sqrt{F(x)[1 - F(x)]}} \right] \to N(0, 1).$$

2.2 Unbiasedness and Sufficiency

Let $X_1, X_2, \ldots, X_n$ be a random sample from a Poisson distribution with parameter λ. Then $T = \sum X_i$ is sufficient for λ. Also $E(X_1) = \lambda$ then X_1 is unbiased for λ but it is not based on T. Moreover, we can say that it is not a function of T.
(i) Let $T_1 = E(X_1|T)$. We will prove that T_1 is better than X_1 as an estimate of λ. The distribution of X_1 given T as

$$f(X_1|T = t) = \begin{cases} \binom{t}{x_1} \left(\frac{1}{n}\right)^{x_1} \left(1 - \frac{1}{n}\right)^{t-x_1} ; & x_1 = 0, 1, 2, \ldots, t \\ 0; & \text{otherwise} \end{cases} \qquad (2.2.1)$$

$E[X_1|T = t] = \frac{t}{n}$ and distribution of T is $P(n\lambda)$

$$V\left(\frac{T}{n}\right) = \frac{1}{n^2} V(T) = \frac{n\lambda}{n^2} = \frac{\lambda}{n}$$

$$V(X_1) > V\left(\frac{T}{n}\right) \qquad (2.2.2)$$

(ii) Let $T_2 = \left(X_n, \sum_{i=1}^{n-1} X_i\right)$ is also sufficient for λ.

$T_0 = \sum_{i=1}^{n-1} X_i$. We have to find the distribution of X_1 given T_2

$$
\begin{aligned}
P[X_1|T_2] &= \frac{P[X_1 = x_1, T_2 = t_2]}{P[T_2 = t_2]} \\
&= \frac{P[X_1 = x_1, X_n = x_n, \sum_{i=2}^{n-1} X_i = t_0 - x_1]}{P[X_n = x_n, \sum_{i=1}^{n-1} X_i = t_0]} \\
&= \frac{e^{-\lambda}\lambda^{x_1}}{x_1!} \frac{e^{-\lambda}\lambda^{x_n}}{x_n!} \frac{e^{-(n-2)\lambda}[(n-2)\lambda]^{t_0-x_1}}{(t_0 - x_1)!} \frac{x_n!}{e^{-\lambda}\lambda^{x_n}} \frac{t_0!}{e^{-(n-1)\lambda}[(n-1)\lambda]^{t_0}} \\
&= \frac{t_0!}{x_1!(t_0 - x_1)!} \frac{(n-2)^{t_0-x_1}}{(n-1)^{t_0}} \\
&= \binom{t_0}{x_1} \left(\frac{n-2}{n-1}\right)^{t_0} \left(\frac{1}{n-2}\right)^{x_1} \\
&= \binom{t_0}{x_1} \left(\frac{1}{n-1}\right)^{x_1} \left(\frac{n-2}{n-1}\right)^{t_0-x_1}; \quad x_1 = 0, 1, 2, \ldots, t_0 \qquad (2.2.3)
\end{aligned}
$$

Now X_1 given T_2 has $B(t_0, \frac{1}{n-1})$

$$
E[X_1|T_2] = \frac{t_0}{n-1} = \frac{\sum_{i=1}^{n-1} X_i}{n-1}
$$

$$
V\left[\frac{T_0}{n-1}\right] = \frac{(n-1)\lambda}{(n-1)^2} = \frac{\lambda}{n-1} \qquad (2.2.4)
$$

We conclude that $\frac{\sum_{i=1}^{n-1} X_i}{n-1}$ is unbiased for λ and has smaller variance than X_1. Comparing the variance of X_1, $\bar{X}$, and $\frac{\sum_{i=1}^{n-1} X_i}{n-1}$, we have

$$
V(X_1) > V\left(\frac{\sum_{i=1}^{n-1} X_i}{n-1}\right) > V(\bar{X})
$$

This implies $\lambda > \frac{\lambda}{n-1} > \frac{\lambda}{n}$.

Hence, we prefer $\bar{X}$ to $\frac{\sum_{i=1}^{n-1} X_i}{n-1}$ and X_1.

Note:

1. One should remember that $E(X_1|T = t)$ and $E(X_1|T_2 = t_2)$ are the unbiased estimators for λ.
2. Even though sufficient statistic reduce the data most we have to search for the minimal sufficient statistic.

Let $T_1(X_1, X_2, \ldots, X_n)$ and $T_2(X_1, X_2, \ldots, X_n)$ be two unbiased estimates of a parameter θ. Further, suppose that $T_1(X_1, X_2 \quad\quad Y_n)$ be sufficient for θ, and $T_1 = f(t)$ for some function f. If sufficiency of T for θ is t_0 have any meaning, we should expect T_1 to perform better than T_2 in the sense that $V(T_1) \le V(T_2)$. More generally, given an unbiased estimate h for θ, is it possible to improve upon h using a sufficient statistics for θ? We have seen in the above example that the estimator is improved. Therefore, the answer is "Yes."

If T is sufficient for θ then by definition, the conditional distribution of $(X_1, X_2, \ldots, X_n)$ given T does not depend on θ.

Consider $E\{h(X_1, X_2, \ldots, X_n) | T(X_1, X_2, \ldots, X_n)\}$. Since T is sufficient then this expected value does not depend on θ.

Set $T_1 = E\{h(X_1, X_2, \ldots, X_n) | T(X_1, X_2, \ldots, X_n)\}$ is itself an estimate of θ.

Using Theorem 5 in "Prerequisite", we can get ET_1

$$E(T_1) = E\left[E\{h(X_1, X_2, \ldots, X_n) | T(X_1, X_2, \ldots, X_n)\}\right]$$

$$= E\{h(X_1, X_2, \ldots, X_n)\} = \theta$$

Since h is unbiased for θ, hence $E(T_1)$ is also unbiased for θ.

Thus, we have found out another unbiased estimate of θ that is a function of the sufficient statistic. What about the variance of T_1?

Using Theorem 6 in "Prerequisite"

$$V[h(X_1, X_2, \ldots, X_n)] = E\{V(h(X_1, X_2, \ldots, X_n) | T(X_1, X_2, \ldots, X_n))\}$$
$$+ V\{Eh(X_1, X_2, \ldots, X_n) | T(X_1, X_2, \ldots, X_n)\}$$

$$= E\{V(h(X_1, X_2, \ldots, X_n) | T(X_1, X_2, \ldots, X_n))\} + V(T_1) \qquad (2.2.5)$$

Since $V(h|T) > 0$ so that $E[V(h|T)] > 0$

From (2.2.5), $V(T_1) < V[h(X)]$

If $T(X)$ is minimal sufficient for θ then T_1 is the best unbiased estimate of θ. Sometimes we face the problem of computations of expectation of h given T.

The procedure for finding unbiased estimates with smaller variance can now be summarized.

1. Find the minimal sufficient statistic.
2. Find a function of this sufficient statistic that is unbiased for the parameter.

Remark If you have a minimal sufficient statistic then your unbiased estimate will have the least variance. If not, the unbiased estimate you construct will not be the best possible but you have the assurance that it is based on a sufficient statistic.

Theorem 2.2.1 *Let $h(X)$ be an unbiased estimator of $g(\theta)$. Let $T(X)$ be a sufficient statistics for θ. Define $\Psi(T) = E(h|T)$. Then $E[\Psi(T)] = g(\theta)$ and $V[\Psi(T)] \leq V(h) \ \forall \ \theta$. Then $\Psi(T)$ is uniformly minimum variance unbiased estimator (UMVUE) of $g(\theta)$.*
This theorem is known as Rao–Blackwell Theorem.

Proof Using Theorem 5 in "Prerequisite",

$$E[h(X)] = E[Eh(X)|T = t] = E[\Psi(T)] = g(\theta) \qquad (2.2.6)$$

Hence $\Psi(T)$ is unbiased estimator of $g(\theta)$
Using Theorem 6 in "Prerequisite",

$$V[h(X)] = V[E(h(X)|T(X))] + E[V(h(X)|T(X))]$$

$$= V[\Psi(T)] + E[V(h(X)|T(X))]$$

Since $V[h(X)|T(X)] \geq 0$ and $E[V(h(X)|T(X))] > 0$
Therefore,

$$V[\Psi(T)] \leq V[h(X)] \qquad (2.2.7)$$

We have to show that $\Psi(T)$ is an estimator,
i.e., $\Psi(T)$ is a function of sample only and independent of θ.
From the definition of sufficiency, we can conclude that the distribution of $h(X)$ given $T(X)$ is independent of θ. Hence $\Psi(T)$ is an estimator.
Therefore, $\Psi(T)$ is UMVUE of $g(\theta)$.

Note: We should remember that conditioning on anything will not result in improving the estimator.

Example 2.2.1 Let X_1, X_2 be iid $N(\theta, 1)$.

Let

$$h(X) = \bar{X} = \frac{X_1 + X_2}{2},$$

$$Eh(X) = \theta \quad \text{and} \quad V[h(X)] = \frac{1}{2},$$

Now conditioning on X_1, which is not sufficient. Let $\Psi(X_1) = E(\bar{X})|X_1)$.
Using Theorem 5 in "Prerequisite", $E[\Psi(X_1)] = E\bar{X} = \theta$. Using Theorem 6 in "Prerequisite", $V[\Psi(X_1)] \leq V(\bar{X})$. Hence $\Psi(X_1)$ is better than $\bar{X}$. But question is whether $\Psi(X_1)$ is an estimator?

$$\Psi(X_1) = E(\bar{X}|X_1)$$

$$= E\left(\frac{Y_1 + Y_2}{2}|X_1\right) = \frac{1}{2}E(X_1|X_1) + \frac{1}{2}E(X_2|X_1)$$

$$= \frac{1}{2}X_1 + \frac{1}{2}E(X_2) \ (X_1 \text{ and } X_2 \text{ are independent})$$

$$= \frac{1}{2}X_1 + \frac{1}{2}\theta$$

Hence $\Psi(X_1)$ is not an estimator. This imply that we cannot say that $\Psi(X_1)$ is better than $\bar{X}$.

Theorem 2.2.2 (Lehmann–Scheffe Theorem) *If T is a complete sufficient statistic and there exists an unbiased estimate h of $g(\theta)$, there exists a unique UMVUE of θ, which is given by $Eh|T$.*

Proof Let h_1 and h_2 be two unbiased estimators of $g(\theta)$ Rao–Blackwell theorem, $E(h_1|T)$ and $E(h_2|T)$ are both UMVUE of $g(\theta)$.
Hence $E[E(h_1|T) - E(h_2|T)] = 0$
But T is complete therefore

$$[E(h_1|T) - E(h_2|T)] = 0$$

This implies $E(h_1|T) = E(h_2|T)$.
Hence, UMVUE is unique.
Even if we cannot obtain sufficient and complete statistic for a parameter, still we can get UMVUE for a parameter. Therefore, we can see the following theorem:

Theorem 2.2.3 *Let T_0 be the UMVUE of $g(\theta)$ and v_0 be the unbiased estimator of 0. Then T_0 is UMVUE if and only if $Ev_0T_0 = 0 \ \forall \ \theta \in \Theta$. Assume that the second moment exists for all unbiased estimators of $g(\theta)$.*

Proof (i) Suppose T_0 is UMVUE and $Ev_0T_0 \neq 0$ for some θ_0 and v_0 where $Ev_0 = 0$. Then $T_0 + \alpha v_0$ is unbiased for all real α. If $Ev_0^2 = 0$ then v_0 is degenerate rv. Hence $Ev_0T_0 = 0$. This implies $P[v_0 = 0] = 1$.
Let $Ev_0^2 > 0$

$$E[T_0 + \alpha v_0 - g(\theta)]^2 = E(T_0 + \alpha v_0)^2 - 2g(\theta)E(T_0 + \alpha v_0) + g^2(\theta)$$
$$= E(T_0 + \alpha v_0)^2 - g^2(\theta)$$
$$= E(T_0)^2 + 2\alpha E(T_0 v_0) + \alpha^2 Ev_0^2 - g^2(\theta) \quad (2.2.8)$$

Choose α such that (2.2.8) is equal to zero, then differentiating (2.2.8) with respect to α, we get

$$= 2E(T_0 v_0) + 2\alpha Ev_0^2 = 0$$

Hence

$$\alpha_0 = -\frac{\mathrm{E}(T_0 v_0)}{\mathrm{E}v_0^2} \tag{2.2.9}$$

$$\begin{aligned}
\mathrm{E}(T_0 + \alpha v_0)^2 &= \mathrm{E}(T_o)^2 + 2\alpha \mathrm{E}(T_0 v_0) + \alpha^2 \mathrm{E}v_0^2 \\
&= \mathrm{E}(T_0)^2 - \frac{(\mathrm{E}(T_0 v_0))^2}{\mathrm{E}v_0^2} \\
&< \mathrm{E}(T_0)^2
\end{aligned} \tag{2.2.10}$$

Because $\frac{(\mathrm{E}T_0 v_0)^2}{\mathrm{E}v_0^2} > 0$ (our assumption $\mathrm{E}(T_0 v_0) \neq 0$)
Then we can conclude that

$$\mathrm{V}(T_0 + \alpha v_0) < \mathrm{E}(T_0)^2$$

which is a contradiction, because T_0 is UMVUE.
Hence $\mathrm{E}v T_0 = 0$
(ii) Suppose that

$$\mathrm{E}v T_0 = 0 \quad \forall \ \theta \in \Theta \tag{2.2.11}$$

Let T be an another unbiased estimator of θ, then $\mathrm{E}(T - T_0) = 0$.
Now T_0 is unbiased estimator and $(T - T_0)$ is unbiased estimator of 0, then by
(2.2.11),

$$\mathrm{E}T_0(T - T_0) = 0$$

$$\mathrm{E}T_0 T - \mathrm{E}T_0^2 = 0$$

This implies $\mathrm{E}T_0^2 = \mathrm{E}T_0 T$
Using Cauchy–Schwarz's inequality

$$\mathrm{E}T_0 T \le (\mathrm{E}T_0^2)^{\frac{1}{2}} (\mathrm{E}T^2)^{\frac{1}{2}}$$

Therefore,

$$\mathrm{E}T_0^2 \le (\mathrm{E}T_0^2)^{\frac{1}{2}} (\mathrm{E}T^2)^{\frac{1}{2}}$$

$$(\mathrm{E}T_0^2)^{\frac{1}{2}} \le (\mathrm{E}T^2)^{\frac{1}{2}} \tag{2.2.12}$$

Now if $ET_0^2 = 0$ then $P[T_0 = 0] = 1$

Then (2.2.12) is true

Next, if $ET_0^2 > 0$ then also (2.2.12) is true

Hence $V(T_0) \leq V(T) \Rightarrow T_0$ is UMVUE.

Remark We would like to mention the comment made by Casella and Berger (2002). "An unbiased estimator of 0 is nothing more than random noise; that is there is no information in an estimator of 0. It makes sense that most sensible way to estimate 0 is with 0, not with random noise. Therefore, if an estimator could be improved by adding random noise to it, the estimator probably is defective."

Casella and Berger (2002) gave an interesting characterization of best unbiased estimators.

Example 2.2.2 Let X be an rv with $\cup(\theta, \theta + 1)$, $EX = \theta + \frac{1}{2}$, then $(X - \frac{1}{2})$ is an unbiased estimator of θ and its variance is $\frac{1}{12}$. For this pdf, unbiased estimators of zero are periodic functions with period 1.

If $h(x)$ satisfies $\int_\theta^{\theta+1} h(x) = 0$

$$\frac{d}{d\theta} \int_\theta^{\theta+1} h(x) = 0$$

$$h(\theta + 1) - h(\theta) = 0 \quad \forall \quad \theta$$

Such a function is $h(x) = \sin 2\pi x$.

Now,

$$\text{Cov}\left[X - \frac{1}{2}, \sin 2\pi X\right] = \text{Cov}[X, \sin 2\pi X] = \int_\theta^{\theta+1} x \sin 2\pi x \, dx$$

$$= -\frac{(\theta + 1)\cos 2\pi(\theta + 1)}{2\pi} + \theta\frac{\cos 2\pi\theta}{2\pi}$$
$$+ \frac{\sin 2\pi(\theta + 1)}{4\pi^2} - \frac{\sin 2\pi\theta}{4\pi^2}$$

Since $\sin 2\pi(\theta + 1) = \sin 2\pi\theta$

$$\cos 2\pi(\theta + 1) = \cos 2\pi\theta \cos 2\pi - \sin 2\pi\theta \sin 2\pi$$

$$= \cos 2\pi\theta \quad (\cos 2\pi = 1, \sin 2\pi = 0)$$

$$\text{Cov}[X, \sin 2\pi X] = -\frac{\cos 2\pi\theta}{2\pi}$$

Hence $\left(X - \frac{1}{2}\right)$ is correlated with an unbiased estimator of zero. Therefore, $\left(X - \frac{1}{2}\right)$ cannot be the best unbiased estimator of θ.

Example 2.2.3   Sometimes UMVUE is not sensible.

Let $X_1, X_2, \ldots, X_n$ be $N(\mu, 1)$. Now X_1 is unbiased estimator for μ and $\bar{X}$ is complete sufficient statistic for μ then $E(X_1|\bar{X})$ is UMVUE. We will show that $E(X_1|\bar{X}) = \bar{X}$. See (ii) of Example 2.2.11
Note that $\bar{X}$ is $N(\mu, \frac{1}{n})$

$$E(X_1\bar{X}) = \frac{1}{n}EX_1[X_1 + X_2 + \cdots + X_n]$$

$$= \frac{1}{n}[E(X_1^2) + E(X_1X_2) + \cdots + E(X_1X_n)]$$

$$= \frac{1}{n}[1 + \mu^2 + \mu^2 + \cdots + \mu^2]$$

$$\mathrm{Cov}(X_1, \bar{X}) = \frac{1 + n\mu^2}{n} - \mu^2 = \frac{1}{n}$$

$$E(X_1|\bar{X}) = EX_1 + \frac{\mathrm{Cov}(X_1, \bar{X})}{\mathrm{V}(\bar{X})}[\bar{X} - E\bar{X}]$$

$$= \mu + \frac{1}{n}n[\bar{X} - \mu]$$

$$= \mu + [\bar{X} - \mu] = \bar{X}$$

$(X_1, \bar{X})$ is a bivariate rv with mean

$$\begin{pmatrix} \mu \\ \mu \end{pmatrix}$$

and covariance matrix

$$\begin{pmatrix} 1 & \frac{1}{n} \\ \frac{1}{n} & \frac{1}{n} \end{pmatrix}$$

In this example, we want to estimate $d(\mu) = \mu^2$ then $\left(\bar{X}^2 - \frac{1}{n}\right)$ is UMVUE for μ^2. One can easily see that $E\bar{X}^2 = \frac{1}{n} + \mu^2$.

Hence $E\left(\bar{X}^2 - \frac{1}{n}\right) = \mu^2$ and $\bar{X}^2$ is sufficient and complete for μ^2.

Now μ^2 is always positive but sometimes $\left(\bar{X}^2 - \frac{1}{n}\right)$ may be negative. Therefore, UMVUE for μ^2 is not sensible, see (2.2.56).

Now, we will find UMVUE for different estimators for different distributions.

Example 2.2.4 Let $X_1, X_2, \ldots, X_n$ are iid rvs with $B(n, p), 0 < p < 1$. In this case, we have to find the UMVUE of $p^r q^s$, $q = 1 - p$, r, $s \neq 0$ and $P[X \leq c]$. Assume n is known.

Binomial distribution belongs to exponential family. So that $\sum_{i=1}^{n} X_i$ is sufficient and complete for p.

(i) The distribution of T is $B(mn, p)$.

Let $U(t)$ be unbiased estimator for $p^r q^s$.

$$\sum_{t=0}^{nm} u(t) \binom{nm}{t} p^t q^{nm-t} = p^r q^s \qquad (2.2.13)$$

$$\sum_{t=0}^{nm} u(t) \binom{nm}{t} p^{t-r} q^{nm-t-s} = 1$$

$$\sum_{t=r}^{nm-s} u(t) \frac{\binom{nm}{t}}{\binom{nm-s-r}{t-r}} \binom{nm-s-r}{t-r} p^{t-r} q^{nm-t-s} = 1$$

Then

$$u(t) \frac{\binom{nm}{t}}{\binom{nm-s-r}{t-r}} = 1$$

Hence

$$u(t) = \begin{cases} \dfrac{\binom{nm-s-r}{t-r}}{\binom{nm}{t}} & ; \ t = r, r+1, r+2, \ldots, nm - s \\ 0 & ; \text{otherwise} \end{cases} \qquad (2.2.14)$$

Note: For $m = n = 1, r = 2$, and $s = 0$, the unbiased estimator of p^2 does not exist, see Example 2.1.9

(ii) To find UMVUE of $P[X \leq c]$

Now

$$P[X \leq c] = \sum_{x=0}^{c} \binom{n}{x} p^x q^{n-x}$$

Then UMVUE of

$$p^x q^{n-x} = \frac{\binom{nm-n}{t-x}}{\binom{nm}{t}}$$

Hence UMVUE of $P[X \le c]$

$$= \begin{cases} \sum_{x=0}^{c} \binom{n}{x} \frac{\binom{nm-n}{t-x}}{\binom{nm}{t}} \; ; \; t = x, x+1, x+2, \dots, nm-n+x, \quad c \le \min(t, n) \\ 1 \qquad\qquad\qquad ; \text{otherwise} \end{cases}$$

(2.2.15)

Note: UMVUE of $P[X = x] = \binom{n}{x} p^x q^{n-x}$ is $\frac{\binom{n}{x}\binom{nm-n}{t-x}}{\binom{nm}{t}}$; $x = 0, 1, 2, \dots, t$

Particular cases:

(a) $r = 1, s = 0$. From (2.2.14), we will get UMVUE of p,

$$u(t) = \frac{\binom{nm-1}{t-1}}{\binom{nm}{t}} = \frac{t}{nm}$$

(2.2.16)

(b) $r = 0, s = 1$. From (2.2.14), we will get UMVUE of q,

$$u(t) = \frac{\binom{nm-1}{t}}{\binom{nm}{t}} = \frac{nm - t}{nm} = 1 - \frac{t}{nm}$$

(2.2.17)

(c) $r = 1, s = 1$. From (2.2.14), we will get UMVUE of pq,

$$u(t) = \left(\frac{t}{nm}\right)\left(\frac{nm - t}{nm - 1}\right)$$

(2.2.18)

Remark We have seen that in (2.2.16), (2.2.17), and (2.2.18),

$$\hat{p} = \frac{t}{nm}; \hat{q} = 1 - \frac{t}{nm} \text{ and } \widehat{pq} = \left(\frac{t}{nm}\right)\left(\frac{nm - t}{nm - 1}\right)$$

Hence, UMVUE of $pq \ne$ (UMVUE of p) (UMVUE of q).

Example 2.2.5 Let $X_1, X_2, \dots, X_m$ are iid rvs with $P(\lambda)$. In this case we have to find UMVUE of (i) $\lambda^r e^{-s\lambda}$ (ii) $P[X \le c]$

Poisson distribution belongs to exponential family. So that $T = \sum_{i=1}^{n} X_i$ is sufficient and complete for λ.

(i) The distribution of T is $P(m\lambda)$.

Let $U(t)$ be unbiased estimator for $\lambda^r e^{-s\lambda}$

$$\sum_{t=0}^{\infty} u(t) \frac{e^{-m\lambda}(m\lambda)^t}{t!} = e^{-s\lambda} \lambda^r$$

(2.2.19)

$$\sum_{t=0}^{\infty} u(t) \frac{e^{-(m-s)\lambda} m^t \lambda^{t-r}}{t!}$$

$$\sum_{t=r}^{\infty} u(t) \frac{m^t}{(m-s)^{t-r}} \frac{(t-r)!}{t!} \frac{e^{-(m-s)\lambda}[(m-s)\lambda]^{t-r}}{(t-r)!} = 1$$

Then

$$u(t) \frac{m^t}{(m-s)^{t-r}} \frac{(t-r)!}{t!} = 1$$

$$u(t) = \begin{cases} \frac{(m-s)^{t-r}}{m^t} \frac{t!}{(t-r)!} & ; \ t = r, r+1, \ldots, s \le m \\ 0 & ; \ \text{otherwise} \end{cases} \tag{2.2.20}$$

(ii) To find UMVUE of $P[X \le c]$

$$P[X \le c] = \sum_{x=0}^{c} \frac{e^{-\lambda} \lambda^x}{x!}$$

Now, UMVUE of $e^{-\lambda} \lambda^x$ is $\frac{(m-1)^{(t-x)}}{m^t} \frac{t!}{(t-x)!}$
UMVUE of $P[X \le c]$

$$= \sum_{x=0}^{c} \frac{t!}{(t-x)! x!} \left(\frac{m-1}{m}\right)^t \left(\frac{1}{m-1}\right)^x$$

$$= \begin{cases} \sum_{x=0}^{c} \binom{t}{x} \left(\frac{1}{m}\right)^x \left(\frac{m-1}{m}\right)^{t-x} & ; c \le t \\ 1 & ; \ \text{otherwise} \end{cases} \tag{2.2.21}$$

Remark UMVUE of $P[X = x] = \frac{e^{-\lambda} \lambda^x}{x!}$ is $\binom{t}{x} \left(\frac{1}{m}\right)^x \left(\frac{m-1}{m}\right)^{t-x}$; $\quad x = 0, 1, \ldots, t$
Particular cases:
(a) $s = 0, r = 1$
From (2.2.20), we will get the UMVUE of λ,

$$u(t) = \frac{m^{t-1} t!}{m^t (t-1)!} = \frac{t}{m} \tag{2.2.22}$$

(b) $s = 1, r = 0$
From (2.2.20), we will get the UMVUE of $e^{-\lambda}$,

$$u(t) = \left(\frac{m-1}{m}\right)^t \tag{2.2.23}$$

(c) $s = 1, r = 1$
From (2.2.20), we will get the UMVUE of $\lambda e^{-\lambda}$

$$u(t) = \frac{(m-1)^{t-1} t!}{m^t (t-1)!} = \left(\frac{m-1}{m}\right)^t \frac{t}{m-1} \qquad (2.2.24)$$

Remark UMVUE of $\lambda e^{-\lambda} \neq$ (UMVUE of λ)(UMVUE of $e^{-\lambda}$)

Example 2.2.6 Let $X_1, X_2, \ldots, X_m$ are iid rvs with $NB(k, p)$. In this case we have to find UMVUE of

1. $p^r q^s (r, s \neq 0)$
2. $P[X \leq c]$

$P[X = x] = $ Probability of getting kth successes at the xth trial

$$= \binom{k+x-1}{x} p^k q^x; \quad x = 0, 1, 2, \ldots, 0 < p < 1, \ q = 1 - p \quad (2.2.25)$$

Negative Binomial distribution belongs to exponential family.
Therefore, $T = \sum_{i=1}^{m} X_i$ is complete and sufficient for p. Distribution of T is $NB(mk, p)$.
Let $U(t)$ be unbiased estimator for $p^r q^s$

$$\sum_{t=0}^{\infty} u(t) \binom{mk+t-1}{t} p^{mk} q^t = p^r q^s.$$

$$\sum_{t=0}^{\infty} u(t) \binom{mk+t-1}{t} p^{mk-r} q^{t-s} = 1$$

$$\sum_{s=0}^{\infty} u(t) \frac{\binom{mk+t-1}{t}}{\binom{mk-r-s+t-1}{t-s}} \binom{mk-r-s+t-1}{t-s} p^{mk-r} q^{t-s} = 1$$

Then

$$u(t) \frac{\binom{mk+t-1}{t}}{\binom{mk-r-s+t-1}{t-s}} = 1$$

Hence,

$$u(t) = \frac{\binom{mk-r-s+t-1}{t-s}}{\binom{mk+t-1}{t}}$$

$$u(t) \begin{cases} \dfrac{\binom{mk-r-s+t-1}{t-s}}{\binom{mk+t-1}{t}} & ; \ t = s, s+1, \ldots, r \leq mk \\ 0 & ; \ \text{otherwise} \end{cases} \qquad (2.2.10)$$

(ii) To find UMVUE of $P[X \leq c]$

$$P[X \leq c] = \sum_{x=0}^{c} \binom{k+x-1}{x} p^k q^x$$

Now UMVUE of $p^k q^x = \dfrac{\binom{mk-k-x+t}{t-x}}{\binom{mk+t-1}{t}}$

UMVUE of $P[X \leq c]$

$$= \begin{cases} \sum_{x=0}^{c} \dfrac{\binom{k+x-1}{x}\binom{mk-k-x+t}{t-x}}{\binom{mk+t-1}{t}} & ; \ t = x, x+1, \ldots \\ 1 & ; \ \text{otherwise.} \end{cases} \qquad (2.2.27)$$

Remark UMVUE of $P[X = x] = \binom{k+x-1}{x} p^k q^x$ is $\dfrac{\binom{k+x-1}{x}\binom{mk-k-x+t}{t-x}}{\binom{mk+t-1}{t}}$

Particular cases:

(a) $r = 1, s = 0$

From (2.2.26), we will get UMVUE of p,

$$u(t) = \frac{\binom{mk+t-2}{t}}{\binom{mk+t-1}{t}} = \frac{mk-1}{mk+t-1} \qquad (2.2.28)$$

(b) $r = 0, s = 1$

From (2.2.26), we will get UMVUE of q,

$$u(t) = \frac{\binom{mk+t-2}{t-1}}{\binom{mk+t-1}{t}} = \frac{t}{mk+t-1} \qquad (2.2.29)$$

(c) $r = 1, s = 1$

From (2.2.26), we will get UMVUE of pq,

$$u(t) = \frac{\binom{mk+t-3}{t-1}}{\binom{mk+t-1}{t}} = \frac{t(mk-1)}{(mk+t-1)(mk+t-2)} \qquad (2.2.30)$$

Remark UMVUE of $pq \neq$ (UMVUE of p)(UMVUE of q)

Example 2.2.7 Let $X_1, X_2, \ldots, X_m$ be iid discrete uniform rvs with parameter $N (N > 1)$. We have to find UMVUE of $N^s (s \neq 0)$.

Then joint distribution of $(X_1, X_2, \ldots, X_m)$ is

$$f(x_1, x_2, \ldots, x_m) = \frac{1}{N^m} I(N - x_{(m)}) I(x_{(1)} - 1)$$

$$I(y) = \begin{cases} 1 \; ; \; y > 0 \\ 0 \; ; \; \text{otherwise} \end{cases}$$

By factorization theorem, $X_{(m)}$ is sufficient for N.
Now, we will find the distribution of $X_{(m)}$.

$$P[X_{(m)} \leq z] = \prod_{i=1}^{m} P[X_i \leq z] = \frac{z^m}{N^m}$$

$$P[X_{(m)} = z] = P[X_{(m)} \leq z] - P[X_{(m)} \leq z - 1]$$

$$= \frac{z^m}{N^m} - \frac{(z-1)^m}{N^m}; \quad z = 1, 2, \ldots, N \tag{2.2.31}$$

We have to show that this distribution is complete, i.e., we have to show if $\mathrm{E}h(z) = 0$
then $h(z) = 0$ with probability 1.

$$\mathrm{E}h(z) = \sum_{z=1}^{N} h(z) \left[\frac{z^m}{N^m} - \frac{(z-1)^m}{N^m} \right] = 0$$

Now $\left(\frac{z^m - (z-1)^m}{N^m} \right)$ is always positive then $h(z) = 0$ with probability 1.
Therefore, $X_{(m)}$ is sufficient and complete for N.
Let $u(z)$ be unbiased estimator of N^s
Then

$$\sum_{z=1}^{N} u(z) \left[\frac{z^m - (z-1)^m}{N^m} \right] = N^s$$

$$\sum_{z=1}^{N} u(z) \left[\frac{z^m - (z-1)^m}{N^{m+s}} \right] = 1$$

$$\sum_{z=1}^{N} u(z) \left[\frac{z^m - (z-1)^m}{z^{m+s} - (z-1)^{m+s}} \right] \left[\frac{z^{m+s} - (z-1)^{m+s}}{N^{m+s}} \right] = 1$$

Hence,

$$u(z) \left[\frac{z^m - (z-1)^m}{z^{m+s} - (z-1)^{m+s}} \right] = 1$$

$$u(z) = \left[\frac{z^{m+s} - (z-1)^{m+s}}{z^m - (z-1)^m} \right]$$

Therefore,

$$u(X_{(m)}) = \left[\frac{X_{(m)}^{m+s} - (X_{(m)} - 1)^{m+s}}{X_{(m)}^m - (X_{(m)} - 1)^m} \right] \tag{2.2.32}$$

Then $u(X_{(m)})$ in (2.2.32) is UMVUE of N^s.
Particular cases:
(a) $s = 1$
From (2.2.32), we get UMVUE of N,

$$\hat{N} = \left[\frac{X_{(m)}^{m+1} - (X_{(m)} - 1)^{m+1}}{X_{(m)}^m - (X_{(m)} - 1)^m} \right] \tag{2.2.33}$$

(b) $s = 5$
From (2.2.33), we get UMVUE of N^5

$$\hat{N}^5 = \left[\frac{X_{(m)}^{m+5} - (X_{(m)} - 1)^{m+5}}{X_{(m)}^m - (X_{(m)} - 1)^m} \right] \tag{2.2.34}$$

(c) To find UMVUE of e^N
Now

$$e^N = \sum_{j=0}^{\infty} \frac{N^j}{j!} \tag{2.2.35}$$

Using (2.2.32), UMVUE of e^N is

$$e^{\hat{N}} = \sum_{j=0}^{\infty} \frac{1}{j!} \left[\frac{X_{(m)}^{m+j} - (X_{(m)} - 1)^{m+j}}{X_{(m)}^m - (X_{(m)} - 1)^m} \right]$$

Remark UMVUE of $e^N \neq e^{\hat{N}}$

Example 2.2.8 Let $X_1, X_2, \ldots, X_m$ be iid rvs with power series distribution.

$$P(X = x) = \frac{a(x)\theta^x}{c(\theta)}; x = 0, 1, 2, \ldots \tag{2.2.36}$$

where $c(\theta) = \sum_{x=0}^{\infty} a(x)\theta^x$.
This distribution belongs to exponential family.

Therefore, $T = \sum X_i$ is sufficient and complete for θ. In this case, we will find UMVUE of $\frac{\theta^r}{[c(\theta)]^s}$ $(r, \ s \neq 0)$.

This distribution of T is again a power series distribution, see Roy and Mitra (1957), and Patil (1962)

$$P(T = t) = \frac{A(t, m)\theta^t}{[c(\theta)]^m},$$ (2.2.37)

where $A(t, m) = \sum_{(x_1, x_2, \ldots, x_m)} \prod_{i=1}^{m} a(x_i)$

Let $U(t)$ be an unbiased estimator of $\frac{\theta^r}{[c(\theta)]^s}$

$$\sum_{t=0}^{\infty} u(t) \frac{A(t, m)\theta^t}{[c(\theta)]^m} = \frac{\theta^r}{[c(\theta)]^s}$$ (2.2.38)

$$\sum_{t=0}^{\infty} u(t) \frac{A(t, m)\theta^{t-r}}{[c(\theta)]^{m-s}} = 1$$

$$\sum_{t=0}^{\infty} u(t) \frac{A(t, m)}{A(t - r, m - s)} \frac{A(t - r, m - s)\theta^{t-r}}{[c(\theta)]^{m-s}} = 1$$

Now

$$u(t) \frac{A(t, m)}{A(t - r, m - s)} = 1$$

This implies

$$U(t) = \begin{cases} \frac{A(t-r, m-s)}{A(t, m)} & ; \ t \geq r, \ m \geq s \\ 0 & ; \ \text{otherwise} \end{cases}$$ (2.2.39)

Example 2.2.9 Let $X_1, X_2, \ldots, X_m$ be iid rvs with $G(p, \frac{1}{\theta})$.
Let

$$f(x, \theta) = \frac{e^{-\frac{x}{\theta}} x^{p-1}}{\theta^p \Gamma(p)}; \ x > 0, \ p > 0, \ \theta > 0$$ (2.2.40)

Now gamma distribution belongs to an exponential family. $T = \sum X_i$ is sufficient and complete for θ.

The distribution of T is

$$f(t) = \frac{e^{-\frac{t}{\theta}} t^{mp-1}}{\theta^{mp} \Gamma(mp)}; \ t > 0, \ p > 0, \ \theta > 0$$ (2.2.41)

We have to find UMVUE of (i) $e^{-\frac{k}{\theta}} \theta^r$ (ii) $P(X \geq k)$
(i) Let $u(t)$ be an unbiased estimator of $e^{-\frac{k}{\theta}} \theta^r$

$$\int\limits_0^\infty u(t)\frac{e^{-\frac{t}{\theta}}t^{mp-1}}{\theta^{mp}\Gamma(mp)} = e^{\frac{k}{\theta}}\theta^{r}$$

$$\int\limits_0^\infty u(t)\frac{e^{-\frac{t-k}{\theta}}t^{mp-1}}{\theta^{mp+r}\Gamma(mp)} = 1$$

$$\int\limits_k^\infty \left(u(t)\frac{t^{mp-1}\Gamma(mp+r)}{(t-k)^{mp+r-1}\Gamma(mp)}\right)\left(\frac{e^{-\frac{t-k}{\theta}}(t-k)^{mp+r-1}}{\theta^{mp+r}\Gamma(mp+r)}\right)dt = 1$$

Then,

$$u(t)\frac{t^{mp-1}\Gamma(mp+r)}{(t-k)^{mp+r-1}\Gamma(mp)} = 1$$

$$u(t) = \begin{cases} \frac{(t-k)^{mp+r-1}\Gamma(mp)}{t^{mp-1}\Gamma(mp+r)} & ;\ t > k,\ mp > -r \\ 0 & ;\ \text{otherwise} \end{cases} \qquad (2.2.42)$$

(ii) We have to find UMVUE of $P[X \geq k]$. Note that

$$P[X \geq k] = \int\limits_k^\infty \frac{e^{-\frac{x}{\theta}}x^{p-1}}{\theta^p\Gamma(p)}dx$$

Let

$$Y = \begin{cases} 1\ ;\ X_1 \geq k \\ 0\ ;\ \text{otherwise} \end{cases}$$

$$E(Y) = P[X_1 \geq k]$$

Hence Y is unbiased estimator for $P[X_1 \geq k]$. We have seen in Sect. 2.2 that $[EY|T = t]$ is an estimator and has minimum variance.

So $E[Y|T = t] = P[X_1 \geq k|T = t]$. Now we will require the distribution of $X_1|T = t$

$$P[X_1|T = t] = \frac{f(x_1)f(t_1)}{f(t)}, \quad \text{where } T_1 = \sum_{i=2}^m X_i$$

Distribution of $(T_1 = t_1) = f(t_1)$

$$f(t_1) = \frac{e^{-\frac{t_1}{\theta}}t_1^{(m-1)p-1}}{\Gamma((m-1)p)\theta^{(m-1)p}}; \quad t_1 \geq 0$$

$$P[X_1|T=t] = \frac{e^{-\frac{x_1}{\theta}} x_1^{p-1}}{\Gamma(p)\theta^p} \frac{e^{-\frac{t_1}{\theta}} t_1^{(m-1)p-1}}{\Gamma((m-1)p)\theta^{(m-1)p}} \frac{\Gamma(mp)\theta^{mp}}{e^{-\frac{t}{\theta}} t^{mp-1}}$$

$$= \frac{\left(\frac{x_1}{t}\right)^{p-1} \left(1 - \frac{x_1}{t}\right)^{(m-1)p-1}}{t\beta(p,(m-1)p)}; \quad 0 \le \frac{x_1}{t} \le 1 \qquad (2.2.43)$$

$$E[Y|T=t] = P[X_1 \ge k|T=t] = \int\limits_{k}^{t} \frac{\left(\frac{x_1}{t}\right)^{p-1} \left(1 - \frac{x_1}{t}\right)^{(m-1)p-1}}{t\beta(p,(m-1)p)} dx_1$$

Let $\frac{x_1}{t} = w$

$$= \int\limits_{\frac{k}{t}}^{1} \frac{w^{p-1}(1-w)^{(m-1)p-1}}{\beta(p,(m-1)p)} dw$$

$$= 1 - \int\limits_{0}^{\frac{k}{t}} \frac{w^{p-1}(1-w)^{(m-1)p-1}}{\beta(p,(m-1)p)} dw \qquad (2.2.44)$$

$$P[X_1 \ge k|T=t] = \begin{cases} 1 - I_{\frac{k}{t}}(p, mp-p) \; ; \; 0 < k < t \\ 0 \qquad\qquad\qquad ; \; k \ge t \end{cases}$$

Now

$$P[X \ge k] = \int\limits_{k}^{\infty} \frac{e^{-\frac{x}{\theta}} x^{p-1}}{\theta^p \Gamma(p)} dx$$

$$1 - I_{\frac{k}{\theta}}(p) = \text{Incomplete Gamma function.} \qquad (2.2.45)$$

Hence UMVUE of $1 - I_{\frac{k}{\theta}}(p)$ is given by incomplete Beta function $1 - I_{\frac{k}{t}}(p, mp-p)$.
Note: Student should use R or Minitab software to calculate UMVUE.

Example 2.2.10 Let $X_1, X_2, \ldots, X_m$ be iid rvs with the following pdfs.

1. $f(x|\lambda) = \frac{\lambda}{(1+x)^{\lambda+1}}; \quad x > 0$
2. $f(x|\lambda) = \lambda x^{\lambda-1}; \quad 0 < x < 1, \; \lambda > 0$
3. $f(x|\lambda) = \frac{1}{2\lambda} e^{\frac{|x|}{\lambda}}; \quad x > 0, \; \lambda > 0$
4. $f(x|\lambda) = \frac{\alpha}{\lambda} x^{\alpha-1} e^{-\frac{x^\alpha}{\lambda}}; \quad x > 0, \; \lambda > 0, \; \alpha > 0$
5. $f(x|\lambda) = \frac{1}{\sqrt{2\pi\lambda}} e^{-\frac{x^2}{2\lambda}}; \quad x > 0, \; \lambda > 0$

(i) Let $Y = \log(1 + r)$ then $f(y|\lambda) = \lambda e^{-\lambda y}$; $y > 0$, $\lambda > 0$

The UMVUE of λ^r is given as

$$u(t) = \frac{t^{-r}\Gamma(m)}{\Gamma(m-r)}; \quad m > r \tag{2.2.46}$$

Consider $r = 1$ then we will get UMVUE of λ,

$$\hat{\lambda} = \frac{m-1}{T} \tag{2.2.47}$$

(ii) Let $Y = -\log X$ then $f(y|\lambda) = \lambda e^{-\lambda y}$; $y > 0$, $\lambda > 0$

We will get the UMVUE of λ^r in (2.2.46) and for $r = 1$, UMVUE of λ is given in (2.2.47)

(iii) Let $|x| = y$ then $f(y|\lambda) = \lambda e^{-\lambda y}$; $y > 0$, $\lambda > 0$

In the same way as (i) and (ii) we can obtain the UMVUE of λ^{-r}.

(iv) Let $x^\alpha = y$ then $f(y|\lambda) = \frac{1}{\lambda}e^{-\frac{y}{\lambda}}$; $y > 0$, $\lambda > 0$

In the same way as (i) and (ii), we can obtain the UMVUE of λ^r (here $\theta = \lambda$).

(v) Let $\frac{x^2}{2} = y$ then $f(y|\lambda) = \frac{e^{-\frac{y}{\lambda}}y^{-\frac{1}{2}}}{\Gamma(\frac{1}{2})\lambda^{\frac{1}{2}}}$; $y > 0$, $\lambda > 0$

In this case $p = \frac{1}{2}$ and $\theta = \lambda$.

Similarly, we can obtain the UMVUE of λ^r.

Example 2.2.11 Let $X_1, X_2, \ldots, X_m$ be iid rvs with $N(\mu, \sigma^2)$. We will consider three cases

$$(i)\ \mu \text{ known}, \sigma^2 \text{ unknown}$$

$$(ii)\ \mu \text{ unknown}, \sigma^2 \text{ known}$$

$$(iii)\ \mu \text{ and } \sigma^2 \text{ both unknown}$$

(i) Normal distribution belongs to exponential family.

$T = \sum_{i=1}^{m}(X_i - \mu)^2$ is complete and sufficient for σ^2.

$$\frac{\sum_{i=1}^{m}(X_i - \mu)^2}{\sigma^2} \text{ has } \chi^2 \text{ with } m\ df \tag{2.2.48}$$

Hence, $E\frac{T}{\sigma^2} = m$. This implies that UMVUE of σ^2 is $\hat{\sigma^2} = \frac{\sum(X_i - \mu)^2}{m}$

Let $\sigma^2 = \theta$ and $Y = \frac{T}{\theta}$

Then

$$f(y) = \frac{e^{-\frac{y}{2}}y^{\frac{m}{2}-1}}{2^{\frac{m}{2}}\Gamma\left(\frac{m}{2}\right)}; y > 0$$

To find the unbiased estimator of θ^r. Let $u(y)$ be an unbiased estimator of θ^r.

$$\int\limits_0^\infty u(y)\frac{e^{-\frac{y}{2}}y^{\frac{m}{2}-1}}{2^{\frac{m}{2}}\Gamma\left(\frac{m}{2}\right)}dy = \theta^r = \frac{t^r}{y^r}$$

$$\int\limits_0^\infty \frac{u(y)}{t^r}\frac{e^{-\frac{y}{2}}y^{\frac{m}{2}+r-1}}{2^{\frac{m}{2}}\Gamma\left(\frac{m}{2}\right)}dy = 1$$

$$\int\limits_0^\infty \left(\frac{u(y)}{t^r}\frac{\Gamma\left(\frac{m}{2}+r\right)2^{\frac{m}{2}+r}}{2^{\frac{m}{2}}\Gamma\left(\frac{m}{2}\right)}\right)\frac{e^{-\frac{y}{2}}y^{\frac{m}{2}+r-1}}{2^{\frac{m}{2}+r}\Gamma\left(\frac{m}{2}+r\right)}dy = 1$$

Now

$$\left(\frac{u(y)}{t^r}\frac{\Gamma\left(\frac{m}{2}+r\right)2^{\frac{m}{2}+r}}{2^{\frac{m}{2}}\Gamma\left(\frac{m}{2}\right)}\right) = 1$$

$$u(y) = \frac{t^r\Gamma\left(\frac{m}{2}\right)}{2^r\Gamma\left(\frac{m}{2}+r\right)}; \quad r = 1,2,\dots \tag{2.2.49}$$

Particular cases: $r = 1$

$$u(y) = \frac{t\Gamma\left(\frac{m}{2}\right)}{2\Gamma\left(\frac{m}{2}+1\right)} = \frac{t}{(2)\left(\frac{m}{2}\right)} = \frac{t}{m}$$

$$= \frac{\sum(X_i-\mu)^2}{m} \tag{2.2.50}$$

Therefore, $\frac{\sum(X_i-\mu)^2}{m}$ is the UMVUE of σ^2.

Next, we will find the UMVUE of $P[X_1 \geq k]$

$$P[X_1 \geq k] = P\left[\frac{X_1-\mu}{\sigma} \geq \frac{k-\mu}{\sigma}\right]$$

$$= 1 - P\left[\frac{X_1-\mu}{\sigma} < \frac{k-\mu}{\sigma}\right]$$

$$= 1 - \Phi\left[\frac{k-\mu}{\sigma}\right] \tag{2.2.51}$$

Define

$$Y_1 = \begin{cases} 1 \; ; \; X_1 \geq k \\ 0 \; ; \text{otherwise} \end{cases}$$

$$EY_1 = P[X_1 \geq k]$$

According to Rao-Blackwell theorem, we have to find $P[X_1 > k|T = t]$.
For this we will have to find the distribution of X_1 given $T = t$. Then it is necessary
to find the joint distribution of X_1 and $T = t$
Let $T = (X_1 - \mu)^2 + Z$ and $\frac{T}{\sigma^2}$ has χ^2_m. So $Z = T - (X_1 - \mu)^2$ then $\frac{Z}{\sigma^2}$ has χ^2_{m-1}. Let
$y = \frac{z}{\sigma^2}$. Then

$$f(y) = \frac{e^{-\frac{y}{2}} y^{\frac{m-1}{2}-1}}{2^{\frac{m-1}{2}} \Gamma\left(\frac{m-1}{2}\right)}$$

$$f(z) = \frac{e^{-\frac{z}{2\sigma^2}} z^{\frac{m-1}{2}-1}}{2^{\frac{m-1}{2}} \Gamma\left(\frac{m-1}{2}\right) \sigma^{m-1}}; \quad z > 0 \qquad (2.2.52)$$

$$f(x_1, t) = f(x_1)f(z)$$

$$= \frac{e^{-\frac{(x_1-\mu)^2}{2\sigma^2}} e^{-\frac{[z-(x_1-\mu)^2]}{2\sigma^2}} [t - (x_1 - \mu)^2]^{\frac{m-1}{2}-1}}{(\sigma\sqrt{2\pi}) 2^{\frac{m-1}{2}} \Gamma\left(\frac{m-1}{2}\right) \sigma^{m-1}}$$

$$= \frac{e^{-\frac{t}{2\sigma^2}} [t - (x_1 - \mu)^2]^{\frac{m-1}{2}-1}}{2^{\frac{m}{2}} \Gamma\left(\frac{m-1}{2}\right) \sigma^m \sqrt{\pi}} \qquad (2.2.53)$$

$$f(t) = \frac{e^{-\frac{t}{2\sigma^2}} t^{\frac{m}{2}-1}}{2^{\frac{m}{2}} \Gamma\left(\frac{m}{2}\right) \sigma^m}$$

$$f(x_1|T = t) = \begin{cases} \frac{\Gamma\left(\frac{m}{2}\right)[t-(x_1-\mu)^2]^{\frac{m-1}{2}-1}}{\Gamma\left(\frac{1}{2}\right) t^{\frac{m}{2}-1} \Gamma\left(\frac{m-1}{2}\right)} & ; \mu - \sqrt{t} < x_1 < \mu + \sqrt{t} \\ 0 & ; \text{otherwise} \end{cases} \qquad (2.2.54)$$

Note that $\sqrt{\pi} = \Gamma\left(\frac{1}{2}\right)$
Consider

$$\frac{\left[t - (x_1 - \mu)^2\right]^{\frac{m-1}{2}-1}}{t^{\frac{m}{2}-1}}$$

$$= \frac{t^{\frac{m-1}{2}-1} \left[1 - \left(\frac{x_1-\mu}{\sqrt{t}}\right)^2\right]^{\frac{m-1}{2}-1}}{t^{\frac{m}{2}-1}}$$

$$= t^{-\frac{1}{2}} \left[1 - \left(\frac{x_1 - \mu}{\sqrt{t}}\right)^2\right]^{\frac{m-1}{2}-1}$$

$$P[X_1 \geq k|T = t] = \int\limits_{k}^{\mu+\sqrt{t}} \frac{t^{-\frac{1}{2}}\left[1 - \left(\frac{x_1-\mu}{\sqrt{t}}\right)^2\right]^{\frac{m-1}{2}-1}}{\beta\left(\frac{1}{2}, \frac{m-1}{2}\right)} dx_1$$

Let $\left(\frac{x_1-\mu}{\sqrt{t}}\right)^2 = v \Rightarrow \frac{2}{\sqrt{t}}\left(\frac{x_1-\mu}{\sqrt{t}}\right) dx_1 = dv$

$\Rightarrow dx_1 = \frac{\sqrt{t}}{2} v^{-\frac{1}{2}} dv$

when $X_1 = k \Rightarrow v = \left(\frac{k-\mu}{\sqrt{t}}\right)^2$ and $X_1 = \mu + \sqrt{t} \Rightarrow v = 1$

$$= \frac{1}{2} \int\limits_{\left(\frac{k-\mu}{\sqrt{t}}\right)^2}^{1} \frac{v^{-\frac{1}{2}}(1-v)^{\frac{m-1}{2}-1}}{\beta\left(\frac{1}{2}, \frac{m-1}{2}\right)} dv$$

Hence

$$2P[X_1 \geq k|T = t] = 1 - \int\limits_{0}^{\left(\frac{k-\mu}{\sqrt{t}}\right)^2} \frac{v^{-\frac{1}{2}}(1-v)^{\frac{m-1}{2}-1}}{\beta\left(\frac{1}{2}, \frac{m-1}{2}\right)} dv$$

$$P[X_1 \geq k|T = t] = \frac{1}{2} - \frac{1}{2}\mathbf{I}_{\left(\frac{k-\mu}{\sqrt{t}}\right)^2}\left(\frac{1}{2}, \frac{m-1}{2}\right)$$

UMVUE of $P[X_1 \geq k] = 1 - \Phi\left(\frac{k-\mu}{\sigma}\right)$ is $P[X_1 \geq k|T = t]$

$$P[X_1 \geq k|T = t] = \begin{cases} \frac{1}{2} - \frac{1}{2}\mathbf{I}_{\left(\frac{k-\mu}{\sqrt{t}}\right)^2}\left(\frac{1}{2}, \frac{m-1}{2}\right) \; ; \; \mu - \sqrt{t} < k < \mu + \sqrt{t} \\ 1 \qquad\qquad\qquad\qquad\qquad ; \; k < \mu - \sqrt{t} \\ 0 \qquad\qquad\qquad\qquad\qquad ; \; k > \mu + \sqrt{t} \end{cases} \qquad (2.2.55)$$

(ii) For σ known, $\sum X_i$ or $\bar{X}$ is complete and sufficient for μ. The distribution of $\bar{X} \sim N\left(\mu, \frac{\sigma^2}{m}\right)$.

Now, $E\bar{X} = \mu$ and $E\bar{X}^2 = \mu^2 + \frac{\sigma^2}{n}$

$$E\left(\bar{X}^2 - \frac{\sigma^2}{m}\right) = \mu^2$$

Hence,

$$\left(\bar{X}^2 - \frac{\sigma^2}{m}\right) \text{ is UMVUE for } \mu^2 \qquad (2.2.56)$$

For (2.2.56), see Example 2.2.3.

$$E(\bar{X}^r) = \int\limits_{-\infty}^{\infty} \bar{x}^r \frac{\sqrt{m}}{\sigma\sqrt{2\pi}} e^{-\frac{m}{2}(\bar{x}-\mu)^2} d\bar{x}$$

Let $w = \frac{(\bar{x}-\mu)\sqrt{m}}{\sigma} \Rightarrow \bar{x} = \mu + \frac{w\sigma}{\sqrt{m}}$

$$= \int\limits_{-\infty}^{\infty} \left(\mu + \frac{w\sigma}{\sqrt{m}}\right)^r \frac{1}{\sqrt{2\pi}} e^{-\frac{w^2}{2}} dw$$

Since odd moments of w are 0

$$= \int\limits_{-\infty}^{\infty} \left[\left(\frac{w\sigma}{\sqrt{m}}\right)^r + \binom{r}{1}\left(\frac{w\sigma}{\sqrt{m}}\right)^{r-1}\mu + \binom{r}{2}\left(\frac{w\sigma}{\sqrt{n}}\right)^{r-2}\mu^2 + \cdots + \mu^r\right] \frac{e^{-\frac{w^2}{2}}}{\sqrt{2\pi}} dw$$

$$= \frac{\sigma^r}{m^{\frac{r}{2}}}\mu_r + \binom{r}{1}\frac{\sigma^{r-1}}{m^{\frac{r-1}{2}}}\mu_{r-1}\mu + \binom{r}{2}\frac{\sigma^{r-2}}{m^{\frac{r-2}{2}}}\mu_{r-2}\mu^2 \cdots + \mu^r$$

$$\mu_r = \begin{cases} 0 & ; \ r \text{ is odd} \\ (r-1)(r-3)\ldots 1 & ; \ r \text{ is even} \end{cases}$$

Particular cases: (a) $r = 3$ (b) $r = 4$
(a) $r = 3$

$$E(\bar{X}^3) = \frac{\sigma^3}{m^{\frac{3}{2}}}\mu_3 + \binom{3}{1}\frac{\sigma^2\mu_2\mu}{m} + \binom{3}{2}\frac{\sigma\mu_1\mu^2}{m^{\frac{1}{2}}} + \mu^3$$

$$= 3\frac{\sigma^2\mu}{m} + \mu^3$$

$$\text{UMVUE of } \mu^3 = \bar{X}^3 - 3\frac{\sigma^2\mu}{m} \tag{2.2.57}$$

(b) $r = 4$

$$E(\bar{X}^4) = \frac{\sigma^4}{m^2}\mu_4 + \binom{4}{1}\frac{\sigma^3}{m^{\frac{3}{2}}}\mu_3(\mu) + \binom{4}{2}\frac{\sigma^2}{m}\mu_2(\mu)^2 + \binom{4}{3}\frac{\sigma}{m^{\frac{1}{2}}}\mu_1(\mu)^3 + \mu^4$$

$$\mu_4 = (4-1)(4-3) = 3, \ \mu_3 = 0, \ \mu_2 = 1$$

$$E(\bar{X}^4) = \frac{3\sigma^4}{m^2} + \frac{6\sigma^2}{m}(\mu)^2 + \mu^4$$

UMVUE of μ^4 is

$$\bar{X}^4 - \frac{3\sigma^4}{m^2} - 6\frac{\sigma^2(\mu)^2}{m}$$

$$= \bar{X}^4 - \frac{3\sigma^4}{m^2} - \frac{6\sigma^2}{m}\left(\bar{x}^2 - \frac{\sigma^2}{m}\right) \tag{2.2.58}$$

Similarly, we can find UMVUE of $\mu^r (r \geq 1)$
Next, find the UMVUE of $P[X_1 \geq k]$
Again define

$$Y_1 = \begin{cases} 1 \; ; \; X_1 \geq k \\ 0 \; ; \; \text{otherwise} \end{cases}$$

$$EY_1 = P[X_1 \geq k]$$

According to Rao–Blackwell theorem, we have to find $P[X_1 \geq k|T = t]$ where $T = \sum_{i=1}^{m} X_i$ and $T_1 = \sum_{i=2}^{m} X_i$. $T \sim N(m\mu, m\sigma^2)$ and $T_1 \sim N\left((m-1)\mu, (m-1)\sigma^2\right)$

$$f(x_1, t) = f(x_1)f(t_1)$$

$$f(x_1, t) = \frac{1}{\sigma\sqrt{2\pi}}e^{-\frac{(x_1-\mu)^2}{2\sigma^2}} \frac{1}{\sigma\sqrt{2\pi(m-1)}}e^{-\frac{[t_1-(m-1)\mu]^2}{2(m-1)\sigma^2}}$$

$$f(t) = \frac{1}{\sigma\sqrt{2\pi m}}\exp\left[-\frac{1}{2m\sigma^2}(t-\mu)^2\right]$$

$$f(x_1|T = t) = \frac{1}{\sigma\sqrt{2\pi}\sqrt{\frac{m-1}{m}}}e^{-\frac{m}{2(m-1)\sigma^2}(x_1-\frac{t}{m})^2} \tag{2.2.59}$$

Therefore, $(X_1|T = t)$ has $N\left(\frac{t}{m}, \frac{(m-1)\sigma^2}{m}\right)$
To find $P[X_1 \geq k|T = t]$

$$= 1 - \Phi\left(\frac{k - \frac{t}{m}}{\sigma\sqrt{\frac{m-1}{m}}}\right)$$

$$= 1 - \Phi\left(\frac{k - \bar{x}}{\sigma\sqrt{\frac{m-1}{m}}}\right) \tag{2.2.60}$$

We conclude that $\Phi\left(\frac{k-\bar{x}}{\sigma\sqrt{\frac{m-1}{m}}}\right)$ is UMVUE of $\Phi\left(\frac{k-\mu}{\sigma}\right)$.

(iii) Both μ and σ are unknown

$(\bar{x}, S^2)$ is jointly sufficient and complete for (μ, σ^2) because normal distribution belongs to exponential family where $S^2 = \sum(x_i - \bar{x})^2$.

Now, $\frac{S^2}{\sigma^2}$ has χ^2 distribution with $m - 1$ df.

Let $\frac{S^2}{\sigma^2} = y$ then $EY^r = \frac{\Gamma(\frac{m-1}{2}+r)}{\Gamma(\frac{m-1}{2})} 2^r$

Hence

$$E(S^2)^r = \frac{\Gamma\left(\frac{m-1}{2}+r\right)}{\Gamma\left(\frac{m-1}{2}\right)} (2\sigma^2)^r \tag{2.2.61}$$

Therefore, $\frac{\Gamma(\frac{m-1}{2})S^2}{\Gamma(\frac{m-1}{2}+r)2^r}$ is UMVUE of σ^{2r}

Particular case: (a) $r = \frac{1}{2}$ (b) $r = 1$

(a)

$$\hat{\sigma} = \frac{\Gamma\left(\frac{m-1}{2}\right) S}{\Gamma\left(\frac{m-1}{2} + \frac{1}{2}\right) 2^{\frac{1}{2}}}$$

(b)

$$\hat{\sigma^2} = \frac{\Gamma\left(\frac{m-1}{2}\right) S^2}{\Gamma\left(\frac{m-1}{2} + \frac{1}{2}\right) 2} = \frac{S^2}{m - 1}$$

$$E(\bar{X}^2) = \mu^2 + \frac{\sigma^2}{m}$$

Then

$$E\left[\bar{X}^2 - \frac{S^2}{m(m - 1)}\right] = \mu^2$$

So that

$$\text{UMVUE of } \mu^2 \text{ is } \bar{X}^2 - \frac{S^2}{m(m - 1)} \tag{2.2.62}$$

Next,

$$E(\bar{X}^3) = \mu^3 + \frac{3\sigma^2}{n}\mu$$

$$E\left[\bar{X}^3 - \frac{3\bar{x}S^2}{m(m - 1)}\right] = \mu^3$$

$$[\bar{X}^3 - \frac{3\bar{x}S^2}{m(m - 1)}] \text{ is UMVUE of } \mu^3 \tag{2.2.63}$$

Similarly, one can obtain UMVUE of $\mu^r (r \geq 1)$
(c) $r = -1$

$$\text{UMVUE of } \frac{1}{\sigma^2} = \widehat{\sigma^{-2}}$$

$$= \frac{\Gamma\left(\frac{m-1}{2}\right)}{\Gamma\left(\frac{m-1}{2} - 1\right)} \frac{S^{-2}}{2^{-1}}$$

$$= \frac{m-3}{S^2}; \quad m > 3 \tag{2.2.64}$$

Next, we will find the UMVUE of $P[X_1 \geq k]$

$$P[X_1 \geq k] = 1 - \Phi\left(\frac{k - \mu}{\sigma}\right)$$

As usual

$$Y = \begin{cases} 1 \; ; \; X_1 \geq k \\ 0 \; ; \text{ otherwise} \end{cases} \tag{2.2.65}$$

$$EY = P[X_1 \geq k] = 1 - \Phi\left(\frac{k - \mu}{\sigma}\right)$$

As we have done earlier,

$$\text{E}(Y|\bar{X}, S^2) = P[X_1 \geq k|\bar{X}, S^2]$$

We need to find the distribution of $(X_1, \bar{X}, S^2)$.
Consider the following orthogonal transformation:

$$z_1 = \frac{1}{\sqrt{m}}(x_1 + x_2 + \cdots + x_m) = \sqrt{m}\bar{x}$$

$$z_2 = \left[\left(1 - \frac{1}{m}\right)x_1 - \frac{x_2}{m} - \cdots - \frac{x_m}{m}\right]\sqrt{\frac{m}{m-1}}$$

$$z_i = c_{i1}x_1 + c_{i2}x_2 + \cdots + c_{im}x_m \quad i = 3, 4, \ldots, m$$

where $\sum_{j=1}^{m} c_{ij} = 0, i = 3, 4, \ldots, m$ and $\sum_{j=1}^{m} c_{jj}^2 = 1$

$$z_1 \sim N(\sqrt{m}\mu, \sigma^2) \tag{2.2.66}$$

$$z_r \sim N(0, \sigma^2) \quad r = 2, 3, \ldots, n$$

Let $Z = PX$, where P is an orthogonal matrix

$$Z'Z = X'P'PX = X'X$$

Hence,

$$\sum_{i=1}^{m} z_i^2 = \sum_{i=1}^{m} x_i^2 \qquad (2.2.67)$$

$$\sum_{i=3}^{m} z_i^2 = \sum_{i=1}^{m} x_i^2 - z_1^2 - z_2^2$$

$$= \sum_{i=1}^{m} x_i^2 - m\bar{x}^2 - z_2^2 = S^2 - z_2^2$$

Let $v = S^2 - z_2^2$,

where $v = \sum_{i=3}^{m} z_i^2$

Let $z_1 = \sqrt{m}\bar{x}$, $z_2 = \sqrt{\frac{m}{m-1}}(x_1 - \bar{x})$, $v = S^2 - z_2^2$

$$J = \frac{\partial(z_1, z_2, v)}{\partial(x_1, \bar{x}, S^2)}$$

$$= \begin{pmatrix} \frac{\partial z_1}{\partial x_1} & \frac{\partial z_1}{\partial \bar{x}} & \frac{\partial z_1}{\partial S^2} \\ \frac{\partial z_2}{\partial x_1} & \frac{\partial z_2}{\partial \bar{x}} & \frac{\partial z_2}{\partial S^2} \\ \frac{\partial v}{\partial x_1} & \frac{\partial v}{\partial \bar{x}} & \frac{\partial v}{\partial S^2} \end{pmatrix}$$

$$J = \begin{pmatrix} 0 & \sqrt{m} & 0 \\ \sqrt{\frac{m}{m-1}} & -\sqrt{\frac{m}{m-1}} & 0 \\ 0 & 0 & 1 \end{pmatrix} = -\frac{m}{\sqrt{m-1}}$$

Therefore,

$$|J| = \frac{m}{\sqrt{m-1}}$$

$$f(z_1, z_2, v) = \frac{e^{-\frac{(z_1 - \sqrt{m}\mu)^2}{2\sigma^2}}}{\sigma\sqrt{2\pi}} \frac{e^{-\frac{(z_2)^2}{2\sigma^2}}}{\sigma\sqrt{2\pi}} \frac{e^{-\frac{v}{2\sigma^2}} v^{\frac{m-2}{2}-1}}{\Gamma\left(\frac{m-2}{2}\right) 2^{\frac{m-2}{2}} \sigma^{m-2}} |J| \qquad (2.2.68)$$

Note that $\frac{v}{\sigma^2} \sim \chi^2_{m-3}$

$$f(x_1|\bar{x}, S^2) = \frac{f(z_1, z_2, v)}{f(\bar{x}, S^2)}$$

$$= |J| \frac{\exp\left[\frac{(z_1 - \sqrt{m}\mu)^2}{2\sigma^2} - \frac{z_2^2}{2\sigma^2} - \frac{v}{2\sigma^2}\right] v^{\frac{m-2}{2}-1} \frac{\sigma}{\sqrt{n}}\sqrt{2\pi} 2^{\frac{m-1}{2}} \sigma^{m-1}\Gamma\left(\frac{m-1}{2}\right)}{\exp\left[-\frac{m}{2\sigma^2}(\bar{x} - \mu)^2 - \frac{s^2}{2\sigma^2}\right] \sigma^m(2\pi)\Gamma\left(\frac{m-2}{2}\right) 2^{\frac{m-2}{2}}(s^2)^{\frac{m-1}{2}-1}}$$

Consider

$$\exp\left[-\frac{m}{2\sigma^2}(\bar{x} - \mu)^2 - \frac{m}{m-1}\frac{(x_1 - \bar{x})^2}{2\sigma^2} - \frac{S^2}{2\sigma^2} + \frac{m}{m-1}\frac{(x_1 - \bar{x})^2}{2\sigma^2} + \frac{m}{2\sigma^2}(\bar{x} - \mu)^2 + \frac{S^2}{2\sigma^2}\right] = 1$$

$$f(x_1|\bar{x}, S^2) = \frac{m}{\sqrt{m-1}} \frac{2^{\frac{m-1}{2}}\Gamma\left(\frac{m-1}{2}\right)}{\sqrt{m}\sqrt{2\pi}} \frac{v^{\frac{m-2}{2}-1}}{\Gamma\left(\frac{m-2}{2}\right) 2^{\frac{m-1}{2}}(S^2)^{\frac{m-1}{2}-1}}$$

$$= \frac{m2^{\frac{1}{2}}}{2^{\frac{1}{2}}\sqrt{m-1}\sqrt{\pi}} \frac{\Gamma\left(\frac{m-1}{2}\right)}{\Gamma\left(\frac{m-2}{2}\right)} \frac{[S^2 - \frac{m}{m-1}(x_1 - \bar{x})^2]^{\frac{m-2}{2}-1}}{(S^2)^{\frac{m-1}{2}-1}\sqrt{m}}$$

$$= \frac{m}{\sqrt{m-1}} \frac{\Gamma\left(\frac{m-1}{2}\right)}{\Gamma\left(\frac{1}{2}\right)\Gamma\left(\frac{m-2}{2}\right)} \frac{[S^2 - \frac{m}{m-1}(x_1 - \bar{x})^2]^{\frac{m-2}{2}-1}}{(S^2)^{\frac{m-1}{2}-1}}$$

$$= \frac{\sqrt{m}}{\sqrt{m-1}} \frac{[S^2 - \frac{m}{m-1}(x_1 - \bar{x})^2]^{\frac{m-2}{2}-1}}{(S^2)^{\frac{m-1}{2}-1}\beta\left(\frac{1}{2}, \frac{m-2}{2}\right)} \qquad (2.2.69)$$

$$= \sqrt{\frac{m}{m-1}} \frac{\sqrt{\frac{m-1}{2}}}{\Gamma\left(\frac{1}{2}\right)\Gamma\left(\frac{m-2}{2}\right)} \frac{[S^2 - \frac{m}{m-1}(x_1 - \bar{x})^2]^{\frac{m-2}{2}-1}}{(S^2)^{\frac{m-1}{2}-1}}$$

$$= \sqrt{\frac{m}{m-1}} \frac{1}{\beta\left(\frac{1}{2}, \frac{m-2}{2}\right)} \left(\frac{1}{S^2}\right)^{\frac{m-1}{2}-1} [S^2 - m(x_1 - \bar{x})^2]^{\frac{m-2}{2}-1}$$

$$= \sqrt{\frac{m}{m-1}} \frac{1}{\beta\left(\frac{1}{2}, \frac{m-2}{2}\right)} (S^2)^{-\frac{1}{2}} \left[1 - \frac{m}{m-1}\left(\frac{x_1 - \bar{x}}{S}\right)^2\right]^{\frac{m-2}{2}-1} \qquad (2.2.70)$$

Now

$$S^2 > \frac{m}{m-1}(x_1 - \bar{x})^2 \Rightarrow \frac{(m-1)S^2}{m} > (x_1 - \bar{x})^2$$

This implies that $|x_1 - \bar{x}| \leq S\sqrt{\frac{m-1}{m}}$

Hence,

$$\bar{x} - S\sqrt{\frac{m-1}{m}} \leq x_1 \leq \bar{x} + S\sqrt{\frac{m-1}{m}} \tag{2.2.71}$$

$$P[X_1 \geq k | T = t] = \int_k^{\bar{x}+S\sqrt{\frac{m-1}{m}}} \frac{1}{\beta\left(\frac{1}{2}, \frac{m-2}{2}\right)} \sqrt{\frac{m}{m-1}} (S^2)^{-\frac{1}{2}} \left[1 - \frac{m}{m-1}\left(\frac{x_1 - \bar{x}}{S}\right)^2\right]^{\frac{m-2}{2}-1} dx_1$$

Let $\frac{m}{m-1}\frac{(x_1-\bar{x})^2}{S^2} = t$, $\frac{2m}{m-1}\frac{(x_1-\bar{x})}{S^2}dx_1 = dt$, and $dx_1 = \frac{m-1}{2m}\frac{S^2}{(x_1-\bar{x})}dt$

$$= \int_{\frac{m}{m-1}\left(\frac{k-\bar{x}}{S}\right)^2}^{1} \frac{1}{2\beta\left(\frac{1}{2}, \frac{m-2}{2}\right)}[1-t]^{\frac{m-2}{2}-1}t^{-\frac{1}{2}}dt \tag{2.2.72}$$

UMVUE of $P[X_1 \geq k]$ is

$$P[X_1 \geq k | \bar{x}, S^2] = \begin{cases} 0 & ; \ k > \bar{x} + S\sqrt{\frac{m-1}{m}} \\ \int_k^{\bar{x}+S\sqrt{\frac{m-1}{m}}} f(x_1|\bar{x}, S^2)dx_1 & ; \ \bar{x} - S\sqrt{\frac{m-1}{m}} \leq x_1 \leq \bar{x} + S\sqrt{\frac{m-1}{m}} \\ 1 & ; \ k > \bar{x} - S\sqrt{\frac{m-1}{m}} \end{cases} \tag{2.2.73}$$

Further, if $\bar{x} - S\sqrt{\frac{m-1}{m}} \leq x_1 \leq \bar{x} + S\sqrt{\frac{m-1}{m}}$

$$\int_k^{\bar{x}+S\sqrt{\frac{m-1}{m}}} f(x_1|\bar{x}, s^2)dx_1 = \frac{1}{2}\left[1 - \mathbf{I}_{\frac{m}{m-1}\left(\frac{k-\bar{x}}{s}\right)^2}\left(\frac{1}{2}, \frac{m-2}{2}\right)\right] \tag{2.2.74}$$

where $\mathbf{I}$ is an incomplete Beta distribution.

2.3 UMVUE in Nonexponential Families

This section is devoted to find UMVUE from right, left, and both truncation families. One can see Tate (1959), Guenther (1978), and Jadhav (1996).

Example 2.3.1 Let $X_1, X_2, \ldots, X_m$ be iid rvs from the following pdf:

$$f(x|\theta) = \begin{cases} Q_1(\theta)M_1(x) & ; \ a < x < \theta \\ 0 & ; \ \text{otherwise} \end{cases} \tag{2.3.1}$$

where $M_1(x)$ is nonnegative and absolutely continuous over (a, θ) and $Q_1(\theta) = \left[\int_a^\theta M_1(x)dx\right]^{-1}$, $Q_1(\theta)$ is differentiable everywhere.

The joint pdf of $X_1, X_2, \ldots, X_m$ is

$$f(x_1, x_2, \ldots, x_m|\theta) = [Q_1(\theta)]^m \prod_{i=1}^m M_1(x_i)\mathbf{I}(\theta - x_{(m)})\mathbf{I}(x_{(1)} - a)$$

where

$$\mathbf{I}(y) = \begin{cases} 1 \; ; \; y > 0 \\ 0 \; ; \; y \le 0 \end{cases}$$

By factorization theorem, $X_{(m)}$ is sufficient for θ. The distribution of $X_{(m)}$ is $w(x|\theta)$, where

$$w(x|\theta) = m[F(x)]^{m-1}f(x) \tag{2.3.2}$$

Now

$$\int_a^\theta Q_1(\theta)M_1(x)dx = 1$$

This implies

$$\int_a^\theta M_1(x)dx = \frac{1}{Q_1(\theta)}$$

Then

$$\int_a^x M_1(x)dx = \frac{1}{Q_1(x)} \tag{2.3.3}$$

This implies $F(x) = \frac{Q_1(\theta)}{Q_1(x)}$
From (2.3.2)

$$w(x|\theta) = \frac{m[Q_1(\theta)]^m M_1(x)}{[Q_1(x)]^{m-1}}, \quad a < x < \theta \tag{2.3.4}$$

Let $h(x)$ be a function $X_{(m)}$. Now, we will show that $X_{(m)}$ is complete.

$$\mathbf{E}[h(x)] = \int_a^\theta h(x)\frac{[Q_1(\theta)]^m M_1(x)}{[Q_1(x)]^{m-1}}dx = 0 \tag{2.3.5}$$

Consider the following result
Let $f = f(x|\theta)$, $a = a(\theta)$, $b = b(\theta)$

$$\frac{d}{d\theta}\left[\int_a^b f\, dx\right] = \int_a^b \frac{df}{d\theta}\, dx + f(b|\theta)\frac{db}{d\theta} - f(a|\theta)\frac{da}{d\theta} \qquad (2.3.6)$$

Now,

$$\int_a^\theta h(x)\frac{M_1(x)}{[Q_1(x)]^{m-1}}\, dx = 0 \qquad (2.3.7)$$

Using (2.3.6),

$$\frac{dh(x)\frac{M_1(x)}{[Q_1(x)]^{m-1}}}{d\theta} = 0 \qquad (2.3.8)$$

Differentiating (2.3.7) with respect to θ

$$\frac{h(\theta)M_1(\theta)}{[Q_1(\theta)]^{m-1}} = 0 \text{ and, } M_1(\theta) \text{ and } Q_1(\theta) \neq 0$$

Hence $h(\theta) = 0$ for $a < x < \theta$.
This implies $h(x) = 0$ for $a < x < \theta$.
We will find UMVUE of $g(\theta)$. Let $U(x)$ be an unbiased estimator of $g(\theta)$.

$$\int_a^\theta u(x)\frac{m[Q_1(\theta)]^m M_1(x)}{[Q_1(x)]^{m-1}}\, dx = g(\theta)$$

$$\int_a^\theta u(x)\frac{M_1(x)}{[Q_1(x)]^{m-1}}\, dx = \frac{g(\theta)}{m[Q_1(\theta)]^m} \qquad (2.3.9)$$

Differentiating (2.3.9) with respect to θ

$$\frac{u(\theta)M_1(\theta)}{[Q_1(\theta)]^{m-1}} = \frac{1}{m}\left[\frac{g^{(1)}(\theta)}{[Q_1(\theta)]^m} + \frac{g(\theta)[Q_1^{(1)}(\theta)](-m)}{[Q_1(\theta)]^{m+1}}\right]$$

$$= \frac{1}{m}\left[\frac{g^{(1)}(\theta)}{[Q_1(\theta)]^m} - \frac{mg(\theta)Q_1^{(1)}(\theta)}{[Q_1(\theta)]^{m+1}}\right] \qquad (2.3.10)$$

where $g^{(1)}(\theta) = $ First derivative of $g(\theta)$
$Q_1^{(1)}(\theta) = $ First derivative of $Q_1(\theta)$
Now

$$\int\limits_a^\theta M_1(x)dx = \frac{1}{Q_1(\theta)} \tag{2.3.11}$$

Differentiating (2.3.11) with respect to θ

$$M_1(\theta) = -\frac{Q_1^{(1)}(\theta)}{Q_1^2(\theta)} \tag{2.3.12}$$

Substitute (2.3.12) in (2.3.10),

$$\frac{u(\theta)M_1(\theta)}{[Q_1(\theta)]^{m-1}} = \frac{1}{m}\left[\frac{g^{(1)}(\theta)}{[Q_1^m(\theta)]} + \frac{mg(\theta)M_1(\theta)}{[Q_1(\theta)]^{m-1}}\right]$$

$$u(\theta) = \frac{g^{(1)}(\theta)}{m[Q_1^m(\theta)]}\frac{[Q_1(\theta)]^{m-1}}{M_1(\theta)} + \frac{g(\theta)M_1(\theta)}{[Q_1(\theta)]^{m-1}}\frac{[Q_1(\theta)]^{m-1}}{M_1(\theta)}$$

$$= \frac{g^{(1)}(\theta)}{mQ_1(\theta)M_1(\theta)} + g(\theta) \quad \forall \ \theta$$

Therefore,

$$u(x_{(m)}) = \frac{g^{(1)}(x_{(m)})}{mQ_1(x_{(m)})M_1(x_{(m)})} + g(x_{(m)}) \tag{2.3.13}$$

We can conclude that $U(x_{(m)})$ is UMVUE of $g(\theta)$.
Particular cases:
(a)

$$f(x|\theta) = \begin{cases} \frac{1}{\theta} \; ; \; 0 < x < \theta \\ 0 \; ; \text{otherwise} \end{cases} \tag{2.3.14}$$

Comparing (2.3.14) with (2.3.1), $Q_1(\theta) = \frac{1}{\theta}$ and $M_1(x) = 1$
In this case we will find UMVUE of $\theta^r (r > 0)$.
Then $g(\theta) = \theta^r$. Using (2.3.13), $g(x_{(m)}) = [x_{(m)}]^r$, $g^{(1)}(x_{(m)}) = r[x_{(m)}]^{r-1}$,
$Q_1(x_{(m)}) = \frac{1}{x_{(m)}}, M_1(x_{(m)}) = 1$

$$u(x_{(m)}) = \frac{r(x_{(m)})^{r-1}}{m\frac{1}{x_{(m)}}(1)} + (x_{(m)})^r$$

$$= x_{(m)}^r\left[\frac{r}{m} + 1\right] \tag{2.3.15}$$

If $r = 1$, then

$$u(x_{(m)}) = \frac{m+1}{m} x_{(m)} \qquad (2.3.16)$$

is UMVUE of θ.

(b)

$$f(x|\theta) = \begin{cases} \frac{a\theta}{(\theta-a)x^2} & ; \ a < x < \theta \\ 0 & ; \ \text{otherwise} \end{cases} \qquad (2.3.17)$$

In comparing (2.3.17) with (2.3.1), $Q_1(\theta) = \frac{a\theta}{(\theta-a)}$, and $M_1(x) = \frac{1}{x^2}$
Let $g(\theta) = \theta^r$ $(r > 0)$, $g^{(1)}(\theta) = r\theta^{r-1}$
Using (2.3.13),

$$u(x_{(m)}) = \frac{rx_{(m)}^{r-1}}{m\left(\frac{ax_{(m)}}{x_{(m)}-a}\right)\left(\frac{1}{x_{(m)}^2}\right)} + x_{(m)}^r \qquad (2.3.18)$$

$$= x_{(m)}^r \left[\frac{r(x_{(m)} - a)}{am} + 1\right] \qquad (2.3.19)$$

Put $r = 1$ in (2.3.19)

$$u(x_{(m)}) = x_{(m)} \left[\frac{x_{(m)} - a}{am} + 1\right] \qquad (2.3.20)$$

is UMVUE of θ

(c)

$$f(x|\theta) = \frac{3x^2}{\theta^3}; \ \ 0 < x < \theta \qquad (2.3.21)$$

In this case $M_1(x) = 3x^2$, $Q_1(\theta) = \frac{1}{\theta^3}$, $g(\theta) = \theta^r$

$$u(x_{(m)}) = \frac{rx_{(m)}^{r-1}}{m\frac{1}{x_{(m)}^3}3x_{(m)}^2} + x_{(m)}^r$$

$$= x_{(m)}^r \left[\frac{r + 3m}{3m}\right] \qquad (2.3.22)$$

Put $r = 1$ in (2.3.22) then $U(x_{(m)}) = x_{(m)}\left(\frac{3m+1}{3m}\right)$ is UMVUE of θ.

(d)

$$f(x|\theta) = \frac{1}{\theta}; \ \ -\theta < x < 0 \qquad (2.3.23)$$

Let

$$Y_i = |X_i|, i = 1, 2, \ldots, m \tag{2.3.24}$$

then $Y_1, Y_2, \ldots, Y_m$ are iid rvs with $\cup(0, \theta)$.
From (2.3.15), UMVUE of θ^r is $u(y_{(m)})$, hence

$$u(y_{(m)}) = y_{(m)}^r \left[\frac{r}{m} + 1 \right] \tag{2.3.25}$$

Example 2.3.2 Let $X_1, X_2, ldots, X_m$ be iid rvs from the following pdf:

$$f(x|\theta) = \begin{cases} Q_2(\theta)M_2(x) & ; \theta < x < b \\ 0 & ; \text{otherwise} \end{cases} \tag{2.3.26}$$

where $M_2(x)$ is nonnegative and absolutely continuous over (θ, b) and $Q_2(\theta) = \left[\int_\theta^b M_2(x)dx \right]^{-1}$, $Q_2(\theta)$ is differentiable everywhere.

The joint pdf of $X_1, X_2, \ldots, X_m$ is

$$f(x_1, x_2, \ldots, x_m|\theta) = [Q_2(\theta)]^m \prod_{i=1}^m M_2(x_i)\mathbf{I}(\theta - x_{(1)})\mathbf{I}(x_{(m)} - b)$$

By factorization theorem, $X_{(1)}$ is sufficient for θ. The distribution of $X_{(1)}$ is $w(x|\theta)$, where

$$w(x|\theta) = m[1 - F(x)]^{m-1}f(x) \tag{2.3.27}$$

Now

$$\int_\theta^b M_2(x)dx = \frac{1}{Q_2(\theta)}$$

This implies then

$$\int_x^b M_2(x)dx = \frac{1}{Q_2(x)} \tag{2.3.28}$$

$$1 - F(x) = P[x \geq x] = \int_x^b Q_2(\theta)M_2(x)dx$$

$$= \frac{Q_2(\theta)}{Q_2(x)} \tag{2.3.29}$$

$$w(x|\theta) = m\frac{[Q_2(\theta)]^m M_2(x)}{[Q_2(x)]^{m-1}}, \theta < x < b \qquad (2.3.30)$$

Using (2.3.6), we can get

$$\frac{h(\theta)M_2(\theta)}{[Q_2(\theta)]^{m-1}} = 0 \text{ and } M_2(\theta), Q_2(\theta) \neq 0$$

Hence $h(\theta) = 0$ for $\theta < x < b$
This implies $h(x) = 0$ for $\theta < x < b$
We conclude that $X_{(1)}$ is complete.
Let $U(x)$ be an unbiased estimator of $g(\theta)$.

$$\int_{\theta}^{b} u(x)\frac{m[Q_2(\theta)]^m M_2(x)}{[Q_2(x)]^{m-1}}dx = g(\theta)$$

Using (2.3.6)

$$-\frac{u(\theta)M_2(\theta)}{[Q_2(\theta)]^{m-1}} = \frac{1}{m}\left[\frac{g^{(1)}(\theta)}{[Q_2(\theta)]^m} - \frac{mg(\theta)[Q_2^{(1)}(\theta)]}{[Q_2(\theta)]^{m+1}}\right] \qquad (2.3.31)$$

Now,

$$\int_{\theta}^{b} M_2(x)dx = \frac{1}{Q_2(\theta)} \qquad (2.3.32)$$

Differentiating (2.3.32) with respect to θ

$$M_2(\theta) = \frac{Q_2^{(1)}(\theta)}{Q_2(\theta)} \qquad (2.3.33)$$

Substituting (2.3.33) into (2.3.31)

$$u(\theta) = g(\theta) - \frac{1}{m}\frac{g^{(1)}(\theta)}{Q_2(\theta)M_2(\theta)}$$

Hence

$$u(x_{(1)}) = g(x_{(1)}) - \frac{1}{m}\frac{g^{(1)}(x_{(1)})}{Q_2(x_{(1)})M_2(x_{(1)})} \qquad (2.3.34)$$

Particular cases:
(a)

$$f(x|\theta) = \frac{\left(\frac{1}{\theta}\right)\left(\frac{\theta}{x}\right)^2}{1 - \frac{\theta}{b}}; \quad \theta < x < b \tag{2.3.35}$$

Here $Q_2(\theta) = \frac{\theta}{1-\frac{\theta}{b}}$ and $M_2(x) = x^{-2}$

We wish to find UMVUE of $g(\theta) = \theta^r$ using (2.3.31),

$$u(x_{(1)}) = x_{(1)}^r - \frac{1}{m}\frac{rx_{(1)}^{r-1}}{\left(\frac{x_{(1)}}{1-\frac{x_{(1)}}{b}}x_{(1)}^{-2}\right)}$$

$$= x_{(1)}^r \left[1 - \frac{1}{m}\frac{r(b - x_{(1)})}{b}\right]$$

For $r = 1$

$$u(x_{(1)}) = x_{(1)}\left[1 - \frac{b - x_{(1)}}{mb}\right] \tag{2.3.36}$$

(b)

$$f(x, \theta) = \begin{cases} \frac{e^{-x}}{e^{-\theta}-e^{-b}} & ; \ \theta < x < b \\ 0 & ; \ \text{otherwise} \end{cases} \tag{2.3.37}$$

Comparing (2.3.37) and (2.3.23)
$Q_2(\theta) = (e^{-\theta} - e^{-b})^{-1}$ and $M_2(x) = e^{-x}$
To find UMVUE of $g(\theta) = \theta^r$ using (2.3.31),

$$u(x_{(1)}) = x_{(1)}^r - \frac{1}{m}\frac{rx_{(1)}^{r-1}(e^{-x_{(1)}} - e^{-b})}{e^{-x_{(1)}}}$$

Put $r = 1$, then UMVUE of θ

$$u(x_{(1)}) = x_{(1)} - \frac{1}{m}e^{x_{(1)}}(e^{-x_{(1)}} - e^{-b}) \tag{2.3.38}$$

In the following example, we will find UMVUE from two-point truncation parameter families. This technique was introduced by Hogg and Craig (1972) and developed by Karakostas (1985).

Example 2.3.3 Let $X_1, X_2, \ldots, X_m$ be iid rvs from the following pdf:

$$f(x|\theta_1, \theta_2) = \begin{cases} Q(\theta_1, \theta_2)M(x) & ; \ \theta_1 < x < \theta_2 \\ 0 & ; \ \text{otherwise} \end{cases} \tag{2.3.39}$$

where $M(x)$ is an absolutely continuous function and $Q(\theta_1, \theta_2)$ is differentiable everywhere.

The joint pdf of $X_1, X_2, \ldots, X_m$ is

$$f(x_1, x_2, \ldots, x_m | \theta_1, \theta_2) = [Q(\theta_1, \theta_2)]^m \prod_{i=1}^{m} M(x_i) \mathbf{I}(x_{(1)} - \theta_1) \mathbf{I}(\theta_2 - x_{(m)}) \quad (2.3.40)$$

By factorization theorem, $(x_{(1)}, x_{(m)})$ is jointly sufficient for (θ_1, θ_2). Suppose we are looking for UMVUE of $g(\theta_1, \theta_2)$ is such that $\frac{dg(x_{(1)}, x_{(m)})}{dx_{(1)}}$ and $\frac{dg(x_{(1)}, x_{(m)})}{dx_{(m)}}$ both exists. The joint pdf of $(x_{(1)}, x_{(m)})$ is

$$f_{(x_{(1)}, x_{(m)})}(x, y) = \begin{cases} m(m-1)[F(y) - F(x)]^{m-2} f(x) f(y) & ; \theta_1 < x < y < \theta_2 \\ 0 & ; \text{otherwise} \end{cases}$$
$$(2.3.41)$$

Now,

$$\int_{\theta_1}^{\theta_2} M(x) dx = \frac{1}{Q(\theta_1, \theta_2)} \quad (2.3.42)$$

Hence

$$\int_{x}^{y} M(t) dt = \frac{1}{Q(x, y)} \quad (2.3.43)$$

$$F(y) - F(x) = \int_{x}^{y} Q(\theta_1, \theta_2) M(t) dt$$

$$= \frac{Q(\theta_1, \theta_2)}{Q(x, y)} \quad (2.3.44)$$

$$f(x, y | \theta_1, \theta_2) = \begin{cases} m(m-1) \frac{[Q(\theta_1, \theta_2)]^m}{[Q(x, y)]^{m-2}} M(x) M(y) & ; \theta_1 < x < y < \theta_2 \\ 0 & ; \text{otherwise} \end{cases} \quad (2.3.45)$$

Assume that $\frac{df(x, y)}{dx}$ and $\frac{df(x, y)}{dy}$ both exists.
To prove the completeness of $f(x, y | \theta_1, \theta_2)$, let

$$R(y, \theta_1) = \int_{\theta_1}^{y} h(x, y)[Q(x, y)]^{-(m-2)} M(x) dx$$

where $h(x, y)$ is any continuous function of (x, y) and

$$R(\theta_1, \theta_2) = \int\limits_{\theta_1}^{\theta_2} M(y)R(y, \theta_1)dy$$

$$R(\theta_1, \theta_2) = \int\limits_{\theta_1}^{\theta_2}\int\limits_{\theta_1}^{y} h(x, y)[Q(x, y)]^{-(m-2)}M(x)M(y)dxdy = 0 \qquad (2.3.46)$$

Hence to prove $h(x, y) = 0$, i.e., to prove $h(\theta_1, \theta_2) = 0$

$$\frac{\partial R(\theta_1, \theta_2)}{\partial \theta_1} = \int\limits_{\theta_1}^{\theta_2} -h(\theta_1, y)[Q(\theta_1, y)]^{-(m-2)}M(\theta_1)M(y)dy \qquad (2.3.47)$$

$$\frac{\partial^2 R(\theta_1, \theta_2)}{\partial \theta_1 \partial \theta_2} = -h(\theta_1, \theta_2)[Q(\theta_1, \theta_2)]^{-(m-2)}M(\theta_1)M(\theta_2) = 0 \qquad (2.3.48)$$

which implies that $h(\theta_1, \theta_2) = 0$. Hence $h(x, y) = 0$.
Completeness of $f(x, y|\theta_1, \theta_2)$ implies that a UMVUE $u(x, y)$ for some function of θ's, $g(\theta_1, \theta_2)$, say, will be found by solving the integral equation.

$$g(\theta_1, \theta_2) = E[u(x, y)]$$

That is,

$$g(\theta_1, \theta_2) = \int\limits_{\theta_1}^{\theta_2}\int\limits_{x}^{\theta_2} u(x, y)m(m-1)M(x)M(y)\frac{[Q(\theta_1, \theta_2)]^m}{[Q(x, y)]^{m-2}}dxdy$$

$$= [Q(\theta_1, \theta_2)]^m \int\limits_{\theta_1}^{\theta_2} m(m-1)M(x)\left\{\int\limits_{x}^{\theta_2} \frac{u(x, y)M(y)}{[Q(x, y)]^{m-2}}dy\right\}dx \qquad (2.3.49)$$

Now, we will have to find the solution of the integral equation (2.3.49). Since

$$\frac{1}{Q(\theta_1, \theta_2)} = \int\limits_{\theta_1}^{\theta_2} M(x)dx$$

$$-\frac{\frac{\partial Q(\theta_1, \theta_2)}{\partial \theta_1}}{[Q(\theta_1, \theta_2)]^2} = -M(\theta_1)$$

$$\frac{\partial Q(\theta_1, \theta_2)}{\partial \eta_1} = [Q(\theta_1, \theta_2)]^2 M(\theta_1) \qquad (2.3.50)$$

Let

$$Q_1(\theta_1, \theta_2) = Q^2(\theta_1, \theta_2) M(\theta_1), \qquad (2.3.51)$$

where $Q_1(\theta_1, \theta_2) = \frac{\partial Q(\theta_1, \theta_2)}{\partial \theta_1}$
Next,

$$\frac{-\frac{\partial Q(\theta_1, \theta_2)}{\partial \theta_2}}{Q^2(\theta_1, \theta_2)} = M(\theta_2)$$

$$-\frac{\partial Q(\theta_1, \theta_2)}{\partial \theta_2} = Q^2(\theta_1, \theta_2) M(\theta_2)$$

Let

$$Q_2(\theta_1, \theta_2) = -Q^2(\theta_1, \theta_2) M(\theta_2) \qquad (2.3.52)$$

where $Q_2(\theta_1, \theta_2) = \frac{\partial Q(\theta_1, \theta_2)}{\partial \theta_2}$

$$\frac{\partial^2 Q(\theta_1, \theta_2)}{\partial \theta_1 \theta_2} = Q_{12}(\theta_1, \theta_2) = -2Q^3(\theta_1, \theta_2) M(\theta_1) M(\theta_2) \qquad (2.3.53)$$

Differentiating (2.3.49) with respect to θ_1,

$$g_1(\theta_1, \theta_2) = [Q(\theta_1, \theta_2)]^m [-m(m-1)M(\theta_1)] \left\{ \int_{\theta_1}^{\theta_2} \frac{u(\theta_1, y)M(y)}{[Q(\theta_1, y)]^{m-2}} dy \right\}$$

$$+ mQ^{m-1}(\theta_1, \theta_2) Q_1(\theta_1, \theta_2) \int_{\theta_1}^{\theta_2} m(m-1)M(x) \left\{ \int_x^{\theta_2} \frac{u(x, y)M(y)}{[Q(x, y)]^{m-2}} dy \right\} dx$$

where $g_1(\theta_1, \theta_2) = \frac{\partial g}{\partial \theta_1}$
Using (2.3.51)

$$= mQ^{m+1}(\theta_1, \theta_2) M(\theta_1) \int_{\theta_1}^{\theta_2} m(m-1)M(x) \left\{ \int_x^{\theta_2} \frac{u(x, y)M(y)}{[Q(x, y)]^{m-2}} dy \right\} dx$$

$$- Q^m(\theta_1, \theta_2) [m(m-1)M(\theta_1)] \left\{ \int_{\theta_1}^{\theta_2} \frac{u(\theta_1, y)M(y)}{[Q(\theta_1, y)]^{m-2}} dy \right\}, \qquad (2.3.54)$$

Using (2.3.49)

$$g_1(\theta_1, \theta_2) = mQ(\theta_1, \theta_2)M(\theta_1)g(\theta_1, \theta_2)$$

$$- m(m-1)Q^m(\theta_1, \theta_2)M(\theta_1)\left[\int_{\theta_1}^{\theta_2} \frac{u(\theta_1, y)M(y)}{[Q(\theta_1, y)]^{m-2}}dy\right] \quad (2.3.55)$$

This equation can be written as

$$\int_{\theta_1}^{\theta_2} \frac{u(\theta_1, y)M(y)}{[Q(\theta_1, y)]^{m-2}}dy = \frac{g_1(\theta_1, \theta_2) - mQ(\theta_1, \theta_2)M(\theta_1)g(\theta_1, \theta_2)}{-m(m-1)[Q(\theta_1, \theta_2)]^m M(\theta_1)}$$

$$= \frac{g(\theta_1, \theta_2)}{(m-1)[Q(\theta_1, \theta_2)]^{m-1}} - \frac{g_1(\theta_1, \theta_2)}{m(m-1)M(\theta_1)[Q(\theta_1, \theta_2)]^m} \quad (2.3.56)$$

Differentiating with respect to θ_2,

$$\frac{u(\theta_1, \theta_2)M(\theta_2)}{[Q(\theta_1, \theta_2)]^{m-2}(\theta_1, \theta_2)} = \frac{g(\theta_1, \theta_2)[-(m-1)]Q[(\theta_1, \theta_2)]^{-(m-1)-1}Q_2(\theta_1, \theta_2)}{m-1}$$

$$+ \frac{g_2(\theta_1, \theta_2)}{(m-1)[Q(\theta_1, \theta_2)]^{m-1}}$$

$$- \left[\frac{g_1(\theta_1, \theta_2)(-m)[Q(\theta_1, \theta_2)]^{-(m+1)}Q_2(\theta_1, \theta_2)}{m(m-1)M(\theta_1)}\right.$$

$$\left. + \frac{g_{12}(\theta_1, \theta_2)}{m(m-1)[Q(\theta_1, \theta_2)]^m M(\theta_1)}\right] \quad (2.3.57)$$

$$\frac{u(\theta_1, \theta_2)M(\theta_2)}{[Q(\theta_1, \theta_2)]^{m-2}} = \frac{g(\theta_1, \theta_2)[-(m-1)]Q_2(\theta_1, \theta_2)}{(m-1)[Q(\theta_1, \theta_2)]^m}$$

$$+ \frac{g_2(\theta_1, \theta_2)}{(m-1)[Q(\theta_1, \theta_2)]^{m-1}}$$

$$- \left[\frac{g_1(\theta_1, \theta_2)(-m)Q_2(\theta_1, \theta_2)}{m(m-1)Q^{m+1}(\theta_1, \theta_2)M(\theta_1)}\right]$$

$$- \frac{g_{12}(\theta_1, \theta_2)}{m(m-1)Q^m(\theta_1, \theta_2)M(\theta_1)} \quad (2.3.58)$$

$$\frac{u(\theta_1, \theta_2)M(\theta_2)}{Q^m \, g(\theta_1, \theta_2)} = \frac{g(\theta_1, \theta_2)M(\theta_2)}{m[Q(\theta_1, \theta_2)]^{m-2}}$$
$$- \frac{g_1(\theta_1, \theta_2)M(\theta_2)}{(m-1)[Q(\theta_1, \theta_2)]^{m-1}M(\theta_1)}$$
$$+ \frac{g_2(\theta_1, \theta_2)}{(m-1)[Q(\theta_1, \theta_2)]^{m-1}}$$
$$- \frac{g_{12}(\theta_1, \theta_2)}{m(m-1)[Q(\theta_1, \theta_2)]^m M(\theta_1)} \qquad (2.3.59)$$

$$u(\theta_1, \theta_2) = g(\theta_1, \theta_2) - \frac{g_1(\theta_1, \theta_2)}{(m-1)[Q(\theta_1, \theta_2)]^m M(\theta_1)}$$
$$+ \frac{g_2(\theta_1, \theta_2)}{(m-1)M(\theta_2)Q(\theta_1, \theta_2)}$$
$$- \frac{g_{12}(\theta_1, \theta_2)}{m(m-1)M(\theta_1)M(\theta_2)[Q(\theta_1, \theta_2)]^2} \qquad (2.3.60)$$

Replacing θ_1 by $X_{(1)}$ and θ_2 by $X_{(m)}$,

$$u(X_{(1)}, X_{(m)}) = g(X_{(1)}, X_{(m)}) - \frac{g_1(X_{(1)}, X_{(m)})}{(m-1)Q(X_{(1)}, X_{(m)})M(X_{(1)})}$$
$$+ \frac{g_2(X_{(1)}, X_{(m)})}{(m-1)M(X_{(m)})Q(X_{(1)}, X_{(m)})}$$
$$- \frac{g_{12}(X_{(1)}, X_{(m)})}{m(m-1)M(X_{(1)})M(X_{(m)})[Q(X_{(1)}, X_{(m)})]^2} \qquad (2.3.61)$$

is UMVUE of $g(\theta_1, \theta_2)$.
Particular cases:
(a)

$$f(x|\theta_1, \theta_2) = \begin{cases} \frac{1}{\theta_2 - \theta_1} & ; \ \theta_1 < x < \theta_2 \\ 0 & ; \ \text{otherwise} \end{cases} \qquad (2.3.62)$$

Comparing (2.3.62) and (2.3.39), $Q(\theta_1, \theta_2) = \frac{1}{\theta_2 - \theta_1}$, $M(x) = 1$
To find UMVUE of (i) θ_1, (ii) θ_2, (iii) $\frac{\theta_1 - \theta_2}{2}$ and (iv) $\frac{\theta_1 + \theta_2}{2}$
(i) $g(\theta_1, \theta_2) = \theta_1$, $g(X_{(1)}, X_{(m)}) = X_{(1)}$, $g_1(X_{(1)}, X_{(m)}) = 1$,
$g_2(X_{(1)}, X_{(m)}) = 0$ and $g_{12}(X_{(1)}, X_{(m)}) = 0$
$M(X_{(1)}) = M(X_{(m)}) = 1$, $Q(X_{(1)}, X_{(m)}) = \frac{1}{X_{(m)} - X_{(1)}}$. Using (2.3.61),

$$u(X_{(1)}, X_{(m)}) = X_{(1)} - \frac{X_{(m)} - X_{(1)}}{(m-1)}$$

$$= \frac{mX_{(1)} - X_{(1)} - X_{(m)} + X_{(1)}}{(m-1)}$$

$$= \frac{mX_{(1)} - X_{(m)}}{(m-1)} \tag{2.3.63}$$

Hence, $\frac{mX_{(1)} - X_{(m)}}{(m-1)}$ is UMVUE of θ_1

(ii) $g(\theta_1, \theta_2) = \theta_2$, $g(X_{(1)}, X_{(m)}) = X_{(m)}$, $g_1(X_{(1)}, X_{(m)}) = 0$, $g_2(X_{(1)}, X_{(m)}) = 1$
and $g_{12}(X_{(1)}, X_{(m)}) = 0$

$M(X_{(1)}) = M(X_{(m)}) = 1$, $Q(X_{(1)}, X_{(m)}) = \frac{1}{X_{(m)} - X_{(1)}}$

$$u(X_{(1)}, X_{(m)}) = X_{(m)} + \frac{X_{(m)} - X_{(1)}}{(m-1)}$$

$$= \frac{mX_{(m)} - X_{(m)} + X_{(n)} - X_{(1)}}{(m-1)}$$

$$= \frac{mX_{(m)} - X_{(1)}}{(m-1)} \tag{2.3.64}$$

Hence, $\frac{mX_{(m)} - X_{(1)}}{(m-1)}$ is UMVUE of θ_2

(iii) UMVUE of $\frac{\theta_1 - \theta_2}{2}$

$$= \frac{mX_{(1)} - X_{(m)} - mX_{(m)} + X_{(1)}}{2(m-1)}$$

$$= \frac{(m+1)}{2(m-1)}[X_{(m)} - X_{(1)}] \tag{2.3.65}$$

(iv) UMVUE of $\frac{\theta_1 + \theta_2}{2}$

$$= \frac{1}{2}\left[\frac{mX_{(1)} - X_{(m)}}{(m-1)} + \frac{mX_{(m)} - X_{(1)}}{(m-1)}\right]$$

$$= \frac{1}{2(m-1)}\left[(m-1)X_{(1)} + (m-1)X_{(m)}\right]$$

$$= \frac{X_{(m)} + X_{(1)}}{2} \tag{2.3.66}$$

(b)

$$f(x, \theta_1, \theta_2) = \begin{cases} \frac{\theta_1 \theta_2}{\theta_2 - \theta_1} x^{-2} & ; \theta_1 < x < \theta_2 \\ 0 & ; \text{otherwise} \end{cases} \tag{2.3.67}$$

Comparing (2.3.67) to (2.3.39)

$$Q(\theta_1, \theta_2) = \frac{\theta_1 \theta_2}{\theta_2 - \theta_1}, M(x) = x^{-2}$$

To find UMVUE of $(\theta_1 \theta_2)^m$

$$g(X_{(1)}, X_{(m)}) = [(X_{(1)} X_{(m)}]^m, g_1(X_{(1)}, X_{(m)}) = m[(X_{(1)} X_{(m)}]^{m-1} X_{(m)}$$

$$g_2(X_{(1)}, X_{(m)}) = m[(X_{(1)} X_{(m)}]^{m-1} X_{(1)},$$

$$g_{12}(X_{(1)}, X_{(m)}) = m(m-1)[X_{(1)} X_{(m)}]^{m-2} X_{(1)} X_{(m)} + m[X_{(1)} X_{(m)}]^{m-1}$$

$$M(X_{(1)}) = X_{(1)}^{-2}, M(X_{(m)}) = X_{(m)}^{-2}, Q(X_{(1)}, X_{(m)}) = \frac{X_{(1)} X_{(m)}}{X_{(m)} - X_{(1)}}$$

$$U(X_{(1)}, X_{(m)}) = (X_{(1)} X_{(m)})^m - \frac{m[X_{(1)} X_{(m)}]^{m-1} X_{(m)} [X_{(m)} - X_{(1)}]}{(m-1) X_{(1)} X_{(m)} X_{(1)}^{-2}}$$

$$+ \frac{m[X_{(1)} X_{(m)}]^{m-1} X_{(1)} [X_{(m)} - X_{(1)}]}{(m-1) X_{(1)} X_{(m)} X_{(m)}^{-2}}$$

$$- \frac{m(m-1)[X_{(1)} X_{(m)}]^{m-2} X_{(1)} X_{(m)} + m[X_{(m)} X_{(1)}]^{m-1}}{m(m-1) X_{(1)}^{-2} X_{(m)}^{-2} X_{(1)}^2 X_{(m)}^2} [X_{(m)} - X_{(1)}]^2$$

$$= (X_{(1)} X_{(m)})^m - \frac{m[X_{(m)} - X_{(1)}][X_{(1)} X_{(m)}]^{m-1}}{(m-1) X_{(1)}^{-1}}$$

$$+ \frac{m[X_{(1)} X_{(m)}]^{m-1} [X_{(m)} - X_{(1)}]}{(m-1) X_{(m)}^{-1}} - \frac{[X_{(1)} X_{(m)}]^{m-1} [X_{(m)} - X_{(1)}]^2}{1}$$

$$- \frac{m[X_{(1)} X_{(m)}]^{m-1} [X_{(m)} - X_{(1)}]^2}{m(m-1) X_{(1)}^{-2} X_{(m)}^{-2} X_{(1)}^2 X_{(m)}^2}$$

$$= (X_{(1)} X_{(m)})^m - \frac{m}{m-1} X_{(1)} [X_{(m)} - X_{(1)}][X_{(1)} X_{(m)}]^{m-1}$$

$$+ \frac{m}{m-1} X_{(m)} [X_{(m)} - X_{(1)}][X_{(1)} X_{(m)}]^{m-1} - [X_{(1)} X_{(m)}]^{m-1} [X_{(m)} - X_{(1)}]^2$$

$$- \frac{[X_{(1)} X_{(m)}]^{m-1} [X_{(m)} - X_{(1)}]^2}{(m-1)}$$

$$= (X_{(1)}X_{(m)})^m + \frac{m}{m-1}[X_{(m)} - X_{(1)}]^2[X_{(1)}X_{(m)}]^{m-1}$$

$$- [X_{(1)}X_{(m)}]^{m-1}[X_{(m)} - X_{(1)}]^2 - \frac{[X_{(1)}X_{(m)}]^{m-1}[X_{(m)} - X_{(1)}]^2}{m-1}$$

$$\doteq (X_{(1)}X_{(m)})^m + [X_{(m)} - X_{(1)}]^2[X_{(1)}X_{(m)}]^{m-1}\left[\frac{m}{m-1} - 1 - \frac{1}{m-1}\right]$$

$$= (X_{(1)}X_{(m)})^m \tag{2.3.68}$$

Hence, $(X_{(1)}X_{(m)})^m$ is UMVUE of $(\theta_1\theta_2)^m$. One should note that MLE of $(\theta_1\theta_2)^m$ is again the same.

Stigler (1972) had obtained an UMVUE for an incomplete family.

Example 2.3.4 Consider the Example 1.5.5.
Further, consider a single observation $X \sim P_N$.

$$P[X = k] = \begin{cases} \frac{1}{N} \; ; \; k = 1, 2, \ldots, N \\ 0 \; ; \text{otherwise} \end{cases}$$

Now X is sufficient and complete.

$$EX = \frac{N+1}{2} \text{ and } E[2X - 1] = N$$

Then, $\Phi_1(X) = (2X - 1)$ is UMVUE of N.

$$V[\Phi_1(X)] = \frac{N^2 - 1}{3} \tag{2.3.69}$$

Now the family $\wp - P_n$ is not complete, see Example 1.5.5.
We will show that for this family the UMVUE of N is

$$\Phi_2(k) = \begin{cases} 2k - 1 \; ; \; k \neq n, k \neq n + 1 \\ 2n \quad ; \; k = n, n + 1 \end{cases} \tag{2.3.70}$$

According to Theorem 2.2.3, we have to show that $\Phi_2(k)$ is UMVUE iff it is uncorrelated with all unbiased estimates of zero.

In Example 1.5.5, we have shown that $g(X)$ is an unbiased estimator of zero, where

$$g(x) = \begin{cases} 0 \; ; \quad x = 1, 2, \ldots, n-1, n+2, n+3 \ldots \\ a \; ; \quad x = n \\ -a \; ; \quad x = n + 1 \end{cases} \tag{2.3.71}$$

where a is nonzero constant.

Case (i) $N < n$

$$Eg(X) = \sum_{k=1}^{N} g(x) \frac{1}{N} = 0$$

Case (ii) $N > n$

$$Eg(X) = \sum_{k=1}^{N} \frac{1}{N} g(x)$$

$$= \frac{1}{N} [0 + \cdots + 0 + (-a) + (a) + 0] = 0$$

Case (iii) $N = n$

$$Eg(X) = \sum_{k=1}^{N} g(x) \frac{1}{N}$$

$$= \frac{1}{N} [0 + \cdots + 0 + (a)] = \frac{a}{N}$$

$$Eg(X) = \begin{cases} 0 & ; \ N = n \\ \frac{a}{N} & ; \ N = n \end{cases}$$

Thus we see that $g(x)$ is an unbiased estimate of zero for the family $\wp - P_n$ and therefore the family is not complete.

Remark: Completeness is a property of a family of distribution rather than the random variable or the parametric form, that the statistical definition of "complete" is related to every day usage, and that removing even one point from a parameter set may alter the completeness of the family, see Stigler (1972).

Now, we know that the family $\wp - \{P_n\}$ is not complete. Hence $\Phi_1(X)$ is not UMVUE of N for the family $\wp - \{P_n\}$. For this family consider the UMVUE of N as $\Phi_2(X)$, where

$$\Phi_2(X) = \begin{cases} 2x - 1 ; & x \neq n, x \neq n+1 \\ 2n & ; x = n, n+1 \end{cases} \tag{2.3.72}$$

According to Theorem 2.2.3, $\Phi_2(X)$ is UMVUE iff it is uncorrelated with all unbiased estimates of zero.

Already, we have shown that $g(x)$ is an unbiased estimator of zero for the family $\wp - \{P_n\}$.

Since $Eg(x) = 0$ for $N \neq n$

Now, we have to show that $\text{Cov}[g(x), \Phi_2(X)] = 0$.

$$\text{Cov}[g(x), \Phi_2(X)] = \text{E}[g(x)\Phi_2(X)]$$

Case (i) $N > n$

$$\text{E}[g(x)\Phi_2(X)] = \frac{1}{N} \sum_{k=1}^{N} g(x)\Phi_2(k)$$

$$= \frac{1}{N} [(0)(2k - 1) + (a)(2n) + (-a)(2n)] = 0$$

Case (ii) $N < n$

$$\text{E}[g(x)\Phi_2(X)] = \frac{1}{N} [(0)(2k - 1)] = 0$$

Thus, $\Phi_2(X)$ is UMVUE of N for the family $\wp - \{P_n\}$.
Note that $\text{E}\Phi_2(X) = N$. We can compute the variance of $\Phi_2(X)$
Case (i) $N < n$

$$\text{E}\Phi_2(x) = \sum_{x=1}^{N} (2x - 1)\frac{1}{N}$$

$$= \frac{1}{N} \left[\frac{2N(N + 1)}{2} - N \right] = N$$

$$\text{E}\Phi_2^{\,2}(x) = \frac{1}{N} \sum_{x=1}^{N} (2x - 1)^2 \frac{1}{N}$$

$$= \frac{1}{N} \left[\sum_{k=1}^{N} (4x^2 - 4x + 1) \right]$$

$$= \frac{1}{N} \left[\frac{4N(N + 1)(2N + 1)}{6} - \frac{4N(N + 1)}{2} + N \right]$$

$$= \frac{2(N + 1)(2N + 1)}{3} - 2(N + 1) + 1$$

$$= \frac{4N^2 - 1}{3}$$

$$\text{Var}[\Phi_2(X)] = \frac{4N^2 - 1}{3^\cdot} - N^2$$

$$= \frac{N^2 - 1}{3}$$

Case (ii) $N > n$

$$E[\Phi_2(x)] = \frac{1}{N}\left[\sum_{x=1}^{N}\Phi_2(x)\right]$$

$$= \frac{1}{N}[\Phi_2(1) + \Phi_2(2) + \cdots + \Phi_2(n-1) + \Phi_2(n) + \Phi_2(n+1)$$
$$+ \Phi_2(n+2) + \cdots + \Phi_2(N)]$$

$$= \frac{1}{N}[1 + 3 + \cdots + 2n - 3 + 2n + 2n + 2n + 3 + 2n + 5 + \cdots + 2N - 1]$$

$$= \frac{1}{N}[1 + 3 + \cdots + 2n - 3 + (2n - 1 + 2n + 1) + 2n + 3 + \cdots + 2N - 1$$
$$+ 2n + 2n - (2n - 1 + 2n + 1)]$$

$$= \frac{1}{N}\left[\frac{N}{2}(1 + 2N - 1) + 0\right] = N$$

$$E\Phi_2{}^2(x) = \frac{1}{N}[\Phi_2{}^2(1) + \Phi_2{}^2(2) + \cdots + \Phi_2{}^2(n-1) + \Phi_2{}^2(n)$$

$$+ \Phi_2{}^2(n+1) + \Phi_2{}^2(n+2) + \cdots + \Phi_2{}^2(N)]$$

$$= \frac{1}{N}[1^2 + 3^2 + 5^2 \cdots + (2n-3)^2 + \{(2n-1)^2 + (2n+1)^2\}$$

$$+ (2n+3)^2 + (2n+5)^2 + \cdots + (2N-1)^2 + (2n)^2 + (2n)^2 - \{(2n-1)^+ (2n+1)^2\}]$$

$$= \frac{1}{N}\left[\sum_{k=1}^{N}(2k-1)^2 + 4n^2 + 4n^2 - 4n^2 + 4n - 1 - 4n^2 - 4n - 1\right]$$

$$= \frac{4N^2}{3} - \frac{1}{3} - \frac{2}{N}$$

$$\mathrm{Var}[\Phi_2(X)] = \frac{4N^2}{3} - \frac{1}{3} - \frac{2}{N} - N^2 = \frac{N^2 - 1}{3} - \frac{2}{N}$$

$$\mathrm{Var}[\Phi_2(X)] = \begin{cases} \frac{N^2-1}{3} & ; N < n \\ \frac{N^2-1}{3} - \frac{2}{N} & ; N > n \end{cases} \tag{2.3.73}$$

Thus $\Phi_2(X)$ is UMVUE for $\wp - \{P_n\}$ but $\Phi_2(X)$ is not unbiased for the family $\wp$. Note that for $N = n$,

$$E[\Phi_2(X)] = \frac{1}{n}\sum_{x=1}^{N}\Phi_2(X)$$

$$= \frac{1}{n}[\Phi_2(1) + \cdots + \Phi_2(n-1) + \Phi_2(n)]$$

$$= \frac{1}{n}[1 + 3 + \cdots + 2n - 3 + 2n]$$

$$= \frac{1}{n}\left[\sum_{x=1}^{N}(2x-1)^2 + 2n - (2n-1)\right] = \frac{n^2+1}{n} \tag{2.3.74}$$

$$E[\Phi_2{}^2(X)] = \frac{1}{n}\left[\sum_{x=1}^{N}(2x-1)^2 + (2n)^2 - (2n-1)^2\right]$$

$$= \frac{4n^2-1}{3} + \frac{4n-1}{n}$$

$$\mathrm{Var}[\Phi_2(X)] = \frac{4n^2-1}{3} + \frac{4n-1}{n} - \left(\frac{n^2+1}{n}\right)^2$$

Example 2.3.5 Let $X_1, X_2, \ldots, X_m$ be iid discrete rvs with following pmf $f(x|N)$. Find the UMVUE of $g(N)$.

$$f(x|N) = \begin{cases} \phi(N)M(x) \; ; \; a \le X \le N \\ 0 \qquad\qquad ; \text{otherwise} \end{cases} \tag{2.3.75}$$

where $\sum_{x=a}^{N} M(x) = \frac{1}{\phi(N)}$.

According to Example 2.2.7, we can show that $X_{(m)}$ is sufficient and complete for N.

$$P[X_{(m)} \le z] = \left[\frac{\phi(N)}{\phi(z)}\right]^m$$

$$P[X_{(m)} \le z-1] = \left[\frac{\phi(N)}{\phi(z-1)}\right]^m$$

$$P[X_{(m)} = z] = \phi^m(N)[\phi^{-m}(z) - \phi^{-m}(z-1)]$$

Let $u(X_{(m)})$ is UMVUE of $g(N)$

$$\sum_{z=a}^{N} u(z)\phi^m(N)[\phi^{-m}(z) - \phi^{-m}(z-1)] = g(N)$$

$$\sum_{z=a}^{N} u(z)\frac{\phi^m(N)}{g(N)}[\phi^{-m}(z) - \phi^{-m}(z-1)] = 1$$

Let $\psi(N) = \frac{\phi(N)}{g^{\frac{1}{m}}(N)}$

$$\sum_{z=a}^{N} u(z)\psi^m(N)[\phi^{-m}(z) - \phi^{-m}(z-1)] = 1$$

$$\sum_{z=a}^{N} \frac{u(z)[\phi^{-m}(z) - \phi^{-m}(z-1)]}{[\psi^{-m}(z) - \psi^{-m}(z-1)]} \psi^{-m}(N)[\psi^{-m}(z) - \psi^{-m}(z-1)] - 1$$

Hence

$$\frac{u(z)[\phi^{-m}(z) - \phi^{-m}(z-1)]}{[\psi^{-m}(z) - \psi^{-m}(z-1)]} = 1,$$

This implies

$$u(z) = \frac{[\psi^{-m}(z) - \psi^{-m}(z-1)]}{[\phi^{-m}(z) - \phi^{-m}(z-1)]},$$

Therefore,

$$u(X_{(m)}) = \frac{[\psi^{-m}(X_{(m)}) - \psi^{-m}(X_{(m)} - 1)]}{[\phi^{-m}(X_{(m)}) - \phi^{-m}(X_{(m)} - 1)]},$$

We conclude that $U(X_{(m)})$ is UMVUE of $g(N)$.
Particular cases:
(a) $g(N) = N^s$, s is a real number.
According to (2.3.75), $\phi(N) = N^{-1}$, $M(x) = 1$,
$\psi(N) = N^{-\frac{(s+m)}{m}}$, $\psi(X_{(m)}) = X_{(m)}^{-\frac{(s+m)}{m}}$, $\phi(X_{(m)}) = X_{(m)}^{-1}$.

$$u(X_{(m)}) = \frac{X_{(m)}^{m+s} - (X_{(m)} - 1)^{m+s}}{X_{(m)}^{m} - (X_{(m)} - 1)^{m}},$$

which is same as (2.2.32).
(b) $g(N) = e^N$

$$\psi(N) = N^{-1}e^{-\frac{N}{m}} \Rightarrow \psi(X_{(m)}) = X_{(m)}^{-1}e^{-\frac{X_{(m)}}{m}}$$

Hence $u(X_{(m)})$ is UMVUE of e^N.
Hence,

$$u(X_{(m)}) = \frac{X_{(m)}^{m}e^{X_{(m)}} - (X_{(m)} - 1)^{m}e^{X_{(m)}-1}}{X_{(m)}^{m} - (X_{(m)} - 1)^{m}},$$

Reader should show that the above UMVUE of e^N is same as in Example 2.2.7.
Now, we will consider some examples which can be solved using R software.

Example 2.3.6 2, 5, 7, 3, 4, 2, 5, 4 is a sample of size 8 drawn from binomial distribution B(10,p). Obtain UMVUE of p, p^2, p^2q, $p(x \leq 2)$, $p(x > 6)$.

```
a=function (r,s)
{
m<-8
n<-10
x<-c(2,5,7,3,4,2,5,4)
t<-sum(x)
umvue=(choose(m*n-r-s,t-r)/choose(m*n,t))
print(umvue)
}
a(1,0) #UMVUE of p
a(2,0) #UMVUE of p^2
a(2,1) #UMVUE of p^2*q
b=function(c)
{
m<-8
n<-10
x<-c(2,5,7,3,4,2,5,4)
t<-sum(x)
g<-array(,c(1,c+1))
for (i in 1:c)
{
g[i]=((choose(n,i)*choose(m*n-n,t-i))/choose(m*n,t))
}
g[c+1]=((choose(n,0)*choose(m*n-n,t))/choose(m*n,t))
umvue=sum(g)
print (umvue)
}
b(2)#UMVUE of P(X<=2)
1-b(6)#UMVUE of P(X<=6) & P(X>6)
```

Example 2.3.7   0, 3, 1, 5, 5, 3, 2, 4, 5, 4 is a sample of size 10 from the Poisson distribution $P(\lambda)$. Obtain UMVUE of λ, λ^2, $\lambda e^{-\lambda}$, and $P(x \geq 4)$.

```
 d=function (s,r) {
m<-10
x<-c(0,3,1,5,5,3,2,4,5,4)
t<-sum(x)
umvue=((m-s)^(t-r)*factorial(t))/(m^t*factorial(t-r))
print (umvue) } d(0,1) #UMVUE of lamda d(0,2) #UMVUE of
lamda^2 d(1,1) #UMVUE of lamda*e^(-lamda) f=function (c) {
m<-10
x<-c(0,3,1,5,5,3,2,4,5,4)
t<-sum(x)
g<-array(,c(1,c+1))
for (i in 1:c)
```

```
{
g[i]<-(choose(t,i)*(1/m)^i*(1-(1/m))^(t-i))
}
g[c+1]=choose(t,0)*(1-(1/m))^t
umvue=sum(g)
print (umvue) } 1-f(3) #UMVUE of P(X<4) & P(X>=4)
```

Example 2.3.8   8, 4, 6, 2, 9, 10, 5, 8, 10, 8, 3, 10, 1, 6, 2 is a sample of size 15 from the following distribution:

$$P[X = k] = \begin{cases} \frac{1}{N} \ ; \ k = 1, 2, \ldots, N \\ 0 \ ; \ \text{otherwise} \end{cases}$$

Obtain UMVUE of N^5.

```
 h<-function (s) {
n<-15
x<-c(8,4,6,2,9,10,5,8,10,8,3,10,1,6,2)
z<-max(x)
umvue=(z^(n+s)-(z-1)^(n+s))/((z^n)-(z-1)^n)
print (umvue) } h(5) #UMVUE of N^5
```

Example 2.3.9 Lots of manufactured articles are made up of items each of which is an independent trial with probability p of it being defective. Suppose that four such lots are sent to a consumer, who inspects a sample of size 50 from each lot. If the observed number of defectives in the ith lot is 0, 1, or 2, the consumer accepts this lot. The observed numbers of defectives are 0, 0, 0, 3. Obtain UMVUE of the probability that a given lot will be accepted.

```
j=function (c) {
m<-4
n<-50
t<-3
g<-array(,c(1,c+1))
for (i in 1:c)
{
g[i]<-(choose(50,i)*choose((m*n)-n,t-i))/(choose(m*n,t))
}
g[c+1]<-(choose(m*n-n,t))/(choose(m*n,t))
umvue=sum(g)
print (umvue) } j(2) #UMVUE of P(X<=2)
```

Example 2.3.10 Let $X_1, X_2, \ldots X_n$ be a sample from $NB(1, \theta)$.
Find the UMVUE of $d(\theta) = P(X = 0)$, for the data 3, 4, 3, 1, 6, 2, 1, 8

```
k=function (r,s) {
m<-8
k<-1
x<-c(3,4,3,1,6,2,1,8)
t=sum(x)
umvue=choose(t-s+m*k-r-1,m*k-r-1)/choose(t+m*k-1,t)
print(umvue) } k(1,0) #UMVUE of P(X=0), i.e., p
```

Example 2.3.11 The following observations were recorded on a random variable X
having pdf:

$$f(x) = \begin{cases} \frac{x^{p-1}e^{-\frac{x}{\sigma}}}{\sigma^p \Gamma(p)} & ; \ x > 0, \ \sigma > 0, \ p = 4 \\ 0 & ; \text{otherwise} \end{cases}$$

7.89, 10.88, 17.09, 16.17, 11.32, 18.44, 3.32, 19.51, 6.45, 6.22.
Find UMVUE of σ^3

```
x1<-function (k,r) {
p<-4
n<-10
y<-c(7.89,10.88,17.09,16.17,11.32,18.44,3.32,19.51,6.45,6.22)
t<-sum(y)
umvue=((gamma(n*p))*(t-k)^(n*p-r-1))/((gamma(n*p-r))*t^(n*p-1))
print (umvue) } x1(0,-3) #UMVUE of sigma^3
```

Example 2.3.12 A random sample of size 10 is drawn from the following pdf:

1.

$$f(x, \theta) = \begin{cases} \frac{\theta}{(1+x)^{\theta+1}} & ; \ x > 0, \ \theta > 0 \\ 0 & ; \text{otherwise} \end{cases}$$

Data: 0.10, 0.34, 0.35, 0.08, 0.03, 2.88, 0.45, 0.49, 0.86, 3.88

2.

$$f(x, \theta) = \begin{cases} \theta x^{\theta-1} & ; \ 0 < x < 1 \\ 0 & ; \text{otherwise} \end{cases}$$

Data: 0.52, 0.79, 0.77, 0.76, 0.71, 0.76, 0.47, 0.35, 0.55, 0.63

3.

$$f(x, \theta) = \begin{cases} \frac{1}{\theta}e^{-\frac{|x|}{\theta}} & ; \ -\infty < x < \infty \\ 0 & ; \text{otherwise} \end{cases}$$

Data: 9.97, 0.64, 3.17, 1.48, 0.81, 0.61, 0.62, 0.72, 3.14, 2.99
Find UMVUE of θ in (i), (ii), and (iii),

(i)

```
x2<-function (k,r) {
n<-10
y<-c(0.10,0.34,0.35,0.08,0.03,2.88,0.45,0.49,0.86,3.88)
x<-array(,c(1,10))
for (i in 1:10)
{
x[i]=log(1+y[i])
}
t<-sum(x)
umvue=(((t-k)^(n-r-1))*gamma(n))/((t^(n-1))*gamma(n-r))
print (umvue) } x2(0,1) #UMVUE of theta
```

(ii)

```
 x3<-function (k,r) {
n<-10
y<-c(0.52,0.79,0.77,0.76,0.71,0.76,0.47,0.35,0.55,0.63)
x<-array(,c(1,10))
for (i in 1:10)
{
x[i]=-log(y[i])
}
t<-sum(x)
umvue=(((t-k)^(n-r-1))*gamma(n))/((t^(n-1))*gamma(n-r))
print (umvue) } x3(0,1) #UMVUE of theta
```

(iii)

```
 x4<-function (k,r) {
n<-10
y<-c(9.97,0.64,3.17,1.48,0.81,0.61,0.62,0.72,3.14,2.99)
t<-sum(y)
umvue=(((t-k)^(n-r-1))*gamma(n))/((t^(n-1))*gamma(n-r))
print (umvue) } x4(0,-1) #UMVUE of theta
```

Example 2.3.13 The following observations were obtained on an rv X following:

1. $N(\theta, \sigma^2)$
   Data: 5.77, 3.81, 5.24, 8.81, 0.98, 8.44, 3.16, 11.27, 4.40, 4.87, 7.28, 8.48, 6.43,
   −0.00, 9.67, 12.04, −5.06, 13.71, 6.12, 4.76
   Find UMVUE of θ, θ^2, ϑ^3 and $P(x \le 2)$

2. $N(6, \sigma^2)$

   Data: 7.26, -0.23, 7.55, 3.09, 7.62, 16.79, 5.27, 8.46, 5.16, -0.66.

   Find UMVUE of $\frac{1}{\sigma}, \sigma, \sigma^2, P(X \geq 2)$

3. $N(\theta, \sigma^2)$

   Data: 10.59, -1.50, 6.40, 7.55, 4.70, 1.63, 0.04, 2.96, 6.47, 6.42

   Find UMVUE of $\theta, \theta^2, \theta + 2\sigma$,

(i)

```
x5<-function (sigsq,n,k)
{x<-c(5.77,3.81,5.24,8.81,0.98,8.44,3.16,11.27,4.4,4.87,7.28,
8.48,6.43,0,9.67,12.04,-5.06,13.71,6.12,4.76)
umvue1=mean(x)
umvue2=umvue1^2-(sigsq/n)
umvue3=umvue1^3-(3*sigsq*umvue1/n)
umvue4=pnorm((k-(mean(x)))/(sqrt((sigsq*((n-1)/n)))))
print (umvue1) #UMVUE of theta
print (umvue2) #UMVUE of theta^2
print (umvue3) #UMVUE of theta^3
print (umvue4) #UMVUE of P(X<=2) } x5(4,20,2)
```

(ii)

```
x6<-function (n,r) {
x<-c(7.26,-0.23,7.55,3.09,7.62,16.79,5.27,8.46,5.16,-0.66)
t<-sum((x-6)^2)
umvue=(((t^r)*gamma(n/2))/((2^r)*gamma((n/2)+r)))
print (umvue) } x6 (10,-0.5) #UMVUE of 1/sigma x6 (10,0.5)
#UMVUE of sigma x6 (10,1) #UMVUE of sigma^2

x7<-function (n,k) {
x<-c(7.26,-0.23,7.55,3.09,7.62,16.79,5.27,8.46,5.16,-0.66)
t<-sum((x-6)^2)
umvue<-(1-pbeta(((k-6)/sqrt(t))^2,0.5, ((n-1)/2)))*0.5
print (umvue) } x7(10,2) #UMVUE of P(X>=2)
```

(iii)

```
x8<-function(n,r) {
x<-c(10.59,-1.5,6.4,7.55,4.7,1.63,0.04,2.96,6.47,6.42)
s<-sum((x-mean(x))^2)
umvue1<-mean(x) #UMVUE of theta
umvue2<-((s^(r))*gamma((n-1)/2))/(gamma(((n-1)/2)+r)*(2^r))
#UMVUE of sigma^2
print (umvue1)
```

```
print (umvue2)
print ((umvue1*2) (umvue2/11) #UMVUE HF FRAFA )
print (umvue1+2*sqrt(umvue2))#UMVUE of theta+2*sigma }
x8(10,1)
```

Example 2.3.14 If rv X is drawn from $U(\theta_1, \theta_2)$. Find the UMVUE of θ_1 and θ_2 from the following data:
3.67, 2.65, 4.41, 3.48, 2.07, 2.91, 2.77, 4.82, 2.73, 2.98.

```
x<-c(3.67,2.65,4.41,3.48,2.07,2.91,2.77,4.82,2.73,2.98)
umvue1<-(max(x)-length(x)*min(x))/(1-length(x)) umvue1 #UMVUE
of theta1 umvue2<-(length(x)*max(x)-min(x))/(length(x)-1)
umvue2 #UMVUE of theta1
```

Example 2.3.15 If rv X is drawn from $U(0, \theta)$ Find the UMVUE of θ, θ^2, and $\frac{1}{\theta}$ from the following data:
1.60, 1.91, 3.68, 0.78, 2.52, 4.34, 1.15, 4.69, 1.53, 4.53

```
x9<-function (n,r) {
x<-c(1.6,1.91,3.68,0.78,2.52,4.34,1.15,4.69,1.53,4.53)
umvue<-((max(x)^r)*((n+r)/n))
print (umvue) } x9(10,1) #UMVUE of theta x9(10,2) #UMVUE of
theta^2 x9(10,-1)#UMVUE of (1/theta)
```

2.4 Exercise 2

1. For the geometric distribution,

$$f(x|\theta) = \theta(1-\theta)^{x-1}; x = 1, 2, 3, \ldots; 0 < \theta < 1$$

Obtain an unbiased estimator of $\frac{1}{\theta}$ for a sample of size n. Calculate it for given data: 6, 1, 1, 14, 1, 1, 6, 5, 2, 2.

2. $X_1, X_2, \ldots, X_n$ is a random sample from an exponential distribution with mean θ. Find an UMVUE of $\exp(-\frac{1}{\theta})$ when $t > 1$, where $T = \sum_{i=1}^{n} X_i$ for the given data: 0.60, 8.71, 15.71, 2.32, 0.02, 6.22, 8.79, 2.05, 2.96, 3.33

3. Let

$$f(x|\mu, \sigma) = \frac{1}{\sigma}\exp\left[-\frac{(x-\mu)}{\sigma}\right]; \ x \geq \mu \in R \text{ and } \sigma > 0$$

For a sample of size n, obtain
(a) an unbiased estimate of μ when σ is known,
(b) an unbiased estimate of σ when μ is known,
(c) Ten unbiased estimators of σ^2 when μ is known.

4. Let $X_1, X_2, \ldots, X_n$ be a random sample of size n from $N(\mu, \sigma^2)$, where μ is known and if $T = \frac{1}{n}\sum_{i=1}^{n} |X_i - \mu|$, examine if T is unbiased for σ and if not obtain an unbiased estimator of σ.

5. If $X_1, X_2, \ldots, X_n$ is a random sample from the population

$$f(x|\theta) = (\theta + 1)x^\theta; \ \ 0 < x < 1, \theta > -1$$

Prove that $\left[-\frac{(n-1)}{\sum \ln X_i} - 1\right]$ is an UMVUE of θ.

6. Suppose X has a truncated Poisson distribution with pmf

$$f(x|\theta) = \begin{cases} \frac{\exp[-\theta]\theta^x}{[1-e^{-\theta}]x!} \ ; & x = 1, 2 \\ 0 & ; \text{otherwise} \end{cases}$$

Prove that the only unbiased estimator of $[1 - e^{-\theta}]$ based on X is the statistic $T(X)$,

$$T(x) = \begin{cases} 0 \ ; \text{when } x \text{ is odd} \\ 2 \ ; \text{when } x \text{ is even} \end{cases}$$

[Hint $\sum_{x=1}^{\infty} \frac{\theta^{2x}}{(2x)!} = \frac{e^{-\theta} + e^{\theta}}{2} - 1$]

7. Let $X_1, X_2, \ldots, X_n$ be iid rvs from $f(x|\theta)$,

$$f(x|\theta) = \begin{cases} \exp[i\theta - x] \ ; & x \ge i\theta \\ 0 & ; \ x < i\theta \end{cases}$$

Prove that

$$T = \min_i [\frac{X_i}{i}]$$

is minimal sufficient statistic for θ. If possible obtain the distribution of X_1 given T. Can you find an unbiased estimator of θ? If "Yes," find and if "No," explain.

8. Let $X_1, X_2, \ldots, X_n$ be iid rvs with $f(x|\mu)$,

$$f(x|\mu) = \begin{cases} \frac{1}{2i\mu} \ ; & -i(\mu - 1) < x_i < i(\mu + 1) \\ 0 & ; \text{otherwise} \end{cases}$$

where $\mu > 0$. Find the sufficient statistic for μ. If T is sufficient for μ then find the distribution of X_1, X_2 given T. If possible, find an unbiased estimator of μ.

9. If X_1, X_2, and X_3 are iid rvs with the following pmfs:

(a)

$$f(x|\lambda) = \frac{e^{-\lambda}\lambda^x}{x!}; \ \ x = 0, 1, 2, \ldots, \lambda > 0$$

(b)

$$f(x|\lambda) = \binom{n}{x}\lambda^x(1-\lambda)^{n-\lambda}; \quad 0 < \lambda < 1, x = 0, 1 ? \qquad n$$

(c)

$$f(x|\lambda) = (1-\lambda)\lambda^x; \quad x = 0, 1, 2, \ldots \lambda > 0$$

Prove that $X_1 + 2X_2$, $X_2 + 3X_3$, and $X_1 + 2X_2 + X_3$ are not sufficient for λ in (a), (b), and (c). Further, prove that $2(X_1 + X_2 + X_3)$ is sufficient for λ in (a), (b), and (c).

10. Let $X_1, X_2, \ldots, X_n$ be iid rvs having $\cup(\theta, 3\theta)$, $\theta > 0$. Then prove that $(X_{(1)}, X_{(n)})$ is jointly minimal sufficient statistic.

11. Let $\{(X_i, Y_i), i = 1, 2, \ldots, n\}$ be n independent random vectors having a bivariate distribution

$$N = \left(\binom{\theta_1}{\theta_2}, \begin{pmatrix} \sigma_1^2 & \rho\sigma_1\sigma_2 \\ \rho\sigma_1\sigma_2 & \sigma_2^2 \end{pmatrix}\right); \quad -\infty < \theta_1, \theta_2 < \infty, \ \sigma_1, \sigma_2 > 0, \ -1 \le \rho \le 1.$$

Prove that

$$\left(\sum X_i, \sum X_i^2, \sum X_i Y_i, \sum Y_i \sum Y_i^2\right)$$

is jointly sufficient $(\theta_1, \sigma_1, \rho, \theta_2, \sigma_2)$.

12. Let the rv X_1 is $B(n, \theta)$ and X_2 is $P(\theta)$ where n is known and $0 < \theta < 1$. Obtain four unbiased estimators of θ.

13. Let $X_1, X_2, \ldots, X_n$ are iid rvs with $\cup(\theta, \theta + 1)$.

(i) Find sufficient statistic for θ

(ii) Show that the sufficient statistic is not complete

(iii) Find an unbiased estimator of θ

(iv) Find the distribution of X_1 given T, where T is sufficient for θ

(v) Can you find UMVUE of θ ? If "No," give reasons.

14. Let X be a rv with pmf

$$f(x|p) = \left(\frac{p}{2}\right)^{|x|}(1-p)^{1-|x|}; \quad x = -1, 0, 1, \ 0 < p < 1$$

(i) Show that X is not complete.

(ii) Show that $|X|$ is sufficient and complete.

15. Let $X_1, X_2, \ldots, X_n$ are iid rvs from the following pdf:
(i)

$$f(x|\alpha) = \frac{\alpha}{(1+x)^{1+\alpha}}; \quad x > 0, \alpha > 0$$

(ii)

$$f(x|\alpha) = \frac{(\ln \alpha)\alpha^x}{\alpha - 1}; \quad 0 < x < \infty, \alpha > 1$$

(iii)

$$f(x|\alpha) = \exp[-(x - \alpha)]\exp[-e^{-(x-\alpha)}]; \quad -\infty < x < \infty, -\infty < \alpha < \infty$$

(iv)

$$f(x|\alpha) = \frac{x^3 e^{-\frac{x}{\alpha}}}{6\alpha^4}; \quad x > 0, \alpha > 0$$

(v)

$$f(x|\alpha) = \frac{kx^{k-1}}{\alpha^k}; \quad 0 < x < \alpha, \alpha > 0$$

Find a complete sufficient statistic or show that it does not exist.
Further if it exists, then find the distribution of X_1 given T, where T is sufficient statistic. Further, find UMVUE of α^r, whenever it exists.
16. Let $X_1, X_2, \ldots, X_N$ are iid rvs with $B(1, p)$, where N is also a random variable taking values $1, 2, \ldots$ with known probabilities $p_1, p_2, \ldots, \sum p_i = 1$.
(i) Prove that the pair (X, N) is minimal sufficient and N is ancillary for p.
(ii) Prove that the estimator $\frac{X}{N}$ is unbiased for p and has variance $p(1 - p)E\frac{1}{N}$.
17. In a normal distribution $N(\mu, \mu^2)$, prove that $(\sum X_i, \sum X_i^2)$ is not complete in a sample of size n.
18. Let $X_1, X_2, \ldots, X_n$ be iid rvs from the following pdf:
(i)

$$f(x|\theta) = \theta x^{\theta-1}; 0 < x < 1, \theta > 0$$

Find UMVUE of (a) $\theta e^{-\theta}$ (b) $\dfrac{\theta}{\theta + 1}$ (c) $\dfrac{1+\theta}{e^{2\theta}}$

(ii)

$$f(x|\theta_1, \theta_2) = \frac{1}{(\theta_2 - \theta_1)}; \theta_1 < x < \theta_2, \theta_1, \theta_2 > 0$$

Find minimal sufficient statistic and show that it is complete, further if possible, find the the distribution of X_1 given T, where T is sufficient statistic. Find UMVUE of $\exp(\theta_2 - \theta_1)$, $\frac{\theta_1}{\theta_1 + \theta_2}$, $\sin(\theta_1 - \theta_2)$, and $\cos(\theta_1 - \theta_2)$

19. Let T_1, T_2 be two unbiased estimates having common variances $a\sigma^2(a > 1)$, where σ^2 is the variance of the UMVUE. Prove that the correlation coefficient between T_1 and T_2 is greater than or equal to $\frac{2-a}{a}$.

20. Let $X_1, X_2, \ldots, X_n$ are iid rvs from discrete uniform distribution

$$f(x|N_1, N_2) = \frac{1}{N_2 - N_1}; \quad x = N_1 + 1, N_1 + 2, \ldots, N_2.$$

Find the sufficient statistic for N_1 and N_2.
If exists, find UMVUE for N_1 and N_2.

21. Let $X_1, X_2, \ldots, X_n$ are iid rvs from $P(\lambda)$. Let $g(\lambda) = \sum_{i=0}^{\infty} c_i \lambda^i$ be a parametric function. Find the UMVUE for $g(\lambda)$. In particular, find the UMVUE for (i)$g(\lambda) = (1 - \lambda)^{-1}$ (ii) $g(\lambda) = \lambda^r (r > 0)$

22. Let $X_1, X_2, \ldots, X_n$ are iid rvs with $N(\theta, 1)$. Show that S^2 is ancillary.

23. In scale parameter family, prove that $\left(\frac{X_1}{X_n}, \frac{X_2}{X_n}, \ldots, \frac{X_{n-1}}{X_n}\right)$ are ancillary.

24. Let X_1, X_2 are iid rvs with $N(0, \sigma^2)$. Prove that $\frac{X_1}{X_2}$ is ancillary.

25. Let $X_1, X_2, \ldots, X_n$ are iid rvs with (i) $N(\mu, \sigma^2)$ (ii)$N(\mu, \mu^2)$. Examine $T = \left(\left(\frac{X_1 - \bar{X}}{S}\right), \left(\frac{X_2 - \bar{X}}{S}\right), \ldots, \left(\frac{X_n - \bar{X}}{S}\right)\right)$ is ancillary in (i) and (ii).

26. Let $X_1, X_2, \ldots, X_m$ are iid rvs with $B(n, p)$, $0 < p < 1$ and n is known. Find the UMVUE of $P[X = x] = \binom{n}{x} p^x q^{n-x}$; $x = 0, 1, 2, \ldots, n$, $q = 1 - p$

27. Let $X_1, X_2, \ldots, X_m$ are iid rvs from Poisson (λ). Find the UMVUE of $P[X = x] = \frac{e^{-\lambda}\lambda^x}{x!}$; $x = 0, 1, 2, \ldots$, $\lambda > 0$

28. Let $X_1, X_2, \ldots, X_m$ are iid rvs from gamma distribution with parameters p and σ. Then find the UMVUE of $\frac{e^{-\frac{x}{\sigma}} x^{p-1}}{\sigma^p \Gamma(p)}$ for p known, $x > 0$, $\sigma > 0$.

29. Let $X_1, X_2, \ldots, X_n$ are iid rvs from $N(\mu, \sigma^2)$, $\mu \in R$, $\sigma > 0$. Find UMVUE of $P[X_1 \leq k], k > 0$.

30. Let $X_1, X_2, \ldots, X_n$ are iid rvs with pdf,

$$f(x|\theta) = \begin{cases} \frac{1}{2\theta}; & -\theta < x < \theta \\ 0 & ; \text{otherwise} \end{cases}$$

Prove that $T(X) = \max\left[-X_{(1)}, X_{(n)}\right]$ is a complete sufficient statistic. Find UMVUE of $\theta^r (r > 0)$. If $Y = |X|$, then find UMVUE of

1. θ^r

2. $\dfrac{\theta}{1+\theta}$

3. $\sin(\theta)$

based on Y.

31. Let $X_1, X_2, \ldots, X_n$ are iid rvs from the pdf,

$$f(x|\mu, \sigma^2) = \frac{1}{\sigma} \exp\left[-\frac{(x-\mu)}{\sigma}\right]; x \geq \mu, \sigma > 0$$

(i) Prove that $[X_{(1)}, \sum_{j=1}^{n}(X_j - X_{(1)})]$ is a complete sufficient statistic for (μ, σ).
(ii) Prove that UMVUE of μ and σ are given by

$$(\hat{\mu} = X_{(1)}) - \frac{n}{(n-1)} \sum_{j=1}^{n}(X_j - X_{(1)})$$

$$\hat{\sigma} = \frac{1}{n-1} \sum_{j=1}^{n}(X_j - X_{(1)})$$

32. Let $X_1, X_2, \ldots, X_n$ are iid rvs from $\cup(\theta_1, \theta_2)$ or $\cup(\theta_1 + 1, \theta_2 + 1)$. Find the UMVUE of $g(\theta_1, \theta_2)$ without using the general result from Example 2.3.3. Further, find the UMVUE of $\theta_1^r \theta_2^s (r, \ s > 0)$.

33. Let $X_1, X_2, \ldots, X_n$ be iid rvs from $\cup(-k\theta, k\theta)$, $k, \ \theta > 0$. Show that the UMVUE of $g(\theta)$ is

$$u(y_{(m)} = g(y_{(m)}) + \frac{y_{(m)} g'(y_{(m)})}{m},$$

where $y_{(m)} = \max_i Y_i, Y_i = \frac{|X_i|}{k} : i = 1, 2, \ldots, n$

34. Let $X_1, X_2, \ldots, X_m$ be iid rvs from discrete uniform distribution where

$$f(x|N) = \begin{cases} \frac{1}{2N} ; & x = -N, -N+1, \ldots, -1, 1, 2, \ldots, N \\ 0 ; & \text{otherwise} \end{cases}$$

Find UMVUE of (i) $\sin N$ (ii) $\cos N$ (iii) e^N (iv) $\frac{N}{e^N}$

35. Let $X_1, X_2, \ldots, X_m$ be iid rvs from $f(x|N)$

$$(a) \ f(x|N) = \frac{2x}{N(N+1)}; \quad x = 1, 2, \ldots, N$$

$$(b) \ f(x|N) = \frac{6x^2}{N(N+1)(2N+1)}; \quad x = 1, 2, \ldots, N$$

Find UMVUE of (i) $\sin N$ (ii) $\cos N$ (iii) e^N (iv) $\frac{N}{e^N}$ (v) $\frac{e^N}{\sin N}$ (vi) $\frac{e^N}{\cos N}$

$X_1, \ldots, X_m$ be iid rvs from $f(x|N_1, N_2)$

$$f(x|N_1, N_2) = \frac{1}{N_2 - N_1 + 1}; x = N_1, N_1 + 1, \ldots, N_2$$

Find UMVUE of (i) N_1 (ii) N_2 (iii) $(N_1 N_2)^2$

37. Let $X_1, X_2, \ldots, X_m$ be iid rvs with $\cup(0, \theta)$.
Then find UMVUE of (i) e^θ (ii) $\sin \theta$ (iii) $\frac{\theta}{1 + \theta}$.

38. Let $X_1, X_2, \ldots, X_m$ be iid rvs with $f(x|\theta)$,

$$f(x|\theta) = \frac{4x^3}{\theta^4} ; \ 0 < x < \theta,$$

Find UMVUE of (i) θ^5 (ii) $\frac{\theta^2}{1 + \theta^3}$ (iii) $\cos \theta$.

References

Casella G, Berger RL (2002) Statistical inference. Duxbury

Guenther WC (1978) Some easily found minimum variance unbiased estimators. Am Stat 32:29–34

Hogg RV, Craig AT (1972) Introduction to mathematical statistics. The Macmillan Company, New York

Jadhav DT (1996) Minimum variance unbiased estimation in some non-regular families. Unpublished M.Phil Thesis, Shivaji University, Kolhapur

Karakostas KX (1985) On minimum variance unbiased estimators. Am Stat 39(4):303–305

Patil GP (1962) Certain properties of the generalized power series distributions. Ann Inst Stat Math 14:179–182

Roy J, Mitra S (1957) Unbiased minimum variance estimation in a class of discrete distributions. Sankhya 18:371–378

Stigler SM (1972) Completeness and unbiased estimation. Am Stat 26(2):28–29

Tate RF (1959) Unbiased estimation: functions of location and scale parameters. Ann Math Stat 30:341–366

Chapter 3
Moment and Maximum Likelihood Estimators

In the previous chapters, we have seen that unbiased estimators are not unique. Using Rao–Blackwell and Lehman–Scheffe theorem, we could find the best estimator among the class of unbiased estimators which has minimum variance. Generally, such estimators are called UMVUE. Is it possible for us to get a biased estimator which is better than UMVUE with respect to MSE? We cannot say 'Yes' with probability one. But sometimes we may get a better estimator than UMVUE. In this chapter, our effort will be to find an alternative estimator which may be better than UMVUE in some cases.

It is an easy work to estimate a parameter in some cases. For example, the sample mean is a good estimate for the population mean. Method of moments (MM) is, the oldest method of finding point estimator. It is very simple to use and always yields some sort of estimate. According to Fisher, MM produces the estimators with large variance.

3.1 Method of Moments

Let $X_1, X_2, \ldots, X_n$ be iid rvs with pdf $f(x|\theta)$. Here, we have to equate the first $r(r \geq 1)$ sample moments to the corresponding r population moments. Then by solving the resulting system of equations, we can obtain the moment estimators. Let

$$E[g(X)] = h(\theta_1, \theta_2, \theta_3, \ldots, \theta_k) \qquad (3.1.1)$$

Suppose $g(X) = X$ then $\mu'_1 = E(X)$.
The corresponding sample moments $m'_1 = \frac{1}{n} \sum_{i=1}^{n} X_i$

© Springer Science+Business Media Singapore 2016
U.J. Dixit, *Examples in Parametric Inference with R*,
DOI 10.1007/978-981-10-0889-4_3

Similarly,

$$\mu_2' = E(X^2) \quad \text{and} \quad m_2' = \frac{1}{n} \sum_{i=1}^{n} X_i^2$$

$$\mu_r' = E(X^r) \quad \text{and} \quad m_r' = \frac{1}{n} \sum_{i=1}^{n} X_i^r$$

One should note that the population moments $\mu_1', \mu_2', \ldots, \mu_r'$ are the functions of $(\theta_1, \theta_2, \theta_3, \ldots, \theta_k)$. According to (3.1.1), we call this function as $h_r(\theta_1, \theta_2, \theta_3, \ldots, \theta_k)$. Therefore,

$$m_1' = h_1(\theta_1, \theta_2, \theta_3, \ldots, \theta_k), \ m_2' = h_2(\theta_1, \theta_2, \theta_3, \ldots, \theta_k) \ \ldots \ m_r' = h_r(\theta_1, \theta_2, \theta_3, \ldots, \theta_k)$$
$$(3.1.2)$$

After solving the Eq. (3.1.2), we get the estimators of θ_i, $(i = 1, 2, \ldots, k)$.

Example 3.1.1 Let $X_1, X_2, \ldots, X_n$ be iid rvs with $P(\lambda)$. Using MM, we will find the estimator of λ.

Population moment: $\mu_1' = E(X) = \lambda$ and Sample moment $m_1' = \frac{1}{n} \sum_{i=1}^{n} X_i$
Hence,

$$\hat{\lambda} = \frac{1}{n} \sum_{i=1}^{n} X_i \qquad (3.1.3)$$

But $\mu_2 = \mu_2' - \mu_1'^2$
Now, $X^2 = X(X - 1) + X$. Hence,

$$E[X(X - 1)] = \lambda^2 \quad \text{and} \quad E(X) = \lambda$$

Population moments:

$$\mu_2' = \lambda^2 + \lambda \quad \text{and} \quad \mu_2 = \lambda$$

sample moments: $m_2 = \frac{1}{n} \sum_{i=1}^{n} (X_i - \bar{X})^2$.
Hence,

$$\hat{\lambda} = \frac{1}{n} \sum_{i=1}^{n} (X_i - \bar{X})^2 \qquad (3.1.4)$$

The reader should think which estimator from (3.1.3) to (3.1.4) should be selected.

Example 3.1.2 Let $X_1, X_2, \ldots, X_m$ be iid rvs with $B(n, p)$

$$P[X = x | n, p] = \binom{n}{x} p^x (1 - p)^{n-x} \; ; \; x = 0, 1, 2, \ldots n, \; 0 < p < 1, \; q = 1 - p.$$

Here, we assume that both the parameters are unknown. Now, we will estimate n and p.

Equating the sample moments to population moments,

$$\bar{X} = np \tag{3.1.5}$$

$$\frac{1}{m} \sum (X_i - \bar{X})^2 = npq \tag{3.1.6}$$

After solving (3.1.5) and (3.1.6),

$$\hat{p} = \frac{m\bar{X} - \sum (X_i - \bar{X})^2}{m\bar{X}} \quad \text{and} \quad \hat{n} = \frac{m\bar{X}^2}{m\bar{X} - \sum (X_i - \bar{X})^2} \tag{3.1.7}$$

In (3.1.7), if $\hat{p} > 0$ then $m\bar{X} > \sum X_i^2 - m\bar{X}^2$.
$\Rightarrow \bar{X} + \bar{X}^2 > m_2'$
$\Rightarrow (\bar{X} + \frac{1}{2})^2 > m_2' + \frac{1}{4}$.
If $|\bar{X} + \frac{1}{2}| < \sqrt{m_2' + \frac{1}{4}}$ then

$-\frac{1}{2} - \sqrt{m_2' + \frac{1}{4}} < \bar{X} < \sqrt{m_2' + \frac{1}{4}} - \frac{1}{2}$. In such situation $\hat{p}$ is negative. Then the value of $\hat{p}$ is not admissible. Hence, value of $\bar{X}$ has to lie outside of $(-\frac{1}{2} - \sqrt{m_2' + \frac{1}{4}}, \sqrt{m_2' + \frac{1}{4}} - \frac{1}{2})$.

In other words, we can say that if sample mean is smaller than the sample variance then it suggests that there is a large degree of variability in the data. Same argument is true for the estimate of n. One needs to reduce the difference between sample mean and variance then we can get the better estimate of n and p.

For more details, see Hamedani and Walter (1988), Draper and Guttman (1971), Feldman and Fox (1968).

Example 3.1.3 Let $X_1, X_2, \ldots, X_m$ be iid rvs with $G(p, \sigma)$. Here, we will try to find the moment estimators of p and σ.

$$\bar{X} = p\sigma \tag{3.1.8}$$

$$\frac{1}{m} \sum (X_i - \bar{X})^2 = p\sigma^2 \tag{3.1.9}$$

After some algebra,

$$\hat{\sigma} = \frac{m^{-1} \sum_{i=1}^{m} (X_i - \bar{X})^2}{\bar{X}} \tag{3.1.10}$$

$$\hat{p} = \frac{\bar{X}^2}{m^{-1} \sum_{i=1}^{m} (X_i - \bar{X})^2}, \tag{3.1.11}$$

One should see Dixit (1982).

Example 3.1.4 Method of moment estimators may not be functions of sufficient or complete statistics.

For example,

$$f(x|\theta_1, \theta_2) = \begin{cases} \frac{1}{2\theta_2} ; & \theta_1 - \theta_2 < X < \theta_1 + \theta_2, \ \theta_2 > 0 \\ 0 & ; \text{otherwise} \end{cases} \tag{3.1.12}$$

Moment estimator for (θ_1, θ_2) are $\left(\bar{X}, \sqrt{\frac{3}{n} \sum_{i=1}^{n} (X_i - \bar{X})^2} \right)$.

But sufficient and complete statistics for (θ_1, θ_2) is $(X_{(1)}, X_{(n)})$.

3.2 Method of Maximum Likelihood

So far we have considered the problem of finding an estimator on the basis of the criteria of unbiasedness and minimum variance. Another more popular principle which is very often used is that of method of maximum likelihood.

Let $X_1, X_2, \ldots, X_n$ be iid rvs with pdf $f(x|\theta), \theta \in \Theta$. Consider the joint pdf of $X_1, X_2, \ldots, X_n$. Treating the X's as if they were constants and looking at this joint pdf as a function of θ, we denote it by $L(\theta|x_1, x_2, \ldots, x_n)$ and this is known as the likelihood function. It is defined as

$$L(\theta|\underline{X}) = L(\theta_1, \theta_2, \ldots, \theta_k|x_1, x_2, \ldots, x_n) = \prod_{i=1}^{n} f(x_i|\theta_1, \theta_2, \ldots, \theta_n)$$

One should note that for likelihood $X_1, X_2, \ldots, X_n$ need not be iid. For example, let X_1, X_2, X_3 are three rvs such that any one has a pdf $g(x|\theta)$ and remaining two rvs have a pdf $f(x|\theta)$. The joint density of (X_1, X_2, X_3) is

$$h(x_1, x_2, x_3|\theta) = \frac{1}{3} \prod_{i=1}^{3} f(x_i) \sum_{i=1}^{3} \frac{g(x_i|\theta)}{f(x_i|\theta)}$$

This can be written as

$$L(\theta|x_1, x_2, x_3) = \frac{1}{3} \prod_{i=1}^{3} f(x_i) \sum_{i=1}^{3} \frac{g(x_i|\theta)}{f(x_i|\theta)} \qquad (3.2.1)$$

For details see Dixit (1987, 1989)
$\hat{\theta} = \hat{\theta}(x_1, x_2, \ldots, x_n)$ is called maximum. Hence, we can obtain MLE of θ if a sample X is given.

Note

1. Intuitively, MLE is a reasonable choice for an estimator.
2. To find a global maximum, one has to verify it.
3. Since the function $y = \log x$ $x > 0$, is strictly increasing, in order to maximize with respect to θ, it is much more convenient to work with log function.

Remark:

1. Suppose that the X's are discrete. Then,

$$L(\theta|x_1, x_2, \ldots, x_n) = P_\theta[X_1 = x_1, X_2 = x_2, \ldots, X_n = x_n]$$

This implies, $L(\theta|x)$ is the probability of observing the X's which were actually observed. Then one should select the estimate of θ which maximizes the probability of observing the X's, which were actually observed if such a θ exists. A similar argument holds true for the continuous case.
2. Generally, MLE is obtained by differentiating the likelihood with respect to θ and equating to zero.
3. Sometimes, it is not differentiable or the derivative tends to get messy and sometimes it is even harder to implement, then one should evaluate the likelihood function for all possible values of parameter and find MLE. The general technique is to find a global upper bound on the likelihood function and then establish that there is a unique point (s) for which the upper bound is attained.
4. Over all, one should remember that we have to maximize the likelihood function with respect to θ by any method. Differentiation is one method while other methods are also there.
5. MLE is often shown to have several desirable properties, We will consider the properties later on.

Example 3.2.1 Let $X_1, X_2, \ldots, X_n$ be iid rvs with $P(\lambda)$.

Then

$$L(\lambda|x_1, x_2, \ldots, x_n) = \frac{e^{-n\lambda} \lambda^{\sum_{i=1}^{n} x_i}}{\prod_{i=1}^{n} x_i!} \qquad (3.2.2)$$

Next,

$$\log L(\lambda|x) = -\log \prod_{i=1}^{n} x_i! - n\lambda + \left(\sum_{i=1}^{n} x_i\right) \log \lambda$$

Therefore the likelihood equation

$$\frac{d \log L(\lambda|x)}{d\lambda} = 0$$

Hence,

$$-n + \frac{\sum_{i=1}^{n} x_i}{\lambda} = 0 \qquad (3.2.3)$$

which gives $\hat{\lambda} = \bar{x}$

Next,

$$\frac{d^2 \log L}{d\lambda^2} = -\frac{\sum_{i=1}^{n} x_i}{\lambda^2} < 0$$

Thus, $\hat{\lambda} = \bar{x}$ is the MLE of λ.

Example 3.2.2 Let $X_1, X_2, \ldots, X_n$ be multinomially distributed rv with parameters $\theta = (p_1, p_2, \ldots, p_n) \in \Theta$, where Θ is the $(n-1)$-dimensional hyperplane in R^n defined by

$$\Theta = \left\{\theta = (p_1, p_2, \ldots, p_n) \in R^n, \ 0 < p_i < 1, \ i = 1, 2, \ldots, n \ \text{and} \ \sum_{i=1}^{n} p_i = 1\right\}$$

Then

$$L(\theta|x_1, x_2, \ldots, x_n) = \frac{n! p_1^{x_1} p_2^{x_2} \cdots p_n^{x_n}}{\prod_{i=1}^{n} x_i!}$$

$$= \frac{n!}{\prod_{i=1}^{n} x_i!} p_1^{x_1} p_2^{x_2} \cdots p_{n-1}^{x_{n-1}} [1 - p_1 - p_2 - \cdots - p_{n-1}]^{x_n},$$

where $n = \sum_{i=1}^{n} x_i$. Then

$$\log L(\theta|x) = \log n! - \log \prod_{i=1}^{n} x_i! + x_1 \log p_1 + x_2 \log p_2 + \cdots + x_{n-1} \log p_{n-1}$$

$$+ x_n \log(1 - p_1 - p_2 - \cdots - p_{n-1}) \qquad (3.2.4)$$

Differentiating with respect to $p_i = i = 1, 2, \ldots, n-1$ and equating it to zero

$$\frac{x_i}{p_i} - \frac{x_n}{1 - p_1 - p_2 - \cdots - p_{n-1}} = 0, i = 1, 2, \ldots, n-1 \qquad (3.2.5)$$

Since $p_n = 1 - p_1 - p_2 - \cdots - p_{n-1}$

$$\frac{x_i}{p_i} - \frac{x_n}{p_n} = 0$$

$$\frac{x_i}{p_i} = \frac{x_n}{p_n}; \quad i = 1, 2, \ldots, n-1 \qquad (3.2.6)$$

Hence

$$\frac{x_1}{p_1} = \frac{x_2}{p_2} = \cdots = \frac{x_{n-1}}{p_{n-1}} = \frac{x_n}{p_n} \qquad (3.2.7)$$

and this common value is equal to

$$\frac{x_1 + x_2 + \cdots + x_n}{p_1 + p_2 + \cdots + p_n} = \frac{n}{1}$$

Therefore

$$\frac{x_i}{p_i} = n \quad \text{and} \quad p_i = \frac{x_i}{n}, i = 1, 2, \ldots, n$$

We can say that these values of p_i's actually maximize the likelihood function, and therefore, $\hat{p}_i = \frac{x_i}{n}, i = 1, 2, \ldots, n$ are the MLE's of the p_i's.

Example 3.2.3 Let $X_1, X_2, \ldots, X_n$ be iid rvs with (i) $N(\mu, \sigma^2)$ (ii) $N(\mu, \mu)$ (iii) $N(\mu, \mu^2)$

(i) $\Theta = \{\theta = (\mu, \sigma^2), \mu \in R, \sigma > 0\}$

The likelihood function is given by

$$L(\theta|x) = \left(\frac{1}{\sigma\sqrt{2\pi}}\right)^n \exp\left[-\frac{1}{2\sigma^2} \sum_{i=1}^{n}(x_i - \mu)^2\right]$$

$$\log L = -\frac{n}{2}\log\sigma^2 - \frac{n}{2}\log 2\pi - \frac{1}{2\sigma^2}\sum_{i=1}^{n}(x_i - \mu)^2$$

$$\frac{\partial \log L}{\partial \mu} = \frac{n(\bar{x} - \mu)}{\sigma^2} = 0 \qquad (3.2.8)$$

$$\frac{\partial \log L}{\partial \sigma^2} = \frac{-n}{2\sigma^2} + \frac{1}{2\sigma^4} \sum_{i=1}^{n} (x_i - \mu)^2 = 0 \qquad (3.2.9)$$

After solving (3.2.8) and (3.2.9), we get,

$$\hat{\mu} = \bar{x} \quad \text{and} \quad \hat{\sigma^2} = \frac{1}{n} \sum_{i=1}^{n} (x_i - \bar{x})^2 \qquad (3.2.10)$$

It can be shown that $\hat{\mu}$ and $\hat{\sigma^2}$ actually maximize the likelihood function and therefore $\hat{\mu} = \bar{X}$ and $\hat{\sigma^2} = n^{-1} \sum_{i=1}^{n} (x_i - \bar{x})^2$ are the MLE of μ and σ^2 respectively.

Remark:

1. If we assume that σ^2 is known then we get $\hat{\mu} = \bar{x}$ as the MLE of μ.
2. If we assume that μ is known then we get $\hat{\sigma^2} = n^{-1} \sum_{i=1}^{n} (x_i - \mu)^2$ is the MLE of σ^2.
3. When both μ and σ^2 are unknown, $\hat{\mu}$ is unbiased estimator of μ and $\hat{\sigma^2}$ is not unbiased for σ^2.

Comment on MLE and UMVUE of (μ, σ).

Since $\hat{\mu} = \bar{X}$ and $\hat{\sigma^2} = \frac{1}{n} \sum_{i=1}^{n} (X_i - \bar{X})^2$

Now, $E(\bar{X}) = \mu$ and $V(\bar{X}) = \frac{\sigma^2}{n}$.

Now the distribution of $\frac{\sum_{i=1}^{n} (x_i - \bar{x})^2}{\sigma^2}$ is χ^2_{n-1}

$$E(\hat{\sigma^2}) = \frac{\sigma^2}{n} E \left[\frac{\sum_{i=1}^{n} (X_i - \bar{X})^2}{\sigma^2} \right] = \frac{(n-1)\sigma^2}{n}$$

$$V(\hat{\sigma^2}) = \frac{\sigma^4}{n^2} V \left[\frac{\sum_{i=1}^{n} (X_i - \bar{X})^2}{\sigma^2} \right] = \frac{2(n-1)\sigma^4}{n^2} \qquad (3.2.11)$$

UMVUE of $\mu = \tilde{\mu} = \bar{X}$ and UMVUE of $\sigma^2 = \tilde{\sigma^2} = \frac{\sum_{i=1}^{n} (X_i - \bar{X})^2}{n-1}$

$$V \left[(n-1)^{-1} \sum_{i=1}^{n} (x_i - \bar{x})^2 \right] = \frac{2\sigma^4}{n-1} \qquad (3.2.12)$$

It may be noted that when we have to compare more than one estimator for two methods, then consider the determinant of the covariance matrix in both methods.

Covariance matrix for $(\hat{\mu}, \hat{\sigma^2})$ in MLE

$$M = \begin{pmatrix} \frac{\sigma^2}{n} & 0 \\ 0 & \frac{2(n-1)\sigma^4}{n^2} \end{pmatrix} \qquad (3.2.13)$$

Covariance matrix for $(\hat{\mu}, \hat{\sigma^2})$ in UMVUE

$$U = \begin{pmatrix} \frac{\sigma^2}{n} & 0 \\ 0 & \frac{2\sigma^4}{n-1} \end{pmatrix} \tag{3.2.14}$$

$$\det |M| = |M| = \frac{2(n-1)\sigma^6}{n^3} \tag{3.2.15}$$

$$\det |U| = |U| = \frac{2\sigma^6}{n(n-1)} \tag{3.2.16}$$

We get,

$$\frac{2(n-1)\sigma^6}{n^3} < \frac{2\sigma^6}{n(n-1)}$$

$$\Rightarrow \frac{(n-1)^2}{n^2} \leq 1,$$

which is always true. Therefore, in this case MLE are more efficient than the, UMVUE.

(ii) $\Theta = \{\theta : (\mu, \mu), \mu > 0\}$,

In this case, mean = variance = μ, then the likelihood function is given by

$$L(\theta|x) = \left(\frac{1}{\sqrt{2\pi\mu}}\right)^n \exp\left\{-\frac{1}{2\mu}\sum_{i=1}^{n}(x_i - \mu)^2\right\}$$

$$\log L(\theta|x) = -\frac{n}{2}\log 2\pi - \frac{n}{2}\log\mu - \frac{1}{2\mu}\sum_{i=1}^{n}(x_i - \mu)^2$$

$$= -\frac{n}{2}\log 2\pi - \frac{n}{2}\log\mu - \frac{1}{2\mu}\left[\sum_{i=1}^{n}x_i^2 - 2\mu\sum_{i=1}^{n}x_i + n\mu^2\right]$$

$$= -\frac{n}{2}\log 2\pi - \frac{n}{2}\log\mu - \frac{\sum_{i=1}^{n}x_i^2}{2\mu} + \sum_{i=1}^{n}x_i - \frac{n\mu}{2}$$

$$\frac{d\log L}{d\mu} = -\frac{n}{2\mu} + \frac{\sum_{i=1}^{n}x_i^2}{2\mu^2} - \frac{n}{2} = 0 \tag{3.2.17}$$

After some algebra,

$$\mu^2 + \mu - m_2' = 0, \quad \text{where } m_2' = \frac{\sum_{i=1}^{n}x_i^2}{n}.$$

Hence

$$\hat{\mu} = \frac{-1 \pm \sqrt{1 + 4m_2}}{2} \qquad (3.2.18)$$

Here we get two MLEs for μ.

We have to select $\hat{\mu} = \frac{-1 + \sqrt{1 + 4m_2}}{2}$. Reader should think why we have to select this value of μ.

(iii) $\Theta = \{\theta : (\mu, \mu^2), \mu > 0\}$

The likelihood function is given by

$$L(\theta|x) = \left(\frac{1}{\mu\sqrt{2\pi}}\right)^n \exp\left\{-\frac{1}{2\mu^2} \sum_{i=1}^{n} (x_i - \mu)^2\right\}$$

$$\log L(\theta|x) = -n \log \mu - \frac{n}{2} \log 2\pi - \frac{1}{2\mu^2} \sum_{i=1}^{n} (x_i - \mu)^2$$

$$= -n \log \mu - \frac{n}{2} \log 2\pi - \frac{1}{2\mu^2} \left[\sum_{i=1}^{n} x_i^2 - 2n\mu\bar{x} + n\mu^2\right]$$

$$= -n \log \mu - \frac{n}{2} \log 2\pi - \frac{\sum_{i=1}^{n} x_i^2}{2\mu^2} + \frac{n\bar{x}}{\mu} - \frac{n}{2}$$

Therefore,

$$\frac{d \log L}{d\mu} = -\frac{n}{\mu} + \frac{\sum_{i=1}^{n} x_i^2}{\mu^3} - \frac{n\bar{x}}{\mu^2} = 0$$

Hence, after some algebra,

$$\mu^2 + \bar{x}\mu - m_2' = 0,$$

which implies

$$\hat{\mu} = \frac{-\bar{x} \pm \sqrt{\bar{x}^2 + 4m_2'}}{2}$$

Since, $\mu > 0$ then $\hat{\mu} = \frac{-\bar{x} + \sqrt{\bar{x}^2 + 4m_2'}}{2}$

Example 3.2.4 Let $X_1, X_2, \ldots, X_m$ be iid rvs from $B(n, p)$, where p is known and n is unknown.

The likelihood function is given by

$$L(n|x, p) = \prod_{i=1}^{m} \binom{n}{x_i} p^{x_i} q^{n-x_i}$$

$$= \prod_{i=1}^{m} \binom{n}{x_i} p^{\sum_{i=1}^{m} x_i} q^{mn - \sum_{i=1}^{m} x_i}$$

Since n is an integer,

$$L(n|\bar{x}, p) = 0 \text{ if } n < x_{(m)} \tag{3.2.19}$$

Hence, MLE is an integer $n > x_{(m)}$ that satisfies

$$\frac{L(n|\bar{x}, p)}{L(n-1|\bar{x}, p)} \geq 1 \quad \text{and} \quad \frac{L(n+1|\bar{x}, p)}{L(n|\bar{x}, p)} < 1 \tag{3.2.20}$$

Then

$$\frac{L(n|\bar{x}, p)}{L(n-1|\bar{x}, p)} = \frac{n^m (1-p)^m}{\prod_{i=1}^{m}(n-x_i)} \geq 1 \tag{3.2.21}$$

and

$$\frac{L(n+1|\bar{x}, p)}{L(n|\bar{x}, p)} = \frac{(n+1)^m (1-p)^m}{\prod_{i=1}^{m}(n+1-x_i)} < 1 \tag{3.2.22}$$

From (3.2.21) to (3.2.22),

$$n^m (1-p)^m \geq \prod_{i=1}^{m}(n-x_i) \quad \text{and} \quad (n+1)^m (1-p)^m < \prod_{i=1}^{m}(n+1-x_i)$$

Dividing n^m and letting $z = \frac{1}{n}$

$$(1-p)^m \geq \prod_{i=1}^{m}(1 - x_i z)$$

Then we have to solve

$$(1-p)^m = \prod_{i=1}^{m}(1 - x_i z) \text{ for } 0 \leq z \leq \frac{1}{x_{(m)}} \tag{3.2.23}$$

Let $g(z) = \prod_{i=1}^{m}(1 - x_i z)$. Then $g(0) = 1$ and $g(\dfrac{1}{x_m}) = 0$. Further g is monotone and convex on $\left[0, \dfrac{1}{x_m}\right]$. Thus there is a unique z (say $\hat{z}$) that solves the equation. The quantity $z = \dfrac{1}{x_{(m)}}$ may. not be an integer but the integer $\hat{n}$ that satisfies the inequalities and is the MLE, is the smallest integer greater than or equal to $x_{(m)}$.

For example, $p = \frac{3}{4}, p = \frac{2}{3}, p = \frac{1}{2}, m = 2, X_1 = 20, X_2 = 25$
From (3.2.23),

p	Quadratic equation (from 3.2.23)	Roots	$\hat{(n)} = \max(\text{Root})^{-1}$
$\frac{3}{4}$	$8000z^2 - 720z + 15$	$0.0572, 0.0327$	31
$\frac{2}{3}$	$4500z^2 - 405z + 8$	$0.0607, 0.0292$	34
$\frac{1}{2}$	$2000z^2 - 180z + 3$	$0.0679, 0.0220$	45

The description of the MLE for n was found by Feldman and Fox (1968), Draper and Guttman (1971). Further see Dixit and Kelkar (2011, 2012). Reader should refer Casella and Berger (2002). Note that for a binomial distribution in the presence of outliers, Dixit and Kelkar (2011) have shown that MM estimator of p is better than MLE of p.

Example 3.2.5 Let X_1 and X_2 be independent exponential random variables with mean λ_1 and λ_2 respectively.

Let $Z_1 = \min(X_1, X_2)$ and

$$Z_2 = \begin{cases} 0 & ; \ Z_1 = X_1 \\ 1 & ; \ Z_1 = X_2 \end{cases}$$

Find the MLE of λ_1 and λ_2 in a sample of size m.
We will have to find the distribution of (Z_1, Z_2)

$$P[Z_1 \leq z_1, Z_2 = 0] = P[\min(X_1, X_2) \leq z_1, Z_2 = 0] = P[X_1 \leq z_1, X_1 < X_2]$$

$$= \int_0^{z_1} \int_{x_1}^{\infty} \frac{1}{\lambda_1} e^{-\frac{x_1}{\lambda_1}} \frac{1}{\lambda_2} e^{-\frac{x_2}{\lambda_2}} dx_1 dx_2$$

Since

$$\int_{x_1}^{\infty} \frac{1}{\lambda_2} e^{-\frac{x_2}{\lambda_2}} dx_2 = e^{-\frac{x_1}{\lambda_2}}$$

Next,

$$\int_0^{z_1} e^{-\frac{x_1}{\lambda_2}} \frac{1}{\lambda_1} e^{-\frac{x_1}{\lambda_1}} dx_1 = \int_0^{z_1} \frac{1}{\lambda_1} e^{-x_1 \theta_{12}} dx_1,$$

$$= \frac{1}{\lambda_1 \theta_{12}} \left[1 - e^{-z_1 \theta_{12}} \right]$$

where

$$\theta_{12} = \frac{1}{\lambda_1} + \frac{1}{\lambda_2}$$

Hence

$$P[Z_1 \le z_1, Z_2 = 0] = \frac{1}{\lambda_1 \theta_{12}} [1 - e^{-z_1 \theta_{12}}] \qquad (3.2.24)$$

Similarly,

$$P[Z_1 \le z_1, Z_2 = 1] = P[\min(X_1, X_2) \le z_1, Z_2 = 1] = P[X_2 \le z_1, X_2 < X_1]$$

$$= \int_0^{z_1} \int_{x_2}^{\infty} \frac{1}{\lambda_1} e^{-\frac{x_1}{\lambda_1}} \frac{1}{\lambda_2} e^{-\frac{x_2}{\lambda_2}} dx_1 dx_2$$

$$= \int_0^{z_1} \frac{1}{\lambda_2} e^{-\frac{x_2}{\lambda_2}} e^{-\frac{x_2}{\lambda_1}} dx_2$$

$$= \int_0^{z_1} \frac{1}{\lambda_2} e^{-\theta_{12} x_2} dx_2$$

$$= \frac{1}{\lambda_2 \theta_{12}} [1 - e^{-\theta_{12} z_1}] \qquad (3.2.25)$$

Now

$$P[Z_2 = 0] = P[X_1 \le X_2]$$

$$= \int_0^{\infty} \int_{x_1}^{\infty} \frac{1}{\lambda_1} e^{-\frac{x_1}{\lambda_1}} \frac{1}{\lambda_2} e^{-\frac{x_2}{\lambda_2}} dx_2 dx_1$$

$$= \int_0^{\infty} \frac{1}{\lambda_1} e^{-\frac{x_1}{\lambda_1}} e^{-\frac{x_1}{\lambda_2}} dx_1$$

$$= \int_0^{\infty} \frac{1}{\lambda_1} e^{-\theta_{12} x_1} dx_1 = \frac{1}{\lambda_1 \theta_{12}} \qquad (3.2.26)$$

Similarly,

$$P[Z_2 = 1] = \frac{1}{\lambda_2 \theta_{12}} \qquad (3.2.27)$$

From (3.2.24), (3.2.25), (3.2.26) and (3.2.27),

$$P[Z_1 \le z_1] = 1 - e^{-\theta_{12} z_1} \quad ; z_1 > 0$$

and

$$f(z_1) = \theta_{12} e^{-\theta_{12} z_1}; \quad z_1 > 0, \theta_{12} > 0 \qquad (3.2.28)$$

This implies that Z_1 and Z_2 are independent.

If we have a sample of size m, let Z_2 takes value zero r times and one $(m - r)$ times. Hence the likelihood function is given by

$$L(\lambda_1, \lambda_2 | z_1 z_2) = (\theta_{12})^m \exp\left[-\theta_{12} \sum_{i=1}^{m} z_{1i}\right] \frac{m!}{r!(m-r)!} (\lambda_1 \theta_{12})^{-r} (\lambda_2 \theta_{12})^{-(m-r)}$$

$$L = \text{const} + \exp\left[-\theta_{12} \sum_{i=1}^{m} z_{1i}\right] (\lambda_1)^{-r} (\lambda_2)^{-(m-r)}$$

$$\log L = \text{const} - \exp\left[-\theta_{12} \sum_{i=1}^{m} z_{1i}\right] - r \log \lambda_1 - (m - r) \log \lambda_2$$

$$\frac{\partial \log L}{\partial \lambda_1} = \frac{\sum_{i=1}^{m} z_{1i}}{\lambda_1^2} - \frac{r}{\lambda_1} \qquad (3.2.29)$$

$$\frac{\partial \log L}{\partial \lambda_2} = \frac{\sum_{i=1}^{m} z_{1i}}{\lambda_2^2} - \frac{m - r}{\lambda_2} \qquad (3.2.30)$$

From (3.2.29), $\hat{\lambda}_1 = \frac{1}{r} \sum_{i=1}^{m} z_{1i}$ and
From (3.2.30), $\hat{\lambda}_2 = \frac{1}{m-r} \sum_{i=1}^{m} z_{1i}$

Example 3.2.6 Let $X_1 \ X_2$ and X_3 be independent exponential random variables with mean $\lambda_i (i = 1, 2, 3)$.

Let $Z_1 = \min(X_1, X_2, X_3)$ and

$$Z_2 = \begin{cases} 0 \ ; \ Z_1 = X_1 \\ 1 \ ; \ Z_1 = X_2 \\ 2 \ ; \ Z_1 = X_3 \end{cases}$$

Find the MLE of λ_1, λ_2 and λ_3 in a sample of size m.

Let θ_{ijk} $\frac{1}{\lambda_i}$ | $\frac{1}{\lambda_j}$ | $\frac{1}{\lambda_k}$; $i \neq j \neq k$

$\theta_{ij} = \frac{1}{\lambda_i} + \frac{1}{\lambda_j}$; $i \neq j$

To find $P[Z_1 \leq z_1, Z_2 = 0]$

$Z_2 = 0$ indicates that $Z_1 = \min(X_1, X_2, X_3) = X_1$

We have two cases (i) $X_1 < X_2 < X_3$ (ii) $X_1 < X_3 < X_2$

Case (i)

$$= \int_0^{z_1} \frac{1}{\lambda_1} e^{-\frac{x_1}{\lambda_1}} dx_1 \left[\int_{x_1}^{\infty} \frac{1}{\lambda_2} e^{-\frac{x_2}{\lambda_2}} \left(\int_{x_2}^{\infty} \frac{1}{\lambda_3} e^{-\frac{x_3}{\lambda_3}} dx_3 \right) dx_2 \right] \qquad (3.2.31)$$

Consider

$$\int_{x_1}^{\infty} \frac{1}{\lambda_2} e^{-\frac{x_2}{\lambda_2}} e^{-\frac{x_2}{\lambda_3}} dx_2 = \int_{x_1}^{\infty} \frac{1}{\lambda_2} e^{-\theta_{23} x_1} dx_2 = \frac{e^{-\theta_{23} x_1}}{\lambda_2 \theta_{23}},$$

Equation (3.2.31) becomes

$$= \int_0^{z_1} \frac{e^{-\theta_{123} x_1}}{\lambda_1 \lambda_2 \theta_{23}} dx_1 = \frac{1 - \exp(-\theta_{123} z_1)}{\lambda_1 \lambda_2 \theta_{23} \theta_{123}} \qquad (3.2.32)$$

Case (ii)

$$= \int_0^{z_1} \frac{1}{\lambda_1} e^{-\frac{x_1}{\lambda_1}} dx_1 \left[\int_{x_1}^{\infty} \frac{1}{\lambda_3} e^{-\frac{x_3}{\lambda_3}} \left(\int_{x_3}^{\infty} \frac{1}{\lambda_2} e^{-\frac{x_2}{\lambda_2}} dx_2 \right) dx_3 \right] \qquad (3.2.33)$$

$$= \int_0^{z_1} \frac{e^{-\theta_{123} x_1}}{\lambda_1 \lambda_3 \theta_{23}} dx_1 = \frac{1 - \exp(-\theta_{123} z_1)}{\lambda_1 \lambda_3 \theta_{23} \theta_{123}} \qquad (3.2.34)$$

Hence from (3.2.32) and (3.2.34),

$$P[Z_1 \leq z_1, Z_2 = 0] = \frac{1 - \exp(-\theta_{123} z_1)}{\lambda_1 \theta_{23} \theta_{123}} \left[\frac{1}{\lambda_2} + \frac{1}{\lambda_3} \right]$$

$$= \frac{1 - \exp(-\theta_{123} z_1)}{\lambda_1 \theta_{123}}, \quad z_1 > 0 \qquad (3.2.35)$$

Similarly,

$$P[Z_1 \le z_1, Z_2 = 1] = \frac{1 - \exp(-\theta_{123}z_1)}{\lambda_2\theta_{123}}, \quad z_1 > 0 \tag{3.2.36}$$

$$P[Z_1 \le z_1, Z_2 = 2] = \frac{1 - \exp(-\theta_{123}z_1)}{\lambda_3\theta_{123}}, \quad z_1 > 0 \tag{3.2.37}$$

Now, we shall find the $P[Z_2 = z_2]$
$(Z_2 = 0)$ implies (i) $X_1 < X_2 < X_3$ (ii) $X_1 < X_3 < X_2$
In case (i)

$$= \int_0^\infty \frac{1}{\lambda_1}e^{-\frac{x_1}{\lambda_1}}dx_1 \left[\int_{x_1}^\infty \frac{1}{\lambda_2}e^{-\frac{x_2}{\lambda_2}} \left(\int_{x_2}^\infty \frac{1}{\lambda_3}e^{-\frac{x_3}{\lambda_3}}dx_3 \right) dx_2 \right]$$

$$= \int_0^\infty \frac{\exp(-\theta_{123}x_1)dx_1}{\lambda_1\lambda_2\theta_{23}} = \frac{1}{\lambda_1\lambda_2\theta_{23}\theta_{123}}$$

In case (ii)

$$= \int_0^\infty \frac{1}{\lambda_1}e^{-\frac{x_1}{\lambda_1}}dx_1 \left[\int_{x_1}^\infty \frac{1}{\lambda_3}e^{-\frac{x_3}{\lambda_3}} \left(\int_{x_3}^\infty \frac{1}{\lambda_2}e^{-\frac{x_2}{\lambda_2}}dx_2 \right) dx_3 \right]$$

$$= \frac{1}{\lambda_1\lambda_3\theta_{23}\theta_{123}}$$

$$P[Z_2 = 0] = \frac{1}{\lambda_1\lambda_2\theta_{23}\theta_{123}} + \frac{1}{\lambda_1\lambda_3\theta_{23}\theta_{123}} = \frac{1}{\lambda_1\theta_{123}} \tag{3.2.38}$$

Similarly,

$$P[Z_2 = 1] = \frac{1}{\lambda_2\theta_{123}} \tag{3.2.39}$$

$$P[Z_2 = 2] = \frac{1}{\lambda_3\theta_{123}} \tag{3.2.40}$$

From (3.2.35) to (3.2.40), Z_1 and Z_2 are independent random variables.
Hence,

$$f(z_1) = \theta_{123}e^{-\theta_{123}z_1}; \quad z_1 > 0 \tag{3.2.41}$$

If we have a sample of size m, assume Z_2 takes value zero r_1 times, Z_2 takes value one r_2 times and Z_2 takes value two $(m - r_1 - r_2)$ times.

The likelihood function is given by

$$L(\lambda_1, \lambda_2, \lambda_3 | Z) = (\theta_{123})^m e^{-\theta_{123} \sum_{i=1}^m z_{1i}} c_{rm} (\lambda_1 \theta_{123})^{-r_1} (\lambda_2 \theta_{123})^{-r_2} (\lambda_3 \theta_{123})^{-(m-r_1-r_2)}$$

$$(3.2.42)$$

where $c_{rm} = \frac{m!}{r_1! r_2! (m-r_1-r_2)!}$

$$\log L = -\theta_{123} \sum_{i=1}^m z_{1i} - r_1 \log \lambda_1 - r_2 \log \lambda_2 - (m - r_1 - r_2) \log \lambda_3$$

$$\frac{\partial \log L}{\partial \lambda_1} = \frac{\sum_{i=1}^m z_{1i}}{\lambda_1^2} - \frac{r_1}{\lambda_1} \tag{3.2.43}$$

$$\frac{\partial \log L}{\partial \lambda_2} = \frac{\sum_{i=1}^m z_{1i}}{\lambda_2^2} - \frac{r_2}{\lambda_2} \tag{3.2.44}$$

$$\frac{\partial \log L}{\partial \lambda_3} = \frac{\sum_{i=1}^m z_{1i}}{\lambda_3^2} - \frac{m - r_1 - r_2}{\lambda_3} \tag{3.2.45}$$

From (3.2.43),

$$\hat{\lambda}_1 = \frac{\sum_{i=1}^m z_{1i}}{r_1} \tag{3.2.46}$$

From (3.2.44),

$$\hat{\lambda}_2 = \frac{\sum_{i=1}^m z_{1i}}{r_2} \tag{3.2.47}$$

From (3.2.45),

$$\hat{\lambda}_3 = \frac{\sum_{i=1}^m z_{1i}}{m - r_1 - r_2} \tag{3.2.48}$$

Remark: Let $X_1, X_2, \ldots, X_n$ be independent exponential rvs with means $\lambda_i (i = 1, 2, \ldots, n)$ respectively, let $Z_1 = \min(X_1, X_2, \ldots X_n)$.
Let

$$Z_2 = \begin{cases} 0 & Z_1 = X_1 \\ 1 & Z_1 = X_2 \\ 2 & Z_1 = X_3 \\ \vdots \\ \vdots \\ n-1 & Z_1 = X_n \end{cases}$$

$$P[Z_2 = i - 1] = \frac{1}{\lambda_i \theta_{123\ldots n}}; \quad i = 1, 2, \ldots, n \tag{3.2.49}$$

$$f(z_1) = \theta_{123...n}[\exp -(z_1\lambda_i\theta_{123...n})] \; ; \; z_1 > 0 \qquad (3.2.50)$$

where $\theta_{123...n} = \sum_{i=1}^{n} \frac{1}{\lambda_i}$

MLE of $\lambda_1, \lambda_2, \ldots, \lambda_n$ is

$$\hat{\lambda}_1 = \frac{1}{r_1} \sum_{i=1}^{m} z_{1i}, \; \hat{\lambda}_2 = \frac{1}{r_2} \sum_{i=1}^{m} z_{1i}, \ldots, \hat{\lambda}_{n-1} = \frac{1}{r_{n-1}} \sum_{i=1}^{m} z_{1i}$$

$$\hat{\lambda}_n = \frac{1}{m - \sum_{i=1}^{n-1} r_i} \sum_{i=1}^{m} z_{1i}$$

Example 3.2.7 Let $X_1, X_2, \ldots, X_n$ be iid rvs satisfying the following regression equation

$$X_i = \alpha z_i + e_i, i = 1, 2, \ldots, n$$

where $z_1, z_2, \ldots, z_n$ are fixed and $e_1, e_2, \ldots, e_n$ are iid rvs $N(0, \sigma^2)$, σ^2 unknown. We will find the MLE of α.

The log likelihood function is given by

$$\log L(\alpha, \sigma^2)|X) = -\frac{n}{2}\log(2\pi) - \frac{n}{2}\sigma^2 - \frac{1}{2\sigma^2}\sum x_i^2 + \frac{\alpha}{\sigma^2}\sum_{i=1}^{n}(x_iz_i) - \frac{\alpha^2}{2\sigma^2}\sum z_i^2$$

For fixed σ^2,

$$\frac{d\log L}{d\alpha} = \frac{1}{\sigma^2}\sum_{i=1}^{n}(x_iz_i) - \frac{\alpha}{2\sigma^2}\sum z_i^2 = 0$$

Then

$$\hat{\alpha} = \frac{\sum_{i=1}^{n}(x_iz_i)}{\sum z_i^2} \quad \text{and} \quad \frac{d^2\log L}{d\alpha^2} < 0$$

MLE of α is $\hat{\alpha} = \frac{\sum_{i=1}^{n}(x_iz_i)}{\sum z_i^2}$. Reader can show that $\hat{\alpha}$ is unbiased for α. Further find the MLE of σ^2.

Example 3.2.8 Let $X_1, X_2, \ldots, X_n$ be iid rvs with the following uniform pdf

1. $\cup(0, \theta)$
2. $\cup(\theta, 2\theta)$
3. $\cup(\theta - 1, \theta + 1)$
4. $\cup(\theta, \theta + 1)$

(i) The pdf of X is given by

$$f(x|\theta) = \begin{cases} \frac{1}{\theta} & ; \ 0 < x < \theta, \\ 0 & ; \ \text{otherwise} \end{cases} \qquad (3.2.51)$$

and the corresponding likelihood function is

$$L(\theta|x) = \theta^{-n} ; \ 0 < x_i < \theta, \ i = 1, 2, \ldots, n$$

Consider the order statistics $X_{(1)} < X_{(2)} < \cdots < X_{(n)}$. Hence $0 < X_{(1)} < X_{(2)} < \cdots < X_{(n)} < \theta < \infty$. Note that the support of θ is $X_{(n)} < \theta < \infty$
We have to maximize $L(\theta|x)$ which is equivalent to finding the minimum value of θ, and it is given by $\hat{\theta} = X_{(n)}$. Thus,

$$\text{MLE of } \theta \text{ is } X_{(n)} \qquad (3.2.52)$$

(ii) The pdf is given by

$$f(x|\theta) = \begin{cases} \frac{1}{\theta} & ; \ \theta < x < 2\theta, \\ 0 & ; \ \text{otherwise} \end{cases}$$

and the corresponding likelihood function is given by

$$L(x|\theta) = \begin{cases} \theta^{-n} & ; \ \theta < X_{(1)} < X_{(n)} < 2\theta, \ i = 1, 2, \ldots, n \\ 0 & ; \ \text{otherwise} \end{cases}$$

$\theta < X_{(1)}$ and $\frac{X_{(n)}}{2} < \theta$
$\Rightarrow \frac{X_{(n)}}{2} < \theta < X_{(1)}$
Maximizing $L(\theta|x)$ occurs at minimum value of θ
That is,

$$\hat{\theta} = \frac{X_{(n)}}{2} \qquad (3.2.53)$$

(iii) The pdf and its corresponding likelihood functions are given by

$$f(x|\theta) = \begin{cases} \frac{1}{2} & ; \ \theta - 1 < x < \theta + 1, \\ 0 & ; \ \text{otherwise} \end{cases}$$

$$L(\theta|x) = \begin{cases} \frac{1}{2^n} & ; \ \theta - 1 < X_{(1)} < X_{(n)} < \theta + 1, \\ 0 & ; \ \text{otherwise} \end{cases}$$

The support of θ is $X_{(n)} - 1 \leq \theta \leq X_{(1)} + 1$.

Here any value of θ is MLE.

Therefore,

$$\hat{\theta} = \alpha(X_{(n)} - 1) + (1 - \alpha)(X_{(1)} + 1), \tag{3.2.54}$$

where $\alpha \in [0, 1]$

(iv) In this case

$$L(\theta|x) = \begin{cases} 1 & ; \ \theta < x < \theta + 1, \\ 0 & ; \ \text{otherwise} \end{cases}$$

and

$$L(\theta|x) = \begin{cases} 1 & ; \ \theta < X_{(1)} < X_{(n)} < \theta + 1, \\ 0 & ; \ \text{otherwise} \end{cases}$$

The support of θ is $X_{(n)} - 1 < \theta < X_{(1)}$. Here also, any value of θ is MLE,

$$\hat{\theta} = \alpha(X_{(n)} - 1) + (1 - \alpha)X_{(1)} \tag{3.2.55}$$

Remark:

1. In (iii) and (iv), from (3.2.54) and (3.2.55), we can conclude that MLE is not a function of sufficient statistics, if $\alpha = 0$ or 1.
2. From (3.2.54) and (3.2.55), we can say that MLE is not unique.

Example 3.2.9 Let X be a rv with $B(1, p)$, $p \in \left[\frac{a}{a+b}, \frac{b}{a+b}\right]$, $b > a$. The likelihood function is given by

$$L(p|x) = \begin{cases} p^x(1 - p)^{1-x} & ; \ x = 0, 1 \\ 0 & ; \ \text{otherwise} \end{cases}$$

Therefore,

$$\log L = x \log p + (1 - x) \log(1 - p)$$

and

$$\frac{d \log L}{dp} = \frac{x}{p} - \frac{(1 - x)}{(1 - p)} = 0$$

From this, $\hat{p} = x$. This value does not lie in $\Theta = \left[\frac{a}{a+b}, \frac{b}{a+b}\right]$. Here, $L(p|x)$ is maximized, if we select

$$\hat{p} = \begin{cases} \frac{a}{a+b} & ; \ x = 0 \\ \frac{b}{a+b} & ; \ x = 1 \end{cases}$$

Hence

$$\hat{p} = \frac{(b-a)x + a}{a+b} \qquad (3.2.56)$$

Now,

$$E(\hat{p}) = \frac{(b-a)p + 1}{a+b} \neq p,$$

which is a biased estimator of p.

$$\begin{aligned}
\text{MSE}(\hat{p}) &= E\left[\frac{(b-a)X + a}{a+b} - p\right]^2 \\
&= \frac{1}{(a+b)^2} E\left[(b-a)X + a - p(a+b)\right]^2 \\
&= \frac{1}{(a+b)^2}\Big[(b-a)^2 E(X^2) + 2(b-a)a E(X) + a^2 + p^2(a+b)^2 \\
&\quad - 2p(a+b)(b-a)EX - 2p(a+b)a\Big]
\end{aligned}$$

Now $E(X^2) = E(X) = p$

$$\begin{aligned}
&= \frac{1}{(a+b)^2}\left[(b-a)^2 p + 2a(b-a)p + a^2 + p^2(a+b)^2 - 2p^2(a+b)(b-a) - 2pa(a+b)\right] \\
&= \frac{1}{(a+b)^2}\left[a^2 + p^2\left\{a^2 + b^2 + 2ab - 2b^2 + 2a^2\right\} + p\left\{b^2 - 2ab + a^2 + 2ab - 2a^2 - 2a^2 - 2ab)\right\}\right] \\
&= \frac{1}{(a+b)^2}\left[a^2 + p^2\left(3a^2 + 2ab - b^2\right) + p\left\{b^2 - 2ab - 3a^2\right\}\right] \\
&= \frac{1}{(a+b)^2}\left[a^2 + p^2(3a - b)(a+b) + p(b - 3a)(a+b)\right] \\
&= \frac{a^2}{(a+b)^2} + \frac{p^2(3a-b)}{(a+b)} + \frac{p(b-3a)}{(a+b)} \qquad (3.2.57)
\end{aligned}$$

In particular if (i) $a = 2, b = 3$ and $p = \frac{2}{5}$

$$E(\hat{p} - p)^2 = 0.016$$

In particular if (i) $a = 2, b = 3$ and $p = \frac{3}{5}$

$$E(\hat{p} - p)^2 = 0.016 \qquad (3.2.58)$$

(i) $\delta(x) = \frac{1}{2}$ and $p = \frac{2}{5}$, where $\delta(x)$ is any trivial estimator of p.

$$E\left(\frac{1}{2} - p\right)^2 < 0.016 \Rightarrow 0.01 < 0.016 \qquad (3.2.59)$$

(ii) $\delta(x) = \frac{1}{2}$ and $p = \frac{3}{5}$

$$E\left(\frac{1}{2} - p\right)^2 < 0.016$$

In the sense of the MSE, the MLE is worse than the trivial estimator $\delta(x) = \frac{1}{2}$.
Remark: We can conclude that MLE can be worst estimator than the trivial estimator (See Rohatagi and Saleh (2001)).

Example 3.2.10 Consider the following example where MLE does not exist.

Let $X_1, X_2, \ldots, X_n$ be iid rvs with $b(1, p)$. Suppose $p \in (0, 1)$. If $(0, 0, \ldots, 0)$, $(1, 1, \ldots, 1)$ is observed. Then $\hat{p} = \bar{x}$, i.e., $\bar{x} = 0, 1$, which is not admissible value of p. Hence MLE does not exist.

Example 3.2.11 Let $X_1, X_2, \ldots, X_n$ be iid rvs from double exponential pdf as follows:

$$f(x|\mu) = \frac{1}{2} \exp[-|x - \mu|]; \quad -\infty < x < \infty, -\infty < \mu < \infty$$

Then the likelihood function is given by

$$L(\mu|x) = \prod_{r=1}^{n} \frac{1}{2} \exp[-|x_r - \mu|]$$

$$L(\mu|x) = 2^{-n} \exp[-\sum_{r=1}^{n} |x_r - \mu|] \tag{3.2.60}$$

In this case we have to maximize, $L(\mu|x)$, which is equivalent to minimize $\sum_{r=1}^{n} |x_r - \mu|$, where $x_{(r)} = r$th order statistics, $1 \leq r \leq n$.
For $x_{(r)} \leq \mu \leq x_{(r+1)}$,

$$\sum_{i=1}^{n} |x_i - \mu| = \sum_{i=1}^{r} (\mu - x_i) + \sum_{i=r+1}^{n} (x_i - \mu)$$

$$= \mu j - (n - r)\mu - \sum_{i=1}^{r} x_{(i)} + \sum_{i=r+1}^{n} x_{(i)}$$

$$= \mu(2r - n) - \sum_{i=1}^{r} x_{(i)} + \sum_{i=r+1}^{n} x_{(i)} \tag{3.2.61}$$

Equation (3.2.61) is a linear function of μ which decreases if $r < \frac{n}{2}$ and increases if $r > \frac{n}{2}$.

If n is even then $r = \frac{1}{2}$ i.e., $2r - n = n - n = 0$.

Then (3.2.60) or (3.2.61) is a constant function between $x_{\left(\frac{n}{2}\right)}$ and $x_{\left(\frac{n}{2}\right)+1}$ and any value in this interval is an MLE of μ.

In general median is the MLE. If n is odd then $\hat{\mu} = x_{\left(\frac{n+1}{2}\right)}$ and if n is even then $\hat{\mu} = \frac{1}{2}\left[x_{\left(\frac{n}{2}\right)} + x_{\left(\frac{n+1}{2}\right)} \right]$.

Example 3.2.12 Due to Basu (1955), MLE of θ is an inconsistent estimator of θ. Let $X_1, X_2, \ldots$ be a sequence of iid rvs with a probability density where $0 < \theta < 1$,

$$P_\theta[X_i = 1] = \begin{cases} \theta & ; \ \theta \text{ is rational} \\ 1 - \theta & ; \ \theta \text{ is irrational,} \end{cases}$$

Further,

$$P_\theta[X_i = 0] = 1 - P_\theta[X_i = 1],$$

Hence,

$$f(x|\theta) = \begin{cases} \theta^x(1-\theta)^{1-x} & ; \ \theta \text{ is rational} \\ (1-\theta)^x \theta^{1-x} & ; \ \theta \text{ is irrational,} \end{cases}$$

The MLE of θ based on first n observations is $\hat{\theta}_n = \frac{\sum X_i}{n}$ since $\hat{\theta}_n$ is rational for all $n = 1, 2, \ldots$
But

$$\hat{\theta}_n \Rightarrow \begin{cases} \theta & ; \ \theta \text{ is rational} \\ (1-\theta) & ; \ \theta \text{ is irrational,} \end{cases}$$

Hence $\hat{\theta}_n$ is an inconsistent estimator of θ.

Example 3.2.13 Let $X_1, X_2, \ldots, X_n$ be a sample from the following pmf

$$f(X = x) = \begin{cases} \frac{1}{N} & ; \ k = 1, 2, \ldots, N \\ 0 & ; \ \text{otherwise} \end{cases}$$

$$L(N = x) = \begin{cases} \frac{1}{N^n} & ; \ 1 \le X_{(n)} \le N \\ 0 & ; \ \text{otherwise} \end{cases}$$

We have to maximize $L(N|x)$ then to find the minimum value of N. The support of N is $x_{(n)} \le N < \infty$.

$$\text{Hence MLE of } N = \hat{N} = x_{(n)} \tag{3.2.62}$$

Example 3.2.14 Let the rv X have the hypergeometric distribution

$$P_N(x) = \begin{cases} \dfrac{\binom{M}{x}\binom{N-M}{n-x}}{\binom{N}{n}} & ; \quad \max(0, n - N + M) \leq x \leq \min(n, M) \\ 0 & ; \text{ otherwise} \end{cases}$$

We have to find the MLE of N where M and n is known.

MLE is $\hat{N}$ if,

$$\frac{P(N \mid x)}{P(N - 1 \mid x)} \geq 1 \quad \text{and} \quad \frac{P(N + 1 \mid x)}{P(N \mid x)} \leq 1$$

Consider,

$$\begin{aligned}
\lambda(N) &= \frac{P(N \mid x)}{P(N - 1 \mid x)} \\
&= \frac{\binom{M}{x}\binom{N-M}{n-x}}{\binom{N}{n}} \frac{\binom{N-1}{n}}{\binom{M}{x}\binom{N-1-M}{n-x}} \\
&= \frac{N - n}{N} \frac{N - M}{N - M - n + x}
\end{aligned}$$

if

$$\begin{aligned}
\lambda(N) \geq 1 &\Leftrightarrow nM \geq Nx \\
&\Leftrightarrow \frac{nM}{N} \geq x \\
&\Leftrightarrow N \leq \frac{nM}{x} \qquad \text{(a)}
\end{aligned}$$

similarly,

$$\lambda(N + 1) = \frac{N + 1 - n}{N + 1} \frac{N + 1 - M}{N + 1 - M - n + x}$$

Therefore,

$$\lambda(N + 1) \leq 1 \Leftrightarrow nM \leq Nx + x$$
$$N \geq \frac{nM}{x} - 1 \qquad \text{(b)}$$

From (a) and (b) we get,

$$N \leq \frac{nM}{x} \quad \text{and} \quad N \geq \frac{nM}{x} - 1$$

Now N is an integer, if $\dfrac{nM}{x}$ is an integer then $\dfrac{nM}{x} - 1$ is also an integer. Therefore,

MLE of $N = \hat{N} = \dfrac{nM}{x}$ or $\dfrac{nM}{x} - 1$.

If $\dfrac{nM}{x}$ is not an integer then MLE of N is $\left[\dfrac{nM}{x}\right]$. $\qquad$ (3.2.63)

Theorem 3.2.1 *Let T be a sufficient statistics for the family of pdf(pmf) $f(x|\theta, \theta \in \Theta)$. If an MLE of θ exists and it is unique then it is a function of T.*

Proof It is given that T is sufficient, from the factorization theorem,

$$f(x|\theta) = h(x)g(T|\theta)$$

Maximization of the likelihood function with respect to θ is therefore equivalent to the maximization of $g(T|\theta)$, which is a function of T alone. $\qquad \square$

Remark: This theorem does not say that a MLE is itself a sufficient statistics. In Example 3.2.8, we have shown that MLE need not be a function of sufficient statistics (see Remark 1).

Example 3.2.15 Find the MLE of the parameter p and σ of the following pdf

$$f(x|p, \sigma) = \frac{1}{\Gamma p}\left(\frac{p}{\sigma}\right)^p e^{-\frac{px}{\sigma}} x^{p-1}; \ x > 0, \ p, \ \sigma > 0$$

For large value of p, one should use $\Psi(p)$,

$$\Psi(p) = \log p - \frac{1}{2p} \quad \text{and} \quad \Psi'(p) = \frac{1}{p} + \frac{1}{2p^2},$$

where $\Psi(p)$ and $\Psi'(p)$ are known as digamma and trigamma functions,

$$\frac{d \log \Gamma p}{dp} = \Psi(p) \text{ and } \frac{d\Psi(p)}{dp} = \Psi'(p) \qquad (3.2.64)$$

For details see, Abramowitz and Stegun (1972), Dixit (1989).
The corresponding likelihood function is given by,

$$L(p, \sigma|x) = \left(\frac{1}{\Gamma p}\right)^n \left(\frac{p}{\sigma}\right)^n p e^{-\frac{p}{\sigma}\sum_{i=1}^{n} x_i} \prod_{i=1}^{n} x_i^{p-1}$$

$$\log L = -n \log \Gamma p + np[\log p - \log \sigma] - \frac{p}{\sigma}\sum_{i=1}^{n} x_i + (p - 1)\sum_{i=1}^{n} \log x_i$$

Let G be the geometric mean of $x_1, x_2, \ldots, x_n$, then

$$\log G = \frac{1}{n} \sum_{i=1}^{n} \log x_i \Rightarrow n \log G = \sum_{i=1}^{n} \log x_i$$

$$\log L = -n \log \Gamma p + np[\log p - \log \sigma] - \frac{pn\bar{x}}{\sigma} + (p-1)n \log G$$

$$\frac{\partial \log L}{\partial \sigma} = -\frac{np}{\sigma} + \frac{np\bar{x}}{\sigma^2} = 0 \Rightarrow \hat{\sigma} = \bar{x}$$

$$\frac{\partial \log L}{\partial p} = -n\Psi(p) + n[\log p - \log \sigma] + \frac{np}{p} - \frac{n\bar{x}}{p} + n \log G$$

$$\Rightarrow \left[-n \log p + \frac{n}{2p} \right] + n[\log p - \log \sigma + 1] - n + n \log G = 0$$

$$\Rightarrow \frac{1}{2p} - \log \bar{x} + 1 - 1 + \log G = 0$$

$$\frac{1}{2p} - \log \frac{\bar{x}}{G} = 0$$

$$\hat{p} = \frac{1}{2 \log \frac{\bar{x}}{G}} \tag{3.2.65}$$

Hence, MLE of p and σ are .

$$\hat{p} = \frac{1}{2 \log \frac{\bar{x}}{G}} \quad \text{and} \quad \hat{\sigma} = \bar{x} \tag{3.2.66}$$

Example 3.2.16 Consider a power series distribution with pmf

$$f(x|\theta) = \frac{a_x \theta^x}{g(\theta)}; x = 0, 1, 2 \ldots$$

where $g(\theta) = \sum_{x=0}^{\infty} a_x \theta^x$, a_x may be nonzero in a sample space of size n.

For some x, we will show that MLE of θ is the root of the equation:

$$\bar{X} = \theta \frac{g'(\theta)}{g(\theta)} = \lambda(\theta),$$

where $\lambda(\theta) = EX$. The likelihood function is given by,

$$L(\theta|r) = \frac{\prod_{i=1}^{n} a_{x_i} \theta^{x_i}}{g^n(\theta)}$$

$$= \frac{\theta^t \prod_{i=1}^{n} a_{x_i}}{g^n(\theta)} \tag{3.2.67}$$

where $T = \sum_{i=1}^{n} x_i$

$$\log L = t \log \theta + \sum_{i=1}^{n} a_{x_i} - n \log g(\theta).$$

Therefore,

$$\frac{d \log L}{d\theta} = \frac{t}{\theta} - \frac{n g'(\theta)}{g(\theta)} = 0$$

$$\Rightarrow \frac{\theta g'(\theta)}{g(\theta)} = \frac{t}{n} = \bar{x}.$$

Hence,

$$\bar{x} = \lambda(\theta) = \frac{\theta g'(\theta)}{g(\theta)}. \tag{3.2.68}$$

Thus, MLE of θ is the root of the equation (3.2.67)
Consider

$$EX = \sum_{x=0}^{\infty} \frac{a_x x \theta^x}{g(\theta)}$$

Now,

$$\sum_{x=0}^{\infty} a_x \theta^x = g(\theta) \tag{3.2.69}$$

Differentiate (3.2.68) with respect to θ;

$$\sum_{x=0}^{\infty} a_x x \theta^{x-1} = g'(\theta)$$

$$\Rightarrow \sum_{x=0}^{\infty} x a_x \theta^x = \theta g'(\theta)$$

$$\Rightarrow \frac{\sum_{x=0}^{\infty} x a_x \theta^x}{g(\theta)} = \frac{\theta g'(\theta)}{g(\theta)}$$

Hence,

$$E(X) = \lambda(\theta) = \bar{X}. \qquad (3.2.70)$$

Example 3.2.17 We will find the MLE of θ in $N(\theta, 1)$ in a sample of size n, where θ is an integer.

$$L(\theta|x) = \left(\frac{1}{\sqrt{2\pi}}\right)^n \exp\left[-\frac{1}{2}\sum(x_i - \theta)^2\right]$$

$$= (2\pi)^{-\frac{n}{2}} \exp\left[-\frac{1}{2}\{S^2 + n(\bar{x} - \theta)^2\}\right]$$

where $S^2 = \sum(x_i - \bar{x})^2$.

In this case, we have to minimize $(\bar{x} - \theta)^2$ with respect to $\theta = 0, \pm 1, \pm 2, \ldots$
Let $\bar{x} = [\bar{x}] + \delta$, where $0 < \delta < 1$, and $[\bar{x}] =$ integer part of $\bar{x}$
Minimize $A = [[\bar{x}] + \delta - \theta]^2$ with respect to $\theta = \pm 1, \pm 2, \ldots$
If $\theta = [\bar{x}] \Rightarrow A = \delta^2$
$\quad = [\bar{x}] + r \Rightarrow A = (r - \delta)^2$ if $r \geq 2$
$\quad = [\bar{x}] - r \Rightarrow A = (r + \delta)^2 > \delta^2$
Note that $(r - \delta)^2 > \delta^2$ then we require the condition $r \geq 2$.
For $r = 1$ then $A = (1 - \delta)^2$
Consider δ^2 and $(1 - \delta)^2$
If $\delta > \frac{1}{2}$ then $(1 - \delta)^2 < \delta^2$
If $\delta < \frac{1}{2}$ then $(1 - \delta)^2 > \delta^2$
If $\delta = \frac{1}{2}$ then $(1 - \delta)^2 = \delta^2$
Therefore, MLE of θ is
If $\delta = \frac{1}{2}$ then $\hat{\theta} = [\bar{x}]$ or $[\bar{x}] + 1$
If $\delta < \frac{1}{2}$ then $\hat{\theta} = [\bar{x}]$
If $\delta > \frac{1}{2}$ then $\hat{\theta} = [\bar{x}] + 1$

Invariance Property of Estimator in Case of MLE

Invariance estimator in case of an MLE is a very useful property. Suppose for some pdf f with parameter θ, $\bar{X}$ is the MLE of θ. Then the MLE of $h(\theta) = \frac{\theta+1}{\theta-1}$ is $h(\hat{\theta}) = \frac{\bar{X}+1}{\bar{X}-1}$. Here we will give a procedure due to Zehna (1966). In other words for a density function f, we are finding MLE for $h(\theta)$. If $\theta \rightarrow h(\theta)$ is one-to-one, there is no problem. In this case, it is easy to see that it makes no difference whether we maximize the likelihood as a function of θ or as a function of $h(\theta)$, in each case we get the same answer.

Let $\Psi = h(\theta)$ then the inverse function $\theta = h^{-1}(\Psi)$ is well defined. The likelihood function of $h(\theta)$, written as a function of Ψ, is given as

$$L^*(h(\theta)|Y) = L^*(\Psi|\theta) = \prod_{i=1}^{n} f(x_i|\theta)$$

$$= \prod_{i=1}^{n} f(x_i|h^{-1}(\Psi)) = L(h^{-1}(\Psi)|X)$$

$$\sup_{h} = L^*(h(\theta)|X) = \sup_{\theta} L(h^{-1}\Psi|X) = \sup_{\theta} L(\theta|X)$$

Hence, the maximum of $L^*(h(\theta)|X)$ is attained at $\Psi = h(\hat{\theta})$, showing the MLE of $h(\theta)$ is $h(\hat{\theta})$.

If $h(\theta)$ is one-to-one, then it is quite possible that θ may take more than one value which satisfies $h(\theta) = \Psi$. We may say that $h(\hat{\theta})$ is also not unique, see Casella and Berger (2002). We state the theorems without proof.

Theorem 3.2.2 (Invariance property of MLE):
If $\hat{\theta}$ is the MLE of θ, then $h(\hat{\theta})$ is the MLE of $h(\theta)$, where $h(\theta)$ is any continuous function of θ.

Further, we state the following theorems on MLE without proof.

Theorem 3.2.3 Let $X_1, X_2, \ldots, X_n$ be iid rvs having common pdf $f(x|\theta), \theta \in \Theta$. Assumption:

1. The derivative $\frac{\partial^i \log f(x|\theta)}{\partial \theta^i}, i = 1, 2, 3$ exist for almost all x and for every θ belonging to a non-degenerate interval in Θ

2. There exists functions $H_1(x), H_2(x)$ and $H_3(x)$ such that $|\frac{\partial f}{\partial \theta}| < H_1(x), |\frac{\partial^2 f}{\partial \theta^2}| < H_2(x), |\frac{\partial^3 f}{\partial \theta^3}| < H_3(x), \forall \theta \in \Theta, \int H_1(x)dx < \infty, \int H_2(x)dx < \infty, \int H_3(x)dx < \infty,$

3.

$$\int \left[\frac{\partial \log f(x|\theta)}{\partial \theta} \right]^2 f(x|\theta)dx$$

is finite and positive for every $\theta \in \Theta$.

If assumptions (a)–(c) are satisfied and true parameter point θ_0 is an inner point then for sufficiently large n,

1.

$$\sum_{j=1}^{n} \frac{\partial \log f(x_j|\theta)}{\partial \theta} = 0$$

has at least one root $\hat{\theta}_n$ which converges in probability to θ_0.

2. $\sqrt{n}(\hat{\theta}_n - \theta_0)$ converges in distribution to $N(0, I^{-1}(\theta))$, where

$$I(\theta) = \int \left(\frac{\partial \log f(x|\theta)}{\partial \theta} \right)^2 f(x|\theta) dx,$$

which is the Fisher information contained in the sample size n.

Theorem 3.2.4 *Huzurbazar (1948): The consistent root is unique.*

Theorem 3.2.5 *Wald (1949): The estimate which maximizes the likelihood absolutely is a consistent estimate.*

3.3 MLE in Censored Samples

In this section, we assume that the life times $X_i (i = 1, 2, \ldots, n)$ are iid rvs. We discuss, two types of censoring mechanisms and describe their corresponding likelihood functions.

Let X be a non-negative rvs representing the lifetime of an individual in some populations. Let X be observed upto time t then cdf of X is $F(t)$, where

$$F(t) = P[X \leq t] = \int_0^t f(x) dx$$

The probability of an individual surviving to time t is given by the survivor function $\bar{F}(t)$, where

$$\bar{F}(t) = P[X > t] = \int_t^\infty f(x) dx$$

In some contexts, involving systems or lifetimes of manufactured items, $\bar{F}(t)$ is a monotone decreasing continuous function. This function exhibits the complementary properties of $F(t)$. Some authors denote $\bar{F}(t)$ as $S(t)$.

Suppose that n individuals have lifetimes represented by rvs $X_1, X_2, \ldots, X_n$. Consider a time t_i which we know is either the lifetime or censoring time. Define a variable $\delta_i = I(X_i = t_i)$, where,

$$\delta_i = \begin{cases} 1 & ; \ X_i = t_i \\ 0 & ; \ X_i > t_i \end{cases}$$

This is called censoring or status indicator for t_i. It implies that if t_i is an observed lifetime, then $\delta_i = 1$ and censoring time then $\delta_i = 0$. The observed data consist of $(t_i, \delta_i), i = 1, 2, \ldots, n$.

I. Type-I Censoring

A type-I censoring mechanism is said to apply when each individual has a fixed potential censoring time $C_i > 0$ such that X_i is observed if $X_i \leq C_i$; otherwise we know only that $X_i > C_i$. Type-I censoring often arises when a study is conducted over a specified time period. For example, termination of a life test on electrical insulation specimens after 100 minutes would mean that $C_i = 100$ for each item. In another example of clinical trials, there is often staggered entry of individuals to the study combined with a specified end of study date. Consider the problem of estimation in the presence of excess of loss reinsurance. Suppose that the claims record shows only the net claims paid by the insurer. A typical claims record might be

$$x_1, x_2, M, x_3, M, x_4, x_5, \ldots \tag{3.3.1}$$

and an estimate of the underlying gross claims distribution is required.

The sample in (3.3.1) is censored. In general, a censored sample occurs when some values are recorded exactly and the remaining values are known only to exceed a particular value, here the retention level is M.

Let $t_i = \text{Min}(X_i, C_i)$, $\delta_i = I(X_i \leq C_i)$ for type-I censoring.

Since C_i are fixed constants and that t_i can take values $\leq C_i$ with

$$P[t_i = C_i, \delta_i = 0] = P[X_i \geq C_i]$$

$$P[t_i, \delta_i = 1] = f(t_i) \text{ when } X_i \leq C_i$$

Assuming that the lifetimes $X_1, X_2, \ldots, X_n$ are stochastically independent, then likelihood function will be

$$L = \prod_{i=1}^{n} [f(x_i)]^{\delta_i} [\bar{F}(C_i)]^{1-\delta_i}$$

If $C_i = M$, then

$$L(\theta) = \prod_{i=1}^{n} [f(x_i)]^{\delta_i} [\bar{F}(M)]^{1-\delta_i}$$

If $\sum_{i=1}^{n} \delta_i = r$,

$$L(\theta) = \prod_{i=1}^{r} [f(x_i)][\bar{F}(M)]^{n-r}$$

Example 3.3.1 Claims in portfolio are believed to arise as a gamma distribution with shape and scale parameter 2 and λ respectively. There is a retention limit of 1000 in force, and claims in excess of 1200 are paid by the insurer. The insurer, wishing to estimate λ, observes a random sample of 120 claims, and finds that the average amount of the 100 claims that do not exceed is 85. There are 20 claims that do exceed the retention limit. Find the MLE of λ. Note that $r = 100$,

$$f(x|\lambda) = xe^{-\lambda x}\lambda^2 \; ; \; x > 0, \; \lambda > 0$$

$$P[X > 1000] = e^{-1000\lambda}[1000\lambda + 1]$$

The likelihood function is

$$L(\lambda|x) = \prod_{i=1}^{100} \lambda^2 x_i e^{-\lambda x_i} \left[e^{-1000\lambda}(1 + 1000\lambda) \right]^{20}$$

$$= \lambda^{200} \left(\prod_{i=1}^{100} x_i \right) e^{-\lambda \sum_{i=1}^{100} x_i} e^{-20000\lambda} (1 + 1000\lambda)^{20}$$

Now $\sum_{i=1}^{100} x_i = 100 \times 85 = 8500$

$$\log L(\lambda|x) = 200 \log \lambda + \sum_{i=1}^{100} \log x_i - \lambda \sum_{i=1}^{100} x_i - 20000\lambda + 20 \log(1000\lambda + 1)$$

$$\frac{\partial \log L(\lambda|x)}{\partial \lambda} = \frac{200}{\lambda} - \sum_{i=1}^{100} x_i - 20000 + \frac{20000}{1000\lambda + 1} = 0.$$

Hence,

$$\frac{200}{\lambda} - 28500 + \frac{20000}{1000\lambda + 1} = 0$$

$$\Rightarrow 285000\lambda^2 - 1915\lambda - 2 = 0$$

$\lambda_1 = 0.00763806, \quad \lambda_2 = -0.00091876$

II. Type-II Censoring

Type-II censoring refers to the situation where only the $r(< n)$ smallest lifetimes $(x_1 < x_2 < \cdots < x_r | r < n)$ in a random sample of n are observed. Here, by censoring at the right, we may be able to obtain reasonable good estimates of the parameters much sooner than by waiting for all times to fail.

With Type-II censoring the value of r is chosen before the data are collected and the data consists of the r smallest lifetimes in a random sample $x_1, x_2, \ldots, x_n$.

The problem considered is: Given the values of $x_{(1)}, x_{(2)}, \ldots, x_{(r)}$ and n, to find the MLE of the parameter(s) as follows;

The joint distribution of the order statistics $x_{(1)}, x_{(2)}, \ldots, x_{(r)}$ in a sample of size n is given as:

$$f(x_{(1)}, x_{(2)}, \ldots, x_{(r)}|\theta) = \frac{n!}{(n-r)!}\left[\prod_{i=1}^{r} f(x_{(i)})\right][1 - F(x_{(r)})]^{n-r}$$

Wilk (1962) derived the MLE of the parameters of gamma distribution based on the order statistics $(x_{(1)} < x_{(2)} < \cdots < x_{(r)}|r < n)$.

Example 3.3.2 Consider the exponential distribution with parameter λ but lifetimes are type-II censored.

The likelihood function is,

$$L(\lambda|x) = \frac{n!}{(n-r)!}\lambda^r e^{-\lambda \sum_{i=1}^{r} x_{(i)}} e^{-(n-r)x_{(r)}\lambda}$$

$$\log L(\lambda|x) = \text{Const} + r\log\lambda - \lambda\sum_{i=1}^{r} x_{(i)} - (n-r)x_{(r)}\lambda$$

Therefore,

$$\frac{\partial \log L}{\partial \lambda} = \frac{r}{\lambda} - \sum_{i=1}^{r} x_{(i)} - (n-r)x_{(r)}$$

and hence

$$\hat{\lambda} = \frac{r}{\sum_{i=1}^{r} x_{(i)} + (n-r)x_{(r)}}$$

For details, see Lawless (2003), Dixit and Nooghabi Jabbavi (2011).

3.4 Newton–Raphson Method

The Newton–Raphson method is a powerful technique for solving equations numerically. Like so much of the differential calculus, it is based on the simple idea of linear approximation.

Let $f(x)$ be a well-behaved function. Let x^* be a root of the equation $f(x) = 0$ which we want to find. To find let us start with an initial estimate x_0. From x_0, we produce to an improved estimate x_1 (if possible) then from x_1 to x_2 and so on. Continue the procedure until two consecutive values x_i and x_{i+1} in ith and $(i+1)$th steps are very close or it is clear that two consecutive values are away from each other. This style of proceeding is called 'iterative procedure'.

Consider the equation $f(x) = 0$ with root x^*. Let x_0 be a initial estimate. Let $x^* = x_0 + h$ then $h = x^* - x_0$, the number h measures how far the estimate x_0 is from the truth. Since h is small, we can use linear approximation to conclude that

$$0 = f(x^*) = f(x_0 + h) \simeq f(x_0) + hf'(x_0)$$

and therefore, unless $f'(x_0)$ is close to 0,

$$h \simeq -\frac{f(x_0)}{f'(x_0)}$$

This implies,

$$x^* = x_0 + h \simeq x_0 - \frac{f(x_0)}{f'(x_0)}$$

Our new improved estimate x_1 of x^* is given by

$$x_1 = x_0 - \frac{f(x_0)}{f'(x_0)}$$

Next estimate x_2 is obtained from x_1 in exactly the same way as x_1 was obtained from x_0

$$x_2 = x_1 - \frac{f(x_1)}{f'(x_1)}$$

Continuing in this way, if x_n is the current estimate, then next estimate x_{n+1} is given by

$$x_{n+1} = x_n - \frac{f(x_n)}{f'(x_n)},$$

See Kale (1962).

Example 3.4.1 Consider the Example 3.2.15

$$\log L = -n \log \Gamma p + np[\log p - \log \sigma] - \frac{np\bar{x}}{\sigma} + n(p-1)\log G$$

$$\frac{\partial \log L}{\partial \sigma} = \frac{np}{\sigma} + \frac{np\bar{x}}{\sigma^2} \Rightarrow \hat{\sigma} = \bar{x}$$

$$\frac{\partial \log L}{\partial p} = -n\Psi(p) + n[\log p - \log \sigma] + \frac{np}{p} - \frac{n\bar{x}}{\sigma} + n\log G$$

$$\Rightarrow \log p - \Psi(p) = \log \frac{\bar{x}}{G}$$

Let $\log \frac{\bar{x}}{G} = C$

Hence $\log n = \Psi(p) = C$. By Newton–Raphson iteration method gives

$$\hat{p}_k = \hat{p}_{k-1} - \frac{\log(\hat{p}_{k-1}) - \Psi(\hat{p}_{k-1}) - C}{(\hat{p}_{k-1})^{-1} - \Psi'(\hat{p}_{k-1})}$$

$\hat{p}_k$ denotes the kth iterate starting with initial trial value $\hat{p}_0$ and $\Psi'(p) = \dfrac{d\Psi(p)}{dp}$.

The function $\Psi(p)$ and $\Psi'(p)$ are tabulated in Abramowitz and Stegun (1972) in the form of digamma and trigamma functions and can be expressed in power series as

$$\Psi(y) = \log y - \frac{1}{2y} + \sum_{i=1}^{\infty} \frac{B_{2i}}{2i} y^{-2i}(-1)^i$$

$$\simeq \log y - \frac{1}{2y} + \sum_{i=1}^{10} a_{2i} y^{-2i}(-1)^i$$

where $a_{2i} = \frac{B_{2i}}{2i}$ for $i = 1, 2, 3, \ldots, 10$.

$$a_2 = \frac{1}{12} \qquad a_{12} = \frac{691}{32760}$$

$$a_4 = \frac{1}{120} \qquad a_{14} = \frac{7}{84}$$

$$a_6 = \frac{1}{252} \qquad a_{16} = \frac{3617}{8160}$$

$$a_8 = \frac{1}{240} \qquad a_{18} = \frac{43867}{14364}$$

$$a_{10} = \frac{5}{660} \qquad a_{20} = \frac{174611}{6600}$$

$$\Psi'(y) = \frac{1}{y} + \frac{1}{2y^2} + \sum_{i=1}^{10} 2i a_{2i} y^{-2i-1}$$

where B_{2i} are Bernoulli numbers. These numbers are obtained from Abramowitz and Stegun (1972, p.810).

Here when y is less than or equal to 2, the values of $\Psi(y)$ and $\Psi'(y)$ are accurate upto 4 decimals and when y is greater than 2, they are accurate upto 7 decimals as compared to Pairman's (1919) table. For details, see Dixit (1989).

Example 3.4.2 Gross and Clark (1975) consider a gamma model following random sample of 20 survival times (in weeks) of male mice exposed to 240 rads of gamma radiation.

152, 152, 115, 109, 137, 88, 94, 77, 160, 83, 165, 125, 40, 128, 123, 136, 101, 62, 153, 69

$\bar{x} = 113.45$, $G = \left(\prod_{i=1}^{n} x_i\right)^{\frac{1}{n}} = 107.0680$

$S^2 = (n-1)^{-1} \sum_{i=1}^{n} (x_i - \bar{x})^2 = 1280.8921$

Moment estimators

$$\bar{p} = \frac{\bar{x}}{S^2} = 10.0484$$

$$\hat{\sigma} = \frac{\bar{x}}{\hat{p}} = 11.2904$$

Consider the moment estimators as initial solutions.
By Newton–Raphson method,

$$L(p, \sigma | x) = \frac{\prod_{i=1}^{n} x_i^{p-1} e^{-\frac{\sum x_i}{\sigma}}}{(\Gamma p)^n \sigma^{np}}$$

$$\log L(p, \sigma | x) = (p-1) \sum_{i=1}^{n} \log x_i - \frac{\sum x_i}{\sigma} - n \log \Gamma p - np \log \sigma$$

$$\frac{\sum x_i}{\sigma^2} - \frac{np}{\sigma} = 0 \quad \Rightarrow \quad \hat{\sigma} = \frac{\sum x_i}{np} = \frac{\bar{x}}{p}$$

$$\sum_{i=1}^{n} \log x_i - n \Psi(p) - n \log \sigma = 0$$

$$\frac{1}{n} \sum_{i=1}^{n} \log x_i - \Psi(p) - (\log \bar{x} - \log p) = 0$$

$$\log G - \log \bar{x} + \log p - \Psi(p) = 0$$

$$\log p - \Psi(p) = \log \frac{\bar{x}}{G}$$

The following program in *R* is given to estimate the parameters.

```
# Given data
  x <- c(152,152,115,109,137,88,94,77,160,83,165,125,40,128,123,136,101,62,
         153,69) # given observations
  n <- length(x) # number of observations.
# to find arithmetic mean, variance and geometric mean.
  mn <- mean(x) # arithmetic mean
  gm <- prod(x)^(1/n) # geometric mean.
```

```
 s2 <- var(x)  # variance
# To initialise parameters.
 sigma <- s2/mn;p <- mn/sigma;
# To define required variables.
 diff <- 1; eps <- 0.0001; p_val <- rep(0,5); p_val[1] <- p; i <- 2;
# Newton-Raphson method.
 while(diff > eps)
 {
 gp <- log(p) - digamma(p) - log(mn/gm);  # function g(p).
 der_gp <- 1/p - trigamma(p);  # derivative of g(p).
 p_new <- p - (gp/der_gp);  # iteration
 diff <- abs(p_new - p);
 p <- p_new; p_val[i] <- p; i <- i+1;
 }
 sig <- mn/p_val;
# OUTPUT
 p_val; sig;
```

k	$\hat{p}_{k-1}$	$\hat{\sigma}_{k-1}$
1	10.0484	11.2904
2	8.6831	13.0656
3	8.8413	12.8318
4	8.8442	12.8276
5	8.8445	12.8275

II. Scoring Method

As we have said earlier, the MLE equations are usually complicated so that the solutions cannot be obtained directly. Here also, we have to assume a trial solution and derive the linear equations for small additive corrections. The process can be repeated until the corrections become negligible. A nice mechanization is introduced by adopting the method known as the scoring method for obtaining the linear equations for the additive corrections.

The quantity $\frac{d \log L}{d\theta}$ is defined as the efficient score for θ. The MLE is the value of θ for which the efficient score vanishes. If θ_0 is the trial value of the estimate, then expanding $\frac{d \log L}{d\theta}$ and retaining only the first power of $\Delta\theta = \theta - \theta_0$ leads to

$$\frac{d \log L}{d\theta} \simeq \frac{d \log L}{d\theta}|_{\theta=\theta_0} + \Delta\theta \frac{d^2 \log L}{d\theta^2}|_{\theta=\theta_0}$$
$$= \frac{d \log L}{d\theta}|_{\theta=\theta_0} + \Delta\theta I(\theta)|_{\theta=\theta_0}$$

where $I(\theta)$ is the fisher information of θ, i.e., the expected value of $-\frac{d^2 \log L}{d\theta^2}$.

In large samples the difference between $-I(\theta_0)$ and $\frac{d^2 \log L}{d\theta^2}|_{\theta=\theta_0}$ will be $o(\frac{1}{n})$, where n is the number of observations, so that the above approximations holds to the first order of small quantities. The correction $\Delta\theta$ is obtained from the equation

$$\Delta\theta I(\theta_0) = \frac{d\log L}{d\theta}\Big|_{\theta=\theta_0},$$

$$\Delta\theta = \frac{d\log L}{d\theta}\Big|_{\theta=\theta_0} \div I(\theta_0)$$

Example 3.4.3 Consider the Example 3.4.2

This method is given by the following iteration scheme

$$\begin{pmatrix}\hat{p}_k \\ \hat{\sigma}_k\end{pmatrix} = \begin{pmatrix}\hat{p}_{k-1} \\ \hat{\sigma}_{k-1}\end{pmatrix} + \begin{pmatrix}\text{Var}(\hat{p}_{k-1}) & \text{Cov}(\hat{p}_{k-1}, \hat{\sigma}_{k-1}) \\ \text{Cov}(\hat{p}_{k-1}, \hat{\sigma}_{k-1}) & \text{Var}(\hat{\sigma}_{k-1})\end{pmatrix} \begin{pmatrix}m(\hat{p}_{k-1}, \hat{\sigma}_{k-1}) \\ h(\hat{p}_{k-1}, \hat{\sigma}_{k-1})\end{pmatrix}$$

where

$$\text{Var}(\hat{p}_{k-1}) = \frac{\hat{p}_{k-1}}{\hat{p}_{k-1}\Psi(\hat{p}_{k-1}) - 1}$$

$$\text{Cov}(\hat{p}_{k-1}, \hat{\sigma}_{k-1}) = \frac{-\hat{\sigma}_{k-1}}{\hat{p}_{k-1}\Psi(\hat{p}_{k-1}) - 1}$$

$$\text{Var}(\hat{\sigma}_{k-1}) = \frac{\hat{\sigma}_{k-1}^2\Psi(\hat{p}_{k-1})}{\hat{p}_{k-1}\Psi(\hat{p}_{k-1}) - 1}$$

$$m(\hat{p}_{k-1}, \hat{\sigma}_{k-1}) = \frac{-n\hat{p}_{k-1}}{\hat{\sigma}_{k-1}} + \frac{n\bar{x}}{\hat{\sigma}_{k-1}^2}$$

$$h(\hat{p}_{k-1}, \hat{\sigma}_{k-1}) = \frac{1}{n}\sum_{i=1}^{n}\log x_i - \Psi(\hat{p}_{k-1}) - \log\hat{\sigma}_{k-1}$$

$$k = 1, \hat{p}_0 = 10.0484, \hat{\sigma}_0 = 11.2904$$

ML scoring method

Iterations k	$\hat{p}_{k-1}$	$\hat{\sigma}_{k-1}$
1	10.0484	11.2904
2	8.6211	13.1596
3	8.7955	12.8986
4	8.7993	12.8931
5	8.7994	12.89312

3.5 Exercise 3

1. Let $X_1, X_2, \ldots, X_n$ be iid rvs with $\cup(a, a + |a|), a \in \Theta$.

(i) Find the MLE of a when $a \in (-\infty, 0)$
(ii) If possible, find UMVUE of a if $a \in (-\infty, 0)$
(iii) Which estimator is more efficient? Why?
(iv) If $a \in (0, \infty)$, then show that there does not exist UMVUE of θ.
(iv) Find the MLE of a if $a \in (0, \infty)$ and further, find an unbiased estimator of a.

2. If the rv X has the following Bernoulli distribution as follows:

$$P(X = x) = \left(\frac{\theta}{2}\right)^{|x|} (1 - \theta)^{1-|x|}; x = -1, 0, 1$$

Find the MLE of θ.

3. Let $X_1, X_2, \ldots, X_m$ are iid rvs (i) $B(n, p)$ (ii) $P(\lambda)$

(i) Find the MLE of $p^2 q^3$ and compare with the UMVUE estimator of $p^2 q^3$. Which is more efficient for $p = 0.3, n = 5, m = 10$ and 20?
(ii) Find the MLE of λ^2 and compare with the UMVUE estimator of λ^2. Which is more efficient for $\lambda = 2, 3$ and $m = 10, 20$?

4. Let $X_1, X_2, \cdots, X_n$ be iid rvs with $G(p, \lambda)$. Find the MLE of $\frac{p}{\lambda^2}$.

5. Let $X_1, X_2, \ldots, X_n$ be a sample from inverse Gaussian pdf

$$f(x|\mu, \lambda) = \left(\frac{\lambda}{2\pi x^3}\right)^{\frac{1}{2}} \exp\left\{-\frac{\lambda(x - \mu)^2}{(2\mu^2 x)}\right\}, x > 0, \mu > 0, \lambda > 0$$

Show that MLE of μ and λ are $\hat{\mu} = \bar{X}$ and $\hat{\lambda} = \frac{n}{\sum_{i=1}^{n} \frac{1}{x_i} - \frac{1}{\bar{X}}}$.

6. Find the estimator of a and b by the method of moments for $\beta(a, b)$ for a sample of size n.

7. For the Problem 7 in Exercise 2, find the MLE of λ.

8. For the Example 3.2.5, find the moment estimator of λ_1 and λ_2.

9. For the Problem 6 in Exercise 2, find the moment estimator of θ in explicit form. (Hint: See Dixit and Nasiri (2008).)

10. For the Problem 6 in Exercise 2, find the estimate of θ using method of moment and MLE. Which estimator is more efficient? Why?

11. For the Problem 20 in Exercise 2, find the MLE of N_1 and N_2.

12. For the Problem 30 in Exercise 2 find the MLE of θ.

13. The weight of ball bearing is assumed to be normally distributed with mean μ and variance σ^2. Fifteen bearings are weighted are found to have weighs satisfying

$$\sum_{i=1}^{n} x_i = 145.104g, \sum_{i=1}^{n} x_i^2 = 1407.441g^2$$

Find MLE of

(i) σ if $\mu=10$
(ii) σ and μ
(iii) Probability that weight of the ball bearing is greater than 12.

14. For a particular experiment the following frequencies were observed for four mutually exclusive and exhaustive classes. Find the MLE of α.

Sr. No	1	2	3	4
Frequency	10	12	15	30
Probability	$\frac{2+\alpha^2}{4}$	$\frac{1-\alpha^2}{4}$	$\frac{1-\alpha^2}{4}$	$\frac{1-\alpha^2}{4}$

15. The crushing strength of concrete samples in kilograms per square centimeter is modeled as gamma distributed rvs with pdf

$$f(x|\theta) = \frac{xe^{-\frac{x}{\theta}}}{\theta^2}; \; x > 0, \theta > 0$$

(i) Find the MLE of θ based on the observations
     5.4, 7.1, 5.6, 6.2, 4.9, 5.8, 6.3, 5.5, 4.8.
(ii) Find the MLE of $P(X > 3)$.

16. The following is a sample of size 10 from a normal distribution with mean = variance = θ:
6.18, 5.96, 3.60, 3.76, 0.0, 5.92, 5.94, 6.22, −0.38, 4.04.

(i) Obtain the MLE of θ
(ii) Find MLE of $P(X \geq 2)$

17. The distribution of length of life in hours of electronic tubes is assumed to have the pdf of the type

$$f(t) = \frac{1}{\theta}e^{-\frac{t}{\theta}}; \; t > 0, \; \theta > 0$$

A sample of 30 tubes was tested for life until 5 tubes failed. The observed failure times were 780, 820, 850, 900, and 980 h.

Estimate θ by maximum likelihood method.

18. Following table gives the probabilities and the observed frequencies in four phenotypic classes AB, Ab, aB, and ab in a genetic experiment. Estimate θ by the maximum likelihood method and its standard error.

Phenotypic class	Probability	Observed frequency
AB	$\frac{2+\theta}{4}$	100
Ab	$\frac{1-\theta}{4}$	20
aB	$\frac{1-\theta}{4}$	25
ab	$\frac{\theta}{4}$	6

19. Let $X_1, X_2, \ldots, X_n$ be iid rvs from the pdf

$$f(x|\theta, \lambda) = \frac{1}{\lambda} e^{-\frac{(x-\theta)}{\lambda}} \; ; \; x > \theta, \; \lambda > 0$$

(i) Find MLE of (θ, λ)

(ii) Find the MLE of $P[X \geq 2]$.

20. Let $X_1, X_2, \ldots, X_n$ be iid rvs with $N(\theta, 1)$. If there are $m(<n)$ observations which are less than zero but these observations are not available. Find the MLE of θ.

21. The following is a sample of size 10 from a normal distribution with mean = variance = θ
6.18, 5.96, 3.6, 3.76, 0, 5.92, 5.94, 6.22, −0.38, 4.04
Obtain the MLE of θ and $P[X \geq 3]$.

22. A potato manufacturer buys potatoes which are either too large or too small. He accepts potatoes which have width between 3 and 8 cm. The width of potato is assumed to follow a normal distribution with mean μ and variance σ^2. From a lot of 100 potatoes, 20 were rejected because their width was less than 3 cm, 40 were rejected because their width was greater than 8 cm and remaining 40 were accepted. Obtain the maximum likelihood estimator of μ and σ^2 (Hint: Use Example 3.2.3).

23. The distribution of length of life in hours of electronic tubes is assumed to have the pdf of the type $\frac{1}{\theta} \exp(-\frac{t}{\theta}) : t \geq 0, \; \theta > 0$.
A sample of 30 tubes were tested for life until 5 bulbs failed. The observed failure times were 780, 820, 850, 900, 980 h.
Estimate θ by MLE (Hint: Use Example 3.3.1)

24. Let X_1 and X_2 be independent rvs with $U(0, \theta_i)$; $i = 1, 2$ respectively.
Let $z_1 = \min(X_1, X_2)$ and

$$Z_2 = \begin{cases} 0 & ; \ Z_1 = X_1 \\ 1 & ; \ Z_1 = X_2 \end{cases}$$

Find the MLE of θ_1 and θ_2 in a sample of size n.

25. Let the rvs X_1 and X_2 be distributed as exponential with mean θ and $N(\theta, 1)$ respectively. Find the MLE of θ.

26. Let the rvs X_1 and X_2 be distributed as $B(n, \theta)$ and $NB(r, \theta)$ respectively, $0 < \theta < 1$. Find MLE of θ if n and r are known.

27. If the rv X has the following probability distribution as follows
$P(X = -2) = \frac{\theta}{4}, \ P(X = -1) = \frac{\theta}{4}, \ P(X = 0) = 1 - \theta$
$P(X = 1) = \frac{\theta}{4}, \ P(X = 2) = \frac{\theta}{4}$
In a sample of size n, find the MLE of θ.

28. Let the rv X_1 is $\cup(0, \theta)$, X_2 be $G(1, \frac{1}{\theta})$ and X_3 is $\cup(0, \theta)$. Find the MLE of θ. Assume X_1, X_2 and X_3 are independent rvs.

29. Let the rv $X_1, X_2, \ldots, X_n$ be independently distributed as $G(p, \frac{1}{\sigma_i})$; ($i = 1, 2, \ldots, n$). For p known, find the MLE of σ_i; ($i = 1, 2, \ldots, n$).

30. Let the rv $X_1, X_2, \ldots, X_{\frac{n}{2}}$ be distributed as $P(\lambda_1)$, $X_{\frac{n}{2}+1}, X_{\frac{n}{2}+2}, \ldots X_{\frac{3n}{4}}$ be distributed as $P(\lambda_2)$ and $X_{\frac{3n}{4}+1}, X_{\frac{3n}{4}+2}, \ldots X_n$ be distributed as $P(\lambda_3)$. Find the MLE of λ_1, λ_2 and λ_3. Note that $X_1, X_2, \ldots, X_n$ are independent rvs. Assume n is divisible by four.

31. Let $X_1, X_2, \ldots, X_n$ are iid rvs with (i) $\cup(0, \theta^2)$ (ii) $\cup(\theta, \theta^2)$, $\theta > 1$. Find the MLE of θ.

References

Abramowitz M, Stegun IA (1972) Handbook of mathematical functions. Dover, New York

Basu D (1955) An inconsistency of the method of maximum likelihood. Ann Math Stat 26:144–145

Casella G, Berger RL (2002) Statistical inference. Duxbury

Dixit UJ (1982) Estimation of parameters of the gamma distribution. Unpublished M.Phil, Thesis, University of Pune, Pune

Dixit UJ (1987) Characterization of the gamma distribution in the presence of outliers. Bull Bombay Math Colloquium 4:54–59

Dixit UJ (1989) Estimation of the parameters of the gamma distribution in the presence of outliers. Commun Stat Theor Meth 18(8):3071–3085

Dixit UJ, Nasiri Parviz F (2008) Estimation of parameters of right truncated exponential distribution. Stat Papers 49:225–236

Dixit UJ, Nooghabi Jabbavi M (2011) Estimation of parameters of gamma distribution in the presence of outliers in right censored samples. Aligarh J Stat 31:19–31

Dixit UJ, Kelkar SG (2011) Estimation of the parameters of binomial distribution in the presence of outliers. J Appl Stat Sci 19(4):489–408

Dixit UJ, Kelkar SG (2012) Semi-Bayes estimation of the parameters of binomial distribution in the presence of outliers. J Stat Theor Appl 11(2):143–163

Draper N, Guttman I (1971) Bayesian estimation of the binomial parameter. Technometrics 13:667–673

Feldman D, Fox M (1968) Estimation of the parameter n in the binomial distribution. J Am Stat Assoc 63:150–158

Gross AJ, Clark VA (1975) Survival distributions: reliability applications in the biomedical sciences. Wiley

Hamedani GG, Walter GG (1988) Bayes estimation of the binomial parameter n. Commun Stat Theor Meth 17(6):1829–1843

Huzurbazar VS (1948) The likelihood equations consistency and maxima of the likelihood function. Ann Eugen (London) 14:185–200

Kale BK (1962) On the solution of the likelihood equations by iteration processes-the multiparametric case. Biometrika 49:479–486

Lawless JF (2003) Statistical models and methods for life time data, 2nd edn. Wiley, New York

Pairman E (1919) Tables of the digamma and trigamma functions. Cambridge University Press, Cambridge

Rohatagi VK, Saleh EAK (2001) An introduction to probability and statistics. Wiley

Wald A (1949) Note on consistency of the maximum likelihood estimate. Ann Math Stat 20:595–601

Wilk MB (1962) Estimation of the parameters of the gamma distributions using order statistics. Biometrika 49(3/4):525–545

Zehna PW (1966) Invariance of maximum likelihood estimation. Ann Math Stat 37:755

Chapter 4
Bound for the Variance

The history of the lower bounds on the variance of the estimators is long and has many contributors. The widely known bound, and the basis of this theory is the so-called Cramer–Rao bound (Cramer 1946; Rao 1945). It is equal to the inverted value of Fisher's information quantity. It is important to know that Frechet (1943) has given the inequality which is now known as the Cramer–Rao inequality in the statistical literature, after its explicit and independent publication by Cramer (1946), Rao (1945). Specifically, Rohatagi and Saleh (2001) called this bound Frechet-Cramer-Rao (FCR) lower bound. But we call this inequality as Cramer–Rao (CR) lower bound, as it is popularly known. Bhattacharya (1946, 1950) generalized Rao's results, under some additional conditions, to give a sequence of sharper bounds. Chapman and Robbins (1951), Kiefer (1952) gave a lower bound for the variance of an estimate which does not require regularity conditions of the CR lower bound. Detailed review is done by Jadhav (1983).

Before considering the lower bound, we will have to consider the Cauchy–Schwarz inequality. Now, the variance is invariant under translation. Therefore, we expect the bound also to be invariant under translation.

4.1 Cramme–Rao Lower Bound

Theorem 4.1.1 *The C-S inequality, which is translation invariant, is given by*

$$Var(U) \geq \frac{Cov(U, V)}{Var(V)} \tag{4.1.1}$$

Theorem 4.1.2 (Frechet 1943, Rao 1945, Cramer 1946)
We call this inequality as CR inequality.
Let $X_1, X_2, \ldots, X_n$ be a sample with pdf $f(x|\theta)$.

© Springer Science+Business Media Singapore 2016
U.J. Dixit, *Examples in Parametric Inference with R*,
DOI 10.1007/978-981-10-0889-4_4

Let $T(X) = T(X_1, X_2, \ldots, X_n)$ be any unbiased estimator of $g(\theta)$. It satisfies the following conditions

$$\frac{\partial}{\partial \theta} \int T(x) f(x|\theta) dx = \int T(x) \frac{\partial}{\partial \theta} f(x|\theta) dx \; if \; f(x|\theta) \; is \; a \; pdf \qquad (4.1.2)$$

$$\frac{\partial}{\partial \theta} \sum_x T(x) f(x|\theta) dx = \sum_x T(x) \frac{\partial}{\partial \theta} f(x|\theta) dx \; if \; f(x|\theta) \; is \; a \; pmf \qquad (4.1.3)$$

Then

$$Var(T(X)) \geq \frac{[g'(\theta)]^2}{E[\{\frac{\partial}{\partial \theta} \log f(x|\theta)\}^2]} \qquad (4.1.4)$$

If equality holds in (4.1.4), then there exists a real number $c(\theta_0) \neq 0$, such that

$$T(X) - ET(X) = c(\theta_0) \frac{\partial \log f(x|\theta)}{\partial \theta} \qquad (4.1.5)$$

with probability 1, provided $T(X)$ is not constant.

Proof The proof is very simple and it is an application of Theorem 4.1.1, i.e., C-S inequality.

From (4.1.1), replace U by $T(X)$ and V by $\frac{\partial \log f(x|\theta)}{\partial \theta}$

Now

$$Cov(U, V) = Cov\left[T(X), \frac{\partial \log f(x|\theta)}{\partial \theta}\right] \qquad (4.1.6)$$

Since $\displaystyle\int_R f(x|\theta) dx = 1 \Rightarrow \int_R \frac{\partial f(x|\theta)}{\partial \theta} dx = 0$

$$\Rightarrow \int_R \frac{1}{f(x|\theta)} \frac{\partial f(x|\theta)}{\partial \theta} f(x|\theta) dx = 0$$

$$\Rightarrow \int_R \frac{\partial \log f(x|\theta)}{\partial \theta} f(x|\theta) = 0 \qquad (4.1.7)$$

$$\Rightarrow E\left[\frac{\partial \log f(x|\theta)}{\partial \theta}\right] = 0 \qquad (4.1.8)$$

Hence

$$\text{Cov}\left[T(X), \frac{\partial \log f(x|\theta)}{\partial \theta}\right] = \text{E}\left[T(X)\frac{\partial \log f(x|\theta)}{\partial \theta}\right] \tag{4.1.9}$$

Differentiating (4.1.7) with respect to θ,

$$\int \frac{\partial^2 \log f(x|\theta)}{\partial \theta^2} f(x|\theta)dx + \int \frac{\partial \log f(x|\theta)}{\partial \theta}\frac{\partial f(x|\theta)}{\partial \theta}dx = 0$$

$$\int \frac{\partial^2 \log f(x|\theta)}{\partial \theta^2} f(x|\theta)dx + \int \left(\frac{\partial \log f(x|\theta)}{\partial \theta}\right)^2 f(x|\theta)dx = 0$$

Therefore

$$\text{E}\left\{\frac{\partial^2 \log f(x|\theta)}{\partial \theta^2} + \left(\frac{\partial \log f(x|\theta)}{\partial \theta}\right)^2\right\} = 0$$

Hence,

$$\text{E}\left(\frac{\partial \log f(x|\theta)}{\partial \theta}\right)^2 = -\text{E}\left[\frac{\partial^2 \log f(x|\theta)}{\partial \theta^2}\right] \tag{4.1.10}$$

Hence, from (4.1.7),

$$\text{V}\left(\frac{\partial \log f(x|\theta)}{\partial \theta}\right) = \text{E}\left(\frac{\partial \log f(x|\theta)}{\partial \theta}\right)^2$$

$$= -\text{E}\left[\frac{\partial^2 \log f(x|\theta)}{\partial \theta^2}\right] \tag{4.1.11}$$

$$\text{E}[T(X)] = \int T(x)f(x|\theta)dx = g(\theta) \tag{4.1.12}$$

Differentiating (4.1.12) with respect to θ,

$$\int T(x)\frac{\partial f(x|\theta)}{\partial \theta}dx = g'(\theta)$$

$$\int T(x)\frac{\partial \log f(x|\theta)}{\partial \theta} f(x|\theta)dx = g'(\theta)$$

$$\Rightarrow \int T(x) \frac{\partial \log f(x|\theta)}{\partial \theta} f(x|\theta) dx = g'(\theta) \qquad (4.1.13)$$

Using (4.1.1), (4.1.8), (4.1.9), (4.1.11) and (4.1.13)

$$V(T(X)) \geq \frac{[g'(\theta)]^2}{E[\{\frac{\partial \log f(x|\theta)}{\partial \theta}\}^2]}$$

Remark:

1. If $X_1, X_2, \ldots, X_n$ are iid rvs then

$$E\left[\left\{\frac{\partial \log f(X|\theta)}{\partial \theta}\right\}^2\right] = nE\left\{\frac{\partial \log f(x_i|\theta)}{\partial \theta}\right\}^2, \quad i = 1, 2, \ldots, n$$

where $X = (X_1, X_2, \ldots, X_n)$

$$f(X|\theta) = f(x_1, x_2, \ldots, x_n|\theta) = \prod_{i=1}^{n} f(x_i|\theta)$$

$$\log f(X|\theta) = \sum_{i=1}^{n} \log f(x_i|\theta)$$

$$\frac{\partial \log f(X|\theta)}{\partial \theta} = \sum_{i=1}^{n} \frac{\partial \log f(x_i|\theta)}{\partial \theta}$$

$$\left(\frac{\partial \log f(X|\theta)}{\partial \theta}\right)^2 = \left(\sum_{i=1}^{n} \frac{\partial \log f(x_i|\theta)}{\partial \theta}\right)^2$$

$$= \sum_{i=1}^{n} \left(\frac{\partial \log f(x_i|\theta)}{\partial \theta}\right)^2 + \sum_{i \neq j} \frac{\partial \log f(x_i|\theta)}{\partial \theta} \frac{\partial \log f(x_j|\theta)}{\partial \theta}$$

From (4.1.7), $E\left[\frac{\partial \log f(x_i|\theta)}{\partial \theta}\right] = 0$

$$E\left(\frac{\partial \log f(X|\theta)}{\partial \theta}\right)^2 = nE\left(\frac{\partial \log f(x_i|\theta)}{\partial \theta}\right)^2 \qquad (4.1.14)$$

since $X_1, X_2, \ldots, X_n$ are iid rvs

$E \left(\frac{\partial \log f(X|\theta)}{\partial \theta} \right)^2$ is called Fisher information

3. We will consider (4.1.5) when equality holds in (4.1.4)
If there exist $a(\theta)$ such that

$$\frac{\partial \log f(X|\theta)}{\partial \theta} = a(\theta)[T(X) - ET(X)] \tag{4.1.15}$$

$$V\left[\frac{\partial \log f(X|\theta)}{\partial \theta}\right] = a^2(\theta)V[T(X)]$$

Hence

$$a^2(\theta) = \frac{V[\frac{\partial \log f(X|\theta)}{\partial \theta}]}{V[T(X)]} \tag{4.1.16}$$

From (4.1.4),

$$V[T(X)]E\left[\left\{\frac{\partial \log f(X|\theta)}{\partial \theta}\right\}^2\right] = [g'(\theta)]^2$$

$$\Rightarrow V[T(X)]V\left[\frac{\partial \log f(X|\theta)}{\partial \theta}\right] = [g'(\theta)]^2 \tag{4.1.17}$$

From (4.1.16) and (4.1.17),

$$a(\theta) = \frac{V[\frac{\partial \log f(X|\theta)}{\partial \theta}]}{g'(\theta)} \tag{4.1.18}$$

From (4.1.15), since $ET(X) = g(\theta)$

$$\frac{\partial \log f(X|\theta)}{\partial \theta} = \frac{V[\frac{\partial \log f(X|\theta)}{\partial \theta}]}{g'(\theta)}[T(X) - g(\theta)] \tag{4.1.19}$$

$$T(X) = g(\theta) + \frac{g'(\theta)}{V[\frac{\partial \log f(X|\theta)}{\partial \theta}]}\frac{\partial \log f(X|\theta)}{\partial \theta}$$

4.

$$E\left(\frac{\partial \log f(X|\theta)}{\partial \theta}\right)^2 = -E\left(\frac{\partial^2 \log f(X|\theta)}{\partial \theta^2}\right)$$

Example 4.1.1 Let $X_1, X_2, \ldots, X_n$ be iid rvs with $N(\theta, 1)$. We will obtain the lower bound for the variance of an unbiased estimator of θ.

$$\log f(x_1, x_2, \ldots, x_n | \theta) = \text{const} - \frac{1}{2} \sum_{i=1}^{n} (x_i - \theta)^2$$

$$\frac{\partial \log f(x_1, x_2, \ldots, x_n | \theta)}{\partial \theta} = \sum_{i=1}^{n} (x_i - \theta) \qquad (4.1.20)$$

$$\frac{\partial^2 \log f(x_1, x_2, \ldots, x_n | \theta)}{\partial \theta^2} = -n$$

$$V \left[\frac{\partial \log f(x_1, x_2, \ldots, x_n | \theta)}{\partial \theta} \right] = n \quad (\text{see } 4.1.11)$$

Hence, $a(\theta) = n$ (Since $g'(\theta) = 1$)
    From (4.1.15) and (4.1.20),

$$\frac{\partial \log f(x_1, x_2, \ldots, x_n | \theta)}{\partial \theta} = -n(\bar{x} - \theta) \qquad (4.1.21)$$

Since $T(X) = \bar{X}$ and $V(\bar{X}) = \frac{1}{n}$
    From (4.1.4),

$$V(\bar{X}) = \frac{1}{n}$$

Moreover,

$$T(X) = \theta + \frac{(1)(n)(\bar{x} - \theta)}{\theta} = \bar{x}$$

It attains lower bound and it is the unbiased estimator of θ. Therefore, $\bar{X}$ is UMVUE for θ.

Further, we will obtain the lower bound for the unbiased estimator of θ^2.
    Let $T(X) = \bar{X}^2$

$$V(\bar{X}) = E\bar{X}^2 - \theta^2 = \frac{1}{n}$$

Hence,

$$E\left(\bar{X}^2 - \frac{1}{n} \right) = \theta^2 \quad \text{and} \quad V\left(\bar{X}^2 - \frac{1}{n} \right) = V(\bar{X}^2) \qquad (4.1.22)$$

$$V(\bar{X}^2) = E(\bar{X}^4 - E\bar{X})^2 \tag{4.1.23}$$

Now, $E(\bar{X} - \theta)^3 = 0$

$$E(\bar{X}^3) - 3(E[\bar{X}^2])\theta + 3\theta^2 E(\bar{X}) - \theta^3 = 0$$

$$E(\bar{X}^3) = 3\left(\theta^2 + \frac{1}{n}\right)\theta - 3\theta^2(\theta) + \theta^3$$

$$= \frac{3\theta}{n} + \theta^3 \tag{4.1.24}$$

Since in standard normal distribution,

$$\beta_2 = \frac{\mu_4}{\mu_2^2} = 3 \Rightarrow \mu_4 = 3\mu_2^2 = \frac{3}{n^2}$$

Hence,

$$E(\bar{X} - \theta)^4 = \frac{3}{n^2} \tag{4.1.25}$$

$$E(\bar{X}^4) - 4E(\bar{X}^3)\theta + 6E(\bar{X}^2)\theta^2 - 4E(\bar{X})\theta^3 + \theta^4 = \frac{3}{n^2}$$

$$E(\bar{X}^4) = \frac{3}{n^2} + 4\theta\left[\frac{3\theta}{n} + \theta^3\right] - 6\theta^2\left[\theta^2 + \frac{1}{n}\right] + 4\theta^3(\theta) - \theta^4$$

$$= \frac{3}{n^2} + \frac{12\theta^2}{n} + 4\theta^4 - 6\theta^4 - \frac{6\theta^2}{n} + 4\theta^4 - \theta^4$$

$$= \frac{3}{n^2} + \theta^4 + \frac{6\theta^2}{n} \tag{4.1.26}$$

Therefore,

$$V(\bar{X}^2) = \frac{3}{n^2} + \theta^4 + \frac{6\theta^2}{n} - \left(\theta^2 + \frac{1}{n}\right)^2$$

$$= \frac{3}{n^2} + \theta^4 + \frac{6\theta^2}{n} - \theta^4 - \frac{2\theta^2}{n} - \frac{1}{n^2}$$

$$= \frac{2}{n^2} + \frac{4\theta^2}{n} \tag{4.1.27}$$

Next, by using CR lower bound, i.e. (4.1.4),
$T(X) = \bar{X}^2, g(\theta) = \theta^2, g'(\theta) = 2\theta$

$$E\left\{\frac{\partial \log f(X|\theta)}{\partial \theta}\right\}^2 = V\left[\frac{\partial \log f(X|\theta)}{\partial \theta}\right] = n \tag{4.1.28}$$

$$V[\bar{X}^2] \geq \frac{4\theta^2}{n} \tag{4.1.29}$$

From (4.1.27) and (4.1.29)

$$\frac{2}{n^2} + \frac{4\theta^2}{n} \geq \frac{4\theta^2}{n}$$

In this, $\left(\bar{X}^2 - \frac{1}{n}\right)$ does not attain CR bound. Moreover, we can say that the estimator with variance $\frac{4\theta^2}{n}$ may not exist.

Example 4.1.2 Let $X_1, X_2, \ldots, X_n$ are iid rvs with $\cup(0, \theta)$.
   Can you obtain CR lower bound for any unbiased estimator of θ?

It does not satisfy the condition as mentioned in (4.1.2).

$$\frac{\partial}{\partial \theta} \int_0^\theta T(x) f(x|\theta) dx = \frac{\partial}{\partial \theta} \int_0^\theta \frac{T(x)}{\theta} dx$$

$$= \frac{T(\theta)}{\theta} + \int_0^\theta T(x) \frac{\partial}{\partial \theta}\left(\frac{1}{\theta}\right)$$

$$\neq \int_0^\theta T(x) \frac{\partial f(x|\theta)}{\partial \theta} dx$$

In this case, we cannot apply CR lower bound.

Example 4.1.3 Let $X_1, X_2, \ldots, X_n$ be iid rvs with $P(\lambda)$.

$$f(x|\lambda) = \frac{e^{-\lambda} \lambda^x}{x!}; \quad x = 0, 1, 2, \ldots, \lambda > 0.$$

In this case, we will find the CR lower bound for $e^{-\lambda}$

Define

$$\epsilon(x) = \begin{cases} 1 \ ; \ x = 0, \\ 0 \ ; \ \text{otherwise} \end{cases}$$

$$E\epsilon(x) = P(X = 0) = e^{-\lambda}$$

$\epsilon(x)$ is an unbiased estimator of $e^{-\lambda}$.

$$V[\epsilon(x)] = e^{-\lambda} - e^{-2\lambda} = e^{-\lambda}(1 - e^{-\lambda})$$

Now, let us compute CR lower bound for $e^{-\lambda}$

Let $\theta = e^{-\lambda} \Rightarrow \lambda = -\log\theta = \log\frac{1}{\theta}$

$$f(x|\theta) = \frac{\theta(\log\frac{1}{\theta})^x}{x!}$$

$$\log f(x|\theta) = x\log\left(\log\frac{1}{\theta}\right) + \log\theta - \log x!$$

$$\frac{\partial \log f(x|\theta)}{\partial \theta} = \frac{-x}{\theta\log\frac{1}{\theta}} + \frac{1}{\theta}$$

$$= \frac{x}{\theta\log\theta} + \frac{1}{\theta}$$

$$E\left(\frac{\partial \log f(x|\theta)}{\partial \theta}\right)^2 = E\left[\frac{x^2}{\theta^2(\log\theta)^2} + \frac{1}{\theta^2} + \frac{2x}{\theta^2(\log\theta)^2}\right]$$

$$= \frac{\lambda^2 + \lambda}{\theta^2(\log\theta)^2} + \frac{1}{\theta^2} + \frac{2\lambda}{\theta^2(\log\theta)^2}$$

Substitute $\theta = e^{-\lambda}$

$$= \frac{\lambda(\lambda + 1)}{e^{-2\lambda}\lambda^2} + \frac{1}{e^{-2\lambda}} - \frac{2\lambda}{\lambda e^{-2\lambda}}$$

$$= e^{2\lambda}\left[\frac{(\lambda + 1)}{\lambda} + 1 - 2\right]$$

$$= e^{2\lambda}\left[1 + \frac{1}{\lambda} + 1 - 2\right]$$

$$= \frac{e^{2\lambda}}{\lambda} \tag{4.1.30}$$

$$E\left(\frac{\partial \log f(x|\theta)}{\partial \theta}\right)^2 = \frac{ne^{2\lambda}}{\lambda} \quad \text{(see Remark 1)}$$

From (4.1.4), CR lower bound for any unbiased estimator of $e^{-\lambda}$ is $\frac{\lambda e^{-2\lambda}}{n}$

$$V[T(X)] \geq \frac{\lambda e^{-2\lambda}}{n} \tag{4.1.31}$$

But actual variance of $\epsilon(x)$, which is unbiased estimator of $e^{-\lambda}$ is $e^{-\lambda}(1 - e^{-\lambda})$

$$e^{-\lambda}(1 - e^{-\lambda}) > \lambda e^{-2\lambda} \tag{4.1.32}$$

Hence, variance of $\epsilon(x)$ is greater than CR lower bound. Here also, we can say that the estimator with variance $\dfrac{\lambda e^{-2\lambda}}{n}$ may not exist.

Note: Without reparametrization, one can get CR lower bound.
Since

$$E\left(\frac{\partial \log f}{\partial \lambda}\right)^2 = \frac{1}{\lambda}$$

$$V[T(X)] \geq \frac{\lambda e^{-2\lambda}}{n}$$

Example 4.1.4 Let $X_1, X_2, \ldots, X_n$ be iid rvs with exponential distribution having mean θ. What is the CR lower bound for the variance of an unbiased estimator of $g(\theta) = \exp\left(-\frac{1}{\theta}\right)$? Does the variance of the UMVUE of $g(\theta)$ attain this lower bound?

Let $X = (X_1, X_2, \ldots, X_n)$

$$f(X|\theta) = (\theta)^{-n} \exp\left(-\frac{t}{\theta}\right), \quad \text{where} \quad t = \sum_{i=1}^{n} x_i$$

$$\log L = -n \log \theta - \frac{t}{\theta}$$

we want to obtain CR lower bound for an unbiased estimator of $e^{-\frac{1}{\theta}}$
Let $\lambda = e^{-\frac{1}{\theta}} \Rightarrow -\frac{1}{\theta} = \log \lambda$

$$\log f(X|\lambda) = n[\log(-\log \lambda)] + t \log \lambda$$

$$\frac{\partial \log f(X|\lambda)}{\partial \lambda} = \frac{1}{\lambda}\left[t + \frac{n}{\log \lambda} \right]$$

$$E\left(\frac{\partial \log f}{\partial \lambda}\right)^2 = \frac{1}{\lambda^2}\left[E\left\{ t^2 + \frac{2nt}{\log \lambda} + \frac{n^2}{(\log \lambda)^2} \right\} \right]$$

Now $E(T^2) = n(n+1)\theta^2$ and $E(T) = n\theta$

$$= e^{\frac{2}{\theta}}[n(n+1)\theta^2 - 2n^2\theta^2 + n^2\theta^2]$$

$$= e^{\frac{2}{\theta}}(n\theta^2)$$

If $T(X)$ is an unbiased estimator of $e^{-\frac{1}{\theta}}$

Then,

$$V(T(X)) \geq \frac{e^{-\frac{2}{\theta}}}{n\theta^2} \tag{4.1.33}$$

Note: Without reparametrization,

$$E\left(\frac{\partial \log f}{\partial \theta}\right) = -\frac{n}{\theta} + \frac{t}{\theta^2}$$

$$E\left(\frac{\partial \log f}{\partial \theta}\right)^2 = \frac{n}{\theta^2},$$

Since $g'(\theta) = \frac{1}{\theta^2}e^{-\frac{1}{\theta}}$.

$$V[T(X)] \geq \frac{e^{-\frac{2}{\theta}}}{\theta^4} \times \frac{\theta^2}{n} = \frac{e^{-\frac{2}{\theta}}}{n\theta^2}$$

The reader can see from (4.1.19) that it does not attain CR lower bound.

Example 4.1.5 Let $X_1, X_2, \ldots, X_n$ are iid rvs with B(n, p). We will show here that variance of $\frac{\bar{X}}{n}$ attains the CR lower bound.

$$f(X|p) = \prod_{i=1}^{m}\binom{n}{x_i}p^{x_i}q^{n-x_i}; \quad q = 1 - p$$

$$= \prod_{i=1}^{m}\binom{n}{x_i}p^t q^{mn-t}; \quad \text{where} \quad T = \sum_{i=1}^{n}x_i$$

$$\log f = \text{const} + t \log p + (mn - t) \log q$$

$$\frac{\partial \log f}{\partial p} = \frac{t}{p} - \frac{mn - t}{q}$$

$$\frac{\partial^2 \log f}{\partial p^2} = -\frac{t}{p^2} - \frac{mn - t}{q^2}$$

$$\mathrm{E}\left[-\frac{\partial^2 \log f}{\partial p^2}\right] = \frac{mnp}{p^2} + \frac{mn - mnp}{q^2}$$

$$= \frac{mn}{p} + \frac{mn}{q} = \frac{mn}{pq}$$

Therefore,

$$\mathrm{V}\left[\frac{\partial \log f}{\partial p}\right] = \frac{pq}{mn}$$

From (4.1.15) and (4.1.18), $g(p) = p$

$$\frac{\partial \log f}{\partial p} = \frac{pq}{mn}\left[\frac{\bar{X}}{n} - p\right]$$

By using (4.1.4),

$$\mathrm{V}\left[\frac{\bar{X}}{n}\right] = \frac{[g'(p)]^2}{\mathrm{E}\left(\frac{\partial \log L}{\delta p}\right)^2} = \frac{1}{\frac{mn}{pq}} = \frac{pq}{mn}$$

Hence, $\frac{\bar{X}}{n}$ attains CR lower bound.

Theorem 4.1.3 *(i) A necessary condition for $V = \frac{\partial \log f(x|\theta)}{\partial \theta}$ to give an inequality of CR lower bound is that V depends on X only through the minimal sufficient statistics.*

(ii) The above condition is also sufficient when the minimal sufficient statistics is complete.

A detailed proof of this theorem is given by Blyth and Roberts (1972).

Here we have given a counter-example where V depends on X only through the minimal sufficient statistics but it does not give an inequality of CR as the minimal sufficient statistics is not complete.

Example 4.1.6 Let $X_1, X_2, \ldots, X_n$ be iid rvs with $N(\theta, \theta^2)$.

$$f(x|\theta) = (2\pi)^{-\frac{n}{2}} \theta^{-n} \exp\left[-\frac{1}{2\theta^2} \sum_{i=1}^{n}(x_i - \theta)^2\right]$$

Then $\left(\sum_{i=1}^{n} X_i, \sum_{i=1}^{n} X_i^2\right)$ is minimal sufficient statistics. Since $(\bar{X}, S^2)$ is one-to-one function of $\left(\sum_{i=1}^{n} X_i, \sum_{i=1}^{n} X_i^2\right)$, it is also minimal sufficient.

Now, $\bar{X}$ has $N\left(\theta, \frac{\theta^2}{n}\right)$. Also $\frac{(n-1)s^2}{\theta^2} = \frac{1}{\theta^2}\sum_{i=1}^{n}(X_i - \bar{X})^2$ has χ^2 distribution with $(n-1)$ degrees of freedom.

Hence $E(S^2) = \theta^2$. Similarly, $E(\bar{X}^2) = \frac{n+1}{n}\theta^2$.

Then $E\left(\frac{n\bar{X}^2}{n+1} - S^2\right) = 0$, but $\frac{n\bar{X}^2}{n+1} - S^2 \neq 0$ for some X, which implies that $(\bar{X}, S^2)$ is not complete.

Let $T_1 = \frac{n\bar{X}^2}{n+1}$, $T_2(X) = S^2$ and $V(X, \theta) = \frac{T_1(X)+T_2(X)}{2\theta^4}$

Then $E[T_1(X)] = E[T_2(X)] = \theta^2$

$$Cov[T_1, T_1 + T_2] = E\{T_1(T_1 + T_2)\} - E(T_1)E(T_1 + T_2)$$

$$= E(T_1^2) + E(T_1 T_2) - [ET_1]^2 - ET_1 ET_2$$

$$= V(T_1) + Cov(T_1, T_2) = V(T_1)$$

Because $(\bar{X}, S^2)$ is independent, then $Cov(T_1, T_2) = 0$

$$V(T_1) = \frac{2\theta^4}{n+1}$$

Now,

$$Cov[T_1, V] = \frac{1}{2\theta^4}Cov[T_1, T_1 + T_2]$$

$$= \frac{1}{n+1}$$

Similarly,

$$Cov[T_2, T_1 + T_2] = V(T_2) = \frac{2\theta^4}{n-1}$$

$$\Rightarrow \text{Cov}[T_2, V] = \frac{1}{n-1}$$

Hence, we do not get CR inequality.

Joshi (1976) shows that CR lower bound may be attained in cases where the underlying densities do not belong to one parameter exponential family.

Example 4.1.7 Select any number $\lambda \epsilon (0, 1)$ and for this determine β from the equation

$$\int_{\lambda}^{\beta} (t^2 - 1) \exp\left[-\frac{t^2}{2}\right] dt = 0 \tag{4.1.34}$$

Define

$$A(t) = \begin{cases} 2 ; & \lambda \le |t| \le \beta \\ 1 ; & \text{otherwise} \end{cases}$$

Let

$$f(x|\theta) = CA(|x - \theta|) \exp\left\{-\frac{(x - \theta)^2}{2}\right\}, \quad -\infty < x < \infty \tag{4.1.35}$$

where

$$C = \left[(2\pi)^{\frac{1}{2}} \{1 + 2\phi(\beta) - 2\phi(\lambda)\}\right]^{-1} \tag{4.1.36}$$

It is clear that $f(x|\theta)$ is not a member of the exponential class of densities.

Since θ is a location parameter, EX exists and the density is symmetric about θ, $EX = \theta$. Thus, $T(X) = X$ is an unbiased estimator of θ, $\forall \theta \epsilon \Theta$.

Therefore,

$$\text{Var}[T(X)] = E(X - \theta)^2$$

$$= C \int_{-\infty}^{\infty} (x - \theta)^2 A(|x - \theta|) \exp\left\{-\frac{(x - \theta)^2}{2}\right\} dx \tag{4.1.37}$$

Let $t = x - \theta$

$$C \int_{-\infty}^{\infty} t^2 A(|t|) u^{-\frac{t^2}{2}} dt$$

$$= 2C \int_{0}^{\infty} t^2 A(t) e^{-\frac{t^2}{2}} dt$$

$$= C \left[\int_{-\infty}^{\infty} t^2 e^{-\frac{t^2}{2}} dt + 2 \int_{\lambda}^{\beta} t^2 e^{-\frac{t^2}{2}} dt \right]$$

$$= C \left[\sqrt{2\pi} + 2 \int_{\lambda}^{\beta} t^2 e^{-\frac{t^2}{2}} dt \right] \tag{4.1.38}$$

$$= C[C^{-1}] = 1$$

Thus, X is an unbiased estimator of θ with unit variance.

Now, we obtain CR lower bound for the variance of an unbiased estimator of θ. From (4.1.37),

$$\log f(x|\theta) = \log C + \log A(|x - \theta|) - \frac{(x - \theta)^2}{2}$$

Therefore,

$$\frac{\partial \log f(x|\theta)}{\partial \theta} = (x - \theta),$$

because C is independent of θ and $\frac{dA(|x-\theta|)}{d\theta} = 0$ for all x and for all θ except at $x - \beta, x - \lambda, x + \lambda$ and $x + \beta$, where $A|x - \theta|$ is not differentiable with respect to θ.

$$E \left[\frac{\partial \log f(x|\theta)}{\partial \theta} \right] = E(x - \theta) = 0$$

and

$$E \left(\frac{\partial \log f(x|\theta)}{\partial \theta} \right)^2 = E(x - \theta)^2 = 1$$

Hence, by (4.1.19), $V(T(X)) = A(\theta)$ which is the CR lower bound.

This shows that a density for which the CR lower bound $A(\theta)$ is attained is not necessarily exponential.

Hence, in the proofs of usually stated result the CR lower bound is attained if and only if the underlying density is a member of exponential family, the fact that the linear relation between $T(X)$ and $\frac{\partial \log f(x|\theta)}{\partial \theta}$ (see 4.1.19) may fail to hold on a null set which may depend on θ is ignored.

Therefore, Wijsman (1973) stated the necessary and sufficient condition for the attainment of CR lower bound and gave a rigorous proof of the result. We will only state his theorem without proof. First, we state here the regularity conditions of Wijsman. Suppose that

  (i) Θ is an open interval

 (ii) $f(x|\theta) > 0$ for every $\theta \epsilon \Theta, x \epsilon R$ and $f(x|\theta)$ is continuously differentiable function of θ for every $x \epsilon R$

(iii) $0 < \mathrm{Var}(\frac{\partial[\log f(x|\theta)}{\partial \theta}] < \infty \; \forall \, \theta \epsilon \Theta$

(iv) $\int_{-\infty}^{\infty} f(x|\theta)dx$ is differentiable under the integral sign with respect to θ

 (v) $\int_{-\infty}^{\infty} T(X)f(x|\theta)dx$ is finite and can be differentiated under the integral sign with respect to θ.

(vi) There is a linear relation between $T(X)$ and $\frac{\partial \log f(x|\theta)}{\partial \theta}$, see (4.1.19)

Theorem 4.1.4 *Let g be a real valued function on Θ, not identically constant; let $T(X)$ be an unbiased estimator of $g(\theta)$ and the above regularity conditions from (i) to (vi) are satisfied. Then the inequality in (4.1.4) is an equality for all $\theta \epsilon \Theta$ if and only if $\forall x$ and for every $\theta \epsilon \Theta$,*

$$f(x|\theta) = \exp[U(\theta)T(X) - V(\theta) + W(x)],$$

as given in Sect. (1.6).

Remark: From the Theorem 1.6.1, $T(X)$ is sufficient and complete and $T(X)$ is the unbiased estimator of $g(\theta)$. Then $T(X)$ is UMVUE of $g(\theta)$.

In the Theorem 2.2.3, we have shown that a necessary and sufficient condition for an estimator to be UMVUE of its expectation is that, it must have zero covariance with every finite variance, unbiased estimator of zero.

We now prove an interesting result similar to Theorem 2.2.3, which is due to Blyth (1974), and provides a necessary and sufficient condition for $\frac{\partial \log f(x|\theta)}{\partial \theta}$ to give an inequality of CR type.

Theorem 4.1.5 *The variance $V = \frac{\partial \log(f(x|\theta)}{\partial \theta}$, with $0 < \mathrm{Var}(V) < \infty$, gives an inequality of Cramer–Rao type if and only if V has zero covariance with every finite variance unbiased estimator of zero.*

Proof Necessity: Assume that V gives an inequality of Cramer–Rao type.

If T_1 and T_2 are any two unbiased estimators of $g(\theta)$ then

$$\mathrm{Cov}(T_1, V) = \mathrm{Cov}(T_2, V)$$

Now,

$$\mathrm{E}T(X) = \mathrm{E}\{T(X) + T_0(X)\},$$

$T_o(X)$ being any unbiased estimator of zero with finite variance. Therefore,

$$Cov(T, V) = Cov(T + T_0, V)$$
$$= Cov(T, V) + Cov(T_0, V)$$

which implies that $Cov(T_0, V) = 0$

Sufficiency:Here, we have $Cov(T_0, V) = 0$. Let $T_1(X)$ and $T_2(X)$ be two unbiased estimators with finite variance.

Then $E(T_1 - T_2) = 0$, which implies $Cov(T_1 - T_2, V) = 0 \Rightarrow Cov(T_1, V) - Cov(T_2, V) = 0$ which means $Cov(T_1, V) = Cov(T_2, V)$. Therefore, V gives an equality of Cramer–Rao type.

Theorem 4.1.6 *A necessary and sufficient condition for the existence of an achievable Cramer–Rao type bound (4.1.4) for the variance of an estimator $T(X)$ having a specified expectation $g(\theta)$ is that $g(\theta)$ possess UMVUE with positive variance.*

Proof Let the variance of an unbiased estimator $T = T^*$ achieve equality in (4.1.4) if V of that inequality is almost surely linearly related to T^*. From (4.1.15) and (4.1.19),

$$V = \frac{Var(V)}{g'(\theta)} T(X) - \frac{Var(V)g(\theta)}{g'(\theta)}$$

$$= d(\theta)T^*(X) + c(\theta),$$

where $c(\theta)$ and $d(\theta)$ being independent of X. The bound given by (4.1.4) is invariant under linear transformations. Therefore, the bound given by $d(\theta)T^*(X) + c(\theta)$ will be the same as that given by $T^*(X)$. Therefore, to achieve equality in (4.1.4), we can write $V = T^*$. Then according to Theorem 4.1.5, we get an CR inequality, we must have zero covariance with every finite variance unbiased estimator of zero. Therefore, T^* must be UMVUE of $g(\theta)$ since this is a necessary and sufficient condition for an estimator to be UMVUE of its expectation (see Theorem 2.2.3).

4.2 Bhattacharya Bound

Theorem 4.2.1 *Let $S_1, S_2, \ldots, S_k$ and $T_1, T_2, \ldots, T_k$ be the two sets of random variables such that with probability one S_i's are linearly independent, i.e.,*

$$P[a_1 S_1 + a_2 S_2 + \cdots + a_k S_k = 0] = 1 \qquad (4.2.1)$$

Further,

$$\Lambda = Covariance\ matrix\ of\ S_i, i = 1, 2, \ldots, k$$

$$M = Covariance\ matrix\ of\ T_j, j = 1, 2, \ldots, k$$

$$N = Covariance\ matrix\ of\ S_i\ and\ T_j, i \neq j$$

Then the matrix $(M - N'\Lambda^{-1}N) \geq 0$ *is positive semi-definite, i.e.,*

$$v'(M - N'\Lambda^{-1}N)v \geq 0$$

This is also known as Hodge's Lemma.

Proof Without loss of generality, assume that $\mathrm{E}S_i = 0, \mathrm{E}T_j = 0$ and if $\mathrm{E}S_i$ and $\mathrm{E}T_j \neq 0$, then let $S_i^* = S_i - \mathrm{E}S_i$ and $T_j^* = T_j - \mathrm{E}T_j$, then $\mathrm{Var}(S_i) = \mathrm{Var}(S_i^*)$ and $\mathrm{Var}(T_j) = \mathrm{Var}(T_j^*)$.

Using Cauchy–Schwarz inequality,

$$Cov^2(u'S, v'T) \leq \mathrm{Var}(u's)\mathrm{Var}(v'T) \tag{4.2.2}$$

$$[u'Cov(S, T)v)]^2 \leq [u'\mathrm{Var}(S)u][v'\mathrm{Var}(T)v]$$

$$(u'Nv)^2 \leq (u'\Lambda u)(v'Mv)$$

Suppose $\Lambda u = Nv \Rightarrow u = \Lambda^{-1}Nv$

$$[(\Lambda^{-1}Nv)'Nv]^2 \leq [(\Lambda^{-1}Nv)'\Lambda\Lambda^{-1}Nv][v'Mv]$$

$$[v'N'\Lambda^{-1}Nv]^2 \leq [v'N'\Lambda^{-1}\Lambda\Lambda^{-1}Nv][v'Mv]$$

$$[v'N'\Lambda^{-1}Nv]^2 \leq [v'N'\Lambda^{-1}Nv][v'Mv]$$

$$[v'N'\Lambda^{-1}Nv] \leq [v'Mv]$$

Therefore,

$$v'(M - N'\Lambda^{-1}N)v \geq 0 \tag{4.2.3}$$

Theorem 4.2.2 *Let* $X_1, X_2, \ldots, X_n$ *be iid rvs with joint pdf* $f(x_1, x_2, \ldots, x_n|\theta)$ *satisfying the regularity conditions.*

Let

$$S_i = \frac{1}{f(x|\theta)} \frac{\partial^i f(x|\theta)}{\partial \theta^i}$$

then $ES_i = 0, i = 1, 2, \ldots, k$

$$\Lambda = Covariance \ matrix \ of \ S_i, i = 1, 2, \ldots, k$$

$$N' = [g^{(1)}(\theta), g^{(2)}(\theta), \ldots, g^{(k)}(\theta)], \quad where \quad g^{(i)}(\theta) = \frac{\partial^i g(\theta)}{\partial \theta^i}; i = 1, 2, \ldots, k$$

$u(x_1, x_2, \ldots, x_n)$ *is an unbiased estimator of* $g(\theta)$, *then*

$$V(u(x)) \geq L_k \quad where \quad L_k = N'\Lambda^{-1}N \tag{4.2.4}$$

(4.2.4) is called Bhattacharya bound.

Proof Let u(x) be an unbiased estimator of $g(\theta)$
    Hence,

$$\int \int \cdots \int u(x) f(x|\theta) dx = g(\theta)$$

$$\frac{\partial}{\partial \theta} \int \int \cdots \int u(x) f(x|\theta) dx = \frac{\partial g(\theta)}{\partial \theta}$$

$$\int \int \cdots \int \frac{u(x)}{f(x|\theta)} \frac{\partial f(x|\theta)}{\partial \theta} f(x|\theta) dx = g^{(1)}(\theta)$$

$$\int \int \cdots \int u(x) S_1 f(x|\theta) dx = g^{(1)}(\theta)$$

$$E[u(x)S_1] = g^{(1)}(\theta)$$

In general,

$$\int \int \cdots \int \frac{u(x)}{f(x|\theta)} \frac{\partial^i f(x|\theta)}{\partial \theta^i} f(x|\theta) dx = g^{(i)}(\theta)$$

$$E[u(x)S_i] = g^{(i)}(\theta)$$

We know that $ES_i = 0 \Rightarrow Cov[u(x), S_i] = g^{(i)}(\theta)$

By using Hodge's Lemma (Theorem 4.2.1),

$$M - N'\Lambda^{-1}N \geq 0$$

In this case $M = Var[u(x)]$

$$Var[u(x)] - N'\Lambda^{-1}N \geq 0$$

Let $L_k = N'\Lambda^{-1}N$ then $Var[U(x)] \geq L_k$
  $Var[u(x)] \geq L_k$

$$\text{Hence } L_k \geq L_{k-1} \geq \cdots \geq L_1. \tag{4.2.5}$$

Note
For $k = 1, \quad Var[u(x)] \geq L_1$

$$L_1 = \frac{[g^{(1)}(\theta)]^2}{Var(S_1)},$$

$$S_1 = \frac{1}{f(x|\theta)} \frac{\partial f(x|\theta)}{\partial \theta} = \frac{\partial \log f(x|\theta)}{\partial \theta}$$

$$Var(S_1) = Var\left[\frac{\partial \log f(x|\theta)}{\partial \theta}\right]$$

CR bound becomes a particular case of Bhattacharya bound for $k = 1$.
  Steps to find Bhattacharya bound:

1. To get N', differentiate the given parametric function $g(\theta)$.

  i.e., $N' = \left[\frac{\partial g(\theta)}{\partial \theta}, \frac{\partial^2 g(\theta)}{\partial \theta^2}, \ldots, \frac{\partial^k g(\theta)}{\partial \theta^k}\right]$
2. Find $S_i = \frac{1}{f(x|\theta)} \frac{\partial^i f(x|\theta)}{\partial \theta^i}$; $i = 1, 2, \ldots, k$ and verify $ES_i = 0$
3. Find $Var(S_i) = E(S_i)^2$ and $Cov(S_i, S_j) = E(S_i S_j)(i \neq j)$. Then obtain the covariance matrix of $(S_i, S_j), (i \neq j)$, i.e., Λ.
4. Calculate $N'\Lambda^{-1}N$.

Example 4.2.1 Let $X_1, X_2, \ldots, X_n$ be iid rvs with $N(\theta, 1)$. We will obtain the Bhattacharya bound for $g(\theta) = \theta^2$

$$N' = [g^{(1)}(\theta), g^{(2)}(\theta), \ldots, g^{(k)}(\theta)] = [2\theta, 2, 0, \ldots, 0] \tag{4.2.6}$$

Here, we can take $N' = [2\theta, 2]$.

$$f(x|\theta) = (2\pi)^{-\frac{n}{2}} \exp\left[-\frac{1}{2}\sum_{i=1}^{n}(x_i - \theta)^2\right]$$

$$\frac{\partial f(x|\theta)}{\partial \theta} = (2\pi)^{-\frac{n}{2}} n(\bar{x} - \theta) \exp\left[-\frac{1}{2}\sum_{i=1}^{n}(x_i - \theta)^2\right]$$

$$S_1 = \frac{1}{f(x|\theta)}\frac{\partial f(x|\theta)}{\partial \theta} = n(\bar{x} - \theta) \qquad (4.2.7)$$

$$\text{Then } ES_1 = 0$$

$$\frac{\partial^2 f(x|\theta)}{\partial \theta^2} = (2\pi)^{-\frac{n}{2}}\left[\left\{-n\exp\left[-\frac{1}{2}\sum_{i=1}^{n}(x_i - \theta)^2\right]\right\} + n^2(\bar{x} - \theta)^2\exp\left[-\frac{1}{2}\sum_{i=1}^{n}(x_i - \theta)^2\right]\right]$$

$$= (2\pi)^{-\frac{n}{2}}\exp\left[-\frac{1}{2}\sum_{i=1}^{n}(x_i - \theta)^2\right][-n + n^2(\bar{x} - \theta)^2]$$

$$S_2 = \frac{1}{f(x|\theta)}\frac{\partial^2 f(x|\theta)}{\partial \theta^2} = -n + n^2(\bar{x} - \theta)^2 \qquad (4.2.8)$$

Similarly, one can find $S_3, S_4, \ldots, S_k$.

$$ES_2 = -n + n = 0$$

$$Var(S_1) = ES_1^2 = n^2\text{E}(\bar{x} - \theta)^2 = n \qquad (4.2.9)$$

$$Var(S_2) = \text{E}[n^2(\bar{x} - \theta)^2 - n]^2$$

$$= n^2\text{E}[n(\bar{x} - \theta)^2 - 1]^2 \qquad (4.2.10)$$

$$= n^2\text{E}[n^2(\bar{x} - \theta)^4 - 2n(\bar{x} - \theta)^2 + 1]$$

Now,

$$\text{E}(\bar{x} - \theta)^4 = \frac{3}{n^2}$$

$$= n^2[n^2\frac{3}{n^2} - 2n\frac{1}{n} + 1]$$

$$= n^2[3 - 2 + 1] = 2n^2$$

$$Cov(S_1, S_2) = ES_1 S_2$$

$$= E[\{n(\bar{x} - \theta)\}\{n^2(\bar{x} - \theta)^2 - n\}]$$

$$= E[n^3(\bar{x} - \theta)^3] - E[n^2(\bar{x} - \theta)] = 0 \qquad (4.2.11)$$

Hence

$$\Lambda = \begin{pmatrix} n & 0 \\ 0 & 2n^2 \end{pmatrix}, \Lambda^{-1} = \begin{pmatrix} \frac{1}{n} & 0 \\ 0 & \frac{1}{2n^2} \end{pmatrix},$$

$$L_2 = N'\Lambda^{-1}N = (2\theta \ 2) \begin{pmatrix} \frac{1}{n} & 0 \\ 0 & \frac{1}{2n^2} \end{pmatrix} \begin{pmatrix} 2\theta \\ 2 \end{pmatrix}$$

$$= \frac{4\theta^2}{n} + \frac{2}{n^2}, \qquad (4.2.12)$$

and $L_1 = \frac{4\theta^2}{n} = $ CR lower bound
   Therefore $L_1 < L_2$.

Example 4.2.2 Let the rv X have a geometric distribution with parameter p. We will find a Bhattacharya bound for $g(p) = p$.

$$P[X = x] = pq^x; x = 0, 1, 2 \ldots, 0 < p < 1, q = 1 - p$$

$$\frac{\partial P(x)}{\partial p} = q^x + xpq^{x-1}(-1) = q^x - xpq^{x-1}$$

$$= \left[\left[\frac{P(x)}{p} - xP(x-1) \right] (-1)^0 \right]$$

$$\frac{\partial^2 P(x)}{\partial p^2} = -2xq^{x-1} + x(x-1)pq^{x-2}$$

$$= \left[\left[\frac{2xP(x-1)}{p} - x(x-1)P(x-2) \right] (-1)^1 \right]$$

$$\frac{\partial^3 P(x)}{\partial \mu^2} = \left[\frac{3x(x-1)P(x-2)}{n} - x(x-1)(x-2)P(x-3) \right](-1)^2$$

$$\frac{\partial^i P(x)}{\partial p^i} = \left[\frac{ix(x-1)\ldots(x-i+2)P(x-i+1)}{p} - x(x-1)(x-2)\ldots(x-i+1)P(x-i) \right](-1)^{i-1}$$

$$S_i = \frac{1}{P(x)}\frac{\partial^i P(x)}{\partial p^i} = \left[\frac{ix(x-1)\cdots(x-i+2)}{pq^{i-1}} - \frac{x(x-1)(x-2)\cdots(x-i+1)}{q^i} \right](-1)^{i-1}$$

$$(4.2.13)$$

We will find L_1 and L_2,

$$\mathrm{E}S_i = \left[\frac{i\mathrm{E}X^{(i-1)}}{pq^{i-1}} - \frac{\mathrm{E}X^{(i)}}{q^i} \right](-1)^{i-1}$$

Note that $\mathrm{E}X^{(i)} = i$th factorial moment.

$$\mathrm{E}X^{(i)} = \left[\frac{i!q^i}{p^i} \right]$$

$$\mathrm{E}S_i = \left[\frac{i(i-1)!q^{i-1}}{(pq^{i-1})(p^{i-1})} - \frac{i!q^i}{q^i p^i} \right](-1)^{i-1}$$

$$= \left[\frac{i!}{p^i} - \frac{i!}{p^i} \right](-1)^{i-1} = 0 \qquad (4.2.14)$$

Now,

$$S_1 = -\frac{1}{q}\left(x - \frac{q}{p} \right) \quad \text{and} \quad S_1^2 = \frac{1}{q^2}\left(x - \frac{q}{p} \right)^2 \qquad (4.2.15)$$

$$\mathrm{E}(S_1^2) = \frac{1}{qp^2}$$

$$S_2 = (-1)\left[\frac{-x^2}{q^2} + \frac{x(q+1)}{pq^2} \right] \qquad (4.2.16)$$

$$S_2^2 = \left[\frac{x^4}{q^4} + \frac{x^2(q+1)^2}{p^2q^4} - \frac{2x^3(q+1)}{pq^4} \right]$$

$$\mathrm{E}(X^2) = \frac{q(1+q)}{p^2}, \quad \mathrm{E}X^3 = \frac{q(1+4q+q^2)}{p^3}, \quad \mathrm{E}X^4 = \frac{q+11q^2+11q^3+q^4}{p^4}$$

$$E(S_2^2) = \frac{4q^2(q+1)}{p^4q^4} = \frac{4(q+1)}{p^4q^2} \tag{4.2.17}$$

$$S_1S_2 = \left[-\frac{x^3}{q^3} - \frac{x^2(q+1)}{pq^3} + \frac{1}{p}\left(-\frac{x^2}{q^2} + \frac{x(q+1)}{pq^2}\right)\right](-1)^1 \tag{4.2.18}$$

$$ES_1S_2 = -\frac{2}{p^3q} \tag{4.2.19}$$

$g(p) = p, N' = (1,0)$

$$\Lambda = \begin{pmatrix} ES_1^2 & ES_1S_2 \\ ES_1S_2 & ES_2^2 \end{pmatrix} = \begin{pmatrix} \frac{1}{p^2q} & -\frac{2}{p^3q} \\ -\frac{2}{p^3q} & \frac{4(q+1)}{p^4q^2} \end{pmatrix} \tag{4.2.20}$$

$L_2 = N'\Lambda^{-1}N$, where

$$\Lambda^{-1} = \frac{p^6q^3}{4}\begin{pmatrix} \frac{4(1+q)}{p^4q^2} & \frac{2}{p^3q} \\ \frac{2}{p^3q} & \frac{1}{p^2q} \end{pmatrix} \tag{4.2.21}$$

$$N'\Lambda^{-1}N = \begin{pmatrix} 1 & 0 \end{pmatrix} \frac{p^6q^3}{4}\begin{pmatrix} \frac{4(1+q)}{p^4q^2} & \frac{2}{p^3q} \\ \frac{2}{p^3q} & \frac{1}{p^2q} \end{pmatrix}\begin{pmatrix} 1 \\ 0 \end{pmatrix}$$

$$= \frac{4(1+q)}{p^4q^2}\frac{p^6q^3}{4} = p^2q(1+q) \tag{4.2.22}$$

Similarly

$$L_3 = p^2q(1+q+q^2) \tag{4.2.23}$$

In general,

$$L_k = p^2q(1+q+q^2+\cdots+q^k)$$

If $T(X)$ is an unbiased estimator of p, then

$$Var(T(x)) \ge p^2q(1+q+q^2+\cdots+q^k)$$

Remark: If k is tending to infinity, then $V(T(X)) \ge pq$.

Example 4.2.3 Let $X_1, X_2, \ldots, X_n$ are iid rvs with Bernoulli distribution

$$P(X = x) = p^x q^{1-x}; \quad x = 0, 1$$

$$P(X_1 = x_1, X_2 = x_2, \ldots, X_n = x_n) = p^t q^{n-t}; \quad \text{where } t = \sum_{i=1}^{n} x_i$$

$$P(X = x) = \left(\frac{p}{q}\right)^t q^n$$

$$\frac{\partial P(X)}{\partial p} = \left(\frac{p}{q}\right)^t q^n \left[\frac{t}{pq} - \frac{n}{q}\right]$$

$$\frac{\partial^2 P(X)}{\partial p^2} = \left(\frac{p}{q}\right)^t q^n \left[-\frac{t}{qp^2} + \frac{t}{pq^2} - \frac{n}{q^2}\right] + \left(\frac{p}{q}\right)^t q^n \left[\frac{t}{qp} - \frac{n}{q}\right]^2$$

$$S_1 = \frac{1}{P(x)} \frac{\partial P(X)}{\partial p} = \frac{t}{pq} - \frac{n}{q} \tag{4.2.24}$$

$$S_2 = \frac{1}{P(x)} \frac{\partial^2 P(X)}{\partial p^2} = \frac{-t}{qp^2} + \frac{t}{pq^2} - \frac{n}{q^2} + \frac{1}{p^2 q^2}(t - np)^2 \tag{4.2.25}$$

Now

$$ES_1 = \frac{np}{pq} - \frac{n}{q} = 0$$

$$ES_2 = -\frac{np}{qp^2} + \frac{np}{pq^2} - \frac{n}{q^2} + \frac{npq}{p^2 q^2}$$

$$= -\frac{n}{qp} + \frac{n}{q^2} - \frac{n}{q^2} + \frac{n}{pq} = 0$$

Note the result:
$K_i = i$th cumulant

$$K_1 = ET = np, \quad K_2 = npq, \quad K_3 = npq(1 - 2p)$$

$$S_1 S_2 = \left(\frac{t}{pq} - \frac{n}{q}\right)\left[t\left(\frac{-1}{qp^2} + \frac{1}{pq^2}\right) - \frac{n}{q^2} + \frac{1}{p^2 q^2}(t - np)^2\right]$$

$$= \frac{1}{pq}[t - np]\left[t\left(\frac{-1}{qp^2} + \frac{1}{pq^2}\right) - \frac{n}{q^2} + \frac{1}{p^2q^2}(t - np)^2\right] \qquad (4.2.26)$$

Note that

$$E[t(t - np)] = E(t - np + np)(t - np) = E(t - np)^2 + npE(t - np) = npq$$

$$\mu_3 = K_3 = E(t - np)^3 = npq(q - p)$$

$$ES_1S_2 = \frac{Et(t - np)}{pq}\left[\frac{-1}{qp^2} + \frac{1}{pq^2}\right] - \frac{n}{q^2} + \frac{E(t - np)}{pq} + \frac{E(t - np)^3}{p^3q^3}$$

$$= \frac{npq}{pq}\left[\frac{-q + p}{p^2q^2}\right] - 0 + \frac{npq(q - p)}{p^3q^3}$$

$$= -\frac{n(q - p)}{p^2q^2} + \frac{n(q - p)}{p^2q^2}$$

$$= 0$$

$$S_1^2 = \left(\frac{1}{pq}\right)^2(t - np)^2$$

$$ES_1^2 = \left(\frac{1}{pq}\right)^2 E(t - np)^2 = \frac{npq}{(pq)^2} = \frac{n}{pq}$$

Next,

$$S_2 = \left(\frac{t}{pq} - \frac{n}{q}\right)^2 + t\left(\frac{p - q}{p^2q^2}\right) - \frac{n}{q^2}$$

$$= \frac{1}{(pq)^2}(t - np)^2 + t\left(\frac{p - q}{p^2q^2}\right) - \frac{n}{q^2}$$

$$(pq)^4 S_2^2 = [(t - np)^2 + t(p - q) - np^2]^2$$

$$= (t - np)^4 + t^2(p - q)^2 + n^2p^4 + 2(p - q)t(t - np)^2 - 2np^2(p - q)t - 2np^2(t - np)^2$$

$$(pq)^4 ES_2^2 = np^2q[10q + nq + 4np] - np^2(2np^2 + np - 2)$$

$$\mathrm{E}S_2^2 = \frac{n}{l^2 q^3}(10q + nq + 4np) - \frac{n}{l^2 q^4}(2np^2 + np - 2)$$

$g(p) = p, N' = (1, 0)$

$$\Lambda = \begin{pmatrix} \frac{n}{pq} & 0 \\ 0 & b \end{pmatrix}$$

Let $\mathrm{E}S_2^2 = b$

$$\Lambda^{-1} = \begin{pmatrix} \frac{pq}{n} & 0 \\ 0 & \frac{1}{b} \end{pmatrix}$$

$$N\Lambda^{-1}N = (1\,0) \begin{pmatrix} \frac{pq}{n} & 0 \\ 0 & \frac{1}{b} \end{pmatrix} \begin{pmatrix} 1 \\ 0 \end{pmatrix}$$

$$= \left(\frac{pq}{n}\ 0 \right) \begin{pmatrix} 1 \\ 0 \end{pmatrix} = \frac{pq}{n}$$

Example 4.2.4 Let $X_1, X_2, \ldots, X_n$ be iid rvs with exponential distribution having mean σ. We will obtain Bhattacharya bound for the unbiased estimator of σ.

$$f(x) = \frac{1}{\sigma}e^{-\frac{x}{\sigma}}; \quad x > 0, \sigma > 0$$

$$f(x_1, x_2, \ldots, x_n | \sigma) = \frac{1}{\sigma^n}e^{-\frac{t}{\sigma}}; t > 0$$

where $T = \sum_{i=1}^{n} x_i$

$$\frac{\partial f}{\partial \sigma} = \sigma^{-n}e^{-\frac{t}{\sigma}}\left[-\frac{n}{\sigma} + \frac{t}{\sigma^2} \right]$$

$$S_1 = \frac{1}{f}\frac{\partial f}{\partial \sigma} = -\frac{n}{\sigma} + \frac{t}{\sigma^2}$$

$$= \frac{1}{\sigma^2}(t - n\sigma)$$

Note that t has $G(n, \frac{1}{\sigma})$

$$\mathrm{E}(S_1) = 0$$

$$E(S_1^2) = \frac{n\sigma^2}{\sigma^4} = \frac{n}{\sigma^2}$$

$$(S_2) = \frac{1}{f}\frac{\partial^2 f}{\partial\sigma^2} = \frac{(t - n\sigma)^2}{\sigma^4} - \frac{(2t - n\sigma)}{\sigma^3}$$

$$E(S_2) = 0$$

$$S_1 S_2 = \frac{1}{\sigma^6}(t - n\sigma)^3 - \frac{1}{\sigma^5}(t - n\sigma)(2t - n\sigma)$$

$$ES_1 S_2 = \frac{1}{\sigma^6}E(t - n\sigma)^3 - \frac{1}{\sigma^5}E(t - n\sigma)(t + t - n\sigma)$$

$$= \frac{1}{\sigma^6}E(t - n\sigma)^3 - \frac{1}{\sigma^5}\left\{E(t - n\sigma)^2 + Et(t - n\sigma)\right\}$$

$$= \frac{2n\sigma^3}{\sigma^6} - \frac{1}{\sigma^5}\left\{n\sigma^2 + E(t - n\sigma + n\sigma)(t - n\sigma)\right\}$$

$$= \frac{2n\sigma^3}{\sigma^6} - \frac{1}{\sigma^5}\{n\sigma^2 + n\sigma^2\}$$

$$= \frac{2n}{\sigma^3} - \frac{2n}{\sigma^3} = 0$$

$$(S_2^2) = \frac{(t - n\sigma)^4}{\sigma^8} + \frac{(2t - n\sigma)^2}{\sigma^6} - \frac{2}{\sigma^7}(t - n\sigma)^2(2t - n\sigma)$$

$$K_i(t) = (i - 1)!n\sigma^i; \quad i = 1, 2, 3, 4$$

$$\mu_4 = K_4 + 3K_2^2$$

$$E(t - n\sigma)^4 = 6n\sigma^4 + 3n^2\sigma^4 = 3\sigma^4 n(n + 2)$$

$$E(2t - n\sigma)^2 = n\sigma^2(n + 4)$$

$$E(t - n\sigma)^2(2t - n\sigma) = n(n + 4)\sigma^3$$

$$E(S_2^2) = \frac{3n(n+2)\sigma^4}{\sigma^8} + \frac{n(n+4)\sigma^2}{\sigma^6} - \frac{2n(n+4)\sigma^3}{\sigma^7}$$

$$= \frac{3n(n+2)}{\sigma^4} + \frac{n(n+4)}{\sigma^4} - \frac{2n(n+4)}{\sigma^4}$$

$$= \frac{3n(n+2)}{\sigma^4} - \frac{n(n+4)}{\sigma^4}$$

$$= \frac{n}{\sigma^4}[3n+6-n-4] = \frac{(2n+2)n}{\sigma^4} = \frac{2n(n+1)}{\sigma^4}$$

Hence

$$\Lambda = \begin{pmatrix} \frac{n}{\sigma^2} & 0 \\ 0 & \frac{2n(n+1)}{\sigma^4} \end{pmatrix}$$

$$\Lambda^{-1} = \begin{pmatrix} \frac{\sigma^2}{n} & 0 \\ 0 & \frac{\sigma^4}{2n(n+1)} \end{pmatrix}$$

$$N'\Lambda^{-1}N = \begin{pmatrix} 1 & 0 \end{pmatrix} \begin{pmatrix} \frac{\sigma^2}{n} & 0 \\ 0 & \frac{\sigma^4}{2n(n+1)} \end{pmatrix} \begin{pmatrix} 1 \\ 0 \end{pmatrix} = \frac{\sigma^2}{n}$$

$$L_1 = \frac{\sigma^2}{n}, \; L_2 = \frac{\sigma^2}{n} \cdots L_k = \frac{\sigma^2}{n}$$

Bhattacharya bound for the unbiased estimator of σ is $\frac{\sigma^2}{n}$.

4.3 Chapman-Robbins-Kiefer Bound

Theorem 4.3.1 *Let the random vector X have a pdf(pmf) $f(x|\theta)$. Let $T(X)$ be an unbiased estimator $g(\theta)$, where $g(\theta)$ defined on Θ. Further, assume that $ET^2 < \infty$ for all $\theta \epsilon \Theta$. If $\theta \neq \alpha$, then assume that $f(x|\theta)$ and $f(x|\alpha)$ are different. Assume that $S(\theta) = \{f(x|\theta) > 0\}$, $S(\alpha) = \{f(x|\alpha) > 0\}$ and $S(\alpha) \subset S(\theta)$.*
   Then,

$$Var[T(X)] \geq \sup_{S(\alpha) \subset S(\theta), \alpha \neq 0} \frac{[g(\alpha) - g(\theta)]^2}{Var\{\frac{f(x|\alpha)}{f(x|\theta)}\}} \; \forall \, \theta \epsilon \Theta \qquad (4.3.1)$$

Proof Under $f(x|\theta)$ and $f(x|\alpha)$

$$\mathrm{E}T(X) = \int T(X)f(x|\theta)dx = g(\theta)$$

$$\mathrm{E}T(X) = \int T(X)f(x|\alpha)dx = g(\alpha)$$

$$g(\alpha) - g(\theta) = \int T(X)\frac{f(x|\alpha) - f(x|\theta)}{f(x|\theta)}f(x|\theta)dx$$

$$= \int T(X)\left[\frac{f(x|\alpha)}{f(x|\theta)} - 1\right]f(x|\theta)dx$$

$$\mathrm{Cov}\left[T(X), \frac{f(x|\alpha)}{f(x|\theta)} - 1\right] = g(\alpha) - g(\theta)$$

Using Cauchy-Schwarz inequality,

$$\mathrm{Cov}^2\left[T(X), \frac{f(x|\alpha)}{f(x|\theta)} - 1\right] \leq \mathrm{Var}[T(X)]\mathrm{Var}\left[\frac{f(x|\alpha)}{f(x|\theta)} - 1\right]$$

$$= \mathrm{Var}T(X)\mathrm{Var}\left[\frac{f(x|\alpha)}{f(x|\theta)}\right]$$

Therefore,

$$[g(\alpha) - g(\theta)]^2 \leq \mathrm{Var}T(X)\mathrm{Var}\left[\frac{f(x|\alpha)}{f(x|\theta)}\right]$$

$$\mathrm{Var}[T(X)] \geq \frac{[g(\alpha) - g(\theta)]^2}{\mathrm{Var}\{\frac{f(x|\alpha)}{f(x|\theta)}\}} \ \forall \ \theta\epsilon\Theta \tag{4.3.2}$$

Then, (4.3.1) follows immediately.

Chapman and Robbins (1951) had given the same above-mentioned theorem in different form.

Theorem 4.3.2 *Let the random vector* $X = (X_1, X_2, \ldots, X_n)$ *have a pdf(pmf)* $f(x|\theta), \theta\epsilon\Theta$. *Let* $T(X)$ *be an unbiased estimator of* θ. *Suppose,* $\theta + h(h \neq 0)$ *be any two distinct values in* Θ *such that*

$$S(\theta + h) \subset S(\theta)$$

Then,

$$\mathrm{Var}(T) \geq \frac{1}{\inf_{h} \mathrm{E}(J|\theta)},$$

where

$$J = J(\theta, h) = \frac{1}{h^2} \left\{ \left[\frac{f(x|\theta + h)}{f(x|\theta)} \right]^2 - 1 \right\}$$

Proof Note that

$$\int_{S(\theta+h)} f(x|\theta + h)dx = \int_{S(\theta)} f(x|\theta)dx = 1$$

Let $t = t(x)$,

$$\int_{S(\theta)} tf(x|\theta)dx = \theta$$

$$\int_{S(\theta+h)} tf(x|\theta + h)dx = \theta + h$$

Consider

$$\frac{1}{h} \int_{S(\theta+h)} (t - \theta)f(x|\theta + h)dx = 1$$

and

$$\frac{1}{h} \int_{S(\theta+h)} (t - \theta)f(x|\theta)dx = 0$$

Therefore,

$$\int_{S} (t - \theta)\sqrt{f(x|\theta)} \frac{f(x|\theta + h) - f(x|\theta)}{hf(x|\theta)} \sqrt{f(x|\theta)}dx = 1$$

By using Cauchy–Schwarz inequality,

$$\left[\int_S (t-\theta)^2 f(x|\theta)dx\right]\int_S \left[\frac{f(x|\theta+h)-f(x|\theta)}{hf(x|\theta)}\right]^2 f(x|\theta)dx \geq 1$$

$$\Rightarrow \text{Var}(t|\theta)\frac{1}{h^2}\int_S \left[\frac{f(x|\theta+h)-f(x|\theta)}{f(x|\theta)}\right]^2 f(x|\theta)dx \geq 1 \qquad (4.3.3)$$

Since

$$\int \left[\frac{f(x|\theta+h)}{f(x|\theta)}-1\right]^2 f(x|\theta)dx = \int \left\{\left[\frac{f(x|\theta+h)}{f(x|\theta)}\right]^2 - 1\right\} f(x|\theta)dx$$

Let

$$J = J(\theta,h) = \frac{1}{h^2}\left\{\left[\frac{f(x|\theta+h)}{f(x|\theta)}\right]^2 - 1\right\}$$

$$\text{Var}(t|\theta)\text{E}(J|\theta) \geq 1$$

Hence,

$$\text{Var}(t|\theta) \geq \frac{1}{\text{E}(J|\theta)} \qquad (4.3.4)$$

$$\Rightarrow \text{Var}(t|\theta) \geq \frac{1}{\inf_h \text{E}(J|\theta)} \qquad (4.3.5)$$

Example 4.3.1 To find CRK lower bound of an unbiased estimator of the mean of a normal distribution based on a random sample of size n

$$f(x|\theta) = \frac{1}{\sigma\sqrt{2\pi}}\exp\left[-\frac{1}{2\sigma^2}(x_i-\theta)^2\right];$$

$$|J| = \frac{1}{h^2}\left\{\left[\frac{f(x|\theta+h)}{f(x|\theta)}\right]^2 - 1\right\}$$

$$= \frac{1}{n^2} \left\{ \exp\left[-\frac{1}{\sigma^2} \sum_{i=1}^{n} [(x_i - \theta - h)^2 - (x_i - \theta)^2] \right] - 1 \right\}$$

$$= \frac{1}{h^2} \left\{ \exp\left[-\frac{nh^2}{\sigma^2} + \frac{2h}{\sigma^2} \sum_{i=1}^{n} (x_i - \theta) \right] - 1 \right\}$$

Let $k = \frac{h\sqrt{n}}{\sigma}$ then $h = \frac{\sigma k}{\sqrt{n}}$

$$= \frac{n}{\sigma^2 k^2} \left\{ \exp\left[-k^2 + 2ku \right] - 1 \right\}, \quad \text{where} \quad u = \frac{\sum_{i=1}^{n}(x_i - \theta)}{\sigma\sqrt{n}}$$

Now,

$$E J | \theta = \frac{nE[e^{-k^2+2ku} - 1]}{\sigma^2 k^2}$$

Since u is $N(0, 1)$ and $Ee^{tu} = e^{\frac{t^2}{2}}$

$$E[e^{-k^2+2ku} - 1] = e^{-k^2+2k^2} - 1 = e^{k^2} - 1$$

$$E J | \theta = \frac{n(e^{k^2} - 1)}{\sigma^2 k^2}$$

$$\inf E(J | \theta) = \lim_{k \to 0} \frac{n(e^{k^2} - 1)}{\sigma^2 k^2} = \frac{n}{\sigma^2} \quad \text{(use L'Hospital's rule)}$$

Hence, if T is any unbiased estimator of θ, it follows from

$$V(T) \geq \frac{\sigma^2}{n}$$

Since the sample mean $\bar{X}$ is UMVUE of θ with $Var(\bar{x}|\theta) = \frac{\sigma^2}{n}$, it follows that $\bar{X}$ has the minimum variance among the class of unbiased estimators.

Example 4.3.2 Unbiased estimator of the variance when the mean is known. Wlog, we assume that mean is zero,

$$J = \frac{1}{h^2} \left\{ \frac{[(\sigma + h)\sqrt{2\pi}]^{2n} \exp[-\frac{1}{(\sigma+h)^2} \sum x_i^2]}{(\sigma\sqrt{2\pi})^{2n} \exp[-\frac{1}{\sigma^2} \sum x_i^2]} - 1 \right\} \quad (4.3.6)$$

$$= \frac{1}{h^2}\left\{\left(\frac{\sigma}{\sigma+h}\right)^{2n}\exp\left[\sum x_i^2\left\{\frac{1}{\sigma^2}-\frac{1}{(\sigma+h)^2}\right\}\right]-1\right\}$$

$$= \frac{1}{h^2}\left\{\left(\frac{\sigma}{\sigma+h}\right)^{2n}\exp\left[\sum x_i^2\left\{\frac{h^2+2\sigma h}{\sigma^2(\sigma+h)^2}\right\}\right]-1\right\}$$

Let $\frac{h^2+2\sigma h}{(\sigma+h)^2}=s$, then to find $Ee^{s\frac{\sum x_i^2}{\sigma^2}}$

Since $\frac{\sum x_i^2}{\sigma^2}$ is χ_n^2, let $u=\frac{\sum x_i^2}{\sigma^2}$,

$$Ee^{su}=(1-2s)^{-\frac{n}{2}}$$

$$EJ=\frac{1}{h^2}\left[\left(\frac{\sigma}{\sigma+h}\right)^{2n}(1-2s)^{-\frac{n}{2}}-1\right]$$

Let $k=\frac{h}{\sigma}$, $\frac{\sigma}{\sigma+h}=(1+k)^{-1}$,

$$(1-2s)=\frac{1-2k-k^2}{(k+1)^2}$$

$$EJ=\frac{1}{\sigma^2k^2}\left[(1+k)^{-n}(1-2k-k^2)^{-\frac{n}{2}}-1\right]$$

$$\inf EJ=\lim_{k\to0}EJ$$

$$\lim_{k\to0}\frac{(1+k)^{-n}(1-2k-k^2)^{-\frac{n}{2}}-1}{\sigma^2k^2}$$

Use L'Hospital's rule,

$$\lim_{k\to0}\frac{-n(1+k)^{-n-1}(1-2k-k^2)^{-\frac{n}{2}}+n(1+k)^{-n+1}(1-2k-k^2)^{-\frac{n}{2}-1}}{2k\sigma^2}$$

$$\lim_{k\to0}\frac{(-n)[-(n+1)A_k^{-n-2}B_k^{-\frac{n}{2}}+nA_k^{-n}B_k^{-\frac{n}{2}-1}]+n[(-n+1)A_k^{-n}B_k^{-\frac{n}{2}-1}+A_k^{-n}B_k^{-\frac{n}{2}}(n+2)]}{2\sigma^2}$$

where $A_k = (1+k)$ and $B_k = 1 - 2k - k^2$

$$= \frac{(n)[-(n+1)+n] + n[(-n+1)+n+2]}{2\sigma^2} = \frac{(-n)[-1]+3n}{2\sigma^2} = \frac{4n}{2\sigma^2} = \frac{2n}{\sigma^2}$$

Remark Cramer–Rao lower bound yields precisely the same bound.

Example 4.3.3 Unbiased estimation of the standard deviation of a normal distribution with known mean

$$E(J|\sigma) = \frac{(1+k)^{-n}(1-2k-k^2)^{-\frac{n}{2}} - 1}{\sigma^2 k^2}$$

Now, the minimum value of $E(J|\sigma)$ is not approached in the neighborhood of $h = k = 0$

Consider $n = 2$,

For $k = \frac{h}{\sigma}$, $(k+1) > 0$ and $1-2k-k^2 > 0$, we find that $-1 < k < \sqrt{2}-1, k \neq 0$.

Set $p = 1 + k$, then $0 < p < \sqrt{2}$

$$E(J|\sigma) = \frac{p^{-2}[2 - p^2]^{-1} - 1}{\sigma^2(p - 1)^2}$$

$$= \frac{(1 - p^2)^2}{p^2(2 - p^2)\sigma^2(p - 1)^2}$$

$$= \frac{(p + 1)^2}{p^2(2 - p^2)\sigma^2}$$

For $p = 0.729 \Rightarrow E(J|\sigma) = \frac{3.830394}{\sigma^2}$

Hence, CRK lower bound $= (0.2610)\sigma^2$

Consider the UMVUE if σ from (2.2.49) with $\mu = 0$

$$T(X) = \hat{\sigma} = \frac{t^{\frac{1}{2}}\Gamma\left(\frac{n}{2}\right)}{2^{\frac{1}{2}}\Gamma\left(\frac{n}{2}+1\right)},$$

where $T = \sum X_i^2$

Now, we will find the variance of T.

$$T = \hat{\sigma} = \frac{\frac{t^{\frac{1}{2}}}{\sigma}\Gamma\left(\frac{n}{2}\right)}{\Gamma\left(\frac{n}{2}+1\right)}\frac{\sigma}{\sqrt{2}}$$

$$T^2 = \left(\frac{t}{\sigma^2}\right)\left(\frac{\Gamma\left(\frac{n}{2}\right)}{\Gamma\left(\frac{n}{2}+1\right)}\right)^2\frac{\sigma^2}{2}$$

$$ET^2 = \left(\frac{n\sigma^2}{2}\right)\left(\frac{\Gamma(\frac{n}{2})}{\Gamma(\frac{n}{2}+1)}\right)^2$$

$$\mathrm{Var}\,T = \sigma^2\left[\frac{n}{2}\left(\frac{\Gamma(\frac{n}{2})}{\Gamma(\frac{n}{2}+1)}\right)^2 - 1\right]$$

For n = 2 and note $\Gamma(\frac{1}{2}) = \sqrt{\pi}$

$$= \sigma^2\left[\frac{4}{\pi} - 1\right] = (0.2732)\sigma^2$$

which is greater than $0.2610\sigma^2$, the CRK bound.

Example 4.3.4 Unbiased estimation of the mean of an exponential distribution

$$f(x|\sigma) = \frac{1}{\sigma}e^{-\frac{x}{\sigma}}; \quad x > 0, \ \sigma > 0$$

$$J = \frac{1}{h^2}\left[\frac{(\sigma+h)^{-2n}\exp\{-\frac{2t}{\sigma+h}\}}{(\sigma)^{-2n}exp\{-\frac{2t}{\sigma}\}} - 1\right]$$

where $t = \sum_{i=1}^{n} x_i$

$$= \frac{1}{h^2}\left[\left(1+\frac{h}{\sigma}\right)^{-2n}\exp\left\{\frac{2ht}{\sigma(\sigma+h)}\right\} - 1\right]$$

Let $\frac{h}{\sigma} = k$ then $(1+\frac{h}{\sigma})^{-2n} = (1+k)^{-2n}$ and $\frac{2h}{\sigma(\sigma+h)} = \frac{2k}{\sigma(k+1)}$
   Note that $T \sim G(n,\sigma)$

$$J = \frac{1}{\sigma^2 k^2}\left[(1+k)^{-2n}\exp\left\{\frac{2kt}{\sigma(k+1)}\right\} - 1\right]$$

To find $\mathrm{E}\exp\left(\frac{2kt}{\sigma(k+1)}\right)$,
   Let $\frac{2k}{\sigma(k+1)} = s$
   Hence, $\mathrm{E}[\exp(st)] = (1-s\sigma)^{-n}$

$$\mathrm{E}J = \frac{1}{\sigma^2 k^2}\left[(1+k)^{-2n}(1-s\sigma)^{-n} - 1\right]$$

Now, $1 - s\sigma = \dfrac{1-k}{1+\lambda}$

$$EJ = \frac{1}{\sigma^2 k^2}\left[(1+k)^{-2n}\left(\frac{1-k}{1+k}\right)^{-n} - 1\right]$$

$$= \frac{1}{\sigma^2 k^2}\left[(1+k)^{-n}(1-k)^{-n} - 1\right]$$

$$= \frac{1}{\sigma^2 k^2}\left[(1-k^2)^{-n} - 1\right]$$

Next,

$$\inf EJ = \lim_{k\to 0}\frac{1}{\sigma^2 k^2}\left[(1-k^2)^{-n} - 1\right]$$

$$\lim_{k\to 0}\frac{1}{2\sigma^2 k}\left[(-n)(1-k^2)^{-n-1}(-2k)\right]$$

$$\lim_{k\to 0}\frac{1}{\sigma^2}\left[n(1-k^2)^{-n-1}\right] = \frac{n}{\sigma^2}$$

One should note that Crammer-Rao lower bound yields the same bound.

Example 4.3.5 Let $X_1, X_2, \ldots, X_n$ be iid with $\cup(0, \theta)$. We will find CRK lower bound for any unbiased estimate of θ, see Kiefer (1952)

$$J = \frac{1}{h^2}\left\{\left(\frac{\theta}{\theta+h}\right)^{2n} - 1\right\}; \quad -\theta < h < 0$$

$$EJ = \frac{1}{h^2}\left[E\left(\frac{\theta}{\theta+h}\right)^{2n} - 1\right]$$

$$= \frac{1}{h^2}\left[\frac{\theta+h}{\theta}\left(\frac{\theta}{\theta+h}\right)^{2n} - 1\right]$$

$$= \frac{1}{h^2}\left[\left(\frac{\theta}{\theta+h}\right)^{2n-1} - 1\right]$$

Let $\frac{h}{\theta} = k \Rightarrow \frac{\theta}{\theta+h} = (1+k)^{-1}$ then $-1 < k < 0$,

$$EJ = \frac{1}{\theta^2 k^2}[(1+k)^{-2n+1} - 1]$$

For n = 1

$$EJ = -\frac{1}{\theta^2 k(k+1)},$$

Then we have to maximize $-k(k+1)$.

$$\inf EJ = \frac{4}{\theta^2}, \quad \text{for} \quad k = -\frac{1}{2}$$

In this case, $Var(T) \geq \frac{\theta^2}{4}$ and Variance of UMVUE of $T = \frac{\theta^2}{n(n+2)}$ is $\frac{\theta^2}{3}$
n = 2

$$EJ = \frac{1}{\theta^2 k^2} \left[(1+k)^{-3} - 1 \right],$$

$$EJ = \frac{1}{\theta^2 k^2} \frac{[-k^3 - 3k^2 - 3k]}{(1+k)^3},$$

$$\inf_k EJ = \inf_k \frac{1}{\theta^2 k^2} \frac{[-k^3 - 3k^2 - 3k]}{(1+k)^3},$$

$$f(k) = \left(-k - 3 - \frac{3}{k} \right) (1+k)^{-3}$$

$$\frac{df(k)}{dk} = \left(-1 + \frac{3}{k^2} \right) (1+k)^{-3} + \left(-k - 3 - \frac{3}{k} \right) (-3)(1+k)^{-4} = 0$$

This implies

$$\left(-1 + \frac{3}{k^2} \right) (1+k) + \left(k + 3 + \frac{3}{k} \right) (3) = 0$$

$$(-k^2 + 3)(1+k) + (3k^3 + 9k^2 + 9k) = 0$$

It implies that we have to solve the equation $2k^3 + 8k^2 + 12k + 3 = 0$
Roots of the equations are $(-0.308586, -1.8557 \pm 1.20592i)$.
Hence, $k = -0.308586$

$$\inf_k EJ = \frac{21.2698}{\theta^2} \Rightarrow Var(T) \geq \frac{\theta^2}{21.2698}$$

Variance of UMVUE of T for $n = 2 = \frac{\theta^2}{8}$
It does not attend the lower bound.

Remark: In $\cup(0, \theta)$, X is sufficient and complete statistics. 2X is unbiased for θ. So that $T(X) = 2X$ is UMVUE. In this case, CRK attains the lower bound.

Example 4.3.6 Let $X_1, X_2, \ldots, X_m$ are iid with B(n,p). We will find CRK lower bound for any unbiased estimator of p.

$$J = \frac{1}{h^2}\left[\left(\frac{p+h}{p}\right)^{2T}\left(\frac{1-p-h}{1-p}\right)^{2(N-T)} - 1\right]$$

where $T = \sum X_i, N = mn$

$$= \frac{1}{h^2}\left[s^{2T}\left(\frac{1-p-h}{1-p}\right)^{2N} - 1\right],$$

where $s = \frac{q(p+h)}{p(q-h)}, q = 1 - p$

$$EJ = \frac{1}{h^2}\left[\left(\frac{q-h}{q}\right)^{2N} Es^{2T} - 1\right],$$

$Es^{2T} = (ps^2 + q)^N$ and let $k = \frac{h}{p}$

$$E(J) = \frac{1}{p^2k^2}\left\{\left[\frac{q^2p(1+k)^2}{(q-pk)^2} + q\right]^N q^{-2N}(q-pk)^{2N} - 1\right\}$$

$$= \frac{1}{p^2k^2}\left\{q^{-N}\left[qp(1+k)^2 + (q-pk)^2\right]^N - 1\right\}$$

$$= \frac{1}{p^2k^2}\left\{q^{-N}(q+k^2p)^N - 1\right\}$$

$$\inf EJ = \lim_{k \to 0}\frac{q^{-N}(q+k^2p)^N - 1}{p^2k^2}.$$

Use L'Hospital's rule,

$$= \lim_{k \to 0}\frac{Nq^{-N}}{2kp^2}\left[(q+k^2p)^{N-1}2kp\right]$$

$$= \lim_{k \to 0}\frac{Nq^{-N}}{p}\left[(q+k^2p)^{N-1}\right] = \frac{N}{pq}$$

Hence

$$\inf EJ = \frac{N}{pq}$$

In this case, $\text{Var}(T_1) \geq \frac{pq}{N}$, where T_1 is any other unbiased estimator of p.

Example 4.3.7 Let $X_1, X_2, \ldots, X_n$ be iid with $P(\lambda)$. We will find CRK lower bound for any unbiased estimator of λ

$$J = \frac{1}{h^2} \left[\left(\frac{\lambda + h}{\lambda} \right)^{2T} \frac{\exp[-2(\lambda + h)m]}{\exp[-2\lambda m]} - 1 \right], \quad T = \sum_{i=1}^{m} X_i$$

$$J = \frac{1}{h^2} \left[\left(\frac{\lambda + h}{\lambda} \right)^{2T} \exp[-2hm] - 1 \right]$$

Let $s = \left(\frac{\lambda + h}{\lambda} \right)^2$,

$$EJ = \frac{1}{h^2} \left[\exp(-2hm) E(s^T) - 1 \right]$$

$$Es^T = [\exp m\lambda(s - 1)]$$

$$EJ = \frac{1}{h^2} \left[\exp(-2hm) \exp[m\lambda(s - 1)] - 1 \right]$$
$$= \frac{1}{h^2} \left[\exp(-2hm) \exp\left[m\lambda \left(\frac{2h}{\lambda} + \frac{h^2}{\lambda^2} \right) \right] - 1 \right]$$
$$= \frac{1}{h^2} \left[\exp\left(\frac{mh^2}{\lambda} \right) - 1 \right]$$

Let $\frac{h}{\lambda} = k$

$$\inf EJ = \lim_{k \to 0} \frac{1}{\lambda^2 k^2} \left[\exp(m\lambda k^2) - 1 \right]$$

Use L'Hospital's rule,

$$= \lim_{k \to 0} \frac{1}{2k\lambda^2} \left[2km\lambda \, \exp(m\lambda k^2) \right]$$

$$= \lim_{k \to 0} \frac{m}{\lambda} \left[\exp(m\lambda k^2) \right] = \frac{m}{\lambda}$$

$$\Rightarrow EJ = \frac{m}{\lambda_0}$$

If T_1 is any unbiased estimator of λ then $Var(T_1) \geq \frac{\lambda}{m}$.

4.4 Exercise 4

1. Find CR, Bhattacharya and CRK lower bound for the variance of any unbiased estimator of the parameter in a sample of size n, wherever possible.

(1) $f(x|\theta) = \theta(1-\theta)^x$; $x = 0, 1, 2 \ldots$; $0 < \theta < 1$

(2) $f(x|p) = \binom{x+k-1}{x} p^k q^x$; $x = 0, 1, 2 \ldots$, $0 < p < 1$, $q = 1 - p$, k known.

(3) $f(x|\sigma) = \frac{1}{\sigma\sqrt{2\pi}} \exp\left[-\frac{x^2}{2\sigma^2}\right]$; $-\infty < x < \infty, \sigma > 0$

(4) $f(x|\mu) = \frac{1}{\sqrt{2\pi}} \exp\left[-\frac{(x-\mu)^2}{2}\right]$; $-\infty < x < \infty, -\infty < \mu < \infty$

(5) $f(x|\sigma) = \frac{e^{-\frac{x}{\sigma}} x^{p-1}}{\sigma^p \Gamma(p)}$; $x > 0, \sigma > 0$ p is known

(6) $f(x|\alpha) = \frac{\alpha}{x^{\alpha+1}}$; $x \geq 1$

(7) $f(x|\theta) = \theta x^{\theta-1}$; $0 < x < 1, \theta > 0$

(8) $f(x|\theta) = \frac{\log \theta}{\theta - 1} \theta^x$; $0 < x < 1, \theta > 0$

2. Let $X_1, X_2, \ldots, X_n$ be iid rvs with $N(\theta, \theta^2)$, $\theta > 0$. Find CR and CRK lower bound for any unbiased estimator of θ.

3. Let $X_1, X_2, \ldots, X_n$ be iid rvs from the pmf,

$$f(x|\theta) = \left(\frac{\theta}{2}\right)^{|x|} (1-\theta)^{1-|x|}; \quad x = -1, 0, 1$$

Find CRK lower bound for an unbiased estimator of θ.

4. A sample of n observations is taken from the pdf $f(x\theta)$ where

$$f(x|\theta) = \begin{cases} \exp[-(x-\theta)] & ; \quad x > \theta \\ 0 & ; \quad \text{otherwise} \end{cases}$$

Find CRK lower bound for θ.

5. Let $X_1, X_2, \ldots, X_n$ be iid rvs with $B(n,p)$. Find CR, Bhattacharya and CRK lower bound for p^2

6. Let $X_1, X_2, \ldots, X_n$ be iid rvs with $P(\lambda)$. Find CR, Bhattacharya and CRK lower bound for (i) λ^2 (ii) $\lambda e^{-\lambda}$

7. Obtain CRR lower bound for θ in (i) $U(\theta, \theta + 1)$ (ii) $U(\theta - 1, \theta + 1)$ (iii) $U(\theta, 2\theta)$

8. Let the rv X have the following pmf

$$P[X = x] = \frac{e^{-m} m^x}{(1 - e^{-m})x!}; \quad x = 1, 2, \ldots, m > 0$$

Obtain CRK lower bound for m.

9. Let X be a rv with the following pdf

$$f(x) = \begin{cases} \frac{1}{3-a} & ; \ a < x < 3, \ a > 0 \\ 0 & ; \ \text{otherwise} \end{cases}$$

Obtain CRK lower bound for a.

10. Obtain CRK lower bound of θ in (i) $\cup(-\theta, \theta)$ (ii) $\cup(0, 2\theta)$ (iii) $\cup(\theta, \theta^2)$

11. The probability density function of the rv X has the following pdf

$$f(x) = \begin{cases} \frac{1}{2\theta} \exp\left[-\frac{|x-\theta|}{\theta}\right] & ; \ -\infty < x < \infty, \ \theta > 0 \\ 0 & ; \ \text{otherwise} \end{cases}$$

Find CR and CRK lower bound for θ.

12. Obtain CRK lower bound for $e^{-\frac{c}{\sigma}}$ if $X_1, X_2 < \ldots, X_n$ are iid rvs from an exponential distribution with mean σ.

13. Let $X_1, X_2, \ldots, X_n$ be iid rvs with $NB(r, \theta)$. Find CRK lower bound for an unbiased estimator θ, θ^2 and e^θ. Assume r known.

14. Let $X_1, X_2, \ldots, X_n$ be iid rvs with $f(x, \theta)$ and $g(x, \theta)$. If m is even, then X_{2k-1} has $f(x, \theta)$ and X_{2k} has $g(x, \theta)$, where $k = 1, 2, \ldots, \frac{m}{2}$.

Assume $f(x, \theta) = \binom{n}{x}\theta^x(1-\theta)^{n-x}; \ x = 0, 1, 2, \ldots, n$ and $g(x, \theta) = \binom{x+r-1}{x}\theta^r(1-\theta)^x; \ x = 0, 1, 2, \ldots$

Find CR and CRK lower bound for the variance of an unbiased estimator of $\theta, \frac{1}{\theta}$, θ^3, and $\frac{e^\theta}{\theta}$.

15. Let The rv X_1 has exponential distribution with mean $\frac{1}{\theta}$ and the rv X_2 has $f(x|\theta)$,

$$f(x|\theta) = \theta x^{\theta-1}; \quad 0 < x < 1, \theta > 0$$

Assume that X_1 and X_2 are independent rvs.

Find CR, CRK, and Bhattacharya lower bound for the variance of any unbiased estimator of θ and θ^2.

16. Let the rv X_1 has $\cup(0, \theta)$ and X_2 has $\cup(\theta, \theta + 1)$. Find CRK lower bound for θ. Assume X_1 and X_2 are independent rvs.

17. Assume that the rvs X_1 and X_2 are distributed as $\cup(0, \theta)$ and exponential with mean θ, respectively. Find CRK lower bound for θ^2.

18. As in problem 14,

$$f(x|\theta) = \frac{1}{\theta}e^{-\frac{x}{\theta}} \ ; x > 0, \ \theta > 0$$

$$g(x|\theta) = \theta x^{\theta-1} \ ; 0 < x < 1, \ \theta > 0$$

Find CR and CRK lower bound for θ and θ^2 in a sample of size m.

19. Let $X_1, X_2, \ldots, X_n$ be iid rvs with $N(\theta, 1)$. Find CR and CRK lower bound for (i) $\frac{\theta}{\theta+1}$ (ii) θe^{θ} (iii) $\frac{e^{\theta}}{\theta+1}$.

20. Let $X_1, X_2, \ldots, X_n$ be iid rvs with exponential and mean θ. Find CR, CRK, and Bhattacharya lower bound for e^{θ}. Further, find UMVUE of e^{θ}. Whether it attains CR lower bound? Justify.

References

Bhattacharya A (1946) On some analogues of the amount of information and their use in statistical estimation. Sankhya 8:1–14

Bhattacharya A (1950) Unbiased statistics with minimum variance. Proc R Soc Edinburgh 63–69

Blyth CR (1974) Necessary and sufficient condition for inequalities of Cramer-Rao type. Ann Stat 2:464–473

Blyth CR, Roberts DM (1972) On inequalities of Cramer-Rao type and admissibility proof. In: Proceedings of sixth Berkley symposium on mathematical statistics and probability, vol 1. University of California Press, pp 17–30

Chapman DG, Robbins H (1951) Minimum variance estimation without regularity assumptions. Ann Math Stat 22:581–586

Cramer H (1946) Mathematical methods of statistics. Princeton University Press

Frechet M (1943) Sur l' extension de certaines evaluating statistques au cas de petits echantillons. Rev Inst Int Stat 11:182–205

Jadhav DB (1983) On inequalities of Cramer-Rao type. Unpublished M.Phil, Thesis, University of Pune, Pune

Joshi VM (1976) On the attainment of the Cramer-Rao lower bound. Ann Stat 4:998–1002

Kiefer J (1952) On minimum variance estimators. Ann Math Stat 23:627–628

Rao CR (1945) Information and accuracy attainable in the estimation of statistical parameters. Bull Calcutta Math Soc 37:81–91

Rohatagi VK, Saleh EAK (2001) An introduction to probability and statistics. Wiley

Wijsman RA (1973) On the attainment of the Cramer-Rao bound. Ann Stat 538–542

Chapter 5
Consistent Estimator

In the previous chapters, we have seen various methods of estimation. Among the class of unbiased estimators, UMVUE is the best estimator in the sense that it has minimum variance. According to Lehmann–Scheffe theorem, if a complete sufficient statistic T exists, all we need to find is a function of T which is unbiased. One should note that if a complete sufficient statistics does not exist, an UMVUE may still exist. Similarly, we have considered moment and maximum likelihood estimators. These estimators may be biased or unbiased. In some cases moment or maximum likelihood estimators may be more efficient than UMVUE in the sense of mse. If n is large, what happens about all these estimators? It is quite possible that for a large n all these estimators may be equally efficient. Hence in this chapter, we will consider the large sample properties of estimators. This property of sequence of estimators is called consistency. Initially, we will consider some theorems on modes of convergence.

5.1 Prerequisite Theorems

Definition 5.1.1 A sequence of rvs $\{X_n\}$ is said to converge to X in probability, denoted as $X_n \xrightarrow{P} X$, if for every $\epsilon > 0$, as $n \to \infty$

$$P[|X_n - X| \geq \epsilon] \to 0. \tag{5.1.1}$$

Equivalently, $X_n \xrightarrow{P} X$, if for every $\epsilon > 0$, as $n \to \infty$

$$P[|X_n - X| < \epsilon] \to 1. \tag{5.1.2}$$

Now we will state the theorem without proof.

© Springer Science+Business Media Singapore 2016
U.J. Dixit, *Examples in Parametric Inference with R*,
DOI 10.1007/978-981-10-0889-4_5

Theorem 5.1.1 *Let* $X_n \xrightarrow{P} X$, *and* $Y_n \xrightarrow{P} Y$. *Then*

(i) $kX_n \xrightarrow{P} kX$, *(k is real)*

(ii) $X_n + Y_n \xrightarrow{P} X + Y$.

(iii) $X_n Y_n \xrightarrow{P} XY$.

(iv) $\frac{X_n}{Y_n} \xrightarrow{P} \frac{X}{Y}$ *if* $P[Y_n = 0] = 0$, $\forall n$ *and* $P[Y = 0] = 0$.

Definition 5.1.2 The sequence of rvs $\{X_n\}$ is said to converge to X almost surely (a.s.) or almost certainly, denoted as $X_n \xrightarrow{a.s.} X$ iff $X_n(w) \to X(w)$ for all w, except those belonging to a null set N.

Thus,

$$X_n \xrightarrow{a.s.} X \text{ iff } X_n(w) \to X(w) < \infty, \text{ for } w\epsilon N^c,$$

where $P(N) = 0$. Hence we can write as

$$P\left[\lim_n X_n = X\right] \to 1 \qquad (5.1.3)$$

Theorem 5.1.2 $X_n \xrightarrow{a.s.} X \Rightarrow X_n \xrightarrow{P} X$
The reader can see the proof of Theorems 5.1.1 and 5.1.2 in Rohatagi and Saleh (2001), and Bhat (2004).

One should note that the converse of Theorem 5.1.2 is not true.

Definition 5.1.3 Let $F_n(x)$ be the df of a rv X_n and $F(x)$, the df of X. Let $C(F)$ be the set of points of continuity of F. Then $\{X_n\}$ is said to converge to X in distribution or in law or weakly, denoted as $X_n \xrightarrow{L} X$ and/or $F_n \xrightarrow{W} F$, for every $x \in C(F)$.

It may be written as $X_n \xrightarrow{d} X$ or $F_n \xrightarrow{d} F$.

Theorem 5.1.3 *Let* $X_n \xrightarrow{L} X$, $Y_n \xrightarrow{L} c$ *then*

(i) $X_n + Y_n \xrightarrow{L} X + c$.

(ii) $X_n Y_n \xrightarrow{L} Xc$.

(iii) $\frac{X_n}{Y_n} \xrightarrow{L} \frac{X}{c}$ *if* $P[Y_n = 0] = 0$ *and* $c \neq 0$.

Theorem 5.1.4 *Let* $X_n \xrightarrow{P} X \Rightarrow X_n \xrightarrow{L} X$.

Theorem 5.1.5 *Let k be a constant,* $X_n \xrightarrow{L} k \Leftrightarrow X_n \xrightarrow{P} k$.

But in Theorem 5.1.4, we cannot say that $X_n \xrightarrow{L} k \Rightarrow X_n \xrightarrow{P} k$

Definition 5.1.4 A sequence of rvs $\{X_n\}$ is said to converge to X in the rth mean if $E|X_n - X|^r \to 0$ as $n \to \infty$. It is denoted as $X_n \xrightarrow{r} X$.

For $r = 2$, it is called convergence in quadratic mean or mean square.

Theorem 5.1.6 $X_n \xrightarrow{r} X \Rightarrow E|X_n|^r \to E|X|^r$.

Theorem 5.1.7 $X_n \xrightarrow{r} X \Rightarrow X_n \xrightarrow{P} X$.

If $X'_n s$ are a.s. bounded, conversely,

$X_n \xrightarrow{P} X \Rightarrow X_n \xrightarrow{r} X$, for all r

The above relationship between these convergence is explained in the following diagram:

$convergence\ a.s. \overset{\Rightarrow}{\underset{\nRightarrow}{}} convergence\ in\ probability \overset{\Rightarrow}{\underset{\nRightarrow}{}} convergence\ in\ distribution$

$convergence\ in\ probability \Leftrightarrow convergence\ in\ the\ rth\ mean$ (5.1.4)

The proofs of the above theorems are available in any text book on probability theory. See Serfling (1980), Billingsley (2012), Athreya and Lahiri (2006), Feller (1970), Rohatagi and Saleh (2001), and Bhat (2004). The reader should refer to different types of examples on the above theorems in Stoyanov (1997).

Next, we will study the limiting behavior of sums of independent rvs.

Definition 5.1.5 Let $\{X_n\}$ be a sequence of rvs. Let $S_n = \sum_{i=1}^{n} X_i, n = 1, 2, \ldots$. We say that $\{X_n\}$ is said to be stable in probability or it obeys the weak law of large numbers (WLLN) with respect to numerical sequences $\{A_n\}$ and $\{B_n\}$, $B_n > 0$, $B_n \uparrow \infty$ if $B_n^{-1}(S_n - A_n) \to 0$ in probability as $n \to \infty$.

Theorem 5.1.8 *Let $\{X_i\}$ be a sequence of pair-wise uncorrelated rvs with* $EX_i = \mu_i$ *and* $EX_i = \sigma_i^2$, $i = 1, 2, \ldots, n$. *If* $\sum_{i=1}^{n} \sigma_i^2 \to \infty$ *as* $n \to \infty$, *we can choose* $A_n = \sum_{i=1}^{n} \mu_i$ *and* $B_n = \sum_{i=1}^{n} \sigma_i^2$ *such that*

$$\frac{S_n - A_n}{B_n} = \sum_{i=1}^{n} \frac{X_i - \mu_i}{\sum_{i=1}^{n} \sigma_i^2} \xrightarrow{P} 0 \quad as\ n \to \infty \tag{5.1.5}$$

Note:

1. If $X_i's$ are iid rvs, then $A_n = n\mu$ and $B_n = n\sigma^2$.
2. Choose $B_n = n$ provided that $n^{-2} \sum_{i=1}^{n} \sigma_i^2 \to 0$ as $n \to \infty$.
3. In iid case, choose $A_n = n\mu$, $B_n = n$, since $\frac{n\sigma^2}{n^2} \to 0$ as $n \to \infty$ then $\frac{S_n}{n} \xrightarrow{P} \mu$.

Theorem 5.1.9 *Let $\{X_i\}$ be a sequence of rvs. Set $Y_n = n^{-1} \sum_{i=1}^{n} X_i$. A necessary and sufficient condition for the sequence $\{X_n\}$ to satisfy the WLLN is that*

$$E\left[\frac{Y_n^2}{1 + Y_n^2}\right] \to 0\ as\ n \to \infty \tag{5.1.6}$$

Theorem 5.1.10 *WLLN holds if and only if the three following conditions hold;*

(i) $\lim \sum_{k=1}^{n} P[|X_k| > n] = \lim \sum_{k=1}^{n} P[|X_k| \neq X_k^n] \to 0,$

(ii) $\frac{1}{n} \sum_{k=1}^{n} EX_k^n \to 0,$

(iii) $\frac{1}{n^2} \sum_{k=1}^{n} Var(X_k^n) \to 0,$

where

$$X_k^n = \begin{cases} X_k & ; |X_k| \leq n \\ 0 & ; \text{otherwise} \end{cases}$$

Definition 5.1.6 Let $\{X_n\}$ be a sequence of rvs. Let $S_n = \sum_{i=1}^{n} X_i, n = 1, 2, \ldots,$. We say that $\{X_n\}$ is said to be stable for a.s. probability or it obeys the strong law of large numbers (SLLN) with respect to numerical sequences $\{A_n\}$ and $\{B_n\}$, $B_n > 0$, $B_n \uparrow \infty$ if $B_n^{-1}(S_n - A_n) \xrightarrow{a.s.} 0$ as $n \to \infty$.

Theorem 5.1.11 *If $\{X_k\}$ are independent and $Var(X_k) = \sigma_k^2 < \infty$, $B_n \uparrow \infty$, $A_n = ES_n$ and $B_n = \sum_{k=1}^{n} \frac{\sigma_k^2}{b_k^2} < \infty$, then*

$$\left(\frac{S_n - A_n}{B_n} \right) \xrightarrow{a.s.} 0 \tag{5.1.7}$$

The proofs of WLLN and SLLN are available in any text book of probability.

Chebychev's Inequality

Theorem 5.1.12 *Let X be a rv with $EX = \mu$ and $Var X = \sigma^2 < \infty$, for any $k > 0$*

$$P[|X - \mu| \geq k\sigma] \leq \frac{1}{k^2} \tag{5.1.8}$$

or equivalently,

$$P[|X - \mu| < k\sigma] \geq 1 - \frac{1}{k^2} \tag{5.1.9}$$

Note: The distribution of X is not specified.

5.2 Definition and Examples

If we collect a large number of observations then we have a lot of information about any unknown parameter θ, and thus we can construct an estimator with a very small mse. We call an estimator consistent if

$$\lim_{n \to \infty} \text{MSE}(T(X)) = 0 \qquad (5.2.1)$$

which means that as the number of observations increase, the mse decreases to zero. For example, if $X_1, X_2, \ldots, X_n \sim N(\theta, 1)$, then $MSE(\bar{X}) = \frac{1}{n}$. Hence $\lim_{n \to \infty} MSE(\bar{X}) = 0$, $\bar{X}$ is consistent estimator of θ.

Definition 5.2.1 Let $X_1, X_2, \ldots, X_n$ be a sequence of iid rvs with pdf(pmf) $f(x|\theta)$. A sequence of point estimates T is called consistent estimator of θ, where $T = T(X_1, X_2, \ldots, X_n)$ if for a given $\epsilon, \delta > 0$, there exists $n_0(\epsilon, \delta, \theta)$ such that $\forall \theta \in \Theta$

$$P[|T - \theta| < \epsilon] \geq 1 - \delta, \forall \ n > n_0 \qquad (5.2.2)$$

or, using the Definition 5.1.1, we can say that $T \xrightarrow{P} \theta$ as $n \to \infty$.

Moreover, we can say that

$$P[|T - \theta| < \epsilon] \to 1 \qquad (5.2.3)$$

Note: Some authors (5.2.3) define as a weak consistency and if we use a Definition 5.1.2 then they define it as a strong consistency.

Example 5.2.1 Let $\{X_i\}_1^m$ be iid $B(n, p)$, then $\dfrac{\bar{X}}{n}$ is consistent estimator for p, where $\bar{X} = \dfrac{\sum_{i=1}^m X_i}{m}$.

Now, $MSE\left(\dfrac{\bar{X}}{n}\right) = \dfrac{pq}{mn}, q = 1 - p$. As $m \to \infty \Rightarrow MSE\left(\dfrac{\bar{X}}{n}\right) \to 0$

Example 5.2.2 Let $\{X_i\}_1^n$ be iid rvs with $P(\lambda)$ $\lambda > 0$ then $\bar{X}$ is a consistent estimator of λ, $\bar{X} = n^{-1} \sum_{i=1}^n X_i$.

Now, $E\bar{X} = \lambda$ and $MSE(\bar{X}) = \frac{\lambda}{n} \to 0$ as $n \to \infty$.

Example 5.2.3 Let $\{X_i\}_1^n$ be iid rvs with $U(0, \theta), \theta > 0$. $\bar{X}$ is not an consistent estimator of θ.

$MSE(\bar{X}) = \frac{(3n+1)\theta^2}{12n}$ and $\lim_{n \to \infty} \frac{(3n+1)\theta^2}{12n} = \frac{\theta^2}{4} \neq 0$
But $X_{(n)}$ is an consistent estimator.

(i) $EX_{(n)} = \frac{n\theta}{n+1}$ and $MSE(X_{(n)}) = \dfrac{2\theta^2}{(n+1)(n+2)} \to 0$ as $n \to \infty$

(ii) Use the Definition (5.1.1) and assume $\epsilon \leq \theta$, from (5.1.2)

$$P[|X_{(n)} - \theta| < \epsilon] = P[\theta - \epsilon < X_{(n)} < \theta + \epsilon]$$

$$= \int_{\theta - \epsilon}^{\theta} \frac{nx^{n-1}}{\theta^n} dx = 1 - \left(\frac{\theta - \epsilon}{\theta}\right)^n \to 1 \text{ as } n \to \infty$$

(iii) Consider a df of $X_{(n)}$, let it be $H_n(x, \theta)$

$$H_n = P[X_{(n)} \leq x] = \begin{cases} 0 & ; \; x < 0 \\ (\frac{x}{\theta})^n & ; \; 0 \leq x < \theta \\ 1 & ; \; x \geq \theta \end{cases}$$

$$\lim_{n \to \infty} H_n(x, \theta) = H(x, \theta),$$

where

$$H(x, \theta) = \begin{cases} 0 & ; \; x < 0 \\ 1 & ; \; x \geq 0 \end{cases}$$

In (5.1.4), we have explained the relationship between convergence, using this relationship,

$$X_{(n)} \xrightarrow{P} \theta \Leftrightarrow H_n \xrightarrow{d} H$$

In this case $H(x, \theta)$ is a df of a singular random variable, i.e., $P[X = \theta] = 1$, then

$$X_{(n)} \xrightarrow{d} X \Rightarrow X_{(n)} \xrightarrow{P} \theta.$$

Example 5.2.4 Consider $\{X_i\}_1^n$ are iid rvs as Cauchy distribution with location parameter θ.

$$f(x|\theta) = \frac{1}{\pi} \left[\frac{1}{1 + (x - \theta)^2} \right]; \quad x \in R, \; \theta \in R$$

then $\bar{X}$ is not a consistent estimator for θ.

The distribution of $\bar{X}$ is Cauchy with parameter θ.
Using the Definition 5.1.1,

$$P[|\bar{X} - \theta| < \epsilon] = P[\theta - \epsilon < \bar{X} < \theta + \epsilon] \qquad (5.2.4)$$

$$= \int_{\theta - \epsilon}^{\theta + \epsilon} \frac{1}{\pi} \left[\frac{dx}{1 + (x - \theta)^2} \right] = \frac{2}{\pi} \tan^{-1} \epsilon$$

This does not tends to 1.
Hence $\bar{X}$ is not a consistent estimator.

Example 5.2.5 Let $X_1, X_2, \ldots, X_n$ be iid $N(\mu, \sigma^2)$ rvs. We have to find the consistent estimator for σ^2.

We know that $\frac{(n-1)S^2}{\sigma^2} \sim \chi^2_{(n-1)}$, where $S^2 = \frac{\sum(x_i - \bar{x})^2}{n-1}$.

From Chebychev's Inequality, i.e., from Theorem 5.1.12

$\Lambda_{l} = \vdash \Rightarrow \Bbbk \equiv \frac{\cdot}{\upsilon}$

$$P[|S^2 - \sigma^2| > \epsilon] \leq \frac{Var(S^2)}{\epsilon^2} = \frac{2\sigma^4}{(n-1)\epsilon^2} \rightarrow 0 \text{ as } n \rightarrow \infty$$

Hence S^2 is consistent estimator for σ^2.

Theorem 5.2.1 *Let T be a consistent estimator for θ and let g be a continuous function then $g(t)$ is consistent for $g(\theta)$.*

Proof Given any $\epsilon > 0$, there exist a $\delta > 0$, such that, $|g(t) - g(\theta)| < \epsilon$ whenever $|T - \theta| < \delta$

Therefore,

$$\{x| \, |T - \theta| < \delta\} \subseteq \{x||g(t) - g(\theta)| < \epsilon\}$$

Then

$$P\{x| \, |g(t) - g(\theta)| < \epsilon\} \geq P\{x||T - \theta| < \delta\},$$

Hence,

$$P\{x| \, |g(t) - g(\theta)| < \epsilon\} \rightarrow 1$$

Because

$$P\{x| \, |T - \theta| < \delta\} \rightarrow 1$$

$g(t)$ is consistent for $g(\theta)$. $\qquad\square$

Example 5.2.6 Let $X_1, X_2, \ldots, X_n$ be iid $p(\lambda)$ rvs. To find the consistent estimator for $g(\lambda) = e^{-s\lambda}\lambda^r$. We know that $\bar{X}$ is consistent for λ.

Using the Theorem 5.2.1, $g\left(\bar{X}\right) = e^{-s\bar{X}}\left(\bar{X}\right)^r$ is consistent for $g(\lambda) = e^{-s\lambda}\lambda^r$.

Example 5.2.7 Let $X_1, X_2, \ldots, X_m$ be iid $B(n, p)$ rvs. We know that $\dfrac{\bar{X}}{n} = (mn)^{-1} \sum_{i=1}^{m} X_i$ is consistent for p.

Now, using Theorem 5.2.1, $\binom{n}{x}\bar{X}^x \left(1 - \bar{X}\right)^{n-x}$ is consistent for $\binom{n}{x}p^x q^{n-x}$, when $m \rightarrow \infty$.

Example 5.2.8 Let $X_1, X_2, \ldots, X_m$ be iid $B(n, p)$ rvs, where p is a function of θ, in Bioassay problem, $p(\theta) = \dfrac{\exp(\theta y)}{1+\exp(\theta y)}$, where $y > 0$ is a given dose level.

Now $\frac{\bar{X}}{n}$ is consistent for p.

$$\frac{\bar{X}}{n} = \frac{\exp(\theta y)}{1 + \exp(\theta y)} \Rightarrow \hat{\theta} = \frac{1}{y} \log \frac{\frac{\bar{X}}{n}}{1 - \frac{\bar{X}}{n}} \ , \quad \bar{X} = \frac{\sum_{i=1}^{m} X_i}{n}$$

Example 5.2.9 Let $X_1, X_2, \ldots, X_n$ be iid with $f(x|\theta)$,

$$f(x|\theta) = \theta x^{\theta-1} \ ; \ 0 < x < 1, \ \theta > 0$$

Let $y = -\log x$

$$g(y|\theta) = \theta e^{-\theta y} \ ; \ y > 0, \ \theta > 0$$

One can easily see that $\frac{-n}{\sum \log x_i}$ is consistent for θ.

Now we will define population quantiles.

Definition 5.2.2 Let X be a rv with its df $F(x|\theta)$, $\theta \epsilon \Theta$ then population quantile q_p is defined as

$$P[X \le q_p] = p, \quad 0 < p < 1$$

See David and Nagaraja (2003).

If $p = \frac{1}{2}$ then $q_{\frac{1}{2}}$ is median.

If $p = \frac{i}{4}$ $(i = 1, 2, 3)$, then $q_{\frac{i}{4}}$ is called as ith Quartile. In many textbooks, Quartiles such as Q_1, Q_2 and Q_3 are defined.

If $D_i = \frac{i}{10}$ $(i = 1, 2, \ldots, 9)$, then $q_{\frac{i}{10}}$ is called as ith Decile. In many textbooks, it is defined as $(D_1, D_2, \ldots, D_9)$.

Definition 5.2.3 Let the rv X have exponential distribution with mean θ, then to find Q_1, Q_2, Q_3, D_1, D_3, and D_8:

$$f(x|\theta) = \frac{1}{\theta} e^{-\frac{x}{\theta}} \ ; \ x > 0, \ \theta > 0$$

By Definition 5.2.2,

$$P[X \le Q_1] = \frac{1}{4}$$
$$1 - e^{-\frac{Q_1}{\theta}} = \frac{1}{4} \Rightarrow Q_1 = -\theta \log \frac{3}{4}$$

Similarly,

$$Q_2 = -\theta \log \frac{1}{2} \text{ and } Q_3 = -\text{`} \log \frac{1}{4}$$

$$D_1 = -\theta \log \frac{9}{10}, D_3 = -\theta \log \frac{7}{10} \text{ and } D_8 = -\text{`} \log \frac{2}{10}.$$

Lemma 5.2.1 *Let X be a random variable with its df $F(x)$. The distribution of $F(x)$ is $\cup(0, 1)$*

Proof Then

$$P[F(X) \le z] = P\left[X \le F^{-1}(z)\right] = F\left[F^{-1}(z)\right] = z$$

Hence $F(x)$ is $\cup(0, 1)$.   □

Theorem 5.2.2 *Sample quantiles are consistent estimators of population quantiles.
  Let $X_{(r)}$ be the rth order statistics of the sample and q_p is the pth quantiles. Hence*
$X_{(r)} \xrightarrow{P} q_p$. *If $r = [np] + 1$, then $X_{[np]+1} \xrightarrow{P} F^{-1}(p)$.*

Proof Let $\{X_{(1)}, X_{(2)}, \ldots, X_{(r)}\}$ be the order statistics of a sample of size n from the
population.   □

Let $Z_{(r)} = F(X_{(r)})$ has the same distribution as that of rth order statistic for a sample
of size n from $\cup(0, 1)$, see Lemma 5.2.1.

$$f_{Z_{(r)}}(x) = \frac{n!}{(r-1)!(n-r)!} x^{r-1}(1-x)^{n-r}; \ 0 < x < 1$$

$$EZ_{(r)} = EF(X_{(r)}) = \frac{r}{n+1}, EZ_{(r)}^2 = \frac{r(r+1)}{(n+1)(n+2)}$$

Let $r = [np] + 1$, where $[a] = $ integer part of a.
  Consider

$$E(Z_{(r)} - p)^2 = \frac{r(r+1)}{(n+1)(n+2)} - \frac{2pr}{n+1} + p^2$$

$$r = [np] + 1 \Rightarrow np \le r \le np + 1$$

$$\frac{np}{n+1} \le \frac{r}{n+1} \le \frac{np+1}{n+1}$$

$$\lim_{n \to \infty} \frac{np}{n+1} \le \lim_{n \to \infty} \frac{r}{n+1} \le \lim_{n \to \infty} \frac{np+1}{n+1}$$

$$p \le \lim_{n \to \infty} \frac{r}{n+1} \le p$$

Hence

$$\lim_{n \to \infty} \frac{r}{n+1} = p$$

Similarly, $np + 1 \leq r + 1 \leq np + 2$

$$\Rightarrow \frac{np+1}{n+2} \leq \frac{r+1}{n+2} \leq \frac{np+2}{n+2}$$

Then

$$\lim_{n \to \infty} \frac{r+1}{n+2} = p$$

Hence $\lim_{n \to \infty} E(Z_{(r)}) = p$ and $E(Z_{(r)} - p)^2 \to 0$.

Then $F(X_{(r)}) \xrightarrow{P} p$.

Now F^{-1} is continuous (which holds if $\frac{dF}{dx} = f(x) > 0$),

We have $F^{-1}F(X_{(r)}) = X_{(r)} \xrightarrow{P} F^{-1}(p)$

This implies that $X_{(r)} \xrightarrow{P} q_p \Rightarrow X_{[np]+1} \xrightarrow{P} q_p$.

Example 5.2.10 Let $X_1, X_2, \ldots, X_n$ are iid rvs with Pareto distribution

$$f(x|\lambda) = \frac{\lambda}{x^{\lambda+1}} \ ; \ \ x \geq 1, \ \lambda > 0$$

$$F(u) = P[X \leq u] = 1 - u^{-\lambda} \ ; \ \ u \geq 1$$

For $0 < p < 1$, $q_p(\lambda)$ is given by $F(q_p) = p$

Then $1 - (q_p)^{-\lambda} = p$

$$q_p = \exp\left[-\frac{\log(1-p)}{\lambda}\right],$$

By Theorem 5.2.2, $X_{[np]+1} \xrightarrow{P} q_p$.

Therefore, $X_{[np]+1}$ is consistent for $\exp[-\frac{\log(1-p)}{\lambda}]$.

Let $\Psi(\lambda) = \exp\left[-\frac{\log(1-p)}{\lambda}\right]$

Now $\frac{d\Psi}{d\lambda} \neq 0$ and Ψ^{-1} exist.

Therefore, $1 - (q_p)^{-\lambda} = p \Rightarrow \tilde{\lambda} = \frac{-\log(1-p)}{\log q_p}$

Now $X_{[np]+1}$ is consistent for q_p.

Then $\tilde{\lambda} = \frac{-\log(1-p)}{\log X_{[np]+1}}$ is consistent for λ.

Note:

$$EX = \int\limits_{1}^{\omega} \frac{x\lambda}{x^{\lambda+1}} dx = \frac{\lambda}{\lambda-1}; \ \lambda > 1$$

If $\lambda > 1$, $\bar{X} \xrightarrow{P} \frac{\lambda}{\lambda-1}$.

Hence, $\tilde{\lambda} = \frac{\bar{X}}{1-\bar{X}}$,

Then $\tilde{\lambda} \to \lambda$ in probability if $\lambda > 1$

Further, if $\lambda \leq 1$ then EX does not exist and $\bar{X}$ is not consistent for $\frac{\lambda}{\lambda-1}$, and therefore $\tilde{\lambda}$ is not consistent for λ.

Example 5.2.11 Let $X_1, X_2, \ldots, X_n$ be Weibull distribution with pdf given by

$$f(x|\alpha) = \alpha x^{\alpha-1} \exp[-x^{\alpha}]; \ x > 0, \ \alpha > 0$$

In this case, $F(u) = 1 - \exp[-u^{\alpha}]$

Hence,

$$1 - \exp[-q_p{}^{\alpha}] = p$$
$$q_p{}^{\alpha} = -\log(1-p)$$
$$\tilde{\alpha} = \frac{\log\left[\log(1-p)^{-1}\right]}{\log q_p}$$

Now $X_{[np]+1}$ is consistent for q_p

$$\tilde{\alpha} = \frac{\log\left[\log(1-p)^{-1}\right]}{\log X_{[np]+1}}$$

Therefore, $\tilde{\alpha} \xrightarrow{P} \alpha$, i.e., consistent estimator for α is given by $\frac{\log[\log(1-p)^{-1}]}{\log X_{[np]+1}}$.

5.3 Consistent Estimator for Multiparameter

In this section, we consider the case when θ is vector valued parameter.

Definition 5.3.1 Let $\theta = (\theta_1, \theta_2, \ldots, \theta_n)$. Then the consistency of a vector valued statistic $T = (T_1, T_2, \ldots, T_n)$ can be defined in two ways

 (i) Marginal consistency: $T_i \xrightarrow{P} \theta_i (i = 1, 2, \ldots, n)$
(ii) Joint consistency

$$\lim_{n \to \infty} P[||T - \theta|| < \epsilon] = 1, \ \forall \epsilon > 0, \ \forall \theta > 0$$

where $||x||$ is a suitable norm for n-dimensional Euclidean space R_n, $||x|| = \max ||x_i||$, which is equivalent to $\sqrt{x'x}$

$$||T - \theta|| = \left[\sum (T_i - \theta_i)^2\right]^{\frac{1}{2}}$$

Theorem 5.3.1 *T is marginally consistent if and only if it is jointly consistent.*

Proof (i) Assume that T is jointly consistent.
Let $A_i = \{x| \ |T_i - \theta_i| < \epsilon\}$
Let $A = \{x| \max_i \ |T_i - \theta_i| < \epsilon\}$
It implies that $A = \bigcap_{i=1}^{n} A_i$
Since T is jointly consistent then $P(A) \to 1$ as $n \to \infty$ and $A \subseteq A_i$.
Hence $P(A_i) \geq P(A)$
Therefore, $P(A_i) \to 1$ as $n \to \infty$,
Thus, T is marginally consistent.
(ii) Assume that T is marginally consistent.
$P(A_i) \to 1 \Rightarrow \lim_{n \to \infty} P(A_i) = 1$
$\lim_{n \to \infty} P(A_i^c) = 0$ for each i
Using De'Morgan's Law, $A^c = \bigcup_{i=1}^{n} A_i^c$

$$P(A^c) \leq \sum_{i=1}^{n} P\left(A_i^c\right) \to 0$$
$$\Rightarrow \quad P(A^c) \to 0$$
$$\Rightarrow \quad P(A) \to 1$$

This implies that T is jointly consistent for θ.   □

Example 5.3.1 Kale (1999) had given a consistent estimator for a location and scale parameter for $f(x|\mu, \sigma)$, $x \in R$, $\mu \in R$, $\sigma > 0$

    Let

$$f(x|\mu, \sigma) = \frac{1}{\sigma} f_0 \left(\frac{x - \mu}{\sigma}\right)$$

Now

$$F(q_p) = \int_{\mu}^{q_p} \frac{1}{\sigma} f_0 \left(\frac{x - \mu}{\sigma}\right) dx = p$$

Let $Z = \frac{X-\mu}{\sigma}$

$$F(q_p) = \int\limits_{0}^{\frac{q_p-\mu}{\sigma}} f(z)dz = p$$

Let $c_p = \frac{q_p-\mu}{\sigma} \Rightarrow q_p = \mu + \sigma c_p$

Consider $0 < p_1 < p_2 < \cdots p_k < 1$, then $q_{p_r} = \mu + \sigma c_{p_r}$, $r = 1, 2, \ldots, k$,

We know that $X_{[np_r]+1}$ is consistent for q_{p_r}

Hence, $X_{[np_r]+1} = \mu + \sigma c_{p_r}$; $r = 1, 2, \ldots, k$

We have k equations and we can estimate parameters by the method of least squares:

$$\hat{\sigma} = \frac{Cov\left[X_{[np_r]+1}, c_{p_r}\right]}{Var[c_{p_r}]}$$

$$= \frac{k \sum\limits_{r=1}^{k} X_{[np_r]+1}\, c_{p_r} - \sum\limits_{r=1}^{k} X_{[np_r]+1} \sum\limits_{r=1}^{k} c_{p_r}}{k \sum c_{p_r}^2 - \left(\sum c_{p_r}\right)^2}$$

$$\hat{\mu} = \frac{\sum\limits_{r=1}^{k} X_{[np_r]+1} - \hat{\sigma} \sum\limits_{r=1}^{k} c_{p_r}}{k}$$

Since $X_{[np_r]+1} \to \mu + \sigma c_{p_r}$ in probability as $n \to \infty$.

We can easily show that $\hat{\sigma} \xrightarrow{P} \sigma$ and $\hat{\mu} \xrightarrow{P} \mu$. Hence $(\hat{\mu}, \hat{\sigma})$ is consistent for (μ, σ).

Example 5.3.2 Let $X_1, X_2, \ldots, X_n$ be iid

$N(\mu, \sigma^2)$ rvs

Using Theorem 5.1.8,

$m_1' = \bar{X} \xrightarrow{P} \mu$, where $m_1' = \frac{\sum X_i}{n}$

$m_2' \xrightarrow{P} \mu^2 + \sigma^2$, where $m_2' = \frac{\sum X_i^2}{n}$

$m_2' - (m_1')^2 \xrightarrow{P} \sigma^2$

Let $T_1 = \bar{X}$ and $T_2 = m_2' - (m_1')^2$,

Using Theorem 5.3.1, (T_1, T_2) are jointly consistent for (μ, σ^2).

Note: If $T_2 = m_2$, then $\sqrt{m_2} \to \sigma$ in probability,

thus $(\bar{X} + Z_\alpha \sqrt{m_2}) \to \mu + Z_\alpha \sigma$ in probability, where Z_α is $100\alpha \%$ point of the normal distribution.

Example 5.3.3 Consider the two-parameter gamma family:

$$f(x|p, \sigma) = \frac{e^{-\frac{x}{\sigma}} x^{p-1}}{\sigma^p \Gamma(p)}; x > 0, p > 0, \sigma > 0$$

$\mathrm{E}X = p\sigma, \mathrm{E}X^2 = p(p+1)\sigma^2, \mathrm{V}(X) = p\sigma^2$

$\bar{X} = m_1' \xrightarrow{P} p\sigma$ and $m_2 \xrightarrow{P} p\sigma^2$

$\frac{m_2}{m_1'} \xrightarrow{P} \sigma$ and $\frac{(m_1')^2}{m_2} \xrightarrow{P} p$

Using Theorem 5.3.1, $\left(\frac{m_2}{m_1'}, \frac{(m_1')^2}{m_2}\right)$ is jointly consistent for (σ, p).

Example 5.3.4 $\{X_i\}_{i=1}^n$ are iid exponential rvs with location and scale parameter μ and σ respectively.

$\mathrm{E}X = \mu + \sigma, \quad Var(X) = \sigma^2$

$m_2 \xrightarrow{P} \sigma^2$ and $\sqrt{m_2} \xrightarrow{P} \sigma$

$\bar{X} \xrightarrow{P} \mu + \sigma$ and $\bar{X} - \sqrt{m_2} \xrightarrow{P} \mu$.

Hence, $(\bar{X} - \sqrt{m_2}, \sqrt{m_2})$ is jointly consistent for (μ, σ).

Alternatively, we can find consistent estimators for (μ, σ) based on order statistics. Let $X_{(1)} < X_{(2)} < \cdots < X_{(n)}$ be the order statistics.

The distribution of $X_{(1)}$ is exponential with location and scale parameter μ and $\frac{\sigma}{n}$, respectively.

$$g_{X_{(1)}}(x|\mu, \sigma) = \frac{n}{\sigma} \exp\{-\frac{n}{\sigma}(x - \mu)\} \; ; \; x > \mu, \; \sigma > 0$$

$\mathrm{E}X_{(1)} = \mu + \frac{\sigma}{n}$ and $\mathrm{Var}(X_{(1)}) = \frac{\sigma^2}{n^2}$

One can easily verify that

$X_{(1)} \xrightarrow{P} \mu$, and $\bar{X} \xrightarrow{P} \mu + \sigma$. Then $\frac{1}{n}\sum_{i=2}^{n}(X_{(i)}) \xrightarrow{P} \sigma$

Hence, by Theorem 5.2.1,

$[X_{(1)}, \frac{1}{n}\sum_{i=2}^{n}(X_{(i)})] \xrightarrow{P} (\mu, \sigma)$.

5.4   Selection Between Consistent Estimators

Let T_1 and T_2 be the two consistent estimators for θ. Assume T_1 and T_2 converges in quadratic mean to θ (see the Definition 5.1.4), i.e., $T_i \to \theta$ in quadratic. We prefer T_1 to T_2 if for sufficiently large n (say $n > n_0$) and $\forall \theta \in \Theta$,

$$MSE(T_1|\theta) \le MSE(T_2|\theta). \tag{5.4.1}$$

Example 5.4.1 Let $X_1, X_2, \ldots, X_n$ are iid $\cup(0, \theta)$ rvs. We know that $T_1 = X_{(n)}$ and $T_2 = 2\bar{X}$ are both consistent estimators for θ (Reader can show this).

$$MSE(T_1|\theta) = \frac{2\theta^2}{(n+1)(n+2)} \text{ and } MSE(T_2|\theta) = \frac{\theta^2}{3n}.$$

In this case $MSE(T_i|\theta) \to 0$ as $n \to \infty$

From (5.4.1), we choose T_1 to T_2 if $MSE(T_1|\theta)$ converges to zero at a faster rate than that of $MSE(T_2|\theta)$.

Now, $MSE(T_1|\theta) \to 0$ at the rate of $\frac{1}{n^2}$

and $MSE(T_2|\theta) \to 0$ at the rate of $\frac{1}{n}$

Thus, $MSE(T_1|\theta)$ converges to zero at a faster rate than that of $MSE(T_2|\theta)$. We prefer T_1 for θ. In such a situation, we will say T_1 is faster than T_2.

Example 5.4.2 Let $X_1, X_2, \ldots, X_n$ are iid rvs with $f(x|\mu)$,

$$f(x|\mu) = \exp[-(x - \mu)]; \; x \geq \mu$$

Let $T_1 = X_{(1)}$ and $T_2 = 2\bar{X} + 1$
$MSE(T_1) = \frac{2}{n^2}, MSE(T_2) = \frac{1}{n}.$
Here, also, T_1 is faster than T_2.

Example 5.4.3 $\{X_i\}_1^{2n+1}$ are iid rvs with $\cup(\mu - 1, \mu + 1)$.

Let $T_1 = $ Median $= X_{(n+1)}$ and $T_2 = \bar{X}_{2n+1} = \frac{1}{2n+1}\sum_{i=1}^{2n+1} X_i = $ sample mean
Here $E(T_1) = E(T_2) = \mu$, $Var(T_1) = \frac{1}{3(2n+1)}$ and $Var X_{(n+1)} = \frac{1}{2n+3}$.

$$Var(T_1) = \frac{1}{2n}\left(1 + \frac{3}{2n}\right)^{-1}$$

$$= \frac{1}{2n}\left[1 - \frac{3}{2n} + o\left(\frac{1}{n}\right)\right]$$

$$= \frac{1}{2n} - \frac{3}{4n^2} + o\left(\frac{1}{n^2}\right) \qquad (5.4.2)$$

$$Var(T_2) = \frac{1}{6n}\left(1 + \frac{1}{2n}\right)^{-1}$$

$$= \frac{1}{6n}\left[1 - \frac{1}{2n} + o\left(\frac{1}{n}\right)\right]$$

$$= \frac{1}{6n} - \frac{1}{12n^2} + o\left(\frac{1}{n^2}\right) \qquad (5.4.3)$$

Comparing the coefficient of $\frac{1}{n}$ in (5.4.2) and (5.4.3)

$$\frac{1}{6} < \frac{1}{2}$$

Hence T_2 is more efficient than T_1. Therefore, sample mean is preferable than median for μ. Further, taking the ratios of MSE and then taking the limit as $n \to \infty$, we conclude that

$$\frac{MSE[X_{(n+1)}]}{MSE[\bar{X}_{(2n+1)}]} \to 3 \text{ as } n \to \infty.$$

The sample mean is three times more efficient than the median.

Example 5.4.4 Let $X_1, X_2, \ldots, X_n$ are iid $N(\mu, \sigma^2)$ rvs.

Let $T_1 = \frac{S^2}{n}$, where $S^2 = \sum(x_i - \bar{x})^2$ and $T_2 = \frac{S^2}{n-1}$
Note that $\frac{S^2}{\sigma^2} \sim \chi^2_{n-1}$, $E(S^2) = (n-1)\sigma^2$ and $Var(S^2) = 2(n-1)\sigma^4$.

$$\begin{aligned}
MSE(T_1) &= E\left(\frac{S^2}{n} - \sigma^2\right)^2 \\
&= Var(\frac{S^2}{n}) + \left[Bias\left(\frac{S^2}{n} - \sigma^2\right)\right]^2 \\
&= \frac{2(n-1)\sigma^4}{n^2} + [\frac{(n-1)\sigma^2}{n} - \sigma^2)]^2 \\
&= \sigma^4[\frac{2(n-1)}{n^2} + \frac{1}{n^2}] \\
&= \sigma^4[\frac{2}{n} - \frac{1}{n^2}]
\end{aligned} \qquad (5.4.4)$$

Similarly,

$$\begin{aligned}
MSE(T_2) &= \frac{2\sigma^4}{n-1} = \frac{2\sigma^4}{n}\left(1 - \frac{1}{n}\right)^{-1} \\
&= \frac{2\sigma^4}{n}\left[1 + \frac{1}{n} + o\left(\frac{1}{n}\right)\right] \\
&= \sigma^4\left[\frac{2}{n} + \frac{2}{n^2} + o\left(\frac{1}{n^2}\right)\right]
\end{aligned} \qquad (5.4.5)$$

By observing (5.4.4) and (5.4.5)

(a) Coefficient of $\frac{1}{n}$ is same in both (5.4.4) and (5.4.5)
(b) Coefficient of $\frac{1}{n^2}$ in (5.4.4) is -1 and in (5.4.5) is 2.

Hence, $MSE(T_1)$ is smaller than $MSE(T_2)$.
    We prefer T_1 than T_2 as a consistent estimator for σ^2.

5.5 Determination of n_0

Kale (1999) has explained the method of determination of n_0 in a very interesting way. In some examples, we will require the following Cramer and Fisz theorem; see Serfling (1980, p. 77)

Theorem 5.5.1 *Let q_p be the $100p\%$ point of F and let $X_{[np]+1}$ be the sample $100p\%$ point then under regularity conditions*

$$X_{[np]+1} \sim AN\left[q_p, \frac{p(1-p)}{n[f(q_p)]^2}\right] \tag{5.5.1}$$

Using the Definition 5.1.4 and Theorem 5.1.12, we can write if $T \xrightarrow{2} \theta$ then

$$P[|T - \theta| < \epsilon] \geq 1 - \frac{MSE(T)}{\epsilon^2} \tag{5.5.2}$$

In (ϵ, δ) terminology, suppose we want that

$$P[|T - \theta| < \epsilon] \geq 1 - \delta \ \forall \, n \geq n_0 \tag{5.5.3}$$

then using (5.5.2) if $1 - \frac{MSE(T)}{\epsilon^2} \geq 1 - \delta$, the desired level of accuracy specified by (ϵ, δ) is achieved if n_0 is determined by

$$MSE(T) \leq \delta\epsilon^2 \tag{5.5.4}$$

Example 5.5.1 Consider the Example 5.4.3,

$$MSE(\bar{X}_{2n+1}) = \frac{1}{3(2n+1)} \tag{5.5.5}$$

$$MSE(X_{(n+1)}) = \frac{1}{2n+3} \tag{5.5.6}$$

Using (5.5.4) and (5.5.5)

$$\frac{1}{3(2n+1)} \leq \delta\epsilon^2$$

$$\Rightarrow n \geq \frac{1}{2}\left(\frac{1}{3\delta\epsilon^2} - 1\right)$$

$$\Rightarrow n_0 = \left[\frac{1}{2}\left(\frac{1}{3\delta\epsilon^2} - 1\right)\right] + 1 \tag{5.5.7}$$

Table 5.1 n_0 for mean $\bar{X}_{2n+1}$

$\delta\backslash\epsilon$	0.2	0.1	0.01	0.001
0.2	21	83	8333	833,333
0.1	42	167	16,667	16,666,667
0.01	417	1667	16,667	16,666,667
0.001	4167	16,667	1,666,667	166,666,667

Table 5.2 n_0 for median $X_{(n+1)}$

$\delta\backslash\epsilon$	0.2	0.1	0.01	0.001
0.2	62	250	25×10^4	25×10^5
0.1	125	500	5×10^4	5×10^6
0.01	1250	5000	5×10^5	5×10^7
0.001	12,500	5×10^4	5×10^6	5×10^8

Similarly, using (5.5.4) and (5.5.6)

$$\frac{1}{2n+3} \leq \delta\epsilon^2$$

$$\Rightarrow n \geq \frac{1}{2}\left(\frac{1}{\delta\epsilon^2} - 3\right)$$

$$\Rightarrow n_0 = \left[\frac{1}{2}\left(\frac{1}{\delta\epsilon^2} - 3\right)\right] + 1 \tag{5.5.8}$$

The Tables 5.1 and 5.2 give the values for n_0 using R,

```
# To enter epsilon (eps) and delta (del).
  eps <- c(0.2,0.1,0.01,0.001); del <- c(0.2,0.1,0.01,0.001);
  ld <- length(del); # length of delta
  le <- length(eps); # length of epsilon
# declaring matrics for n0t and n0t2.
  n0t1 <- matrix(data=0,nrow=ld,ncol=le);  n0t2 <- matrix(data=0,nrow=ld,ncol=le)
# To find n0 for T1=X(n+1).
  for(i in 1:ld)
    {
      for(j in 1:le)          {  n0t1[i,j] <- (((1/(del[i]*(eps[j]^2)))-3)/2)+1 }
    }
  colnames(n0t1) <- c("[0.2]","[0.1]","[0.01]","[0.001]")
  rownames(n0t1) <- c("[0.2]","[0.1]","[0.01]","[0.001]")
# To find n0 for T2=X_bar(2n+1).
  for(i in 1:ld)
    {
      for(j in 1:le)          {  n0t2[i,j] <- (((1/(3*del[i]*(eps[j]^2)))-1)/2)+1 }
    }
  colnames(n0t2) <- c("[0.2]","[0.1]","[0.01]","[0.001]")
  rownames(n0t2) <- c("[0.2]","[0.1]","[0.01]","[0.001]")
# OUTPUT
  print("n0 for T1"); n0t1; print("n0 for T2"); n0t2
```

Here $\epsilon = 0.2, \ 0.1, \ 0.01$ and 0.001, and $\delta = 0.2, \ 0.1, \ 0.01$ and 0.001

It is interesting to see that n_0 is increasing faster in row-wise than in column wise. Hence, we will have to be more careful in selecting ϵ than δ. Hence, one should select δ smaller than ϵ.

The above analysis of determination of n_0 and the minimum sample size required to attain a given level of accuracy specified by (ϵ, δ) is the basic idea underlying the concept of efficiency or asymptotic relative efficiency. In this case, we can define relative efficiency of T_1 and T_2 as we prefer T_1 to T_2 if $n_0(T_1) \leq n_0(T_2) \; \forall \; \theta \in \Theta$.

Since this efficiency depends on Chebychev's inequality and it gives a very crude lower bound to the probability. Therefore, $n_0(T)$ determined by this inequality may actually be an overestimate.

The CLT plays an important role in statistical theory. We generally make the assumption that underlying observations follow normal distribution at least approximately. The theory of errors used by physicists or astronomers can be justified on the basis of CLT.

Theorem 5.5.2 *Let $X_1, X_2, \ldots, X_n$ be iid rvs with $EX_i = \mu$ and $Var X_i = \sigma^2 < \infty$. Let $Y_n = \sqrt{n}(\frac{\bar{X}-\mu}{\sigma})$, where $\bar{X} = n^{-1} \sum_{i=1}^{n} X_i$. Then Y_n converges to a standard normal variate.*

Note:

1. CLT gives the probability bound for $|\bar{X} - \mu|$, while WLLN gives only the limiting value.
2. Reader should refer Bhat (2004), where CLT's are given with different conditions.

Example 5.5.2 Consider the Example 5.5.1

Let $T_1 = \bar{X}_{2n+1}$ and $T_2 = X_{(n+1)}$
By Theorem 5.1.1

$$\sqrt{3(2n+1)} \left(\bar{X}_{2n+1} - \mu \right) \xrightarrow{d} N(0, 1) \tag{5.5.9}$$

Thus

$$P\left[|\bar{X}_{2n+1} - \mu| < \epsilon\right] = P\left[-\epsilon\sqrt{3(2n+1)} \leq \sqrt{3(2n+1)}(\bar{X}_{2n+1} - \mu) \leq \epsilon\sqrt{3(2n+1)}\right]$$

$$= \Phi[\epsilon\sqrt{3(2n+1)}] - \Phi[-\epsilon\sqrt{3(2n+1)}],$$

where

$$\Phi(y) = \int_{-\infty}^{y} \frac{1}{\sqrt{2\pi}} e^{-\frac{x^2}{2}} dx$$

Using (5.5.3)

$$\Phi[\epsilon\sqrt{3(2n+1)}] - \Phi[-\epsilon\sqrt{3(2n+1)}] \geq 1 - \delta$$

By symmetry,

$$2\Phi\left[\epsilon\sqrt{3(2n+1)}\right] - 1 \geq 1 - \delta$$

$$\epsilon\sqrt{3(2n+1)} \geq \Phi^{-1}\left[1 - \frac{\delta}{2}\right]$$

$$n \geq \frac{1}{6\epsilon^2}\left[\Phi^{-1}\left(1 - \frac{\delta}{2}\right)\right]^2 - \frac{1}{2}$$

Thus

$$n_0(T_1) = \left[\frac{1}{6\epsilon^2}\left\{\Phi^{-1}\left(1 - \frac{\delta}{2}\right)\right\}^2 - \frac{1}{2}\right] + 1 \qquad (5.5.10)$$

Consider the estimator $T_2 = X_{(n+1)}$

$$P[|X_{(n+1)} - \mu| < \epsilon] = \frac{1}{\beta(n+1, n+1)} \int\limits_{\mu-\epsilon}^{\mu+\epsilon} \left(\frac{y - \mu + 1}{2}\right)^n \left(1 - \frac{y - \mu + 1}{2}\right)^n \frac{dy}{2}$$

$$(5.5.11)$$

Substitute $w = y - \mu$

$$= \frac{1}{\beta(n+1, n+1)} \int\limits_{-\epsilon}^{\epsilon} \left(\frac{1 + w}{2}\right)^n \left(\frac{1 - w}{2}\right)^n \frac{dw}{2}$$

$$= \frac{1}{\beta(n+1, n+1)2^{2n}} \int\limits_{0}^{\epsilon} (1 - w^2)^n dw$$

Let $w^2 = t$

$$= \frac{1}{2^{2n+1}\beta(n+1, n+1)} \int\limits_{0}^{\epsilon^2} (1 - t)^n t^{-\frac{1}{2}} dt$$

This is an incomplete Beta type distribution. In such a situation, it is difficult to find n_0. Now, we will use Cramer–Fisz theorem, i.e., Theorem 5.5.1 to find the distribution of $X_{(n+1)}$. In this case, $p = \frac{1}{2}$, $q_p = \mu$, $f(q_p) = \frac{1}{2}$ and n is replaced by $2n + 1$.

$$X_{(n+1)} \sim AN\left[\mu, \frac{1}{(4)(2n+1)(\frac{1}{2})^2}\right]$$

$$\Rightarrow X_{(n+1)} \sim AN\left[\mu, \frac{1}{2n+1}\right]$$

$$\Rightarrow \sqrt{2n+1}[X_{(n+1)} - \mu] \sim AN(0, 1)$$

Therefore

$$P\left[\left|X_{(n+1)} - \mu\right| < \epsilon\right] = \Phi\left[\epsilon\sqrt{2n+1}\right] - \Phi\left[-\epsilon\sqrt{2n+1}\right]$$

Similarly as before,

$$n_0(T_2) = \left[\frac{1}{2\epsilon^2}\left\{\Phi^{-1}\left(1 - \frac{\delta}{2}\right)\right\}^2 - \frac{1}{2}\right] + 1 \qquad (5.5.12)$$

Tables 5.4 and 5.5 give the values of n_0 using R

```
# To enter epsilon (eps) and delta (del).
  eps <- c(0.2,0.1,0.01,0.001); del <- c(0.2,0.1,0.01,0.001);
  ld <- length(del); # length of delta
  le <- length(eps); # length of epsilon
# declaring matrices for n0t1 and n0t2.
  n0t1 <- matrix(data=0,nrow=ld,ncol=le); n0t2 <- matrix(data=0,nrow=ld,ncol=le)
# To fnd n0 for T1.
  for(i in 1:ld)
  {
     for(j in 1:le)
     {
       x <- 1-(del[i]/2); y <- qnorm(x,0,1);
       n0t1[i,j] <- ((y^2/(6*eps[j]^2))-2)+1 }
  }
  colnames(n0t1) <- c("[0.2]","[0.1]","[0.01]","[0.001]")
  rownames(n0t1) <- c("[0.2]","[0.1]","[0.01]","[0.001]")
# To fnd n0 for T2.
  for(i in 1:ld)
  {
     for(j in 1:le)
     {
       x <- 1-(del[i]/2); y <- qnorm(x,0,1);
       n0t2[i,j] <- ((y^2/(2*eps[j]^2))-0.5)+1 }
  }
  colnames(n0t2) <- c("[0.2]","[0.1]","[0.01]","[0.001]")
  rownames(n0t2) <- c("[0.2]","[0.1]","[0.01]","[0.001]")
# OUTPUT
  print("n0 for T1"); n0t1
  print("n0 for T2"); n0t2
```

Here $\epsilon = 0.2$, 0.1, 0.01, and 0.001, and $\delta = 0.2$, 0.1, 0.01, and 0.001. One can use these following values to find $n_0(T_1)$ and $n_0(T_2)$, if you are not using R. If we have to calculate $n_0(T_1)$ and $n_0(T_2)$ without using R, then one can use Table 5.3.

In view of the asymptotic normality of both $\bar{X}_{2n+1}$ and $X_{(n+1)}$, it has been observed that $n_0(T_1) < n_0(T_2)$. The relative efficiency of T_1 is better than T_2.

Table 5.3 Percentile points of standard normal distribution

δ	0.2	0.1	0.01	0.001
$\Phi^{-1}(1 - \frac{\delta}{2})$	1.28155	1.64485	2.57583	3.29053

Table 5.4 n_0 for $\bar{X}_{2n+1}$ (by using CLT)

$\delta \backslash \epsilon$	0.2	0.1	0.01	0.001
0.2	6	27	273	273,728
0.1	11	45	4509	450,921
0.01	27	110	11,058	1,105,816
0.001	44	179	18,044	1,804,596

Table 5.5 n_0 for $X_{(n+1)}$ (by using Theorem 5.5.1)

$\delta \backslash \epsilon$	0.2	0.1	0.01	0.001
0.2	21	82	8212	821,185
0.1	34	135	13,528	1,352,766
0.01	83	332	33,175	3,317,450
0.001	135	541	54,138	5,413,794

Example 5.5.3 For double exponential distribution with mean = median = θ. Use CLT to obtain asymptomatic distribution of $\bar{X}_n$. Use Theorem 5.5.1 to obtain asymptotic distribution of Median = M_n. Determine $n_0(\epsilon, \delta, \theta)$ for $\bar{X}_n$ and M_n for the given $\epsilon = 0.01, 0.1$ and $\delta = 0.01, 0.1$.

In this case,

$$f(x|\theta) = \begin{cases} \frac{1}{2}\exp[-|x - \theta|] & ; \quad -\infty < x < \infty, \ \theta > 0 \\ 0 & \quad \text{otherwise} \end{cases}$$

$$\mathrm{E}(X^r) = \frac{1}{2} \int\limits_{-\infty}^{\infty} (z + \theta)^r e^{-|z|} dz, \quad \text{where} \quad z = |x - \theta|$$

$$= \frac{1}{2} \int\limits_{-\infty}^{\infty} \sum_{k=0}^{r} \binom{r}{k} z^k \theta^{r-k} e^{-|z|} dz, \quad \text{where} \quad z = |x - \theta|$$

$$= \frac{1}{2} \sum_{k=0}^{r} \binom{r}{k} \theta^{r-k} [\int\limits_{-\infty}^{0} (-1)^k z^k e^z dz + \int\limits_{0}^{\infty} z^k e^{-z} dz]$$

$$-\frac{1}{2}\sum_{k=0}^{r}\binom{r}{k}\theta^{r-k}[(-1)^{k}k! + k!]$$

$$=\frac{1}{2}\sum_{k=0}^{r}\binom{r}{k}\theta^{r-k}k![1 + (-1)^{k}]$$

$\mu_1' = \theta, \mu_2' = \theta^2 + 2, V(X) = 2, E\bar{X}_n = \theta$ and $V(\bar{X}_n) = \frac{2}{n}$.

The exact distribution of median is very complicated. Hence using Cramer–Fisz theorem

$$X_{[np]+1} \sim AN\left(q_p, \frac{p(1-p)}{n[f(q_p)]^2}\right)$$

In this case, $p = \frac{1}{2}, q_p = \theta$

$f(q_p) = f(\theta) = \frac{1}{2}$

Median $= M_n \sim AN(\theta, \frac{1}{n})$

$\Rightarrow EM_n = \theta$ and $V(M_n) \to 0$ as $n \to \infty$ $M_n \xrightarrow{P} \theta$

To find n_0,

Consider $P[|M_n - \theta| < \epsilon] \geq 1 - \delta$

$$P[|M_n - \theta| < \epsilon] \geq 1 - \delta$$

$$\Rightarrow \phi[\epsilon\sqrt{n}] \geq 1 - \frac{\delta}{2}$$

$$\Rightarrow n \geq \frac{1}{\epsilon^2}\left[\phi^{-1}\left(1 - \frac{\delta}{2}\right)\right]^2$$

$$\Rightarrow n_0(T_1) = \left[\frac{1}{\epsilon^2}\left\{\phi^{-1}\left(1 - \frac{\delta}{2}\right)\right\}^2\right] + 1 \quad \text{see Table 5.6}$$

Consider the another estimator $\bar{X}_n, E\bar{X}_n = \theta$ and $\lim_{n\to\infty} V\left(\bar{X}_n\right) = 0$.

Hence $\bar{X}_n \xrightarrow{P} \theta$

Using CLT $\bar{X}_n \sim N(\theta, \frac{2}{n})$

Hence,

$$P[|\bar{X}_n - \theta| < \epsilon] \geq 1 - \delta$$

$$P\left[-\epsilon\sqrt{\frac{n}{2}} < z < \epsilon\sqrt{\frac{n}{2}}\right] \geq 1 - \delta, \quad \text{where } z \sim N(0, 1)$$

Table 5.6 n_0 for median

$\delta\backslash\epsilon$	0.1	0.01
0.1	271	664
0.01	27,061	66,307

Table 5.7 n_0 for mean

$\delta\backslash\epsilon$	0.1	0.01
0.1	542	1327
0.01	54,121	132,613

$$2\phi\left[\epsilon\sqrt{\frac{n}{2}}\right] - 1 \geq 1 - \delta$$

$$\phi\left[\epsilon\sqrt{\frac{n}{2}}\right] \geq 1 - \frac{\delta}{2}$$

Therefore,

$$\Rightarrow n \geq \frac{2}{\epsilon^2}\left[\phi^{-1}(1 - \frac{\delta}{2})\right]^2$$

$$\Rightarrow n_0(T_1) = \left[\frac{2}{\epsilon^2}\left\{\phi^{-1}\left(1 - \frac{\delta}{2}\right)\right\}^2\right] + 1 \quad \text{see Table 5.7}$$

n_0 for $\epsilon = 0.1, 0.01$ and $\delta = 0.1, 0.01$

Tables 5.6 and 5.7 give the values of n_0 using R

```
# To enter epsilon (eps) and delta (del).
  eps <- c(0.1,0.01); del <- c(0.1,0.01);
  ld <- length(del); # length of delta
  le <- length(eps); # length of epsilon
# declaring matrices for n0t1 = median and n0t2 = mean.
  n0t1 <- matrix(data=0,nrow=ld,ncol=le)
  n0t2 <- matrix(data=0,nrow=ld,ncol=le)
# To find n0 for T1.
  for(i in 1:ld)
    {
    for(j in 1:le)
      {
      x <- 1-(del[i]/2); y <- qnorm(x,0,1);
      n0t1[i,j] <- ((y^2)/(eps[j]^2))+1
      }
    }
  colnames(n0t1) <- c("[0.1]","[0.01]")
  rownames(n0t1) <- c("[0.1]","[0.01]")
# To fnd n0 for T2.
  for(i in 1:ld)
    {
```

```
for(j in 1:le)
{
    x <- 1-(del[i]/2); y <- qnorm(x,0,1);
    n0t2[i,j] <- (2*(y^2)/(eps[j]^2))+1
}
}
colnames(n0t2) <- c("[0.1]","[0.01]")
rownames(n0t2) <- c("[0.1]","[0.01]")
# OUTPUT
print("n0 for T1"); n0t1
print("n0 for T2"); n0t2
```

Note: In the above examples, $n_0(T)$ is independent of μ.

Practical advantage of $n_0(T)$ being independent of θ is that the minimum sample size required to achieve (ϵ, δ) level accuracy does not depend on the unknown parameter θ, then such a minimum sample size cannot be determined as θ is unknown. We can take $N_0 = \sup_{\theta \in \Theta} n_0(T)$ if such N_0 is finite, which is usually the case if Θ is bounded or closed compact set and $n_0(\epsilon, \delta, \theta)$ for a given T is a continuous function of θ.

Example 5.5.4 Let $X_1, X_2, \ldots, X_n$ are iid $\cup(0, \theta)$ rvs.

Let $T_1 = X_{(n)}$ be a consistent estimator of θ.

$$P[|X_{(n)} - \theta| < \epsilon] = \begin{cases} 1 - (\frac{\theta - \epsilon}{\theta})^n & ; \epsilon < \theta \\ 0 & ; \epsilon \geq \theta \end{cases}$$

If $\epsilon < \theta$ then

$$1 - \left(\frac{\theta - \epsilon}{\theta}\right)^n \geq 1 - \delta,$$

Then

$$n \geq \frac{\log \delta}{\log(1 - \frac{\epsilon}{\theta})}$$

Thus

$$n_0(T_1) = \begin{cases} \left[\frac{\log \delta}{\log(1 - \frac{\epsilon}{\theta})} \right] + 1 & ; \epsilon < \theta \\ 1 & ; \epsilon \geq \theta \end{cases} \tag{5.5.13}$$

Let $T_2 = 2\bar{X}$ be a consistent estimator of θ.

By CLT, $2\bar{X} \sim AN(\theta, \frac{\theta^2}{3n})$,

Then $\frac{\sqrt{3n}}{\theta}(2\bar{X} - \theta) \sim AN(0, 1)$

$$P[|2\bar{X} - \theta| < \epsilon] = \Phi\left(\frac{\epsilon\sqrt{3n}}{\theta}\right) - \Phi\left(-\frac{\epsilon\sqrt{3n}}{\theta}\right)$$

$$= 2\Phi\left(\frac{\epsilon\sqrt{3n}}{\theta}\right) - 1$$

$n_0(T_2)$ is given by

$$2\Phi\left(\frac{\epsilon\sqrt{3n}}{\theta}\right) - 1 \geq 1 - \delta,$$

$$n_0(T_2) = \left[\frac{\theta^2}{3\epsilon^2}\left\{\Phi^{-1}\left(1 - \frac{\delta}{2}\right)\right\}^2\right] + 1 \qquad (5.5.14)$$

The following table gives the values of $n_0(T_1)$ and $n_0(T_2)$ for a given θ, $\epsilon = \delta = 0.1$ using R

```
# Given data
  eps <- 0.1; del <- 0.1; theta <- c(0.5,1,2,4);
  lt <- length(theta);
# Declaring vectors for n0 for T1, T and T2.
  n0t1 <- rep(0,lt); n0t <- rep(0,lt); n0t2 <- rep(0,lt);
# To fnd n0(T1).
  for(i in 1:lt)
    {
      x <- 1-(eps/theta[i]);
      n0t1[i] <- log(del)/log(x) +1;
    }
# To fnd n0(T).
  for(i in 1:lt)
    {
      x <- theta[i]/eps;
      n0t[i] <- -x*log(del) +1;
    }
# To fnd n0(T2).
  for(i in 1:lt)
    {
      x <- 1-(del/2); y <- qnorm(x,0,1);
      n0t2[i] <- ((theta[i]^2)*(y^2)/(3*eps^2))+1
    }
# OUTPUT
  print("n0 for T1 using Eq.\,5.5.13"); n0t1
  print("n0 for T1"); n0t;
  print("n0 for T2"); n0t2
```

Note: We will show that $Y_n = n(\theta - X_{(n)}) \xrightarrow{d} Y$, where Y has exponential distribution with mean θ.

$$
\begin{aligned}
H_n(y) &= P[n(\theta - X_{(n)}) \leq y] \\
&= P\left[X_{(n)} \geq \theta - \frac{y}{n}\right] \\
&= \int_{\theta - \frac{y}{n}}^{\theta} \frac{nt^{n-1}}{\theta^n} \\
&= 1 - \left(1 - \frac{y}{n\theta}\right)^n
\end{aligned}
$$

$$
\lim_{n \to \infty} H_n(y) = \lim_{n \to \infty} 1 - \left(1 - \frac{y}{n\theta}\right)^n = 1 - e^{-\frac{y}{\theta}}
$$

$\Rightarrow$ Y has exponential distribution with mean θ.

Therefore,

$$
P[n(\theta - X_{(n)}) \leq n\epsilon] = P[Y_n \leq n\epsilon]
$$

This tend to

$$
P[Y_n \leq n\epsilon] = 1 - e^{\frac{-n\epsilon}{\theta}},
$$

Then

$$
1 - e^{\frac{-n\epsilon}{\theta}} \geq 1 - \delta
$$

$$
\Rightarrow n_0(T) = \left[\frac{\theta}{\epsilon}(-\log \delta)\right] + 1 \tag{5.5.15}
$$

From the following Table 5.9, we will get almost same value of $n_0(T)$ as in the Table 5.8.

Table 5.8 n_0 for $X_{(n)}$ and $2\bar{X}$

θ	$\frac{1}{2}$	1	2	4
$n_0(T_1)$	11	22	45	91
$n_0(T_2)$	23	90	359	1435

Table 5.9 n_0 for $X_{(n)}$

θ	$\frac{1}{2}$	1	2	4
$n_0(T)$	12	24	47	93

5.6 Exercise 5

1. Let $X_1, X_2, \ldots, X_n$ are iid rvs with $\cup(0, \theta)$. Show that $T(X) = (\prod_{i=1}^{n} X_i)^{\frac{1}{n}}$ is consistent estimator for $\frac{\theta}{e}$.

2. Let $X_1, X_2, \ldots, X_n$ are iid rvs with the pdf $f(x|\theta)$, where $f(x|\theta) = \theta x^{\theta-1}$; $0 < x < 1$. Find the consistent estimator for θ and e^θ. Further obtain n_0 for both the estimators.

3. Let $X_1, X_2, \ldots, X_n$ are iid rvs with $EX_i = \mu$ and $EX_i^2 < \infty$. Show that $T(X) = \frac{2}{n(n+1)} \sum_{i=1}^{n} i X_i$ is a consistent estimator for μ.

4. If $X_1, X_2, \ldots, X_n$ are random observations on a Bernoulli variate X such that $P(X = 1) = p$ and $P(X = 0) = 1 - p$, $0 < p < 1$. Show that (i) $\bar{X}$ is consistent estimator of p.

(ii) $\bar{X}(1 - \bar{X})$ is a consistent estimator of $p(1 - p)$. Find if possible n_0 in both cases.

5. Let $X_1, X_2, \ldots, X_n$ are iid rvs with geometric distribution as,

$$P[X = x] = pq^x, \ x = 0, 1, 2 \ldots$$

Find the consistent estimator for p^{-1} and $\frac{q}{p}$. Further, find n_0 for both estimators.

6. Let $X_1, X_2, \ldots, X_n$ are iid rvs with $\cup(0, \theta)$. Prove or disprove $X_{(1)}$ is a consistent estimator of θ.

7. If $X_1, X_2, \ldots, X_n$ is a random sample obtained from the density function:

$$f(x|\theta) = \begin{cases} 1 \ ; \ \theta < x < \theta + 1 \\ 0 \ ; \ \text{otherwise} \end{cases}$$

Show that $\bar{X}$ is a consistent estimator of $\theta + \frac{1}{2}$. Find n_0 for $\theta = 2, 3$.

8. Show that in sampling from Cauchy distribution

$$f(x|\mu) = \frac{1}{\pi\left[1 + (x - \mu)^2\right]} \ ; \ -\infty < x < \infty, \mu > 0,$$

(i) Sample mean $\bar{X}$ is not a consistent estimator of θ.
(ii) Sample median is a consistent estimator of θ.

9. If T_1 and T_2 are consistent estimators of $g(\theta)$, prove that $\alpha_1 T_1 + \alpha_2 T_2$, such that $\alpha_1 + \alpha_2 = 1$, is also consistent for $g(\theta)$.

10. For a Poisson distribution with mean λ, show that $\bar{X}$ is a consistent estimator for λ. Find n_0. Further prove that $\frac{1}{\bar{X}}$ is consistent estimator of $\frac{1}{\lambda}$. Can you find n_0 in case of an estimator $\frac{1}{\bar{X}}$? Give reasons.

11. Let $X_1, X_2, \ldots, X_n$ be a random sample from a population with pdf

$$f(x|\theta) = \frac{1}{2\theta}; \quad -\theta < x < \theta, \ \theta > 0$$

Find, if exists, a sufficient consistent estimator of θ and its corresponding n_0. (Hint: Consider $\max_i |X_i|$)

12. Let $X_1, X_2, \ldots, X_n$ be independent random sample with pdf

$$f(x_i|\theta) = \frac{1}{2i\theta}; \quad -i(\theta - 1) < x_i < i(\theta + 1), \ \theta > 0,$$

Find a sufficient consistent estimator of θ. Further, if exists, find n_0.

13. Let $X_1, X_2, \ldots, X_n$ be iid rvs with the pdf

$$f(x|\mu, \sigma) = \frac{1}{\sigma} \exp\left[-\frac{(x-\mu)}{\sigma}\right]; \quad x > \mu, \sigma > 0,$$

Find the sufficient and consistent estimator for μ and σ. If possible, find n_0 for both consistent estimators of μ and σ.

14. Let $X_1, X_2, \ldots, X_m$ be iid rvs with $B(n, p)$. Find the consistent estimator of p and the minimum sample size.

15. Let $X_1, X_2, \ldots, X_n$ be iid exponential rvs with mean λ. Find a consistent estimator for λ and its n_0.

16. Let $X_1, X_2, \ldots, X_n$ be iid rvs with $NB(r, \theta)$. Find the consistent estimator of θ, θ^2 and e^θ.

17. Let the rv X_1 be $B(n, \theta)$ and rvs $X_2, X_3 \ldots, X_n$ be exponentially distributed with mean θ. Find the consistent estimator of θ, θ^2, $\frac{1}{\theta}$ and e^θ.

18. Let $X_1, X_2, \ldots, X_n$ are iid rvs with (i) $\cup(-\theta, 0)$, $\theta > 0$ (ii) $\cup(\theta, \theta^2)$, $\theta > 1$ (iii) $\cup(0, \theta^2)$

Find the consistent estimator of θ. Further, find its n_0.

References

Athreya KB, Lahiri SN (2006) Measure theory and probability theory. Springer
Bhat BR (2004) Modern probability theory and its applications. Wiley

Billingsley P (2012) Probability and measure. Wiley, New York
David HA, Nagaraja HN (2003) Order statistics. Wiley
Feller W (1970) An introduction to probability theory and its applications, vol 1. Wiley
Kale BK (1999) A first course on parametric inference. Narosa Publishing House, New Delhi
Rohatagi VK, Saleh EAK (2001) An introduction to probability and statistics. Wiley
Serfling RJ (1980) Approximation theorems of mathematical statistics. Wiley
Stoyanov JM (1997) Counter examples in probability. Wiley

Chapter 6
Bayes Estimator

In all the previous chapters, we have considered the moment, maximum likelihood and uniformly minimum variance unbiased estimators. Generally, mle are better than moment estimators with reference to their mse. More recently, Dixit and Kelkar (2011) had shown that for a binomial distribution, moment estimators are better than mle with reference to the generalized variance in the presence of outliers. In general, it is not possible to decide which estimation procedure is better among mle and UMVUE.

Strictly speaking, there is no such thing as an unconditional probability. However, it often happens that many probability statements are made conditional on everything that is part of an individual's knowledge at a particular time. When many statements are to be made conditional on the same event, it makes for cumbersome notation to refer to this same conditioning event every time.

The Bayesian philosophy involves a completely different approach to statistics. The Bayesian version of estimation is considered here for the basic situation concerning the estimation of a parameter, given a random sample from a particular distribution.

The fundamental difference between Bayesian and classical methods is that the parameter θ is a fixed unknown quantity. This leads to difficulties such as interpreting the classical confidence intervals, because the interval is random in Bayesian interval estimation. For example, if we have a random sample of size n from $N(\mu, 1)$ then 95 % confidence interval is, $\left(\bar{X} - 1.96\frac{1}{\sqrt{n}}, \bar{X} + 1.96\frac{1}{\sqrt{n}} \right)$. We can interpret in classical methods but if μ is random then it is difficult to interpret. Once the data are observed then one can give the confidence interval for θ, because probability is not involved in such a situation.

In the beginning, we will consider some examples based on Bayes theorem.

1. In 2011, there are three candidates for the position of vice-chancellor Dr. Joshi, Dr. Sawant and Dr. Rege whose chances of getting the appointment are in the proportions 5:6:9 respectively. The probability that Dr.Joshi if selected would introduce credit system in the university is 0.6. The probabilities of Dr. Sawant and Dr. Rege doing the same are respectively 0.5 and 0.4.

© Springer Science+Business Media Singapore 2016
U.J. Dixit, *Examples in Parametric Inference with R*,
DOI 10.1007/978-981-10-0889-4_6

If credit system has been introduced, what is the probability that Dr. Joshi, Dr. Sawant and Dr. Rege is the vice-chancellor?

Here, we are not solving the problem.

If c=credit system, given probabilities are $p(c|J), p(c|S)$ and $p(c|R)$, where J, S, and R denotes for Joshi, Sawant and Rege. But we want to find the probability $p(J|c), p(S|c)$ and $p(R|c)$.

2. Lee (1997) had given the following interesting example.

A case of alleged discrimination on the basis of a test to determine eligibility for promotion was considered. It turned out that, of those taking the test, 48 were Black(B) and 259 were White(W), so that if we consider the test

$$p(B) = \frac{48}{307} = 0.16, p(W) = \frac{259}{307} = 0.84.$$

Of the Blacks taking the test, 26 passed(P) and the rest failed(F), whereas of the Whites, 206 passed and the rest failed, so that altogether 232 people passed.

Hence $p(B|P) = \frac{26}{232} = 0.11, p(W|P) = \frac{206}{232} = 0.89.$

One may think that these figures indicate the possibility of discrimination. But instead of the figures that should be considered are $p(P|B) = \frac{26}{48} = 0.54, p(P|W) = \frac{206}{259} = 0.80.$

One should see the fact that $p(B|P)$ is less than $p(W|P)$ is irrelevant to the real question as to whether $p(P|B)$ is less than $p(P|W)$. Therefore, it might or might not be depending on the rest of the relevant information, that is, $p(B)$ and $p(W)$.

It is easily checked that the probabilities are related by Bayes theorem in both the examples. In the first example, given probabilities were $p(c|J), p(c|S)$, and $p(c|R)$.

Further, $p(J) = \frac{5}{20}, p(S) = \frac{6}{20}$, and $p(R) = \frac{9}{20}$, these probabilities of becoming vice-chancellor are given. In this case, posterior probabilities are $p(J|c), p(S|c)$, and $p(R|c)$. Similarly, from $p(B|P)$ and $p(W|P)$, the figures indicate serious discrimination. But from the figures $p(P|B)$ and $p(P|W)$, moderate discrimination may be there.

6.1 Bayes Theorem

Bayes theorem was given by a British mathematician Thomas Bayes in 1763. Given the new information, he updated the prior probabilities by calculating revised probabilities and referred to them as posterior probabilities.

Theorem 6.1.1 (Bayes Theorem) *If $E_1, E_2, \ldots, E_n$ are mutually disjoint events with $P(E_i) > 0 (i = 1, 2, \ldots, n)$ then for any arbitrary event H which is a subset of $\cup_{i=1}^{n} E_i$ such that $P(H) > 0$, we have*

$$P(E_i|H) = \frac{P(E_i)P(H|E_i)}{\sum P(E_i)P(H|E_i)} = \frac{P(E_i)P(H|E_i)}{P(H)}; \quad i = 1, 2, \ldots, n \qquad (6.1.1)$$

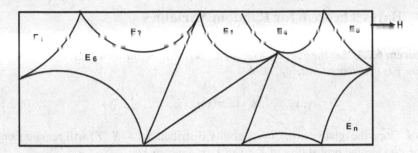

Fig. 6.1 Partition of events

Proof See Fig. 6.1.
  Since

$$H \subseteq \cup_{i=1}^n E_i \Rightarrow H = H \cap (\cup_{i=1}^n E_i) = \cup_{i=1}^n (H \cap E_i)$$

This is true to distributive law. Since $H \cap E_i (i = 1, 2, \ldots, n)$ are mutually disjoint events, using addition theorem of probability

$$P(H) = P[\cup_{i=1}^n (H \cap E_i)] = \sum_{i=1}^n P(H \cap E_i) = \sum_{i=1}^n P(E_i)P(H|E_i)$$

Now $P(H \cap E_i) = P(H)P(E_i|H)$

$$P(E_i \mid H) = \frac{P(H \cap E_i)}{P(H)} = \frac{P(H)P(E_i|H)}{\sum_{i=1}^n P(E_i)P(H|E_i)}$$

Remark 1. The probabilities $P(E_i)$, $i = 1, 2, \ldots, n$ are called as the prior probabilities, because they exist before we gain any information from the experiment itself.
2. The probabilities $P(H|E_i)$, $i = 1, 2, \ldots, n$ are called likelihoods because they indicate how likely the event H under consideration is to occur, given each and every prior probability $E_i (i = 1, 2, \ldots, n)$.
3. The probabilities $P(E_i|H)$, $i = 1, 2, \ldots, n$ are called posterior probabilities, because they are determined after the results of the experiments are known.
4. From the figure, we can conclude:
  If the events $E_1, E_2, \ldots, E_n$ constitute a disjoint partition of the sample space Ω and $P(E_i) > 0$, $i = 1, 2, \ldots, n$ then for every event H in Ω, we have

$$P(H) = \sum_{i=1}^n P(H \cap E_i) = \sum_{i=1}^n P(E_i)P(H|E_i)$$

5. Bayes theorem is extensively used in statistical inference, and by business and management executives in arriving at valid decisions in the face of uncertainty.

6.2 Bayes Theorem for Random Variables

Theorem 6.2.1 *Let the rvs $X_1, X_2, \ldots, X_n$ given Y be iid with $f(x_1, x_2, \ldots, x_n|y)$, where y is distributed as $\pi(y)$. Then*

$$f(y|x_1, x_2, \ldots, x_n) \propto \pi(y)f(x_1, x_2, \ldots, x_n|y) \qquad (6.2.1)$$

Proof Since the relative size of probability distribution of (X, Y) will remain same in the conditional probability of Y given X or X given Y.

Since $f(X|Y) \geq 0$, $\int f(X|Y)dx = 1$

Similarly $f(Y|X) \geq 0$, $\int f(Y|X)dy = 1$

Hence, $f(X, Y) = f(X|Y)\pi(y)$

$f(x) = \int f(X, Y)dy = \int f(X|Y)\pi(y)dy$,

$\pi(y)$ is the pdf of a rv y

It is clear that

$$f(Y|X) = \frac{f(X, Y)}{f(X)} = \frac{f(X|Y)\pi(y)}{f(X)}$$

This implies that

$$f(Y|X) \propto f(X|Y)\pi(y) \qquad (6.2.2)$$

Note: 1. If we replace y for θ then we have our density function $f(x|\theta)$. In this case, pdf or pmf of θ is called prior density of θ.

2. The conditional distribution of θ is given as X i.e., $f(\theta|X)$ is called the posterior probability distribution of θ, given the sample.

6.3 Bayesian Decision Theory

In statistical inference, decision about the population parameter is taken from the sample data. Consider the following example:

A statistician is told that a coin has either a head on one side and a tail on the other side or it has two heads.

A statistician cannot inspect the coin but can observe a single toss of the coin and see whether it shows a head or tail. The statistician must then decide whether or not the coin is two-headed. If the statistician makes the wrong decision there is a penalty of Rs. 1 and otherwise there is no penalty.

Ignoring the fact that the statistician can observe the toss of coin, the problem could be regarded as follows:

Statistician
(Player A)

		a_1	a_2
Nature	θ_1	0	1
(Player B)	θ_2	1	0

$\theta_1 =$ The state of nature is that the coin is two-headed.
$\theta_2 =$ The state of nature is that the coin is balanced.
$a_1 =$ statistician's decision is that the coin is two-headed.
$a_2 =$ Statistician's decision is that the coin is balanced.
Let X be a random variable taking values 0 (heads) and 1 (tails). Consider the following decision function

$$d_1(X) = \begin{cases} a_1; & \text{when X=0} \\ a_2; & \text{when X=1} \end{cases}$$

one can write $d_1(0) = a_1$ and $d_1(1) = a_2$.
    Note that if head occurs, then $X = 0$ and if tail occurs then $X = 1$. Other decision functions are

$$d_2(0) = a_1, d_2(1) = a_1$$

It implies that choose a_1 whatever may be outcome of the experiment.

$$d_3(0) = a_2, \ d_3(1) = a_2 \ d_4(0) = a_2, \ d_4(1) = a_1$$

Some of the decision functions may not be very sensible in practice.
    Consider the loss function in the table

Statistician (Player A)

		a_1	a_2
Nature	θ_1	$L(a_1,\theta_1)$	$L(a_2, \theta_1)$
(Player B)	θ_2	$L(a_1, \theta_2)$	$L(a_2, \theta_2)$

Our option is to choose a_1 when $X = 0$ and a_2 when $X = 1$.
Consider the risk function as $R(d_i, \theta_j)$, $i = 1, 2, 3, 4$ and $j = 1, 2$

$$R(d_i, \theta_j) = E[L(d_i, \theta_j)]$$

Under θ_1, $P[X = 0] = 1$ and $P[X = 1] = 0$.
    Note that the coin is two-headed, i.e., the toss resulting out in head is certain event.
Under θ_2, $P[X = 0] = \frac{1}{2}$, and $P[X = 1] = \frac{1}{2}$.
    This gives

$$R(d_1, \theta_1) = 1L(a_1, \theta_1) + 0L(a_2, \theta_1) = 1(0) + 0(1) = 0$$

$$R(d_1, \theta_2) = \frac{1}{2}L(a_1, \theta_2) + \frac{1}{2}L(a_2, \theta_2) = \frac{1}{2}(1) + \frac{1}{2}(0) = \frac{1}{2}$$

$$R(d_2, \theta_1) = 1L(a_1, \theta_1) + 0L(a_1, \theta_1) = 1(0) + 0(0) = 0$$

$$R(d_2, \theta_2) = \frac{1}{2}L(a_2, \theta_2) + \frac{1}{2}L(a_2, \theta_2) = \frac{1}{2}(1) + \frac{1}{2}(0) = \frac{1}{2}$$

$$R(d_3, \theta_1) = 1L(a_2, \theta_1) + 0L(a_2, \theta_1) = 1(1) + 0(0) = 1$$

$$R(d_3, \theta_2) = \frac{1}{2}L(a_2, \theta_2) + \frac{1}{2}L(a_2, \theta_2) = \frac{1}{2}(0) + \frac{1}{2}(0) = 0$$

$$R(d_4, \theta_1) = 1L(a_2, \theta_1) + 0L(a_1, \theta_1) = 1(1) + 0(0) = 1$$

$$R(d_4, \theta_2) = \frac{1}{2}L(a_2, \theta_2) + \frac{1}{2}L(a_1, \theta_2) = \frac{1}{2}(0) + \frac{1}{2}(1) = \frac{1}{2}$$

The Minimax criteria

We are assuming the minimax criteria, i.e., each player chooses the strategy that minimizes their maximum loss or it otherwise maximizes their minimum gain. We can say that each player will consider the worst possible outcome they could get for each move they make and then select the move for which this works out to be the least worst.

i	$R(d_i, \theta_1)$	$R(d_i, \theta_2)$	$\text{Max}_{\theta_1, \theta_2} R(d_i, \theta)$
1	0	$\frac{1}{2}$	$\frac{1}{2}$
2	0	$\frac{1}{2}$	$\frac{1}{2}$
3	1	0	1
4	1	$\frac{1}{2}$	1

$\text{Min}_i \text{Max}_{\theta_1, \theta_2} R(d_i, \theta) = \frac{1}{2}$. The minimax solution is,

$$d_1(X) = \begin{cases} a_1 \ ; \ X = 0 \\ a_2 \ ; \ X = 1 \end{cases}$$

or

$$d_2(X) = \begin{cases} a_1 \ ; \ X = 0 \\ a_1 \ ; \ X = 1 \end{cases}$$

If θ is regarded as a random variable, under the Bayes criterion, the decision function chosen is that for which $E[R(d, \theta)]$ is minimum where the expectation is taken with respect to θ.

Different Loss Functions

Loss function is a function that maps an event into a real number intuitively representing some "cost" associated with the event. According to the earlier examples, a statistician will suffer loss of $L(\theta, a)$ if he takes action 'a' when the true state nature is θ. If we use the decision function $d(x)$, when L is the loss function and θ is the true parameter value, the loss is the r.v. $L(\theta, d(x))$ for all loss function, $\theta \in \Theta, a \in A$.

1. Quadratic Loss Function

$$L(\theta, a) = (\theta - a)^2, \tag{6.3.1}$$

2. Weighted Quadratic Loss Function

$$L(\theta, a) = w(\theta)(\theta - a)^2 \tag{6.3.2}$$

3. Absolute Loss Function

$$L(\theta, a) = |\theta - a| \tag{6.3.3}$$

4. Zero-One Loss Function

$$L(\theta, a) = \begin{cases} 0 \; ; \; |\theta - a| \leq \epsilon \\ 1 \; ; \; |\theta - a| > \epsilon \end{cases} \tag{6.3.4}$$

On the basis of data, the statistician chooses an action $d(X) \in A$, resulting in a random loss $L(\theta, d(x))$. Then the risk function $R(\theta, d)$ is defined as

$$R(\theta, d) = EL(\theta, d(x))$$

$$= \int L(\theta, d(x)) f(x|\theta) dx \tag{6.3.5}$$

Let the prior density of θ is $p(\theta)$. The Bayes risk $r(d)$ of the decision rule d can be defined as $ER(\theta, d)$ over all possible values of θ, where θ and X are continuous rvs.

$$r(d) = ER(\theta, d) = \int R(\theta, d) p(\theta) d\theta$$

$$= \int \left[\int L(\theta, d(x)) f(x|\theta) dx \right] p(\theta) d\theta$$

$$= \int \int L(\theta, d(x)) f(x, \theta) dx d\theta$$

$$= \int \left\{ \int L(\theta, d(x)) f(\theta|x) d\theta \right\} f(x) dx \tag{6.3.6}$$

Note that $f(x) = \int f(x|\theta)p(\theta)d\theta$ and $h(x) = \int L(\theta, d(x))f(\theta|x)d\theta$.

$$r(d) = \int h(x)f(x)dx$$

Then the Bayes risk is minimized if the decision rule d^* is chosen such that $r(d)$ is minimum for all x.

If θ and X are discrete rvs, then

$$r(d) = \sum_{\theta} \sum_{x} L(\theta, d(x))f(x, \theta) \tag{6.3.7}$$

Definition 6.3.1 A decision rule d^* is known as a Bayes rule if it minimizes $r(d)$ or (6.3.6) or (6.3.7)

$$r(d^*) = \inf_{d} r(d) \tag{6.3.8}$$

Theorem 6.3.1 *Consider a problem of estimation of a parameter $\theta \in \Theta$ with respect to a quadratic loss function $L(\theta, d) = (\theta - d)^2$. A bayes rule is given by $d^*(x) = E(\theta|X = x)$, where $d^*(x)$ is known as an Bayes estimate.*

Proof From (6.3.6), we have to minimize

$$R(\theta, d) = \int L(\theta, d)f(\theta|x)d\theta$$

$$= \int (\theta - d)^2 f(\theta|x)d\theta$$

$$= \int (\theta^2 - 2\theta d + d^2)f(\theta|x)d\theta \tag{6.3.9}$$

Differentiating (6.3.9) with respect to d and put it equal to zero, then

$$\int (-2\theta + 2d)f(\theta|x)d\theta = 0$$

$$\Rightarrow \int df(\theta|x)d\theta = \int \theta f(\theta|x)d\theta$$

$$\Rightarrow d^* = E(\theta|X) \tag{6.3.10}$$

Theorem 6.3.2 *In Theorem 6.3.1, if the loss function is weighted quadratic then $L(\theta, d) = W(\theta)(\theta - d)^2$. A Bayes rule is given by*

$$d^*(X) = \frac{E\theta W(\theta)|X}{EW(\theta)|X} \tag{6.3.11}$$

Proof From (6.3.6), we have to minimize

$$R(\theta, d) = \int L(\theta, d(X)) f(\theta|x) d\theta$$

$$= \int W(\theta)(\theta - d)^2 f(\theta|x) d\theta$$

$$= \int W(\theta)(\theta^2 - 2\theta d + d^2) f(\theta|x) d\theta \qquad (6.3.12)$$

Differentiating (6.3.12) with respect to d and equating it to zero, then

$$\int W(\theta)(-2\theta + 2d) f(\theta|x) d\theta = 0$$

$$\Rightarrow \int \theta W(\theta) f(\theta|x) d\theta = d \int W(\theta) f(\theta|x) d\theta$$

$$\Rightarrow d^* = \frac{\int \theta W(\theta) f(\theta|x) d\theta}{\int W(\theta) f(\theta|x) d\theta}$$

$$\Rightarrow d^*(X) = \frac{E\theta W(\theta)|X}{EW(\theta)|X}$$

Theorem 6.3.3 *In Theorem 6.3.1, if the loss function is absolute error loss function then* $L(\theta, d) = |\theta - d|$. *A Bayes rule is given by*

$$d^*(X) = \text{Median of the posterior distribution of } \theta \text{ given } X. \qquad (6.3.13)$$

Proof Let M be the median of the posterior distribution given X.
   Hence $P(\theta \leq M|X) \geq \frac{1}{2}$ and $P(\theta \geq M|X) \geq \frac{1}{2}$.
   Let d_1 be any rule and for definiteness, $d_1 > M$ for some particular value of X.

There are three cases: (i) $\theta < M < d_1$ (ii) $M < \theta < d_1$ (iii) $M < d_1 < \theta$
   In case (i)

$$L(\theta, M) = -(\theta - M) \text{ and}$$

$$L(\theta, d_1) = -(\theta - d_1)$$

In case (ii)

$$L(\theta, M) = (\theta - M) \text{ and}$$

$$L(\theta, d_1) = -(\theta - d_1)$$

In case (iii)

$$L(\theta, M) = (\theta - M) \text{ and}$$

$$L(\theta, d_1) = (\theta - d_1)$$

$$L(\theta, M) - L(\theta, d_1) = \begin{cases} M - d_1 & ; \ \theta < M < d_1 \\ 2\theta - (M + d_1) & ; \ M < \theta < d_1 \\ d_1 - M & ; \ M < d_1 < \theta \end{cases}$$

For $M < \theta < d_1$

$$2\theta - (M + d_1) < \theta - M < d_1 - M$$

So that

$$L(\theta, M) - L(\theta, d_1) \leq \begin{cases} M - d_1 & ; \ \theta \leq M \\ d_1 - M & ; \ \theta > M \end{cases}$$

Hence

$$\mathrm{E}[L(\theta, M) - L(\theta, d_1)] \leq (M - d_1)P(\theta \leq M|X) + (d_1 - M)P(\theta > M|X)$$

$$= (M - d_1)P(\theta \leq M|X) + (d_1 - M)[1 - P(\theta \leq M|X)]$$

$$= (d_1 - M)\{1 - 2P(\theta \leq M|X)]\}$$

Hence,

$$\mathrm{E}[L(\theta, M) - L(\theta, d_1)] \leq 0$$

Similarly, we can show that if $d_1 < M$.

Hence, we can conclude that posterior median is the approximate Bayes rule for this loss function.

Note: For zero-one loss function, $\mathrm{E}L(\theta, a) = P[|\theta - a| > \epsilon] = 1 - P[|\theta - a| \leq \epsilon]$ A modal interval of length 2ϵ is defined as $(a - \epsilon, a + \epsilon)$, where a is mode of the observation, which has the highest probability for given t. Then the mid point of Mode(X) of this interval is a Bayes estimate for this loss function. For details, see Lee (1997).

Bayes Risk
In calculation of Bayes risk d^*, one can use the following procedure:
First procedure
1. Find

$$\int L(\theta, d^*(x))f(x|\theta)dx \quad ; \text{if } X \text{ is continuous}$$

$$\sum_X L(\theta, d^*(x))f(x|\theta) \quad ; \text{if } X \text{ is discrete} \qquad (6.3.14)$$

2. Let

$$q(\theta) = \int\limits_X L(\theta, d^*(x)) f(x|\theta) dx; \quad \text{if } X \text{ is continuous}$$

$$q(\theta) = \sum\limits_X L(\theta, d^*(x)) f(x|\theta); \quad \text{if } X \text{ is discrete}$$

Then

$$r(d^*) = \int q(\theta) p(\theta) d\theta; \quad \text{if } \theta \text{ is continuous}$$

$$r(d^*) = \sum\limits_\theta q(\theta) p(\theta); \quad \text{if } \theta \text{ is discrete} \tag{6.3.15}$$

Second procedure
1. Find $h(x)$, where

$$h(x) = \int\limits_\theta L(\theta, d^*(x)) f(\theta|x) d\theta \; ; \quad \text{if } \theta \text{ is continuous}$$

$$h(x) = \sum\limits_\theta L(\theta, d^*(x)) f(\theta|x); \quad \text{if } \theta \text{ is discrete rv} \tag{6.3.16}$$

2.

$$r(d^*) = \int\limits_X h(x) f(x) dx; \quad \text{if } X \text{ is continuous}$$

$$r(d^*) = \sum\limits_X h(x) f(x) \; ; \text{if } X \text{ is discrete rv} \tag{6.3.17}$$

6.4 Limit Superior and Limit Inferior

If $\{x_n\}_{n=1}^\infty$ is a convergent sequence, then $\lim_{n\to\infty} x_n$ measures, roughly, "the size of x_n when n is large". In this section, we introduce the concepts of limit superior and limit inferior which can be applied to all sequences.

Definition 6.4.1 Let $\{x_n\}_{n=1}^\infty$ be a sequence of real numbers that is bounded above and let $M_n = $ Least upper bound (l.u.b) $\{x_n, x_{n+1}, x_{n+2}, \ldots\}$

(i) If $\{M_n\}_{n=1}^\infty$ converges, we define
$\lim\sup_{n\to\infty} x_n$ to be $\lim_{n\to\infty} M_n$

(ii) If $\{M_n\}_{n=1}^{\infty}$ diverges to $-\infty$, then we write

$$\limsup_{n \to \infty} x_n = -\infty$$

For example, $x_n = (-1)^n$ then $M_n = 1$
Hence $\lim_{n \to \infty} M_n = 1 \Rightarrow \limsup_{n \to \infty}(-1)^n = 1$

Definition 6.4.2 If $\{x_n\}_{n=1}^{\infty}$ be a sequence of real numbers that is not bounded above, we write

$$\limsup_{n \to \infty} x_n = \infty$$

Theorem 6.4.1 *If $\{x_n\}_{n=1}^{\infty}$ is a convergent sequence of real numbers, then*

$$\limsup_{n \to \infty} x_n = \lim_{n \to \infty} x_n$$

Definition 6.4.3 Let $\{x_n\}_{n=1}^{\infty}$ be a sequence of real numbers that is bounded below and let $m_n = $ greatest lower bound $\{x_n, x_{n+1}, x_{n+2}, \ldots\}$
(i) If $\{m_n\}_{n=1}^{\infty}$ converges, we define
$\liminf_{n \to \infty} x_n$ to be $\lim_{n \to \infty} M_n$
(ii) If $\{x_n\}_{n=1}^{\infty}$ diverges to ∞ then we write

$$\liminf_{n \to \infty} x_n = \infty$$

For example, if $x_n = (-1)^n$ then $m_n = -1$
$\liminf(-1)^n = -1$

Theorem 6.4.2 *If $\{x_n\}_{n=1}^{\infty}$ is a convergent sequence of real numbers, then*

$$\liminf_{n \to \infty} x_n = \lim_{n \to \infty} x_n$$

Theorem 6.4.3 *If $\{x_n\}_{n=1}^{\infty}$ is a sequence of real numbers, then*

$$\liminf_{n \to \infty} x_n \le \limsup_{n \to \infty} x_n$$

Remark

1. For a sequence $\{x_n\}_{n=1}^{\infty}$ of real numbers

$$\limsup_{n \to \infty} x_n = \inf_{n \ge 1} \sup_{k \ge n} x_k \tag{6.4.1}$$

$$\liminf_{n \to \infty} x_n = \sup_{n \ge 1} \inf_{k \ge n} x_k \tag{6.4.2}$$

2. From the Theorem 6.4.3

$$\sup_{n\geq 1} \inf_{k\geq n} x_k \leq \inf_{n\geq 1} \sup_{k\geq n} x_k \tag{6.4.3}$$

Theorem 6.4.4 *Let* $\{f(x|\theta); \theta \in \Theta\}$ *be a family of pdf(pmf). Suppose that an estimate* d^* *of* θ *is a Bayes estimate corresponding to a priori distribution* $\pi(\theta)$, $\theta \in \Theta$. *If the risk function* $R(\theta, d^*)$ *is constant on* Θ, *then* d^* *is a minimax estimate for* θ *(see Berger (1985)).*

Proof Since d^* is the Bayes estimator of θ with constant risk r^* (free of θ), we have

$$r^* = R(\pi, d^*) = \int_{-\infty}^{\infty} R(\theta, d^*)\pi(\theta)d\theta$$

Using Definition (6.3.1), or (6.3.8)

$$r^* = R(\pi, d^*) = \inf_d R(\pi, d)$$

Using (6.3.5)

$$= \inf_{d\in D} \int R(\theta, d)\pi(\theta)d\theta$$

$$\leq \sup_{\theta\in\Theta} \inf_{d\in D} R(\theta, d)$$

Using (6.4.3)

$$\leq \inf_{d\in D} \sup_{\theta\in\Theta} R(\theta, d)$$

Since $r^* = R(\theta, d^*)$ $\forall \theta \in \Theta$

$$r^* = \sup_{\theta\in\Theta} R(\theta, d^*) \geq \inf_{d\in D} \sup_{\theta\in\Theta} R(\theta, d)$$

Together, we have,

$$\sup_{\theta\in\Theta} R(\theta, d^*) = \inf_{d\in D} \sup_{\theta\in\Theta} R(\theta, d)$$

which means d^* is minimax.

Example 6.4.1 Let the rv $X \sim B(n, p)$ and $L(p, d(x)) = [p - d(x)]^2$. We will obtain the Bayes estimate for p, where p is $\cup(0, 1)$.

Let $\pi(p) = 1; 0 < p < 1$

$$f(x) = \int_0^1 f(x|p)\pi(p)dp$$

$$= \int_0^1 \binom{n}{x} p^x q^{n-x} dp = \binom{n}{x} \beta(x+1, n-x+1)$$

$$= \frac{1}{n+1} \tag{6.4.4}$$

$$f(p|x) = \frac{f(x,p)}{f(x)} = \frac{f(x|p)\pi(p)}{f(x)}$$

$$= \frac{\binom{n}{x} p^x q^{n-x}}{\binom{n}{x} \beta(x+1, n-x+1)}$$

Bayes estimate of of p

$$E(p|x) = \int_0^1 pf(p|x)dp$$

$$= \int_0^1 \frac{p^{x+1} q^{n-x}}{\beta(x+1, n-x+1)} dp$$

$$= \frac{\beta(x+2, n-x+1)}{\beta(x+1, n-x+1)} = \frac{x+1}{n+2}$$

From (6.3.10),

$$d^*(x) = \frac{x+1}{n+2}$$

Bayes Risk $= r(d^*)$

Using the first procedure

$$h(p) = \sum_{x=0}^n \left(d^*(x) - p\right)^2 f(x|p)$$

$$= \sum_{x=0}^n \left(\frac{x+1}{n+2} - p\right)^2 \binom{n}{x} p^x q^{n-x}$$

$$= \left(\frac{1}{n+2}\right)^2 \sum_{x=0}^n [x+1 - (n+2)p]^2 \binom{n}{x} p^x q^{n-x}$$

$$= \left(\frac{1}{n+2}\right)^2 \sum_{x=0}^{n} \left[(x-np)+(1-2p)\right]^2 \binom{n}{x} p^x q^{n-x}$$

$$= \left(\frac{1}{n+2}\right)^2 \left[\sum_{x=0}^{n} [x-np]^2 \binom{n}{x} p^x q^{n-x} + (1-2p)^2 \sum_{x=0}^{n} \binom{n}{x} p^x q^{n-x}\right.$$

$$\left. + 2(1-2p)\sum_{x=0}^{n} (x-np)\binom{n}{x} p^x q^{n-x}\right]$$

$$= \frac{npq + (1-2p)^2}{(n+2)^2} \qquad \text{(since other terms are zero)}$$

$$r(d^*) = \int_0^1 \frac{npq + (1-2p)^2}{(n+2)^2} \pi(p)dp$$

$$= \int_0^1 \frac{n}{(n+2)^2} p(1-p)dp + \int_0^1 \frac{1-4p+4p^2}{(n+2)^2} dp$$

$$= \frac{n\beta(2,2)}{(n+2)^2} + \frac{1-2+\frac{4}{3}}{(n+2)^2} = \frac{1}{6(n+2)}$$

Second procedure

$$h(x) = \int_0^1 \left(p - \frac{x+1}{n+2}\right)^2 f(p|x)dp$$

$$= \int_0^1 \left(p - \frac{x+1}{n+2}\right)^2 \frac{p^x q^{n-x}}{\beta(x+1, n-x+1)dp} = \frac{(x+1)(n-x+1)}{(n+2)^2(n+3)}$$

Reader should note that it is the variance of a Beta distribution with parameters $(x+1)$ and $(n-x+1)$.

Next step is to find $r(d^*)$

$$r(d^*) = \sum_{x=0}^{n} \frac{(x+1)(n-x+1)}{(n+2)^2(n+3)} f(x),$$

where $f(x) = \frac{1}{n+1}$; $x = 0, 1, \ldots, n$, which is given in (6.4.4)

$$r(d^*) = \frac{1}{(n+1)(n+2)^2(n+3)} \sum_{x=0}^{n} \left[nx + (n+1) - x^2\right]$$

$$= \frac{1}{(n+1)(n+2)^2(n+3)} \left[\frac{n^2(n+1)}{2} + (n+1)^2 - \frac{n(n+1)(2n+1)}{6} \right]$$

$$= \frac{(n+1)(n+2)(n+3)}{6(n+1)(n+2)^2(n+3)} = \frac{1}{6(n+2)}$$

Example 6.4.2 Let $X_1, X_2, \ldots, X_n$ be iid rvs with $N(\mu, 1)$ and μ is $N(0, 1)$. Find Bayes estimate of μ and its risk.

$$f(x_1, x_2, \ldots, x_n | \mu) = (2\pi)^{-\frac{n}{2}} \exp \left[-\frac{1}{2} \sum_{i=1}^{n} (x_i - \mu)^2 \right],$$

$$\pi(\mu) = \frac{1}{\sqrt{2\pi}} \exp \left[-\frac{\mu^2}{2} \right]$$

$$f(x_1, x_2, \ldots, x_n) = \int_{-\infty}^{\infty} (2\pi)^{-\left(\frac{n+1}{2}\right)} \exp \left[-\frac{1}{2} \left\{ \sum_{i=1}^{n} (x_i - \mu)^2 + \mu^2 \right\} \right] d\mu$$

Consider

$$-\frac{1}{2} \left[\sum_{i=1}^{n} (x_i^2 - 2x_i\mu + \mu^2) + \mu^2 \right]$$

$$= -\frac{1}{2} \left[\sum_{i=1}^{n} x_i^2 - 2n\bar{x}\mu + n\mu^2 + \mu^2 \right]$$

$$= -\frac{1}{2} \sum_{i=1}^{n} x_i^2 - \frac{\mu^2(n+1)}{2} + \frac{2n\mu\bar{x}}{2}$$

$$= -\frac{1}{2} \sum_{i=1}^{n} x_i^2 - \frac{(n+1)}{2} \left[\mu^2 - \frac{2n\mu\bar{x}}{n+1} \right]$$

$$= -\frac{1}{2} \sum x_i^2 - \frac{(n+1)}{2} \left[\mu^2 - 2\frac{2n\mu\bar{x}}{n+1} + \frac{n^2\bar{x}^2}{(n+1)^2} \right] + \frac{n^2\bar{x}^2}{2(n+1)}$$

$$= -\frac{1}{2} \sum x_i^2 + \frac{n^2\bar{x}^2}{2(n+1)} - \frac{(n+1)}{2} \left[\mu - \frac{n\bar{x}}{n+1} \right]^2$$

Therefore,

$$f(x_1, x_2, \ldots, x_n) = (2\pi)^{-\left(\frac{n+1}{2}\right)} (2\pi)^{\frac{1}{2}} (n+1)^{-\frac{1}{2}} \exp \left[-\frac{1}{2} \sum x_i^2 + \frac{n^2\bar{x}^2}{2(n+1)} \right]$$

$$= (2\pi)^{-\left(\frac{n}{2}\right)}(n+1)^{-\frac{1}{2}} \exp\left[-\frac{1}{2}\sum x_i^2 + \frac{n^2\bar{x}^2}{2(n+1)}\right]$$

$$f(\mu|x_1, x_2, \ldots, x_n) = \frac{f(x_1, x_2, \ldots, x_n, \mu)}{f(x_1, x_2, \ldots, x_n)}$$

$$f(\mu|x_1, x_2, \ldots, x_n) = \left(\frac{2\pi}{n+1}\right)^{-\frac{1}{2}} \exp\left\{-\frac{(n+1)}{2}\left(\mu - \frac{n\bar{x}}{n+1}\right)^2\right\}$$

In this case $(\mu|x)$ is $N\left(\frac{n\bar{x}}{n+1}, \frac{1}{n+1}\right)$.

Therefore, Bayes estimate $d^*(x) = \frac{n\bar{x}}{n+1}$.

Now $\bar{x} \sim N\left(\mu, \frac{1}{n}\right)$.

Using first procedure

$$h(\mu) = \int \left(\frac{n\bar{x}}{n+1} - \mu\right)^2 f(\bar{x}|\mu)d\bar{x}$$

Consider

$$= \left(\frac{n\bar{x}}{n+1} - \frac{n\mu}{n+1} + \frac{n\mu}{n+1} - \mu\right)^2$$

$$= \left(\frac{n}{n+1}\right)^2 (\bar{x} - \mu)^2 + \left(\frac{n\mu}{n+1} - \mu\right)^2 + 2\left(\frac{n}{n+1}\right)(\bar{x} - \mu)\left(\frac{n\mu}{n+1} - \mu\right)$$

Since the third term is zero after expectation

$$h(\mu) = \left(\frac{n}{n+1}\right)^2 V(\bar{x}) + \frac{\mu^2}{(n+1)^2}$$

$$h(\mu) = \left(\frac{n}{n+1}\right)^2 \frac{1}{n} + \frac{\mu^2}{(n+1)^2} = \frac{\mu^2 + n}{(n+1)^2}$$

Bayes Risk $= r(d^*)$

$$r(d^*) = \int_{-\infty}^{\infty} \frac{\mu^2 + n}{(n+1)^2} \frac{1}{\sqrt{2\pi}} \exp\left(-\frac{\mu^2}{2}\right)d\mu \qquad (6.4.5)$$

$$= \frac{n + E\mu^2}{(n+1)^2}$$

and $E\mu^2 = 1$

$$r(d^*) = \frac{1}{n+1} \qquad (6.4.6)$$

Using second procedure

$$h(x_1, x_2, \ldots, x_n) = \int \left(\mu - \frac{n\bar{x}}{n+1}\right)^2 \exp\left\{-\frac{n+1}{2}\left(\mu - \frac{n\bar{x}}{n+1}\right)^2\right\} d\mu$$

$$= V(\mu|X) = \frac{1}{n+1}$$

Bayes Risk $= r(d^*)$

$$r(d^*) = \int_{x_1} \cdots \int_{x_n} \left(\frac{1}{n+1}\right)(2\pi)^{-\left(\frac{n}{2}\right)}(n+1)^{-\frac{1}{2}} \exp\left[-\frac{1}{2}\sum x_i^2 + \frac{n^2\bar{x}^2}{2(n+1)}\right] dx_1 \ldots dx_n$$

$$r(d^*) = \frac{1}{n+1}$$

Note:

In both the examples, we get the same Bayes risk. Reader should select one procedure according to simplicity of integral or summation.

Example 6.4.3 Let $X_1, X_2, \ldots, X_n$ be iid rvs with pdf

$$f(x|\theta) = \exp[-(x - \theta)]; \quad x > \theta$$

Consider the prior distribution of θ is $\pi = e^{-\theta}$; $\theta > 0$. Find the Bayes estimator of θ under quadratic loss. In this problem range of θ is very important.

$$f(x_{(1)}, x_{(2)}, \ldots, x_{(n)}|\theta) = \begin{cases} n!\exp\left[-\left(\sum_{i=1}^n x_{(i)} - \theta\right)\right] & ; \theta < x_{(1)} < x_{(2)} < \cdots < x_{(n)} < \infty \\ 0 & ; \text{otherwise} \end{cases}$$

$$\pi = e^{-\theta}; \theta > 0$$

$$f(x_{(1)}, x_{(2)}, \ldots, x_{(n)}, \theta) = \begin{cases} n!\exp\left[-\left(\sum_{i=1}^n x_{(i)} - n\theta\right)\right]e^{-\theta} & ; 0 < \theta < x_{(1)} \\ 0 & ; \text{otherwise} \end{cases}$$

$$f(x_{(1)}, x_{(2)}, \ldots, x_{(n)}) = n!\exp\left[\sum_{i=1}^n x_{(i)}\right]\int_0^{x_{(1)}} e^{\theta(n-1)}d\theta$$

$$= n!\exp\left[-\sum_{i=1}^n x_{(i)}\right]\frac{e^{x_{(1)}(n-1)} - 1}{n-1}$$

$$f(\theta|x) = \frac{(n-1)e^{\theta(n-1)}}{e^{x_{(1)}(n-1)} - 1}; \quad 0 < \theta < x_{(1)}$$

In this case Bayes estimator is $d^*(x)$

$$E(\theta|X) = d^*(x)$$

$$d^*(x) = \left(\frac{x_{(1)}}{n-1}\right)\left(\frac{e^{x_{(1)}(n-1)}}{e^{x_{(1)}(n-1)} - 1}\right) - \frac{1}{(n-1)^2} \tag{6.4.7}$$

Example 6.4.4 Let $X_1, X_2, \ldots, X_n$ be iid $\cup(0, \theta)$. Suppose that the prior distribution of θ is a Pareto with pdf

$$\pi(\theta) = \begin{cases} \frac{\alpha\beta^\alpha}{\theta^{\alpha+1}} & ; \ \theta > \beta \\ 0 & ; \ \text{otherwise} \end{cases}$$

Using the quadratic loss function find the Bayes estimator of θ.

$$f(x_{(1)}, x_{(2)}, \ldots, x_{(n)}|\theta) = \begin{cases} \frac{n!}{\theta^n} & ; \ 0 < x_{(1)} < x_{(2)} < \cdots x_{(n)} < \theta \\ 0 & ; \ \text{otherwise} \end{cases}$$

$$\pi(\theta) = \begin{cases} \frac{\alpha\beta^\alpha}{\theta^{\alpha+1}} & ; \ \theta > \beta \\ 0 & ; \ \text{otherwise} \end{cases}$$

$$f(x_{(1)}, x_{(2)}, \ldots, x_{(n)}, \theta) = \begin{cases} \frac{n!\alpha\beta^\alpha}{\theta^{n+\alpha+1}} & ; \ \max(\beta, X_{(n)}) < \theta < \infty \\ 0 & ; \ \text{otherwise} \end{cases}$$

Case (1) $X_{(n)} < \beta < \theta < \infty$

$$f(x_{(1)}, x_{(2)}, \ldots, x_{(n)}) = \int\limits_{\beta}^{\infty} \frac{n!\alpha\beta^\alpha}{\theta^{n+\alpha+1}} d\theta = \frac{n!\alpha}{(n+\alpha)\beta^n}$$

Case (2) $\beta < X_{(n)} < \theta < \infty$

$$f(x_{(1)}, x_{(2)}, \ldots, x_{(n)}) = \int\limits_{x_{(n)}}^{\infty} \frac{n!\alpha\beta^\alpha}{\theta^{n+\alpha+1}} d\theta = \frac{n!\alpha\beta^\alpha}{(n+\alpha)x_{(n)}^{n+\alpha}}$$

$$f(\theta|x_{(1)}, x_{(2)}, \ldots, x_{(n)}) = \begin{cases} \frac{(n+\alpha)\beta^{\alpha+n}}{\theta^{n+\alpha+1}} & ; \ \beta < \theta < \infty \\ \frac{(n+\alpha)x_{(n)}^{\alpha+n}}{\theta^{n+\alpha+1}} & ; \ x_{(n)} < \theta < \infty \end{cases}$$

Bayes estimate of $\theta = d^*(x)$

$$d^*(x) = \begin{cases} \frac{(n+\alpha)\beta}{n+\alpha-1} & ; \ x_{(n)} < \beta \\ \frac{(n+\alpha)x_{(n)}}{n+\alpha-1} & ; \ \beta < x_{(n)} \end{cases} \tag{6.4.8}$$

Example 6.4.5 Let $X_1, X_2, \ldots, X_n$ be iid rvs from an exponential distribution with mean $\frac{1}{\sigma}$. Let the prior distribution of σ is $\pi(\sigma) = \exp(-\sigma)$, $\sigma > 0$. Find the Bayes estimate of σ and its risk using squared error loss function.

$$f(x_1, x_2, \ldots, x_n | \sigma) = \sigma^n \exp\left[-\sigma \sum_{i=1}^{n} x_i\right]; \quad x_i > 0$$

$$\pi(\sigma) = e^{-\sigma}; \; \sigma > 0$$

$$f(x_1, x_2, \ldots, x_n) = \int_{0}^{\infty} \sigma^n \exp[-\sigma(t+1)] \, d\sigma, \quad t = \sum_{i=1}^{n} x_i$$

$$= \frac{\Gamma(n+1)}{(t+1)^{n+1}}$$

Note that

$$\int \cdots \int \frac{\Gamma(n+1)}{(t+1)^{n+1}} dx_1 \ldots dx_n = 1$$

$$\Rightarrow \int \cdots \int \frac{dx_1 dx_2 \ldots dx_n}{(t+1)^{n+1}} = \frac{1}{n!} \tag{6.4.9}$$

$$f(\sigma | x) = \frac{(t+1)^{n+1}}{\Gamma(n+1)} \sigma^n \exp[-\sigma(t+1)]; \; \sigma > 0$$

Bayes estimate is

$$d^*(x) = E(\sigma | x) = \frac{n+1}{t+1} \tag{6.4.10}$$

$$V(\sigma | x) = \frac{n+1}{(t+1)^2}$$

Using second procedure

$$h(\sigma) = \int \left(\frac{n+1}{(t+1)} - \sigma\right)^2 \frac{(t+1)^{n+1}}{\Gamma(n+1)} \sigma^n e^{-\sigma(t+1)} d\sigma$$

$$= \frac{n+1}{(t+1)^2} = V(\sigma | x)$$

Bayes risk $= r(d^*)$

$$r(d^*) = \int \int \cdots \int \frac{n+1}{(t+1)^2} \frac{\Gamma(n+1)}{(t+1)^{n+1}} dx_1 dx_2 \cdots dx_n$$

$$= (n+1)\Gamma(n+1) \int \int \cdots \int \frac{dx_1 dx_2 \cdots dx_n}{(t+1)^{n+3}}$$

Using (6.4.9)

$$= \frac{(n+1)\Gamma(n+1)}{\Gamma(n+3)} = \frac{(n+1)n!}{(n+2)!} = \frac{1}{n+2}$$

Reader should obtain the above risk using the first procedure. Generally, it is felt that the first procedure is more complicated than the second. Therefore, reader should select the procedure according to his understanding.

Following steps should be remembered to obtain Bayes estimator and its risk.

1. Find the joint distribution of X and θ.
i.e., $f(x, \theta) = f(x|\theta)\pi(\theta)$
2. Find the marginal distribution of X from $f(x, \theta)$. Denote it by $g(x)$.
3. Find the posterior distribution of θ given X.
i.e., $f(\theta|x) = \frac{f(x,\theta)}{g(x)}$
4. According to loss function, find mean or median of θ given X.
5. Use procedure I or II to find Bayes risk.
One should note that it is not always easy to go through these steps in practice.

Definition 6.4.4 Let X be a rv with $f(x|\theta)$ and θ be a rv with $\pi(\theta)$. Then π is said to be conjugate prior family, if the corresponding posterior distribution $f(\theta|x)$ belongs to the same family as $\pi(\theta)$.

Example 6.4.6 Let the rv X be $N(\mu, \sigma^2)$ and the rv μ is $N(\theta, b^2)$.

$$f(x|\mu, \sigma^2) = \frac{1}{\sigma\sqrt{2\pi}} \exp\left[-\frac{1}{2\sigma^2}(x-\mu)^2\right]$$

$$f(x, \mu|\sigma^2) = \frac{1}{\sigma\sqrt{2\pi}} \exp\left[-\frac{1}{2\sigma^2}(x-\mu)^2\right] \frac{1}{b\sqrt{2\pi}} \exp\left[-\frac{1}{2b^2}(\mu-\theta)^2\right]$$

$$= \frac{1}{b\sigma(2\pi)} \exp\left[-\frac{1}{2\sigma^2}(x^2 - 2\mu x + \mu^2) - \frac{1}{2b^2}(\mu^2 - 2\mu\theta + \theta^2)\right]$$

Consider

$$\frac{x^2 - 2\mu x + \mu^2}{\sigma^2} + \frac{\mu^2 - 2\mu\theta + \theta^2}{b^2}$$

$$= \frac{x^2}{\sigma^2} + \frac{\theta^2}{b^2} + \frac{\mu^2}{\sigma^2} + \frac{\mu^2}{b^2} - 2\mu \left(\frac{x}{\sigma^2} + \frac{\theta}{b^2} \right)$$

$$= \frac{x^2 b^2 + \theta^2 \sigma^2}{\sigma^2 b^2} + \mu^2 \left(\frac{\sigma^2 + b^2}{\sigma^2 b^2} \right) - 2\mu \left(\frac{xb^2 + \theta\sigma^2}{\sigma^2 b^2} \right)$$

$$= \frac{x^2 b^2 + \theta^2 \sigma^2}{\sigma^2 b^2} + \mu^2 \left(\frac{\sigma^2 + b^2}{\sigma^2 b^2} \right) - 2\mu \left(\frac{xb^2 + \theta\sigma^2}{\sigma^2 b^2} \right)$$

$$+ \frac{(xb^2 + \theta\sigma^2)^2}{\sigma^2 b^2 (\sigma^2 + b^2)} - \frac{(xb^2 + \theta\sigma^2)^2}{\sigma^2 b^2 (\sigma^2 + b^2)}$$

$$= \frac{x^2 b^2 + \theta^2 \sigma^2}{\sigma^2 b^2} - \frac{(xb^2 + \theta\sigma^2)^2}{\sigma^2 b^2 (\sigma^2 + b^2)} + \left[\mu \left(\frac{(\sigma^2 + b^2)^{\frac{1}{2}}}{\sigma^2 b^2} \right) - \frac{xb^2 + \theta\sigma^2}{\sigma b (\sigma^2 + b^2)^{\frac{1}{2}}} \right]^2$$

$$= \frac{b^2 \sigma^2 (x^2 - 2x\theta + \theta^2)}{(b^2 + \sigma^2) b^2 \sigma^2} + \left[\mu \left(\frac{\sigma^2 + b^2}{\sigma^2 b^2} \right)^{\frac{1}{2}} - \frac{xb^2 + \theta\sigma^2}{\sigma b (\sigma^2 + b^2)^{\frac{1}{2}}} \right]^2$$

$$= \frac{(x - \theta)^2}{b^2 + \sigma^2} + \frac{\sigma^2 + b^2}{\sigma^2 b^2} \left[\mu - \frac{xb^2 + \theta\sigma^2}{\sigma^2 + b^2} \right]^2$$

$$f(x, \mu | \sigma^2) = \frac{1}{b\sigma(2\pi)} \exp \left[-\frac{(x - \theta)^2}{2(b^2 + \sigma^2)} \right] \exp \left[-\frac{\sigma^2 + b^2}{2\sigma^2 b^2} \left\{ \mu - \frac{xb^2 + \theta\sigma^2}{\sigma^2 + b^2} \right\}^2 \right]$$

$$g(x | \sigma^2) = \frac{1}{b\sigma(2\pi)} \exp \left[-\frac{(x - \theta)^2}{2(b^2 + \sigma^2)} \right] \sqrt{2\pi} \frac{b\sigma}{(b^2 + \sigma^2)^{\frac{1}{2}}}$$

$$= \frac{1}{\sqrt{2\pi(b^2 + \sigma^2)}} \exp \left[-\frac{(x - \theta)^2}{2(b^2 + \sigma^2)} \right]$$

Therefore, posterior distribution of μ given x i.e.,

$$f(\mu | x) = \frac{\sqrt{\sigma^2 + b^2}}{\sqrt{2\pi}\sigma b} \exp \left[-\frac{\sigma^2 + b^2}{2\sigma^2 b^2} \left(\mu - \frac{xb^2 + \theta\sigma^2}{\sigma^2 + b^2} \right)^2 \right]$$

In this case μ is $N(\theta, b^2)$ and $\mu \mid x$ is $N \left(\frac{xb^2 + \theta\sigma^2}{\sigma^2 + b^2}, \frac{\sigma^2 b^2}{\sigma^2 + b^2} \right)$.

Hence $\pi(\mu)$ is said to be conjugate prior.

Definition 6.4.5 (*Minimax estimate*) An estimator $d : X \to \theta$ is called minimax with respect to risk function $R(\theta, d)$ if it achieves the smallest maximum risk among all estimators, meaning if it satisfies

$$\sup_{\theta \in \Omega} R(\theta, d) = \inf_{\mathcal{H}} \sup_{\theta \in \Omega} R(\theta, d)$$

Note that if a Bayes' estimator has a constant risk, then it is minimax. In the following example, we try to find a minimax estimate.

Example 6.4.7 Let the rv X is $B(n, p), 0 \le p \le 1$.
Find a minimax estimate of p of the form $\alpha X + \beta$, using squared error loss function.

Consider

$$
\begin{aligned}
R(p, d) &= \mathrm{E}(\alpha X + \beta - p)^2 \\
&= \mathrm{E}[\alpha X + \beta + \alpha np - \alpha np - p]^2 \\
&= \mathrm{E}[\alpha(X - np) + \beta + p(\alpha n - 1)]^2 \\
&= \alpha^2 \mathrm{E}(X - np)^2 + \beta^2 + p^2(\alpha n - 1)^2 + 2\beta\alpha\mathrm{E}(X - np) + 2\beta p(\alpha n - 1) \\
&\quad + 2\alpha p(\alpha n - 1)\mathrm{E}(X - np) \\
&= \alpha^2 \mathrm{V}(x) + \beta^2 + p^2(\alpha n - 1)^2 + 2\beta p(\alpha n - 1) \\
&= \alpha^2 npq + \beta^2 + p^2(\alpha n - 1)^2 + 2\beta p(\alpha n - 1) \\
&= \alpha^2 np(1 - p) + \beta^2 + p^2(\alpha n - 1)^2 + 2\beta p(\alpha n - 1) \\
&= p^2[(\alpha n - 1)^2 - \alpha^2 n] + p[\alpha^2 n + 2\beta(\alpha n - 1)] + \beta^2
\end{aligned}
$$

Let d^* is a minimax estimator of p if $R(p, d)$ is constant.
   Therefore, to find α and β such that coefficient of p^2 and p equal to 0, then $R(p, d)$ is equal to β^2.

$$(\alpha n - 1)^2 - \alpha^2 n = 0 \tag{6.4.11}$$

$$\alpha^2 n + 2\beta(\alpha n - 1) = 0 \tag{6.4.12}$$

From (6.4.11),

$$\alpha = \frac{1}{\sqrt{n}(1 + \sqrt{n})} \quad \text{or} \quad \frac{1}{\sqrt{n}(\sqrt{n} - 1)}$$

$$\beta = \frac{1}{2(1 + \sqrt{n})} \quad \text{or} \quad \frac{-1}{2(\sqrt{n} - 1)}$$

By omitting second set of roots, we get

$$d^*(X) = \frac{X}{\sqrt{n}(1 + \sqrt{n})} + \frac{1}{2(1 + \sqrt{n})} \tag{6.4.13}$$

To show that $d^*(X)$ is a Bayes estimate for some prior $\pi(p)$, where

$$\pi(p) = \frac{p^{a-1}(1-p)^{b-1}}{\beta(a,b)}; \quad 0 \le p \le 1, \quad a, b > 0$$

The posterior pdf of p given X is

$$h(p|X) = \frac{p^{x+a-1}(1-p)^{n-x+b+1}}{\beta(a,b)}; \quad 0 < p < 1$$

$$E(p|X) = \frac{x+a}{n+a+b} \tag{6.4.14}$$

Hence

$$d^*(X) = \frac{x+a}{n+a+b} \tag{6.4.15}$$

From (6.4.13) and (6.4.14)

$$\frac{X}{n+\sqrt{n}} + \frac{1}{2(\sqrt{n}+1)} = \frac{X}{n+a+b} + \frac{a}{n+a+b}$$

$a+b = \sqrt{n}$ and $\frac{1}{2(\sqrt{n}+1)} = \frac{a}{n+a+b}$

$$\Rightarrow \frac{a}{n+\sqrt{n}} = \frac{1}{2(\sqrt{n}+1)} \Rightarrow a = \frac{\sqrt{n}}{2} \quad \text{and} \quad b = \frac{\sqrt{n}}{2}$$

For this choice of a and b, the estimate $d^*(X)$ is minimax with constant risk

$$R(\theta, d^*) = \beta^2 = \frac{1}{4(\sqrt{n}+1)^2} \tag{6.4.16}$$

Now, compare (6.4.16) with the variance of the UMVUE of p.
In this case $\hat{p} = \frac{X}{n}$

$$Var(\hat{p}) = \frac{p(1-p)}{n} \tag{6.4.17}$$

From (6.4.16) and (6.4.17) we will see the following table for $n = 5(5)50(50)200$ and $p = 0.2, 0.5$.

n	$R(\theta, \delta^*)$	$Var(\hat{p})(p = 0.2)$	$Var(\hat{p})(p = 0.5)$
5	1.7387	1.320	
10	0.01443	0.0160	0.025
15	0.01053	0.0107	0.0167
20	0.00834	0.008	0.0125
25	0.00694	0.0064	0.01
30	0.00596	0.0053	0.0083
35	0.00523	0.0046	0.0071
40	0.0047	0.0040	0.00625
45	0.0042	0.0036	0.056
50	0.0038	0.0032	0.005
100	0.0021	0.0016	0.0025
200	0.0011	0.0008	0.0012

$$\frac{p(1-p)}{n} \le \frac{1}{4(1+\sqrt{n})^2} \quad \text{iff} \quad \left|p - \frac{1}{2}\right| \ge \frac{\sqrt{1+2\sqrt{n}}}{2(1+\sqrt{n})} = a_n$$

$$\text{If} \quad \frac{p(1-p)}{n} \ge \frac{1}{4(1+\sqrt{n})^2}$$

$$\Rightarrow p^2 - p + \frac{n}{4(1+\sqrt{n})^2} < 0$$

$$\Rightarrow \frac{1}{2} - \frac{\sqrt{1+2\sqrt{n}}}{2(1+\sqrt{n})} \le p \le \frac{1}{2} + \frac{\sqrt{1+2\sqrt{n}}}{2(1+\sqrt{n})}$$

$$\Rightarrow \frac{1}{2} - a_n \le p \le \frac{1}{2} + a_n$$

$$\Rightarrow p \in \left[\frac{1}{2} - a_n, \frac{1}{2} + a_n\right]$$

$$\frac{\sup V(\hat{p})}{\sup R(P, d^*)} = \frac{\frac{1}{4n}}{\frac{1}{4(1+\sqrt{n})^2}} = \frac{n + 2\sqrt{n} + 1}{n}$$

As $n \to \infty$

$$\Rightarrow \frac{\sup V(\hat{p})}{\sup R(P, d^*)} \to 1$$

We can numerically see the above result.

We can conclude that one should prefer the minimax estimate if n is small and would prefer UMVUE if n is large. Moreover, it is simple.

Example 6.4.8 Let the rv X is $N(\theta, 1)$ and the rv θ is $\pi(\theta)$, where

$$\pi(\theta) = \frac{\exp[-(\theta - \alpha)]}{[1 + \exp[-(\theta - \alpha)]^2]},$$

where α is the location parameter.

Marginal pdf of X is $g(X)$, where

$$g(X) = \frac{e^\alpha}{\sqrt{2\pi}} \int\limits_{-\infty}^{\infty} \frac{e^{-\theta} e^{-\frac{(X-\theta)^2}{2}}}{[1 + e^{-(\theta-\alpha)}]^2} d\theta, \qquad (6.4.18)$$

It is difficult to integrate (6.4.18). Hence the closed form of $g(X)$ is not known. Then we cannot get the closed form of $f(\theta|x)$. Due to mathematical convenience, statisticians use conjugate prior. Naturally, posterior distributions also belong to the same family.

Example 6.4.9 Consider an urn with N balls, M of which are white and $N - M$ are red. Suppose that we draw a sample of n balls at random (without replacement) from the urn. Then the probability of getting k white balls out of n is

$$P(X = k|M) = \frac{\binom{M}{k}\binom{N-M}{n-k}}{\binom{N}{n}}; \quad k = 0, 1, 2, \ldots \min(n, M)$$

Here, we wish to find minimax estimate of M

Note that

$$E(X) = \frac{nM}{N}, V(X) = \frac{nM(N-n)(N-M)}{N^2(N-1)}$$

We seek a minimax estimator of M of the form $\alpha X + \beta$ using squared error loss function.

$$R(M, d) = E[\alpha X + \beta - M]^2$$

$$= E\left[\alpha X + \beta + \frac{\alpha nM}{N} - \frac{\alpha nM}{N} - M\right]^2$$

$$= E\left[\alpha\left(X - \frac{nM}{N}\right) + M\left(\frac{\alpha n}{N} - 1\right) + \beta\right]^2$$

$$= \alpha^2 E\left(X - \frac{nM}{N}\right)^2 + \beta^2 + M^2\left(\frac{\alpha n}{N} - 1\right)^2 + 2\beta M\left(\frac{\alpha n}{N} - 1\right)$$

(other terms are equal to 0 because $EX = \frac{nM}{N}$)

$$= \frac{\alpha^2 nM(N-n)(N-M)}{N^2(N-1)} + M^2\left(\frac{\alpha n}{N} - 1\right)^2 + 2\beta M\left(\frac{\alpha n}{N} - 1\right) + \beta^2$$

Let $Q = \frac{n(N-n)}{N^2(N-1)}$

$$R(M,d) = \alpha^2 QM(N-M) + M^2\left(\frac{\alpha n}{N} - 1\right)^2 + 2\beta M\left(\frac{\alpha n}{N} - 1\right) + \beta^2$$

$$= M^2\left[\left(\frac{\alpha n}{N} - 1\right)^2 - \alpha^2 Q\right] + M\left[\alpha^2 QN + 2\beta\left(\frac{\alpha n}{N} - 1\right)\right] + \beta^2$$

For a minimax estimator, $R(M,d) = \beta^2$

$$\Rightarrow \left(\frac{\alpha n}{N} - 1\right)^2 - \alpha^2 Q = 0 \tag{6.4.19}$$

and

$$\alpha^2 QN + 2\beta\left(\frac{\alpha n}{N} - 1\right) = 0$$

$$\Rightarrow \alpha^2 Q = \left(\frac{\alpha n}{N} - 1\right)^2 \Rightarrow N\left(\frac{\alpha n}{N} - 1\right)^2 + 2\beta\left(\frac{\alpha n}{N} - 1\right) = 0$$

$$N\left(\frac{\alpha n}{N} - 1\right) + 2\beta = 0$$

Therefore,

$$\beta = \frac{N}{2}\left(1 - \frac{\alpha n}{N}\right)$$

Further, from (6.4.19)

$$\left(\frac{\alpha n}{N} - 1\right) = \pm\frac{\alpha}{N}\sqrt{\frac{n(N-n)}{N-1}}$$

$$\alpha\left[\frac{n \pm \sqrt{\frac{n(N-n)}{N-1}}}{N}\right] = 1$$

$$\alpha = \left[\frac{N}{n \pm \sqrt{\frac{n(N-n)}{N-1}}}\right]$$

We will consider

$$\alpha = \left[\frac{N}{n + \sqrt{\frac{n(N-n)}{N-1}}} \right]$$

Next we show that $\alpha X + \beta$ is the Bayes estimator corresponding to the prior pmf.

$$P(M = m) = \int_0^1 \binom{N}{m} p^m q^{N-m} \frac{p^{a-1}(1-p)^{b-1}}{\beta(a, b)} dp$$

$$= \frac{\binom{N}{m}\beta(a+m, N-m+b)}{\beta(a, b)}; \quad m = 0, 1, 2, \ldots, N$$

$$P(X, M) = P(X|M) \times P(M = m)$$

$$= \frac{\binom{m}{k}\binom{N-m}{n-x}}{\binom{N}{n}} \frac{\binom{N}{m}\beta(a+m, N-m+b)}{\beta(a, b)}$$

$$P(X = k) = \sum_{m=k}^{N-n+k} \frac{\binom{m}{k}\binom{N-m}{n-x}\binom{N}{m}\beta(a+m, N-m+b)}{\binom{N}{n}\beta(a, b)}$$

The Bayes estimate is given by

$$d^*(k) = \frac{\sum_{m=k}^{N-n+k} m\binom{m}{k}\binom{N-m}{n-x}\binom{N}{m}\beta(a+m, N-m+b)}{\sum_{m=k}^{N-n+k} \binom{m}{k}\binom{N-m}{n-x}\binom{N}{m}\beta(a+m, N-m+b)}$$

$$\binom{m}{k}\binom{N-m}{n-x}\binom{N}{m}\frac{n!(N-n)!}{n!(N-n)!} = \binom{N-n}{m-k}\binom{N}{n}\binom{n}{k}$$

Set m as $m + a - a$, let $i = m - k$, if $m = k \Rightarrow i = 0$ and $m = N - n + k \Rightarrow i = N - n$

$$d^*(k) = \frac{\sum_{i=0}^{N-n} \binom{N-n}{i}\Gamma(k+a+i+1)\Gamma(N+b-k-i)}{\sum_{i=0}^{N-n} \binom{N-n}{i}\Gamma(k+a+i)\Gamma(N+b-k-i)} - a$$

Consider

$$\sum_{m=0}^{N} \binom{N}{m}\frac{\beta(a+m, N-m+b)}{\beta(a, b)} = 1$$

$$\sum_{m=0}^{N} \binom{N}{m} \Gamma(a+m)\Gamma(N-m+h) - \frac{\Gamma(a)\Gamma(b)\Gamma(N+a+b)}{\Gamma(a\mid b)}$$

Using this

$$\sum_{i=0}^{N-n} \binom{N-n}{i} \Gamma(k+a+1+i)\Gamma(N+b-k-i) = \frac{\Gamma(k+a+1)\Gamma(b-k+N)\Gamma(N+a+b+1)}{\Gamma(a+b+N+1)}$$

Similarly,

$$\sum_{i=0}^{N-n} \binom{N-n}{i} \Gamma(k+a+i)\Gamma(N+b-k-i) = \frac{\Gamma(k+a)\Gamma(b-k+N)\Gamma(N+a+b)}{\Gamma(a+b+N)}$$

Therefore, the Bayes estimator of M is

$$d^*(k) = \frac{(k+a)(N+a+b)}{a+b+n} - a$$

6.5 Exercise 6

1. Let $X_1, X_2, \ldots, X_m$ are iid with $B(n, p)$ and $L(p, d(x)) = [p - d(x)]^2$. Obtain the Bayes estimate for p if p has Beta distribution with parameters a and b. Find the Bayes estimate of p and its risk.

From the following data, obtain the estimate of p for $a = 3$ and $b = 5$.
3 5, 3, 4, 4, 3, 4, 5, 2, 4.

Compare the variance of Bayes estimate of p and UMVUE of p.

Further, obtain the Bayes estimate for weighted loss function.

2. Let $X_1, X_2, \ldots, X_n$ are iid rvs from $N(\mu, 1)$ and μ is $N(\alpha, \beta^2)$. Obtain the Bayes estimate of μ under squared error and weighted quadratic loss function.

From the following data obtain the estimate of μ for $\alpha = 2$ and $\beta = 3$.

0.4384 6.8281 40.0148 29.3679
−10.3823 0.0871 −9.5146 19.8065
12.9548 32.6523 −2.0395 −15.8874
8.9464 −0.2844 11.0987 −10.8222
40.6232 14.3904 −8.7655 −4.4608.

Obtain the Bayes risk of $\hat{\mu}$ and compare it with the variance of UMVUE of μ.

3. In example 2, if μ is $U(0, 1)$, then find Bayes estimate of μ under squared error and weighted quadratic loss function.

4. Let $X_1, X_2, \ldots, X_n$ are iid rvs with pdf

$$f(x|\alpha) = \begin{cases} \exp[-(x - \alpha)] & ; x \ge \alpha \\ 0 & ; \text{otherwise} \end{cases}$$

Obtain the Bayes estimate of α if

$$\pi(\alpha) = \frac{e^{-\frac{\alpha}{a}}\alpha^{b-1}}{a^b\Gamma(b)}; \ \alpha > 0, \ a, b > 0$$

and the loss function is squared error and weighted quadratic loss function.

From the following data obtain the Bayes and UMVUE estimate of α.

6.1398 5.5978 6.4957 6.5645 6.2387
10.7251 7.2395 5.0859 5.4681 5.3441
6.1722 6.4479 7.2601 9.0449 8.2572
5.4054 6.9218 7.3457 5.3869 5.7536
6.1015 6.3037 6.9928 6.4762 5.8694

Assume $a = 4$ and $b = 6$. Find Bayes risk.

5. Suppose that the vector $X = (X_1, X_2, X_3)$ has a trinomial distribution depending on the index n and the parameter $P = (p_1, p_2, p_3)$, where $p_1 + p_2 + p_3 = 1$ and $n = x_1 + x_2 + x_3$, that is,

$$f(x|p) = \frac{n!}{x_1!x_2!x_3!}p_1{}^{x_1}p_2{}^{x_2}p_3{}^{x_3}$$

Obtain the Bayes estimate of p if p_i is $\cup(0, 1)$.

6. Let the rv X is $B(n, p)$. Suppose that your prior for p is $\frac{3}{4} : \frac{1}{4}$ mixture of $\cup(0.1, 0.5)$ and $\cup(0.3, 0.9)$. Obtain the Bayes estimate of p, if the data is $n = 10$, $X = 4$ under squared error and weighted quadratic loss function and further obtain the minimax estimate of p.

7. Let the rv X is $NB(r, p)$, r is known. Suppose that the prior distribution of p is $\cup(0, 1)$. Find Bayes estimate of p and its risk.

8. In problem 1, if the prior distribution of p is $\pi(p)$, where

$$\pi(p) = \begin{cases} \alpha p^{\alpha-1} & ; \ 0 < p < 1 \\ 0 & ; \ \text{otherwise} \end{cases}$$

Obtain the Bayes estimate of p and its risk.

For $\alpha = 3$, calculate the Bayes estimate of p (use the data given in problem 1).

9. Let $X_1, X_2, \ldots, X_n$ is $NB(r, p)$, r is known. Suppose that your prior for p is $\frac{2}{3} : \frac{1}{3}$, mixture of $\cup(0.2, 0.6)$ and $\cup(0.3, 0.9)$. Obtain the Bayes estimate of p and its risk under squared error loss function from the following data.

14 10 11 25 21 13 16 17 7
16 20 20 16 14 22 17 19 26
14 17

Assume $r = 5$

10. A random sample of size n is taken from $N(\mu, 1)$. The prior distribution of μ is

$$\pi(\mu) = \begin{cases} 1 & ; \mu > 0 \\ 0 & , \text{otherwise} \end{cases}$$

Find the Bayes estimate of μ under squared error loss function.

11. Let the rv X be $N(\mu, 1)$. Suppose that your prior for μ is $\frac{1}{4} : \frac{3}{4}$, mixture of $N(1, 1)$ and $N(2, 1)$. Find the posterior probability of $\mu > 2$, if the observation from X is 1.5.

12. Let $X_1, X_2, \ldots, X_n$ are iid from $f(x|\theta)$,

$$f(x|\theta) = \frac{(\alpha + 1)x^\alpha}{\theta^{\alpha+1}}; \ 0 < x < \theta$$

Find the Bayes estimate of θ and its risk using conjugate prior.

13. Let the rv X is $B(n, \theta)$ and the rv Y is $NB(n, \theta)$. Suppose the prior distribution of θ is $\pi(\theta)$,

$$\pi(\theta) = \frac{\theta^{a-1}(1 - \theta)^{b-1}}{\beta(a, b)}; \ 0 < \theta < 1$$

Find the Bayes estimate of θ and its risk.

14. Let $X_1, X_2, \ldots, X_n$ are iid rvs from an exponential distribution with mean $\frac{1}{\sigma}$. Let the prior distribution of σ is $\frac{2}{3} : \frac{1}{3}$, mixture of exponential with mean α and β. That is

$$\pi(\sigma) = \frac{2}{3\alpha}e^{-\frac{\sigma}{\alpha}} + \frac{1}{3\beta}e^{-\frac{\sigma}{\beta}}$$

Find the Bayes estimate of σ and its risk using squared error loss function.

15. Let $X_1, X_2, \ldots, X_n$ be iid $U(0, \theta)$. Suppose that the prior distribution of θ is Pareto with pdf $\pi(\theta)$,

$$\pi(\theta) = \begin{cases} \frac{\alpha\beta^\alpha}{\theta^{\alpha+1}} & ; \theta > \beta \\ 0 & ; \text{otherwise} \end{cases}$$

Using squared error loss function find the risk of the Bayes estimator of θ. Calculate the Bayes estimator from the following data for $\alpha = 2$, $\beta = 5$.

3.92 0.44 9.97 7.68 2.45 2.03 1.34
5.26 3.05 6.46 6.56 3.02 8.73 6.93
7.06 4.51 1.67 0.73 6.13 0.83

16. Let $X_1, X_2, \ldots, X_n$ be iid $U(\theta, \theta+1)$. Suppose that prior distribution of θ is Pareto with pdf $\pi(\theta)$,

$$\pi(\theta) = \begin{cases} \frac{\alpha\beta^\alpha}{\theta^{\alpha+1}} & ; \theta > \beta \\ 0 & ; \text{otherwise} \end{cases}$$

Using quadratic and weighted loss function, find the Bayes estimate of θ. Calculate the Bayes estimator from the following data for $\alpha = 3$ and $\beta = 8$

    3.72 3.56 5.89 3.91 4.27 5.45 8.12
    6.94 2.87 6.72 3.84 8.25 3.70 6.78
    5.30 3.38 5.46 5.45 6.69 8.99

17. Let $X_1, X_2, \ldots, X_n$ be iid rvs with pdf

$$f(x|\theta) = \exp[-(x - \theta)]; \quad x > \theta$$

Find the Bayes risk of estimator if $\pi(\theta) = e^{-\theta} : \theta > 0$
18. Let $X_1, X_2, \ldots, X_n$ be iid rvs with pdf

$$f(x|\theta) = \exp[-(x - \theta)]; \quad x > \theta$$

Assume squared error loss function and find the Bayes estimator of θ if prior distribution of θ is $\pi(\theta)$,

$$\pi(\theta) = \frac{\theta^{k-1} e^{-\frac{\theta}{\alpha}}}{\Gamma(k)\alpha^k}; \quad \theta > 0, \ \alpha, k > 0$$

Calculate the Bayes estimator for the following data.

    3.03 2.34 2.25 3.07 2.79 2.50
    2.08 2.40 2.24 2.26 2.51 3.16
    2.37 3.11 2.25

Assume $k = 4$ and $\alpha = 5$.
19. In problem 17, if θ is $U(0, 1)$ then find the Bayes estimator under squared error and quadratic loss function. Further find its risk. Calculate the Bayes estimator of θ using the data given in problem 18.
20. Let $X_1, X_2, \ldots, X_n$ be iid rvs with $U(\theta, 2\theta)$. Suppose that the prior distribution of θ is Pareto with pdf

$$\pi(\theta) = \begin{cases} \frac{\alpha\beta^\alpha}{\theta^{\alpha+1}} & ; \theta > \beta \\ 0 & ; \text{otherwise} \end{cases}$$

Using quadratic loss function, find Bayes estimator of θ and its risk. Calculate Bayes estimator from the following data for $\alpha = 2$ and $\beta = 4$,

    4.16 2.41 4.01 2.52 2.76 2.30 2.89
    5.60 5.47 2.85 4.89 5.15 2.71 3.51
    3.15

21. Let $X_1, X_2, \ldots, X_m$ be iid rvs with $f(x|\theta)$ and $g(x|\theta)$. If m is even then X_{2k-1} has $f(x|\theta)$ and X_{2k} has $g(x|\theta)$, where $k = 1, 2, \ldots, \frac{m}{2}$. Assume $f(x|\theta) = \binom{n}{x}\theta^x(1 - \theta)^{n-x}; x = 0, 1, 2, \ldots, n$ and $g(x|\theta) = \binom{x+r-1}{x}\theta^r(1-\theta)^x; x = 0, 1, 2, \ldots$. Assuming

prior distribution of θ as $\beta(a, b)$, find Bayes estimate of θ under squared error loss and weighted quadratic loss function.

22. Let the rv X_1 has exponential distribution with mean θ and the rv X_2 has $g(x, \theta)$,

$$g(x|\theta) = \theta x^{\theta-1}; \ 0 < x < 1 \ \theta > 0$$

Find the Bayes estimate of θ under squared error and weighted loss function. Assume that θ has improper prior distribution.

23. Let the rv X_1 has $\cup(0, \theta)$ and X_2 has $\cup(\theta, \theta+1)$. Suppose that the prior distribution of θ is a Pareto distribution with pdf

$$\pi(\theta) = \begin{cases} \frac{\alpha \beta^\alpha}{\theta^{\alpha+1}} & ; \ \theta > \beta \\ 0 & ; \text{otherwise} \end{cases}$$

Using quadratic loss function, find the Bayes estimate of θ.

24. Assume that rvs X_1 and X_2 are distributed as $\cup(0, \theta)$ and exponential with mean θ, respectively. Find the Bayes estimate of θ under the quadratic loss function. Further, assume that prior distribution of θ as specified in problem 23 and obtain the Bayes' estimate of θ.

25. Let $X_1, X_2, \ldots, X_n$ be iid (i) $\cup(-\theta, 0)$ (ii) $\cup(0, \theta^2)$ (iii) $\cup(\theta, k\theta)$, $k > 1$ (iv) $\cup(-2\theta, 2\theta)$.

Suppose the prior distribution of θ is $\pi(\theta)$

$$\pi(\theta) = \begin{cases} \frac{\alpha \beta^\alpha}{\theta^{\alpha+1}} & ; \ \theta > \beta \\ 0 & ; \text{otherwise} \end{cases}$$

Using the squared error loss function, find the risk of the bayes estimator of θ (use the data given in problem 20).

References

Berger JO (1985) Statistical decision theory and bayesian analysis, 2nd edn. Springer

Dixit UJ, Kelkar SG (2011) Estimation of the parameters of binomial distribution in the presence of outliers. J Appl Stat Sci 19(4):489–498

Lee PM (1997) Bayesian statistics: an introduction. Oxford University Press, New York

Chapter 7
Most Powerful Test

Suppose that your parliament is considering a proposal for uniform civil code for all religions. To gather information, a group surveys 500 randomly selected individuals from their district and learns that 65 % of these people favor the proposal.

Can the Member of Parliament conclude that a majority of all adults in their district favor this proposal? Because the result is based on the sample, there is a possibility that the observed majority might have occurred just by the "Luck of the draw." If the majority of the whole population actually opposes the proposal, how likely is it that 65 % of a random sample would favor the proposal?

In this chapter, we will learn how to use the method of statistical hypothesis testing to analyze this type of issue. The hypothesis testing method uses data from a sample to judge whether or not a statement about a population may be true. A hypothesis test is used to answer questions about particular values for a population parameter, or particular relationship in a population, based on information in the sample data. The five steps for any hypothesis test follow.

Step 1: Determine the null and alternative hypothesis.
Step 2: Verify necessary data conditions, and if met summarize the data into an appropriate test statistics.
Step 3: Assuming the null hypothesis, find the p-value.
Step 4: Decide whether or not the result is statistically significant based on the p-value.
Step 5: Report the conclusion in the context of the situation.

We shall learn these steps one by one.

7.1 Type-1 and Type-2 Errors

Let $(X_1, X_2, \ldots, X_n)$ be iid from $f(x|\theta)$, $\theta \in \Theta$ and $\Theta \subseteq \Re$. Further, we assume that the functional form of $f(x|\theta)$ is known except the parameter θ. Also, we assume that θ contains at least two points.

© Springer Science+Business Media Singapore 2016
U.J. Dixit, *Examples in Parametric Inference with R*,
DOI 10.1007/978-981-10-0889-4_7

Any statement about the population or population parameter from which a given random sample $(x_1, x_2, \ldots, x_n)$ may have been drawn is called a null hypothesis. Further, a parametric hypothesis is an assertion about the unknown parameter θ. It is denoted by H_0. Moreover, we can say that no statistical significance exists in a set of given observations. The specific null hypothesis varies from problem to problem, but generally it can be thought of as the status quo, or no relationship, or no difference. In many instances, the statistician hopes to disprove or reject the null hypothesis.

The alternative hypothesis is denoted by H_1. It is a statement that something is happening. In most situations, this hypothesis is what the statistician hopes to prove. It may be a statement that the assumed status quo is false, or that there is a relationship, or that there is a difference.

Consider the following example of null hypothesis:

- Men and women have same I.Q.
- There is no difference between the mean pulse rates of men and women.
- The accused is not guilty.

Some examples of alternative hypothesis

- Men and women do not have the same I.Q.
- There is a difference between the mean pulse rates of men and women.
- The accused is guilty.

In notation, we can write:

$H_0 : \theta \in \Theta_0$, where $\Theta_0 \subset \Theta$

$H_1 : \theta \in \Theta_1$, where $\Theta_1 \subset \Theta$

Definition 7.1.1 If Θ_0 or Θ_1 contains only one point, we say that Θ_0 or Θ_1 is simple hypothesis, otherwise it is composite hypothesis.

If the hypothesis is simple, the probability distribution of X is completely specified.

(1) For example, if the rv X is $N(\mu, 1)$ and $H_0 : \mu = 4$ and $H_1 : \mu = 5$. In this case under H_0, X is $N(4, 1)$ and under H_1, X is $N(5, 1)$. Hence it is a simple hypothesis.

(2) Let the rv X is $N(\mu, \sigma^2)$ and both μ and σ^2 are unknown.

$\Theta = \{(\mu, \sigma^2) : -\infty < \mu < \infty, \sigma^2 > 0\}$

Let $H_0 : \mu \le \mu_0, \sigma^2 > 0$, where μ_0 is known constant against $H_1 : \mu > \mu_0$, $\sigma^2 > 0$.

In this case, both null and alternative hypothesis are composite.

Given the sample point $X = (x_1, x_2, \ldots, x_n)$, we have to find a decision rule which will lead to a decision to accept or reject the null hypothesis. Further partition the n-dimensional Euclidean space $\Re_n$ into two disjoint sets A and A^c.

If $x \in A$, reject H_0 and $x \in A^c$, accept H_0.

Definition 7.1.2 Let $X_1, X_2, \ldots, X_n$ are iid rvs from $f(x|\theta)$, $\theta \in \Theta$. The subset 'A' of $\Re_n$ such that if $x \in A$ then H_0 is rejected with probability 1 is called the critical region or rejection region.

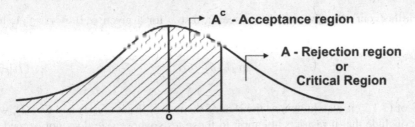

Fig. 7.1 Acceptance and rejection region

In notation, we can write,

$A = \{x \in \Re_n, H_0$ is rejected if $x \in A\}$; see Fig. 7.1.

We can make two possible types of error:

(i) We may reject the hypothesis when we ought to accept it; i.e., when it is true
(ii) We may accept the hypothesis when we ought to reject it; i.e., when it is false

	Decision	
Test	Reject H_0	Accept H_0
H_0 True	Type-1 error	Correct
H_0 False	Correct	Type-2 error

Let α = Probability of Type-1 error and

β = Probability of Type-2 error.

Choose the critical region so as to minimize both types of errors simultaneously, but this is, in general, not possible for a sample of fixed size. In fact, decreasing one type of error may very likely increase the other type. Thus, by deciding to always accept the hypothesis, we can reduce type-1 error to zero, but in that case β would have its largest value, i.e., 1. In practice, we keep type-1 error fixed at a specified value and then, out of these critical regions all of which give this type-1 error, we choose that region which minimizes the type II error. The type-1 error, which is the same for all these regions, is sometimes called the size of the critical regions.

Definition 7.1.3 The test function ϕ is defined as $\phi : \Re \to [0, 1]$

Examples of test function

(i) $\phi(x) = 1 \, \forall x \in \Re_n$
(ii) $\phi(x) = 0 \, \forall x \in \Re_n$
(iii) $\phi(x) = \delta, 0 \leq \delta \leq 1, \, \forall x \in \Re_n$

Definition 7.1.4 The test function ϕ is said to be a test of hypothesis $H_0 : \theta \in \Theta_0$ against the alternative $H_1 : \theta \in \Theta_1$ with the error probability α (it is also called level of significance) if

$$E_\theta \phi(x) \leq \alpha \quad \forall \quad \theta \in \Theta_0 \tag{7.1.1}$$

Further, our objective will be to seek a test ϕ for a given $\alpha, 0 \le \alpha \le 1$, let $\beta_\phi(\theta) = E_\theta \phi(x)$, such that

$$\sup_{\theta \in \Theta_0} \beta_\phi(\theta) \le \alpha \qquad (7.1.2)$$

LHS of (7.1.2) is also known as the size of the test ϕ. From (7.1.1) and (7.1.2), we can conclude that it restricts attention to those tests whose size does not exceed a given level of significance α.

In the hypothesis testing problems involving discrete distributions, it is usually not possible to choose a critical region consisting of realizable values of the statistics of size exactly α, where α is some prescribed value. In simple hypothesis, we can write α = Probability of type I error,

$\quad = P[\text{Reject } H_0 | H_0 \text{ is true}]$, and

β = Probability of type II error,

$\quad = P[\text{Accept } H_0 | H_0 \text{ is false}]$

For a nonrandomized test with rejection region A, ϕ for a region A is just an indicator function. That is,

$$\phi(x) = \begin{cases} 1 \; ; \; x \in A \\ 0 \; ; \; x \in A^c \end{cases}$$

We will extend this to allow for some different action (other than reject and accept) if the outcome X is on the boundary of the critical region. The other action effectively is performing an auxiliary experiment such as tossing a coin with $P[\text{heads}] = p$; if head results, reject H_0; if tail results, H_0 is accepted. The value of p is chosen to make $P[\text{Reject } H_0 | H_0 \text{ is true}]$, the desired value.

More formally, for a test with critical region A and a value of $X = x_0$ on the boundary, we may define

$$\phi(x) = \begin{cases} 1 \; ; \; x < x_0 \\ \gamma \; ; \; x = x_0 \\ 0 \; ; \; \text{otherwise,} \end{cases}$$

where $0 < p < 1$.

Such a test is known as "Randomized Test."

Example 7.1.1 Suppose X has a Poisson distribution with mean λ. A sample of size $n = 10$ is used to test $H_0 : \lambda = 0.1$ against $H_1 : \lambda > 0.1$.

Note that $Y = \sum_{i=1}^{10} X_i$ has a Poisson distribution with mean 10λ.

The test is to reject H_0 for large values of Y. Suppose we wish to have a significance level of $\alpha = 0.05$.

Now, $P[Y \geq 3] = 0.08$ and $P[Y \geq 4] = 0.019$. The desired level of significance can be achieved by the test

$$\phi(Y) = \begin{cases} 1 \; ; \; Y \geq 4 \\ \gamma \; ; \; Y = 3 \\ 0 \; ; \; Y < 3, \end{cases}$$

$P[\text{Reject } H_0 | H_0 \text{ is true}] = 0.05$

$$
\begin{aligned}
E\phi(y) &= 1 \times P[Y \geq 4] + \gamma \times P[Y = 3] + 0 \\
&= 0.019 + \gamma[P(Y \geq 3) - P(Y \geq 4)] \\
&= 0.019 + \gamma[0.08 - 0.019] \\
&= 0.019 + \gamma(0.061)
\end{aligned}
$$

Therefore,

$$= 0.019 + \gamma(0.061) = 0.05$$

$$\Rightarrow \gamma = \frac{31}{61}$$

Hence, the randomized test can be written as

$$\phi(Y) = \begin{cases} 1 \; ; \; Y > 3 \\ \frac{31}{61} \; ; \; Y = 3 \\ 0 \; ; \; Y < 3, \end{cases}$$

Example 7.1.2 Let the random variable X has $\cup(\theta_1, \theta_2)$. We wish to test $H_0 : \theta_1 = 2, \theta_2 = 5$ against $H_1 : \theta_1 = 3, \theta_2 = 8$, using a sample of size 1. Find randomized test of size $\alpha = 0.05$.

Clearly any sensible decision rule would include if $x \in (2, 4)$, H_0 should be accepted, if $x \in (3, 8)$, H_0 should be rejected. But if $x \in (4, 5)$, we will reject H_0 sometimes and accept H_0 sometimes (Fig. 7.2).

Our ϕ function will be

$$\phi(X) = \begin{cases} 1 \; ; \; X \in (5, 8) \\ \gamma \; ; \; X \in (4, 5) \\ 0 \; ; \; X \in (2, 4) \end{cases}$$

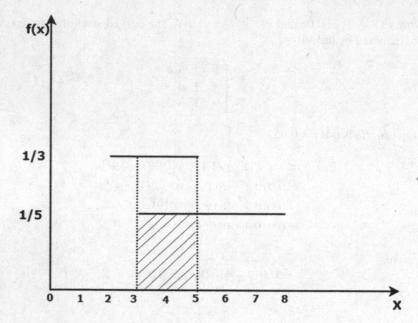

Fig. 7.2 Acceptance-rejection region

$$\alpha = P[\text{Reject } H_0 | H_0 \text{ is true}]$$

$$0.05 = 1 \times P[X \in (5, 8)] + \gamma \times P[X \in (4, 5)]$$

$$= 0 + \gamma \int_4^5 \frac{1}{3} dx$$

$$= \frac{\gamma}{3}$$

$$\Rightarrow \gamma = 0.15$$

Our randomized test is

$$\phi(X) = \begin{cases} 1 & ; X \in (5, 8) \\ 0.15 & ; X \in (4, 5) \\ 0 & ; X \in (2, 4) \end{cases}$$

Definition 7.1.5 Let ϕ be any test function for $H_0 : \theta \in \Theta_0$ and $H_1 : \theta \in \Theta_1$. For every $\theta \in \Theta_1$, define

$$\beta_\phi(\theta) = E_\theta \phi(X); \quad \theta \in \Theta_1 \tag{7.1.3}$$

As a function of θ, $\beta_\phi(\theta)$ is called the power function of the test ϕ, for $\theta \in \Theta_1$. Further in simple hypothesis if β is Type II error then power of the test is $1 - \beta$. Moreover in composite hypothesis $\beta_\phi(\theta)$ for $\theta \in \Theta_1$ is a power function.

Problem in Testing of Hypothesis

Let $X_1, X_2, \ldots, X_n$ be iid rvs from $f(x|\theta)$, $\theta \in \Theta$. Assume $0 < \alpha < 1$ be given. Given a sample point X, find a test $\phi(x)$ such that

$$\sup_{\theta \in \Theta_0} \beta_\phi(\theta) \leq \alpha$$

and $\beta_\phi(\theta)$ for $\theta \in \Theta_1$ is maximum.

Evidently, we can find in general many and often even an infinity of subregions A of the sample space, all satisfying (7.1.2). Which of them should we prefer to the others? This is the problem of the theory of testing of hypothesis. We can put it in different words as to which set of observations are to regard as favoring and which is disfavoring a given hypothesis?

Once the question is put in this way, we are directed to the heart of the problem. Knowing H_o alone, along with the properties of the critical region does not suffice. It is essential to know the alternative hypothesis also. What happens when some other hypothesis holds? In other words, we cannot say whether a given body of observations favors a given hypothesis unless we know to what alternatives this hypothesis is being compared. The hypothesis being used to study the set of observations should be meaningful and appropriate hypothesis should be stated. For example, if our objective is to test the quantity of milk in 500 ml packet, it is obvious that testing should be done around 500 ml and must not vary from far 500 ml, like say 300 or 600 ml. Mathematically, it may be correct but statistically it is incorrect. The problem of testing a hypothesis is essentially one of choice between it and some others. It follows immediately that whether or not we accept the original hypothesis depends crucially upon the alternatives against which is being tested.

Here, we seek a critical region A such that its power defined in (7.1.3) is as large as possible. Then, in addition to having controlled the probability of type-1 error defined in (7.1.1) or (7.1.2), we have to minimize the probability of type-2 error.

Now, we will consider some examples of type-1 and type-2 errors.

Example 7.1.3 Let the rv X is distributed as $\cup(0, \theta)$. We are testing the hypothesis $H_0 : \theta = 1$ against $H_1 : \theta = 2$ based on a single observation. Calculate the type-1 and type-2 errors based on the following critical regions. Also obtain the power of the test.

(i) $A_1 = \{x|0.9 < x\}$
(ii) $A_2 = \{x|1 < x < 1.5\}$

(i) Type-1 error $= P[\text{Reject } H_0 | H_0 \text{ is true}]$

$$= P[X > 0.9 | \theta = 1]$$

$$= \int_{0.9}^{1} dx = 0.10$$

Type-2 error $= P[\text{Accept } H_0 | H_1 \text{ is true}]$

$$= P[0 \le X < 0.9 | \theta = 2]$$

$$= \int_{0}^{0.9} \frac{dx}{2} = 0.45$$

Power $= P[\text{Reject } H_0 | H_1 \text{ is true}]$

$$= P[X > 0.9 | \theta = 2]$$

$$= P[0.9 < X < 2 | \theta = 2]$$

$$= \int_{0.9}^{2} \frac{dx}{2} = 0.55$$

(ii) $A_2 = \{x : 1 \le x \le 1.5\}$

Type-1 error $= P[1 \le x \le 1.5 | \theta = 1]$

$$= \int_{1}^{1} dx = 0$$

Type-2 error $= P[\text{Accept } H_0 | H_1 \text{ is true}]$

$$= P[x \in A_2{}^c | \theta = 2]$$

$$= 1 - P[x \in A_2 | \theta = 2]$$

$$1 \quad \int\limits_{1}^{1.5} \frac{dx}{2} - 0.75$$

Power $= P[\text{Reject } H_0 | H_1 \text{ is true}] = P[x \in A_2 | \theta = 2] = 0.25$

Example 7.1.4 If $X \geq 1$ is the critical region for testing $H_0 : \theta = 2$ against $H_1 : \theta = 1$ on the basis of a single observation from the following population

$$f(x|\theta) = \begin{cases} \theta e^{-\theta x} ; & x > 0, \ \theta > 0 \\ 0 & ; \ \text{otherwise} \end{cases}$$

Obtain Type-1 and Type-2 errors. Further obtain the power of this test.
$A = \{x | x \geq 1\}, A^c = \{x | x < 1\}$
We can say that A is the critical region
(i) Type-1 error $= P[\text{Reject } H_0 \mid H_0 \text{ is true}]$

$$= P[x \geq 1 | \theta = 2]$$

$$= \int\limits_{1}^{\infty} 2e^{-2x} dx = e^{-2}$$

Type-2 error $= P[\text{Accept } H_0 | H_1 \text{ is true}]$

$$= P[X < 1 | \theta = 1]$$

$$= \int\limits_{0}^{1} e^{-x} dx = 1 - e^{-1}$$

Power $= P[\text{Reject } H_0 | H_1 \text{ is true}]$

$$= P[X \geq 1 | \theta = 1] = e^{-1}$$

The probability of Type-1 error α depends on the choice of the critical region A and on the hypothesis H_0, while probability of Type II error β depends on both the hypothesis H_0 and H_1. An increase in α results in decrease in β and a decrease in α results in an increase in β.

Example 7.1.5 Research and development department of a pharmaceutical company has recently developed a dietary supplement to reduce cholesterol level. Company applied for license to FDA, a drug regulatory body, to market the drug. The FDA authorities asked the company to administer the supplement on 100 people and check hypothesis.

H_0: The drug does nothing to reduce the cholesterol, i.e., $p = 0.10$.

H_1: The drug reduces the cholesterol in 20% people, i.e., $p = 0.20$.

FDA knew from the past experience that in about 10% of the people cholesterol declines naturally. FDA decided to issue license for marketing the drug only when the experiment shows strong evidence that the drug reduces cholesterol 20% of all the people. FDA also supplied the decision rule

$$\phi(\hat{p}) = \begin{cases} 1 \; ; \; \hat{p} > 0.15 \\ 0 \; ; \; \hat{p} \le 0.15 \end{cases}$$

where $\hat{p}$ is observed proportion and $\hat{p} \sim N(p, \frac{pq}{n})$.

Probability of type-I error

$$P[\hat{p} > 0.15 | H_0 \text{ is true}]$$

$$= P[\hat{p} > 0.15 | p = 0.10]$$

$$= P\left[\frac{\hat{p} - p}{\sqrt{\frac{pq}{n}}} > \frac{0.15 - 0.10}{\sqrt{0.0009}} \right] = P[z > 1.67], \quad \text{where} \quad z \sim N(0, 1), \quad q = 1 - p,$$

$$= 0.0475,$$

Probability of type II error $= P[\text{Accept } H_0 | H_1 \text{ is true}]$

$$= P[\hat{p} < 0.15 | p = 0.20]$$

$$= P[z \le -1.25] = 0.1056 \tag{7.1.4}$$

Example 7.1.6 Let X and Y be two independent rvs with $\cup(0, \theta)$. We are testing the hypothesis $H_0 : \theta = 1$ against $H_1 : \theta = 2$. Calculate the probability of type-1 error and power of the test based on the following critical regions

(i) $(X + Y) > 0.75$ (ii) $XY > 0.75$ (iii) $\frac{X}{Y} > 0.75$

X and Y are independent random variables.

(i) $X \sim \cup(0, 1)$, $Y \sim \cup(0, 1)$,

To find $P(X + Y > 0.75)$ (ii) $P(XY > 0.75)$ (iii) $P\left(\frac{X}{Y} > 0.75\right)$

Here $X \sim \cup(0, 1)$ and $Y \sim \cup(0, 1)$,

Also X and Y are independent

$$f_{X,Y}(x, y) = f_X(x)f_Y(y)$$

$$f_X(x) = \begin{cases} 1 \; ; \; 0 < x < 1 \\ 0 \; ; \; \text{otherwise} \end{cases}$$

$$f_Y(y) = \begin{cases} 1 \; ; \; 0 < y < 1 \\ 0 \; ; \; \text{otherwise} \end{cases}$$

$$f_{X,Y}(x, y) = \begin{cases} 1 \; ; \; \forall \, 0 < x < 1, 0 < y < 1 \\ 0 \; ; \; \text{otherwise} \end{cases}$$

We will find probability of type-1 error in all the three critical regions.

Here, as the value of pdf is 1, for all $0 < x < 1, 0 < y < 1$, the probability required will actually be the area bounded by $x + y > 0.75, x < 1$ and $y < 1, x > 0$, $y > 0$. Required region is A.

Now Area (A) + Area (B) = Area of Squares

Area (A) = 1-Area (B) = $1 - \dfrac{1}{2} \times \dfrac{3}{4} \times \dfrac{3}{4}$ Since B is a right angled triangle.

$= 1 - \dfrac{9}{32} = \dfrac{23}{32}$

$P(X + Y > 0.75) = 0.71875$; see Fig. 7.3

(ii) $P(XY > 0.75)$ Here, $XY = 0.75$ is a hyperbola.
Required region is A

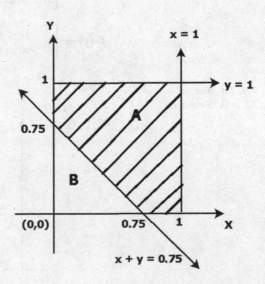

Fig. 7.3 Area for $X + Y > 0.75$ under H_0

$$\text{Area}(A) = \int_{\frac{3}{4}}^{1} \left(1 - \frac{0.75}{x}\right) dx$$

$$= \left[x - 0.75 \log x\right]_{\frac{3}{4}}^{1}$$

$$= (1 - 0.75 \log 1) - \left(\frac{3}{4} - 0.75 \log 0.75\right) = 1 - 0.9657 = 0.034238$$

$\therefore P(XY > 0.75) = 0.034238$; see Fig. 7.4

(iii) $P\left(\dfrac{X}{Y} > 0.75\right)$

$X = \frac{3}{4}Y$. Required region is A.

Area (A) + Area (B) = 1

Area (A) = 1-Area (B) $= 1 - \dfrac{1}{2} \times \dfrac{3}{4} \times 1$

$= 1 - \dfrac{3}{8} = \dfrac{5}{8} = 0.625$

$P\left(\dfrac{X}{Y} > 0.75\right) = 0.625$; see Fig. 7.5.

Next we will find power of the test in all the three regions.

Here $X \sim U(0, 2)$ and $Y \sim U(0, 2)$,

Also X and Y are independent.

$$f_{X,Y}(x, y) = f_X(x) f_Y(y)$$

$$f_X(x) = \begin{cases} \frac{1}{2} \; ; \; 0 < x < 2 \\ 0 \; ; \; \text{otherwise} \end{cases}$$

$$f_Y(y) = \begin{cases} \frac{1}{2} \; ; \; 0 < y < 2 \\ 0 \; ; \; \text{otherwise} \end{cases}$$

Fig. 7.4 Area for $XY > 0.75$ under H_0

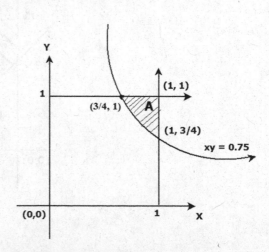

Fig. 7.5 Area for $\frac{X}{Y} > 0.75$ under H_0

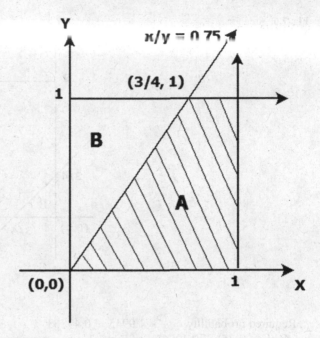

$$f_{X,Y}(x, y) = \begin{cases} \frac{1}{4} & ; \ 0 < x < 2, 0 < y < 2 \\ 0 & ; \text{ otherwise} \end{cases}$$

Here, value of pdf is $\frac{1}{4}$, for all $0 < x < 2, 0 < y < 2$. The required probability will be the area bounded by given lines or curves multiplied by $\frac{1}{4}$.

(i) $P(X + Y > 0.75)$

Area (A) $= 4$-Area (B) $= 4 - \frac{1}{2} \times \frac{3}{4} \times \frac{3}{4}$

$= 4 - \frac{9}{32} = 3.71875$

$P(X + Y > 0.75) = \frac{1}{4} \times 3.71875 = 0.92968$

$\therefore P(X + Y > 0.75) = 0.92968$; see Fig. 7.6

(ii) $P(XY > 0.75)$

$$\text{Area (A)} = \int_{\frac{3}{8}}^{2} \left(2 - \frac{0.75}{x}\right) dx$$

$$= \left[2x - 0.75 \log x\right]_{\frac{3}{8}}^{2}$$

$$= (4 - 0.75 \log 2) - \left(\frac{3}{4} - 0.75 \log \frac{3}{8}\right) = 3.480139 - 1.48562 = 1.9945$$

Fig. 7.6 Area for
$X + Y > 0.75$ under H_1

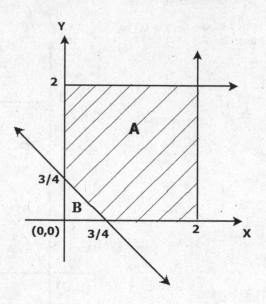

∴ Required probability $= \frac{1}{4} \times 1.9945 = 0.49863$
   $P(XY > 0.75) = 0.49863$; see Fig. 7.7

(iii) $P\left(\dfrac{X}{Y} > 0.75\right)$

  Area (A) $= 4$-Area (B) $= 4 - \dfrac{1}{2} \times 1.5 \times 2 = 4 - 1.5 = 2.5$

$P\left(\dfrac{X}{Y} > 0.75\right) = \dfrac{2.5}{4} = 0.625$; see Fig. 7.8.

Fig. 7.7 Area for
$XY > 0.75$ under H_1

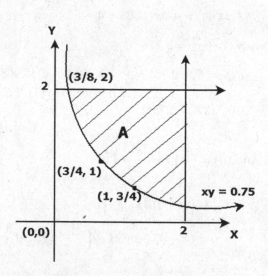

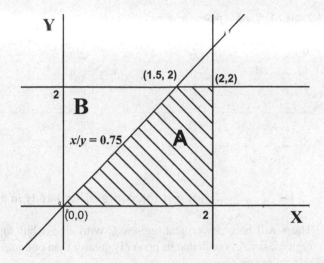

Fig. 7.8 Area for $\frac{X}{Y} > 0.75$ under H_1

7.2 Best Critical Region

A critical region whose power is no smaller than that of any other region of the same size for testing a hypothesis H_0 against the alternative H_1 is called a best critical region (BCR) and a test based on BCR is called a most powerful (MP) test. We will illustrate this BCR in notation.

Let A denote a subset of the sample space S. Then A is called a BCR of size α for testing H_0 against H_1 if for every subset C of the sample space S for which $P[(X_1, X_2, \ldots, X_n) \in C] \le \alpha$.

Let $X = (X_1, X_2, \ldots, X_n)$

$$
\begin{aligned}
&(i) \quad P[X \in A | H_0 \text{ is true}] \le \alpha \\
&(ii) \quad P[X \in A | H_1 \text{ is true}] \le P[X \in C | H_1 \text{ is true}]
\end{aligned}
\tag{7.2.1}
$$

A test based on A is called the MP test.

Definition 7.2.1 Let ϕ_α be the class of all test whose size is α. A test $\phi_0 \in \phi_\alpha$ is said to be most powerful (MP) test against an alternative $\theta \in \Theta_1$ if

$$
\beta_{\phi_0}(\theta) \ge \beta_\phi(\theta), \quad \forall \phi \in \phi_\alpha
\tag{7.2.2}
$$

Note: If Θ_1 contains only one point, this definition suffices. If on the other hand Θ_1 contains at least two points as will usually be the case, we will have a MP test corresponding to each Θ_1.

Definition 7.2.2 A test $\phi_0 \in \phi_\alpha$ for testing $H_0 : \theta \in \Theta_0$ against $H_1 : \theta \in \Theta_1$ of size α is said to be uniformly most powerful (UMP) test if

Table 7.1 Ratio of probabilities under H_1 to H_0

X	$P_0(X)$	$P_1(X)$	$\frac{P_1(X)}{P_0(X)}$
0	$\frac{1}{32}$	$\frac{1}{1024}$	$\frac{1}{32}$
1	$\frac{5}{32}$	$\frac{15}{1024}$	$\frac{3}{32}$
2	$\frac{10}{32}$	$\frac{90}{1024}$	$\frac{9}{32}$
3	$\frac{10}{32}$	$\frac{270}{1024}$	$\frac{27}{32}$
4	$\frac{5}{32}$	$\frac{405}{1024}$	$\frac{81}{32}$
5	$\frac{1}{32}$	$\frac{243}{1024}$	$\frac{243}{32}$

$$\beta_{\phi_0}(\theta) \geq \beta_\phi(\theta), \ \forall \, \phi \in \phi_\alpha \text{ uniformly in } \theta \in \Theta_1 \tag{7.2.3}$$

There will be many critical regions C with size α but suppose one of the critical regions, say A is such that its power is greater than or equal to the power of any such critical region C.

Example 7.2.1 Let the rv X is $B(5, \theta)$. We have to test $H_0 : \theta = \frac{1}{2}$ against $H_1 : \theta = \frac{3}{4}$

Consider the Table 7.1:
(i) $\alpha = \frac{1}{32}$
We can have two regions $A_1 = \{0\}, A_2 = \{5\}$
One can see

$$\alpha = P\left[X = 0 | \theta = \frac{1}{2}\right] = P\left[X = 5 | \theta = \frac{1}{2}\right]$$

Power $(A_1) = P_{A_1}[\text{Reject } H_0 | H_1 \text{ is true}]$

$$= P_{A_1}\left[X = 0 | \theta = \frac{3}{4}\right] = \frac{1}{1024}$$

$$\text{Power } (A_2) = P_{A_2}\left[X = 5 | \theta = \frac{3}{4}\right] = \frac{243}{1024}$$

In this case A_2 is BCR.

One can observe that the BCR is found by observing the points in P_0 at which P_0 is small among other points. Hence, we have two sets A_1 and A_2.

Select BCR, such that $\dfrac{P_1(X)}{P_0(X)}$ is maximum.

At $X = 0, \dfrac{P_1(X)}{P_0(X)} = \dfrac{1}{32}$ and At $X = 5, \dfrac{P_1(X)}{P_0(X)} = \dfrac{243}{32}$

$$\max\left\{\frac{P_1(X = 0)}{P_0(X = 0)}, \frac{P_1(X = 5)}{P_0(X = 5)}\right\} = \frac{243}{32}.$$

Hence, BCR $= A_2 = \{X = 5\}$.

(II) $\alpha = \frac{11}{32}$

We can have the following four regions:

$A_1 = \{0, 1\}, A_2 = \{4, 5\}, A_3 = \{0, 4\}, A_4 = \{1, 5\}$

Critical region	Power
A_1	$\frac{16}{1024}$
A_2	$\frac{648}{1024}$
A_3	$\frac{406}{1024}$
A_4	$\frac{258}{1024}$

Power is maximum at A_2. In this case BCR is A_2.

One can observe that $\frac{P_1}{P_0}$ has largest and next largest values at $X = 4$ and $X = 5$.

Thus, if H_0 and H_1 are both composite, the problem is to find a UMP test ϕ a test very frequently does not exist. Then we will have to put further restriction on the class of the tests ϕ_α.

Remark: If ϕ_1 and ϕ_2 are both tests and λ is a real number, $0 < \lambda < 1$, then $\lambda\phi_1 + (1-\lambda)\phi_2$ is also a test function and it follows that the class of all test functions ϕ_α is convex.

Example 7.2.2 Let $X_1, X_2, \ldots, X_n$ be iid rvs with $N(\mu, 1)$, where μ is unknown. But it is known that $\mu \in \Theta = \{(\mu_0, \mu_1), \mu_1 > \mu_0\}$. We have to find critical regions for given α under $H_0 : \mu = \mu_0$ against $H_1 : \mu = \mu_1 (\mu_1 > \mu_0)$.

Assume that the critical region $A = \{\bar{X} > k\}$. Define the test $\phi(X)$ as

$$\phi(x) = \begin{cases} 1 \; ; \; \bar{X} > k \\ 0 \; ; \; \bar{X} \le k \end{cases}$$

$$\begin{aligned} E_{H_0}\phi(X) &= P[\text{Reject } H_0 | \mu = \mu_0] \\ &= P[\bar{X} > k | \mu = \mu_0] \\ &= P[\sqrt{n}(\bar{X} - \mu_0) > \sqrt{n}(k - \mu_0)] = \alpha \end{aligned}$$

Let $Z = \sqrt{n}(\bar{X} - \mu_0)$ and $Z_\alpha = \sqrt{n}(k - \mu_0)$

Hence $P[Z > Z_\alpha] = \alpha$ then $k = \mu_0 + \frac{Z_\alpha}{\sqrt{n}}$

Hence our test is

$$\phi(x) = \begin{cases} 1 \; ; \; \bar{X} > \mu_0 + \frac{Z_\alpha}{\sqrt{n}} \\ 0 \; ; \; \bar{X} \le \mu_0 + \frac{Z_\alpha}{\sqrt{n}} \end{cases}$$

In this case $\bar{X}$ is known as test statistic.

Power of the test $= E_{H_1}\phi(X) = P[\text{Reject } H_0 | H_1 \text{ is true}]$

$$= P\left[\bar{X} > \mu_0 + \frac{Z_\alpha}{\sqrt{n}} | \mu = \mu_1\right]$$

$$= P\left[\sqrt{n}(\bar{X} - \mu_1) > \sqrt{n}\left(\mu_0 + \frac{Z_\alpha}{\sqrt{n}} - \mu_1\right)\right]$$

$$= P\left[\sqrt{n}(\bar{X} - \mu_1) > Z_\alpha - \sqrt{n}(\mu_1 - \mu_0)\right]$$

$$= 1 - \Phi\left[z_\alpha - \sqrt{n}(\mu_1 - \mu_0)\right]$$

Theorem 7.2.1 (Neyman–Pearson Lemma)

Let the rv X have pdf(pmf) $f(x|\theta_0)$ *and* $f(x|\theta_1)$.

(a) Existence: For testing $H_0 : X \sim f(x|\theta_0)(=f_0)$ against $H_1 : X \sim f(x|\theta_1)(=f_1)$ and for every $0 < \alpha \leq 1$, there exists a test $\phi, k \geq 0$ and $\gamma(x) = \gamma$ such that

$$E_{H_0}\phi(X) = \alpha, \tag{7.2.4}$$

where

$$\phi(X) = \begin{cases} 1 & ; f_1 > kf_0 \\ \gamma(x) & ; f_1 = kf_0 \\ 0 & ; f_1 < kf_0 \end{cases} \tag{7.2.5}$$

and for $\alpha = 0$, there exists a test $\phi(x), k = \infty$ such that

$$E_{H_0}\phi(X) = 0, \tag{7.2.6}$$

where

$$\phi(X) = \begin{cases} 1 & ; f_0 = 0 \\ 0 & ; f_0 > 0 \end{cases} \tag{7.2.7}$$

(b) Sufficient condition for MP test: Any test of the form (7.2.5) for some $k \geq 0$ and $0 \leq \gamma(x) \leq 1$, is most powerful of its size in A_α for testing $H_0 : X \sim f_1$,

$$A_\alpha = \{\phi | E_{H_0}\phi(X) \leq \alpha\} \tag{7.2.8}$$

Further, if $k = \infty$ the test of the form (7.2.7) is most powerful of size o for testing H_0 against H_1.

(c) Necessary condition for MP test:

If ϕ^* is the MP test of its size in A_α, then it has the form as given in (7.2.5) or (7.2.7) except for a set of X with probability γ ~~~~~~~~~ H_0 against H_1

Proof If $k = \infty$ then from (7.2.7)

$$E_{H_0}\phi(X) = 0, \text{ i.e. } \alpha = 0$$

(Note that when k is very large then almost there is no rejection region)

Next, if $k = 0$, then the test

$$\phi(X) = \begin{cases} 1 \; ; f_0 = 0 \\ 0 \; ; f_0 > 0 \end{cases} \tag{7.2.9}$$

$$E_{H_0}\phi(X) = 1$$

Next, consider $0 < \alpha < 1$, let $\gamma(x) = \gamma$,
    From (7.2.5),

$$E_{H_0}\phi(X) = P_{H_0}[f_1 > kf_0] + \gamma P_{H_0}[f_1 = kf_0]$$

$$= 1 - P_{H_0}[f_1 \leq kf_0] + \gamma P_{H_0}[f_1 = kf_0] \tag{7.2.10}$$

Let $Y = \frac{f_1}{f_0}$ then $P_{H_0}[Y \leq k]$ is a df, i.e., it is nondecreasing and right continuous.

If, for a fixed α, there exists a k_0 such that

$$P_{H_0}[Y \leq k_0] = 1 - \alpha$$

then $E_{H_0}\phi(X) = \alpha, \gamma = 0$ and the test is

$$\phi(X) = \begin{cases} 1 \; ; f_1 > kf_0 \\ 0 \; ; f_1 \leq kf_0 \end{cases}$$

It implies that randomization on the boundary with probability $\gamma(x) = \gamma$ is not required when the distribution of Y given H_0 is continuous. If the above situation is not met then from (7.2.10),

$$E_{H_0}\phi(X) = \alpha \quad \text{and} \quad k = k_0$$

$$\alpha = 1 - P_{H_0}[Y \leq k_0] + \gamma P_{H_0}[Y = k_0] \tag{7.2.11}$$

$$P[Y < k_0] + (1 - \gamma)P_{H_0}[Y = k_0] = 1 - \alpha \qquad (7.2.12)$$

It implies

$$P[Y < k_0] \leq 1 - \alpha \qquad (7.2.13)$$

and

$$P[Y \leq k_0] = (1 - \alpha) + \gamma P[Y = k_0] \qquad (7.2.14)$$

$$(1 - \alpha) < P[Y \leq k_0] \qquad (7.2.15)$$

From (7.2.13) and (7.2.15),

$$P[Y < k_0] \leq (1 - \alpha) < P[Y \leq k_0] \qquad (7.2.16)$$

Hence, from (7.2.14),

$$\gamma = \frac{P_{H_0}[Y \leq k_0] - (1 - \alpha)}{P_{H_0}[Y = k_0]} \qquad (7.2.17)$$

Then the test $\phi(X)$ for $k = k_0$ is defined as

$$E_{H_0}\phi(X) = \alpha$$

and

$$\phi(X) = \begin{cases} 1 & ; f_1 > kf_0 \\ \gamma(x) & ; f_1 = kf_0 \\ 0 & ; f_1 < kf_0 \end{cases}$$

(b) Let the sample space S is divided into three regions

$$S^+ = \{x \in S | f_1(x) > kf_0(x)\}$$

$$S^0 = \{x \in S | f_1(x) = kf_0(x)\}$$

$$S^- = \{x \in S | f_1(x) < kf_0(x)\}$$

Let $\phi(x)$ be the test satisfying (7.2.5) and $\phi^*(x)$ be any test with

$$E_{H_0}\phi^*(X) \leq E_{H_0}\phi(X) \qquad (7.2.18)$$

In continuous case,

$$\int [\phi(X) - \phi^*(X)][f_1(x) - kf_0(x)]dx$$

$$= \int_{S^+} [\phi(X) - \phi^*(X)][f_1(x) - kf_0(x)]dx + \int_{S^0} [\phi(X) - \phi^*(X)][f_1(x) - kf_0(x)]dx$$

$$+ \int_{S^-} [\phi(X) - \phi^*(X)][f_1(x) - kf_0(x)]dx$$

For any $x \in S^+$, $\phi(X) = 1 \Rightarrow \phi(X) - \phi^*(X) = 1 - \phi^*(X)$

$$\Rightarrow [1 - \phi^*(X)][f_1(x) - kf_0(x)] > 0 \qquad (7.2.19)$$

Next, $x \in S^0$

$$[1 - \phi^*(X)][f_1(x) - kf_0(x)] = 0 \qquad (7.2.20)$$

Next $x \in S^-$, $\phi(X) = 0 \Rightarrow \phi(X) - \phi^*(X) = -\phi^*(X)$

$$\Rightarrow -\phi^*(X)[f_1(x) - kf_0(x)] > 0 \qquad (7.2.21)$$

From (7.2.19), (7.2.20) and (7.2.21), the complete integral gives a quantity which is greater than or equal to zero and we have,

$$\int_{S} [\phi(X) - \phi^*(X)][f_1(x) - kf_0(x)]dx > 0$$

$$\Rightarrow \int \phi(X)[f_1(x) - kf_0(x)]dx > \int \phi^*(X)[f_1(x) - kf_0(x)]dx$$

$$\Rightarrow E_{H_1}\phi(X) - kE_{H_0}\phi(X) > [E_{H_1}\phi^*(X) - kE_{H_0}\phi^*(X)]$$

$$\Rightarrow E_{H_1}\phi(X) - E_{H_1}\phi^*(X) > k[E_{H_0}\phi(X) - E_{H_0}\phi^*(X)]$$

From (7.2.18), LHS is positive.

$$\Rightarrow E_{H_1}\phi(X) > E_{H_1}\phi^*(X)$$

i.e., $\phi(X)$ is MP test of size α in A_α.

Further, consider a test $\phi(X)$ with size 0 that satisfies the Eqs. (7.2.6) and (7.2.7). Let $\phi^*(X)$ be any other test in A_0 where

$$A_0 = \{\phi | E_{H_0}\phi(X) = 0\}$$

A test $\phi^* \in A_0$, which implies that

$$\int \phi^*(X)f_0(x)dx = 0.$$

This implies that $\phi^* = 0$ with probability 1 on the set $\{x : f_0(x) > 0\}$.
    Consider

$$E_{H_1}\phi(X) - E_{H_1}\phi^*(X) = \int\limits_{x:f_0(x)=0} [\phi(X) - \phi^*(X)]f_1(x)dx$$

$$+ \int\limits_{x:f_0(x)>0} [\phi(X) - \phi^*(X)]f_1(x)dx$$

$$= \int\limits_{x:f_0(x)=0} [1 - \phi^*(X)]f_1(x)dx \geq 0$$

$$E_{H_1}\phi(X) \geq E_{H_1}\phi^*(X)$$

Hence, the test ϕ in (7.2.7) is MP of size 0.
    (c) Necessary condition for MP test
    Let $\phi(X)$ be a MP test given in (7.2.4) and (7.2.5). Let $\phi^*(X)$ be some other MP test of size α in $A_\alpha \alpha > 0$, for testing $H_0 : X \sim f_0$ against $H_1 : X \sim f_1$.
    Since $S = S^+ \cup S^- \cup S^0$,
    Assume that the tests ϕ and ϕ^* takes on different values on S^+ and S^-.
    Consider a set

$$W = S \cap \{x : f_1(x) \neq kf_0(x)\} = \{x : f_1(x) \neq kf_0(x)\}$$

Assume that $P[W] > 0$, consider the integral

$$\int\limits_{S} [\phi(X) - \phi^*(X)][f_1(x) - kf_0(x)]dx = \int\limits_{W} [\phi(X) - \phi^*(X)][f_1(x) - kf_0(x)]dx$$

$$+ \int\limits_{S^0} [\phi(X) - \phi^*(X)][f_1(x) - kf_0(x)]dx$$

Since $\phi(X)$ and $\phi^*(X)$ are both MP test of size α then
$$E_{H_0}\phi(X) = E_{H_0}\phi^*(X) \text{ and } E_{H_1}\phi(X) = E_{H_1}\psi^*(X)$$

$$\Rightarrow \int_W [\phi(X) - \phi^*(X)][f_1(x) - kf_0(x)]dx = 0$$

with $[\phi(X) - \phi^*(X)][f_1(x) - kf_0(x)] > 0$.

It gives $P[W] = 0$. It is a contradiction to the assumption $P[W] > 0$. Therefore $P[W] = 0$. Let P be such that $P(W) = P(W|H_0) + P(W|H_1)$. But $P[W] = 0$, it implies that $P(W|H_0)$ and $P(W|H_1)$ are both zero. Thus the probability on the set W on which ϕ and ϕ^* are different, become zero under the H_0 and H_1 hypothesis.

In other words, we can say that $\phi(X) \neq \phi^*(X)$ on the set $S^0 = \{x|f_1(x) = kf_0(x)\}$. It implies that ϕ and ϕ^* are of the same form except on the set S^0 under H_0 and H_1.

Let ϕ be a test given in Eqs. (7.2.6) and (7.2.7). Further, let ϕ^* be an MP test of size 0 in A_0. The test ϕ and ϕ^* differ on the set

$$W = \{x : f_0(x) = 0, f_1(x) > 0\} \cup \{x : f_0(x) > 0, f_1(x) > 0\}$$

Assume $P(S) > 0$. Since $[\phi - \phi^*]f_1(x) > 0$ on W and the integral

$$\int_W [\phi - \phi^*]f_1(x)dx = 0$$

Because $E_{H_1}\phi(X) = E_{H_1}\phi^*(X)$. This implies that $P(W) = 0$.

This shows that MP tests $\phi(X)$ and $\phi^*(X)$ have the same form as in Eqs. (7.2.6) and (7.2.7) except perhaps on the set $\{x : f_0(x) > 0\} \cup \{x : f_1(x) > 0\}$.

7.3 P-Value

In many experimental situations and financial decisions, we conveniently use type I error as 1 or 5%. This is because we do not know how much one could tolerate the first kind of error. In fact, the choice of significance level should be such that the power of the test against the alternative must not be low.

Under H_0, the distribution of $\frac{f_1(x)}{f_0(x)}$ is continuous. Then, the MP level α test is nonrandomized and reject if $\frac{f_1(x)}{f_0(x)} > k$, where $k = k(\alpha)$ is determined by (7.2.4). For varying α, the resulting tests provide an example of the typical situation in which the rejection regions A_α are nested in the sense that

$$A_\alpha \subset A_{\alpha'} \text{ if } (\alpha < \alpha')$$

When this is the case, it is a good practice to determine not only whether the hypothesis is accepted or rejected at the given significance level, but also to determine the smallest significance level, or more formally.

$$\hat{p} = \hat{p}(X) = \inf\{\alpha : X \in A_\alpha\}$$

at which the hypothesis would be rejected for the given observation. This number, the so-called *p*-value gives an idea of how strongly the data contradicts the hypothesis. It also enables others to reach a verdict based on the significance level of their choice.

Example 7.3.1 Suppose that $H_0 : \mu = \mu_0$ against $H_0 : \mu = \mu_1 > \mu_0$ and $\alpha = 0.05$.
Suppose our test $\phi(X)$ is defined as

$$\phi(X) = \begin{cases} 1 \; ; \; X > k \\ 0 \; ; \; X \le k \end{cases}$$

Let $k = 1.64$. Further, we assume that Z is $N(0,1)$; see Fig. 7.9. Let the calculated value of z from the sample is 1.86. Hence, our test is reject H_0 if $z > 1.64$. In this case 1.86 is greater than 1.64, so we reject H_0; see Fig. 7.10.

Now, we can find *p*-value

$$P[z > 1.86] = 0.0314,$$

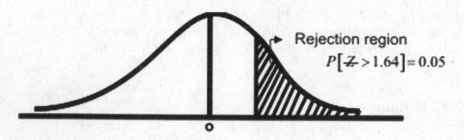

Fig. 7.9 5% upper tail area for standard normal distribution

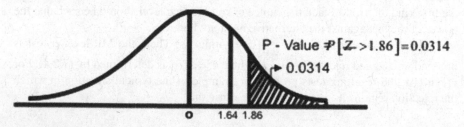

Fig. 7.10 *p*-value

In this case p-value is 0.0314 and it is less than 0.05. One should observe that we reject H_0 if p-value $< \alpha$. Now consider the equivalence between the two methods. Calculated value from sample of a test statistics is more than the table value of the distribution of a test statistics is equivalent that p-value less than α in rejecting H_0. One can see this in the above Fig. 7.10. Hence $1.86 > 1.64 \Leftrightarrow 0.0314 < 0.05$.

If $H_0 : \mu \geq \mu_0$ against $H_1 : \mu < \mu_0$. The test is

$$\phi(X) = \begin{cases} 1 \; ; \; X < -1.64 \\ 0 \; ; \; X > -1.64 \end{cases}$$

$P[z < -1.64] = 0.05$ and $P[z < -1.86] = 0.0314$.

P-value, i.e., probability of observing the value of z in the sample as small as -1.86 and $0.0314 < 0.05$.

P-value in two-sided test

Determining p-value in the two-sided tests presents a problem. The most common practice in two-sided test is to report the p-value as twice as the one-sided p-value.

This procedure is satisfactory for cases in which the sampling distribution of test statistic is symmetric under H_0.

However, if the distribution is not symmetric under H_0, doubling the p-value may lead to get $p > 1$ and other problems. Hence Gibbson and Pratt (1975) suggested that in the case of two-sided tests, we should report the one-sided p-value and state the direction of the observed departure from the null hypothesis.

7.4 Illustrative Examples on the MP Test

Example 7.4.1 Let X be a random variable with pmf under H_0 and H_1. Find MP test of size 0.03 and further find its power.

(i)

X:	1	2	3	4	5	6
$f_0(x)$:	0.01	0.01	0.01	0.01	0.01	0.95
$f_1(x)$:	0.05	0.04	0.03	0.02	0.01	0.85
$\frac{f_1(x)}{f_0(x)}$:	5	4	3	2	1	0.89

Let $\lambda(x) = \frac{f_1}{f_0}$
One can observe that $\lambda(x)$ is decreasing function of X.

If $E\phi(X) \leq 0.03$ then MP test of size 0.03,

$$\phi(X) = \begin{cases} 1 \; ; \; \lambda(X) \geq 3 \\ 0 \; ; \; \lambda(X) < 3 \end{cases}$$

$$\Rightarrow \phi(X) = \begin{cases} 1 \; ; \; X \leq 3 \\ 0 \; ; \; X > 3 \end{cases}$$

Power $= P_{H_1}(X \leq 3) = 0.05 + 0.004 + 0.03 = 0.12$.

One should note that according to NP lemma, if we take any other region of size 0.03, it will have less power.

For example:

$$\Rightarrow \phi_1(X) = \begin{cases} 1 \; ; \; X = 1, 4, 5 \\ 0 \; ; \; \text{otherwise} \end{cases}$$

or

$$\phi_2(X) = \begin{cases} 1 \; ; \; X = 3, 4, 5 \\ 0 \; ; \; \text{otherwise} \end{cases}$$

$$E_{H_1}\phi_1(X) = P_{H_1}(X = 1) + P_{H_1}(X = 4) + P_{H_1}(X = 5) = 0.05 + 0.02 + 0.01 = 0.08$$

$$E_{H_1}\phi_2(X) = P_{H_1}(X = 3) + P_{H_1}(X = 4) + P_{H_1}(X = 5) = 0.03 + 0.02 + 0.01 = 0.06$$

Example 7.4.2 Consider a random sample of size one and let $H_0 : X \sim N(0, 1)$ against $H_1 : X \sim C(0, 1)$. We want to obtain MP test of size α for the given H_0 against H_1

$$\lambda(x) = \frac{f_1}{f_0} = \sqrt{\frac{2}{\pi}} \frac{\exp(\frac{x^2}{2})}{1 + x^2}$$

$$\log \lambda(x) = \text{const} + \frac{x^2}{2} - \log(1 + x^2)$$

$$\frac{\partial \log \lambda(x)}{\partial x} = x - \frac{2x}{1 + x^2} \tag{7.4.1}$$

$$\frac{\partial^2 \log \lambda(x)}{\partial x^2} = 1 - \frac{2(1 - x^2)}{(1 + x^2)^2} \tag{7.4.2}$$

From (7.4.1), $x[x^2 - 1] = 0$

Hence $x = 0$, or $x = \pm 1$

From (7.4.2), at $x = 0$, $\dfrac{\partial^2 \log \lambda(x)}{\partial x^2} < 0$,

at $x = \pm 1$, $\dfrac{\partial^2 \log \lambda(x)}{\partial x^2} > 0$,

Therefore, at $x = 0$, $\lambda(x)$ has a local maximum, i.e., $\lambda(0) = 0.7979 = \sqrt{\dfrac{2}{\pi}}$.

Further at $x = \pm 1$, $\lambda(x)$ has a minimum, i.e., $\lambda(\pm 1) = 0.6577 = \sqrt{\dfrac{2}{\pi}}(\dfrac{e^{0.5}}{2})$

We can say that $\lambda(x)$ is decreasing in $(0, 1)$ and increasing in $(1, \infty)$.

The horizontal line $\lambda(x) = k$ intersects the graph of $\lambda(x)$; see Fig. 7.11

  (i) in two points if $k > 0.7979$ (points a, b)
 (ii) in three points if $k = 0.7979$ (points c, d, e)
(iii) in four points if $k \in (0.6577, 0.7979)$ (points f, g, h, i)
 (iv) in two points if $k = 0.6577$ (points j, k)
  (v) in no points if $k < 0.6577$

A program in R for the function $\lambda(x)$ is written.

```
f=function(x) sqrt(2/pi)*exp((1/2)*x^2)/(1+x^2)
x=seq(-2,2,0.01) plot(y=f(x),x,type='l',col='solid line',lwd=4)
abline(h=0.7979,col='dashed line',lty=2,lwd=3)
abline(h=0.6577,col='dotted line',lty=4,lwd=3) legend("top",
c("-lambda(x)","-lambda(0)","-lambda(+1,-1)"),text.col=c('solid line','dashed line','dotted line'))
```

(i) if $0 \le k \le 0.6577$, then

$$E_{H_0} \phi(x) = 1$$

Fig. 7.11 Graph of $\lambda(x)$

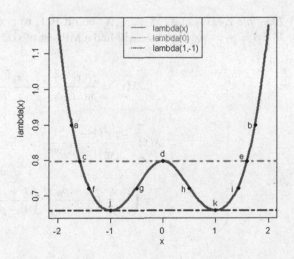

(ii) if $k \in (0.6527, 0.7979)$ then

$$E_{H_0}\phi(x) = P_{H_0}[-c_1 < X < c_1] + P_{H_0}[X < -c_2] + P_{H_0}[X > c_2]$$

(iii) if $k > 0.7979$ then

$$E_{H_0}\phi(x) = P[X < -d_1] + P[X > d_1]$$

For $\alpha = 0.05$, the MP test will have critical region $|x| > d_1$ provided $\lambda(d_1) > 0.7979$.
From tables, $\alpha = 0.05$, $d_1 = 1.96$ and $\lambda(d_1) = 1.2124$ and $\lambda(d_1) > 0.7979$.
We can write the MP test as

$$\phi(X) = \begin{cases} 1 ; & |x| > d_1 \\ 0 ; & |x| \le d_1 \end{cases}$$

For $\alpha = 0.05$

$$= \begin{cases} 1 ; & |x| > 1.96 \\ 0 ; & |x| \le 1.96 \end{cases}$$

Power of the test

$$E_{H_0}\phi(x) = 1 - \int\limits_{-1.96}^{1.96} \frac{1}{\pi}\frac{1}{1+x^2}dx$$

$$= 1 - 2\int\limits_{0}^{1.96} \frac{1}{\pi}\frac{1}{1+x^2}dx = 1 - \frac{2\tan^{-1}1.96}{\pi} = 0.3003$$

Example 7.4.3 Let $X_1, X_2, \ldots, X_n$ be iid B(1, p) rvs and let $H_0 : p = p_0$ against $h_1 : p = p_1 > p_0$. Now, we will find a MP test of size α.

$$\lambda(x) = \frac{f_1(x_1, x_2, \ldots, x_n|p)}{f_0(x_1, x_2, \ldots, x_n|p)}$$

$$\lambda(x) = \frac{p_1^t(1-p_1)^{n-t}}{p_0^t(1-p_0)^{n-t}}; \quad t = \sum_{i=1}^{n} x_i$$

$$= \left(\frac{p_1}{p_0}\right)^t \left(\frac{1-p_0}{1-p_1}\right)^t \left(\frac{1-p_1}{1-p_0}\right)^n$$

$$= \left(\frac{p_1(1-p_0)}{p_0(1-p_1)}\right)^t \left(\frac{1-p_1}{1-p_0}\right)^n$$

For $p_1 > p_0$, $\lambda(x)$ is nondecreasing function of t. It implies that $\lambda(x) > k$,

$$\lambda(x) > k \Leftrightarrow t > k$$

(the reader should note that after many algebraic operations, we will call the constant as k only)

Our MP test is as

$$\phi(X) = \begin{cases} 1 & ; \ t > k \\ \gamma & ; \ t = k \\ 0 & ; \ t < k \end{cases}$$

Now, k and γ are determined from $E_{H_0}\phi(x) = \alpha$

$$\alpha = P_{H_0}[t > k] + \gamma P[t = k]$$

Now $T \sim B(n, p)$

In particular $n = 5, p_0 = \frac{1}{2}, p_1 = \frac{3}{4}$ and $\alpha = 0.05$

$$0.05 = \sum_{r=k+1}^{5} \binom{5}{r}\left(\frac{1}{2}\right)^5 + \gamma\binom{5}{k}\left(\frac{1}{2}\right)^5$$

Let $k = 4$

$$= 0.0312 + \gamma(0.1562)$$

$$\gamma = \frac{0.0188}{0.15620} = 0.12$$

The MP test of size $\alpha = 0.05$ is given as

$$\phi(X) = \begin{cases} 1 & ; \ t > 4 \\ 0.12 & ; \ t = 4 \\ 0 & ; \ t < 4 \end{cases}$$

Thus the MP size $\alpha = 0.05$ test is to reject $p = \frac{1}{2}$ in favor of $p = \frac{3}{4}$ if $\sum x_i = 5$ and reject $p = \frac{1}{2}$ with probability 0.12 if $\sum_{i=1}^{5} x_i = 4$.

For this a program in *R* is also given below.

```
# Given data
  alpha = 0.05; n = 5; p0 = 0.5; p1 = 3/4.
# To find k such that first term is < alpha
  a = seq(from=0,to=(n-1),by=1); # possible values for k
  la = length(a)
# to find cumulative probability, i.e., P(t1 > k)
  cpk <- rep(0,la) # declaring variable to find cumulative probability.
  for(i in 1:la)
    {
      for(j in (a[i]+1):n)
        {
          cpk[i] = cpk[i] + dbinom(j,n,p0);
        }
    }
  ind = min(which(cpk < alpha))  # gives cumulative probability < alpha
# To find gamma
  k = ind-1; b <- dbinom(k,n,p0);
  gamma = (alpha-cpk[k+1])/b
# To check the answer
  check <- cpk[k+1]+(gamma*dbinom(k,n,p0))
# OUTPUT
  print(c("k=",k))
  print(c("gamma =",gamma))
  print(c("check=",check))
```

Example 7.4.4 A sample size of 10 is obtained from a Poisson distribution with parameter m. Obtain a MP test of size $\alpha = 0.01$ to test $H_0 : m = 3$ against $H_1 : m > 3$

$$\lambda(x) = \frac{f_1(x_1, x_2, \ldots, x_n|p)}{f_0(x_1, x_2, \ldots, x_n|p)} = \frac{\prod_{i=1}^{n} e^{-m_1} m_1{}^{x_i}(x_i!)^{-1}}{\prod_{i=1}^{n} e^{-m_0} m_0{}^{x_i}(x_i!)^{-1}}$$

$$= e^{-n(m_0-m_1)} \left(\frac{m_1}{m_0}\right)^t; \quad t = \sum_{i=1}^{n} x_i$$

For $m_1 > m_0$, $\lambda(x) > k \Leftrightarrow t > k$

The MP test is

$$\phi(X) = \begin{cases} 1 \; ; \; t > k \\ \gamma \; ; \; t = k \\ 0 \; ; \; t < k \end{cases}$$

Now, $E_{H_0}\phi(x) = 0.01$ and the distribution of T under H_0 is P(30).

$$P_{H_0}[t > k] + \gamma P_{H_0}[t = k] = 0.01$$

$$P[t > k] \leq 0.01$$

For large m $T \sim N(30, 30)$

$$P[t > k] = P\left[z > \frac{k - 30}{\sqrt{30}}\right] = 0.01$$

$$\Rightarrow \frac{k - 30}{\sqrt{30}} = 2.325 \Rightarrow k = 42.74 \approx 43$$

$$P_{H_0}[t > 43] + \gamma P_{H_0}[t = 43] = 0.01$$

$$\gamma = \frac{0.01 - P[t > 43]}{p[t = 43]}$$

$$= \frac{0.01 - 0.009735}{0.00508432} = 0.05212$$

Hence MP test is given as

$$\phi(X) = \begin{cases} 1 & ; t > 43 \\ 0.05212 & ; t = 43 \\ 0 & ; t < 43 \end{cases}$$

Note: If you are using R then the above method of calculation is not required. A program in R is also given to calculate k and γ.

```
# Given data
  n = 10; m = 3; lambda = n*m;
# To find k such that P(T <= k) > 1-alpha
# Defining function
  fun = function(upper,lambda)
  {
    alpha = 0.01;
    a = seq(from=0,to=(upper-1),by=1); # possible values for k
    la = length(a)
    # to find cumulative probability, i.e., P(t1 <= k)
    cpk <- rep(0,la) # declaring variable to find cumulative probability.
    for(i in 1:la)
    {
      cpk[i] = ppois(a[i],lambda);
    }
    if(cpk[la] < (1-alpha))
      { print("increase the value of k") }
    if(cpk[la] > (1-alpha))
      {
        ind = min(which(cpk > (1-alpha)))  # gives cumulative probability > 1-alpha
        # To find gamma
        k = ind-1; b <- dpois(k,lambda);
        gamma = (cpk[k+1]-1+alpha)/b;
        return(c(k,gamma));
      }
  }
  upper = 10; ans = fun(upper,lambda)
"increase the value of k"
```

```
    upper = 50;
    ans = fun(upper,lambda)
    k = ans[1]; gamma = ans[2];
# To check the answer
    check <- (1-ppois((k),lambda))+(gamma*dpois(k,lambda))
# OUTPUT
    print(c("k=",k))
    print(c("gamma =",gamma))
    print(c("check=",check))
```

Example 7.4.5 Let X be a rv with $C(1, \theta)$. Obtain a most powerful test of level of significance of α to test $H_0 : \theta = 0$ against $H_1 : \theta = 1$

$$f(x|\theta) = \frac{1}{\pi[1 + (x - \theta)^2]}$$

$$\lambda(x) = \frac{1 + x^2}{1 + (x - 1)^2}$$

From NP lemma,

$$\lambda(x) > k \Rightarrow \frac{1 + x^2}{1 + (x - 1)^2} > k$$

$$\Rightarrow x^2(1 - k) + 2kx + (1 - 2k) > 0 \tag{7.4.3}$$

if $k > 1$

$$\Rightarrow x^2 + \frac{2kx}{1 - k} + \frac{1 - 2k}{1 - k} < 0 \tag{7.4.4}$$

Let $k_1 = \frac{2k}{1-k}$ and $k_2 = \frac{1-2k}{1-k}$

$$x = \frac{-k_1 \pm \sqrt{k_1^2 - 4k_2}}{2}$$

Let $\lambda_1(k) = \frac{1}{2}\left[-k_1 - \sqrt{k_1^2 - 4k_2}\right]$ and $\lambda_2(k) = \frac{1}{2}\left[-k_1 + \sqrt{k_1^2 - 4k_2}\right]$.
  From (7.4.4), $\lambda_1(k) < x < \lambda_2(k)$ if $k > 1$
  The MP test is given by

$$\phi(X) = \begin{cases} 1 \; ; \; \lambda_1(k) < x < \lambda_2(k) \\ 0 \; ; \; \text{otherwise} \end{cases}$$

$$E_{H_0}\phi(x) = \alpha$$

$$\Rightarrow h(k) = \int\limits_{\lambda_1(k)}^{\lambda_2(k)} \frac{dx}{\pi[1+x^2]} = \alpha$$

$$\Rightarrow \tan^{-1}\lambda_2(k) - \tan^{-1}\lambda_1(k) = \pi\alpha \tag{7.4.5}$$

From (7.4.3) if $k < 1$

$$x^2 + \frac{2kx}{1-k} + \frac{1-2k}{1-k} > 0$$

The MP test is given by

$$\phi(X) = \begin{cases} 1 \; ; \; x > \lambda_2(k) \text{ or } x < \lambda_1(k) \\ 0 \; ; \text{ otherwise} \end{cases}$$

$$E_{H_0}\phi(x) = \alpha$$

$$\Rightarrow \int\limits_{\lambda_2(k)}^{\infty} \frac{dx}{\pi[1+x^2]} + \int\limits_{-\infty}^{\lambda_1(k)} \frac{dx}{\pi[1+x^2]}$$

$$\Rightarrow \int\limits_{\lambda_1(k)}^{\lambda_2(k)} \frac{dx}{\pi[1+x^2]} = 1 - \alpha$$

$$\Rightarrow \tan^{-1}\lambda_2(k) - \tan^{-1}\lambda_1(k) = \pi(1-\alpha) \tag{7.4.6}$$

$$\lambda_1(k) = \frac{-k - \sqrt{-k^2 + 3k - 1}}{(1-k)}$$

and

$$\lambda_2(k) = \frac{-k + \sqrt{-k^2 + 3k - 1}}{(1-k)}$$

Assume $(-k^2 + 3k - 1)$ is positive.

If $k > 1$ and $\frac{3-\sqrt5}{2} < k < \frac{3+\sqrt5}{2}$ then $1 < k < \frac{3+\sqrt5}{2}$ and if $k < 1$ then $\frac{3-\sqrt5}{2} < k < 1$.

$\pi\alpha = 0.314139$ if $\alpha = 0.10$

k	$\lambda_2(k)$	$\lambda_1(k)$	$\tan^{-1}\lambda_2(k) - \tan^{-1}\lambda_1(k)$
0.3819	−0.618	−0.618	0
0.50	0	−2	1.1071
0.60	0.1583	−3.5183	1.4211541
0.80	0.3589	−8.3589	1.7963
0.99	0.4937	−198.4937	2.0244
1.50	5.2361	0.7639	0.7297
2	3	1	0.4636
2.2	2.5598	1.1069	0.3623
2.3	2.37	1.1684	0.3086
2.29	2.3885	1.1619	0.31415

$k = 2.29$
UMP test

$$\phi(X) = \begin{cases} 1 \; ; \; 1.1619 < X < 2.3885 \\ 0 \; ; \text{otherwise} \end{cases}$$

A program is written for the function $\lambda(x)$ in R.

Assume $f(k) = \tan^{-1}\lambda_2(k) - \tan^{-1}\lambda_1(k)$. If $k = 2.29$, then $f(k) = 0.31$ see Fig. 7.12. Note: One can use the following formula for (7.4.5) and (7.4.6)

$$\tan^{-1}(A) - \tan^{-1}(B) = \tan^{-1}\frac{A - B}{1 + AB}, \text{ if } A \text{ and } B \text{ are positive.}$$

Fig. 7.12 Graph of $f(k)$

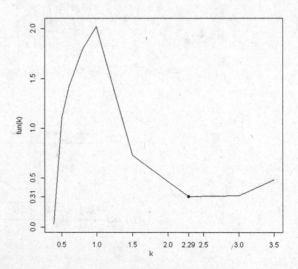

```
# TO FIND VALUES OF LAMBDA1 AND LAMBDA2.
  # To define function lambda1
    lambda1 = function(k)
    {
       k1 = 2*k/(1-k); k2 = (1-2*k)/(1-k);
       l1 = (-k1-sqrt(abs((k1^2)-(4*k2))))/2;
       return(l1)
    }
  # To define function lambda2
  lambda2 = function(k)
  {
     k1 = 2*k/(1-k); k2 = (1-2*k)/(1-k);
     l2 = (-k1+sqrt(abs((k1^2)-(4*k2))))/2;
     return(l2)
  }
# To define function h(k)
  fun <- function(k)
  { f1 <- atan(lambda2(k))-atan(lambda1(k)); return(f1) }
# To obtain values and to plot h(k)
  # Given values of k.
    k <- c(0.3819,0.50,0.60,0.80,0.99,1.50,2,2.2,2.3,2.29,3.0,3.5)
    result <- data.frame(lambda1(k),lambda2(k),fun(k))
    plot(k,fun(k),type="l"); points(2.2899,fun(2.2899),pch=19)
    axis(1,2.29); axis(2,0.31)
  # Result
  result
# To solve h(k)
  # Defining h(k)
    f =function(k)
    {
       alpha <- 0.10; k1 = 2*k/(1-k); k2 = (1-2*k)/(1-k);
       l1 = (-k1-sqrt(abs((k1^2)-(4*k2))))/2; l2 = (-k1+sqrt(abs((k1^2)-(4*k2))))/2;
       fun <- atan(lambda2(k))-atan(lambda1(k))-pi*alpha;
       return(fun)
    }
# To solve h(k)
  y <- uniroot(f,c(1.01,2.98))    print(y$root)
```

Example 7.4.6 Let $X_1, X_2, \ldots, X_n$ be iid rvs $N(\mu, \sigma^2)$, where both μ and σ^2 are unknown. We wish to test $H_0 : \mu = \mu_0, \sigma^2 = \sigma_0^2$ against $H_1 : \mu = \mu_1 > \mu_0$, $\sigma^2 = \sigma_0^2$,

$$\lambda(X) = \frac{\left(\frac{1}{\sigma_0\sqrt{2\pi}}\right)^n \exp\left[-\frac{1}{2\sigma_0^2}\sum_{i=1}^{n}(x_i-\mu_1)^2\right]}{\left(\frac{1}{\sigma_0\sqrt{2\pi}}\right)^n \exp\left[-\frac{1}{2\sigma_0^2}\sum_{i=1}^{n}(x_i-\mu_0)^2\right]}$$

$$= \exp\left[\frac{1}{2\sigma_0^2}\left\{\sum_{i=1}^{n}(x_i-\mu_0)^2 - \sum_{i=1}^{n}(x_i-\mu_1)^2\right\}\right]$$

$$= \exp\left[\frac{1}{2\sigma_0^2}\left\{2\sum_{i=1}^{n}x_i(\mu_1-\mu_0) + n(\mu_0^2-\mu_1^2)\right\}\right]$$

If $\mu_1 > \mu_0 \Rightarrow \lambda(x) > k$ if and only if $\sum_{i=1}^{n} x_i > k$.

The distribution of $\sum X_i$ is $N(n\mu, n\sigma^2)$.

Now, k is determined such that

$$\alpha = P_{H_0}\left[\sum_{i=1}^{n} x_i > k\right] = P_{H_0}\left[z > \frac{k - n\mu_0}{\sigma_0\sqrt{n}}\right],$$

where

$$z = \frac{\sum_{i=1}^{n} x_i - n\mu_0}{\sigma_0\sqrt{n}} \sim N(0, 1)$$

$$Z_\alpha = \frac{k - n\mu_0}{\sigma_0\sqrt{n}} \Rightarrow k = \sigma_0\sqrt{n}Z_\alpha + n\mu_0$$

The MP test is

$$\phi(X) = \begin{cases} 1 \; ; \; \sum X_i > \sigma_0\sqrt{n}Z_\alpha + n\mu_0 \\ 0 \; ; \; \text{otherwise} \end{cases}$$

Similarly, for testing $H_0 : \mu = \mu_0, \sigma^2 = \sigma_0^2$ against $H_1 : \mu = \mu_1 < \mu_0, \sigma^2 = \sigma_0^2$. The MP test is

$$\phi(X) = \begin{cases} 1 \; ; \; \sum X_i < \sigma_0\sqrt{n}Z_\alpha + n\mu_0 \\ 0 \; ; \; \text{otherwise} \end{cases}$$

Note: 1. If $\sigma = \sigma_0$, i.e., σ is known, the test determined is independent of μ_1 as long as $\mu_{1.} > \mu_0$ and it follows that the test is UMP against $H_1 : \mu = \mu_1 > \mu_0$. Similarly the test is UMP for $H_1 : \mu = \mu_1 < \mu_0$. Further, we can say that the test does not depend on H_1.

2. If σ is not known and H_0 is composite hypothesis. Then the test determined above depends on σ^2. Hence the above test will not be an MP test.

Example 7.4.7 Let $X_1, X_2, \ldots, X_n$ be iid random sample of size n from exponential distribution with mean θ. Find MP test for testing $H_0 : \theta = \theta_0$ against $H_1 : \theta = \theta_1 < \theta_0$

$$\lambda(x) = \frac{f_1(x_1, x_2, \ldots, x_n|\theta_1)}{f_0(x_1, x_2, \ldots, x_n|\theta_0)}$$

$$= \left(\frac{\theta_0}{\theta_1}\right)^n \exp\left[\sum_{i=1}^{n} x_i\left(\frac{1}{\theta_0} - \frac{1}{\theta_1}\right)\right]$$

By NP lemma, $\lambda(x) > k$ and $\lambda(x)$ is nonincreasing in t, where $t = \sum_{i=1}^{n} x_i$ because $\theta_1 < \theta_{11}$.

Hence $\lambda(x) > k \Leftrightarrow t < k$

The MP test is given by

$$\phi(X) = \begin{cases} 1 \; ; \; t < k \\ 0 \; ; \; \text{otherwise} \end{cases}$$

Under H_0, $T = t$ has gamma distribution with parameters n and $\frac{1}{\theta_0}$

$$f(t) = \frac{e^{-\frac{t}{\theta_0}} t^{n-1}}{\theta_0^n \Gamma(n)} ; t > 0, \theta_0 > 0$$

Let $V = \frac{2t}{\theta_0}$ then $V \sim \chi_{2n}^2$

$$E_{H_0} \phi(x) = \alpha \Rightarrow P[t < k] = \alpha$$

$$\Rightarrow P\left[\frac{2t}{\theta_0} < \frac{2k}{\theta_0} \right] = \alpha$$

$$\Rightarrow P\left[V < \frac{2k}{\theta_0} \right] = \alpha$$

$$\Rightarrow \chi_{2n,1-\alpha}^2 = \frac{2k}{\theta_0} \Rightarrow k = \frac{\theta_0}{2} \chi_{2n,1-\alpha}^2$$

The MP test is

$$\phi(X) = \begin{cases} 1 \; ; \; T < \frac{\theta_0}{2} \chi_{2n,1-\alpha}^2 \\ 0 \; ; \; \text{otherwise} \end{cases}$$

Note: 1. This test is UMP because it does not depend on H_1.
2. Similarly, one can find a UMP test when $H_1 : \theta = \theta_1 > \theta_0$.

Example 7.4.8 Let $X_1, X_2, \ldots, X_n$ be iid random sample of size n from $N(0, \sigma^2)$. Find MP test for testing $H_0 : \sigma = \sigma_0$ against $H_1 : \sigma = \sigma_1 > \sigma_0$

According to earlier examples, the MP test is given as

$$\phi(X) = \begin{cases} 1 \; ; \; \sum x_i^2 > k \\ 0 \; ; \; \text{otherwise} \end{cases}$$

Note that, under H_0, $\frac{\sum x_i^2}{\sigma_0^2} \sim \chi_n^2$

$$E_{H_0}\phi(x) = \alpha \Rightarrow P\left[\frac{\sum x_i^2}{\sigma_0^2} > \frac{k}{\sigma_0^2}\right] = \alpha$$

$$\frac{k}{\sigma_0^2} = \chi_{n,\alpha}^2 \Rightarrow k = \sigma_0^2 \chi_{n,\alpha}^2$$

Hence, the UMP test is given by

$$\phi(X) = \begin{cases} 1 \; ; \; \sum x_i^2 > \sigma_0^2 \chi_{n,\alpha}^2 \\ 0 \; ; \text{otherwise} \end{cases}$$

Example 7.4.9 Let X be a rv with pdf $f(x|\theta)$

$$f(x|\theta) = \begin{cases} \frac{2(\theta - x)}{\theta^2} \; ; \; 0 < x < \theta \\ 0 \qquad \; ; \text{otherwise} \end{cases}$$

Obtain a MP test of size α to test

(i) $H_0 : \theta = \theta_0$ against $H_1 : \theta = \theta_1 > \theta_0$
(ii) $H_0 : \theta = \theta_0$ against $H_1 : \theta = \theta_1 < \theta_0$

(i) There are three cases:
(a) $0 < x < \theta_0$ (b) $\theta_0 < x < \theta_1$ (c) $\theta_1 < x < \infty$

$$\lambda(x) = \begin{cases} \frac{\theta_0^2(\theta_1 - x)}{\theta_1^2(\theta_0 - x)} \; ; \; 0 < x < \theta_0 \\ \frac{2\theta_1^{-2}(\theta_1 - x)}{0} \; ; \; \theta_0 < x < \theta_1 \\ \frac{0}{0} \qquad \; ; \; \theta_1 < x < \infty \end{cases}$$

$$\lambda'(x) = \frac{\theta_0^2}{\theta_1^2} \frac{(\theta_1 - \theta_0)}{(\theta_0 - x)^2} > 0$$

Hence $\lambda(x)$ is nondecreasing sequence in x
   Hence

$$\lambda(x) > k \Leftrightarrow x > k$$

Hence, MP test of size α is given as

$$\phi(X) = \begin{cases} 1 \; ; \; x > k \\ 0 \; ; \text{otherwise} \end{cases}$$

$$E_{H_0}\phi(x) = \alpha \Rightarrow \int\limits_k^{\theta_0} \frac{2(\theta_0 - x)}{\theta_0^2}\,dx = \alpha$$

$$\frac{k - \theta_0}{\theta_0} = \pm\sqrt{\alpha}$$

$$k = \theta_0 \pm \theta_0\sqrt{\alpha},\ 0 < \alpha < 1$$

Since X lies between 0 and θ_0,
  i.e., $0 < x < \theta_0 \Rightarrow k < \theta_0$
$\Rightarrow k = \theta_0(1 - \sqrt{\alpha})$
This test is UMP because it does not depend on H_1.
Hence, UMP test of size α is given by

$$\phi(X) = \begin{cases} 1 \ ; \ x > \theta_0(1 - \sqrt{\alpha}) \\ 0 \ ; \ \text{otherwise} \end{cases}$$

(ii) $H_0 : \theta = \theta_0$ against $H_1 : \theta_1 < \theta_0$
  There are three cases (a) $0 < x < \theta_1$ (b) $\theta_1 < x < \theta_0$ (c) $\theta_0 < x < \infty$

$$\lambda(x) = \begin{cases} \frac{\theta_0^2(\theta_1 - x)}{\theta_1^2(\theta_0 - x)} & ; \ 0 < x < \theta_1 \\ \frac{0}{2\theta_0^{-2}(\theta_0 - x)} & ; \ \theta_1 < x < \theta_0 \\ \frac{0}{0} & ; \ \theta_0 < x < \infty \end{cases}$$

$$\lambda'(x) = \left(\frac{\theta_0}{\theta_1}\right)^2 \frac{(\theta_1 - \theta_0)}{(\theta_0 - x)^2} < 0,$$

Hence $\lambda(x)$ is nonincreasing in x

$$\lambda(x) > k \Leftrightarrow x < k$$

$$E_{H_0}\phi(x) = \alpha \Rightarrow \int\limits_0^k \frac{2}{\theta_0^2}(\theta_0 - x) = \alpha$$

$$\Rightarrow k = \theta_0 \pm \theta_0\sqrt{1 - \alpha}$$

We can take the value of $k = \theta_0(1 - \sqrt{1 - \alpha})$ because $k < \theta_0$

The UMP test is given as

$$\phi(X) = \begin{cases} 1 \; ; \; x < \theta_0(1 - \sqrt{1-\alpha}) \\ 0 \; ; \; \text{otherwise} \end{cases}$$

Example 7.4.10 Obtain the MP test of size α to test $H_0 : X \sim f_0(x)$, where

$$f_0(x) = \begin{cases} 1 \; ; \; 0 < x < 1 \\ 0 \; ; \; \text{otherwise} \end{cases}$$

against $H_1 : X \sim f_1(x)$ where

$$f_1(x) = \begin{cases} 4x \quad\;\; ; \; 0 < x < \frac{1}{2} \\ 4 - 4x \; ; \; \frac{1}{2} \le x < 1 \end{cases}$$

$$\lambda(x) = \frac{f_1(x)}{f_0(x)} = \begin{cases} 4x \quad\;\; ; \; 0 < x < \frac{1}{2} \\ 4 - 4x \; ; \; \frac{1}{2} \le x < 1 \end{cases}$$

$$\lambda'(x) = \begin{cases} 4 \;\; ; \; 0 < x < \frac{1}{2} \Rightarrow \lambda(x) \text{ is } \uparrow \text{ in}(0, \frac{1}{2}) \\ -4 \; ; \; \frac{1}{2} \le x < 1 \Rightarrow \lambda(x) \text{ is } \downarrow \text{ in}(\frac{1}{2}, 1) \end{cases}$$

$$\lambda\left(\frac{3}{4}\right) = \lambda\left(\frac{1}{4}\right) = 1, \lambda\left(\frac{1}{2}\right) = 2$$

$\lambda(x)$ is symmetric about $x = \frac{1}{2}$

$$\lambda(x) > k \Leftrightarrow k_1 < x < k_2$$

$$\Leftrightarrow \frac{1}{2} - k < x < \frac{1}{2} + k$$

The MP test of size α is given as

$$\phi(x) = \begin{cases} 1 \; ; \; \frac{1}{2} - k < x < \frac{1}{2} + k \\ 0 \; ; \; \text{otherwise} \end{cases}$$

$$E_{H_0} \phi(x) = \alpha \Rightarrow \int\limits_{\frac{1}{2}-k}^{\frac{1}{2}+k} dx = \alpha \Rightarrow k = \frac{\alpha}{2}$$

$$\phi(x) = \begin{cases} 1 \; ; \; \frac{1-\alpha}{2} < x < \frac{1+\alpha}{2} \\ 0 \; ; \; \text{otherwise} \end{cases}$$

Example 7.4.11 Let X be a rv with pdf $f(x|\theta)$,

$$f(x|\theta) = \begin{cases} 2\theta x + 2(1-\theta)(1-x) \; ; \; 0 < x < 1, \; \theta \in [0, 1] \\ 0 \hspace{5.5cm} ; \; \text{otherwise} \end{cases}$$

Find the MP test of size α to test $H_0 : \theta = \theta_0$ against $H_1 : \theta = \theta_1 > \theta_0$

$$\lambda(x) = \frac{2\theta_1 x + 2(1-\theta_1)(1-x)}{2\theta_0 x + 2(1-\theta_0)(1-x)}$$

$$= \frac{x[2\theta_1 - 1] + (1-\theta_1)}{x[2\theta_0 - 1] + (1-\theta_0)}$$

$$\lambda'(x) = \frac{\theta_1 - \theta_0}{[x(2\theta_0 - 1) + (1-\theta_0)]^2} \quad (\text{since } \theta_1 > \theta_0)$$

$\Rightarrow \lambda(x)$ is nondecreasing function in x.
$\quad \Rightarrow \lambda(x) > k \Leftrightarrow x > k$
The MP test is

$$\phi(X) = \begin{cases} 1 \; ; \; X > k \\ 0 \; ; \; \text{otherwise} \end{cases}$$

$$E_{H_0}\phi(x) = \alpha \Rightarrow \int_k^1 [x(2\theta_0 - 1) + (1-\theta_0)]dx$$

$$= (2\theta_0 - 1)(1 - k^2) + 2(1 - \theta_0)(1 - k)$$

Now

$$= (2\theta_0 - 1)(1 - k^2) + 2(1 - \theta_0)(1 - k) = \alpha$$

$$k^2(2\theta_0 - 1) + 2(1 - \theta_0)k + \alpha - 1 = 0$$

$$k = \frac{-(1-\theta_0) + \sqrt{\theta_0^2 - \alpha(2\theta_0 - 1)}}{(2\theta_0 - 1)}$$

The UMP test of size α is given as

$$\phi(X) = \begin{cases} 1 \; ; \; X > \frac{-(1-\theta_0) + \sqrt{\theta_0^2 - \alpha(2\theta_0 - 1)}}{(2\theta_0 - 1)} \\ 0 \; ; \; \text{otherwise} \end{cases}$$

This test is UMP because it does not depend on H_1.

Example 7.4.12 Let X be a rv with $\beta(1, b)$. Find a MP test of size α to test $H_0 : b = 1$ against $H_1 : b = b_1 > 1$

$$f(x) = \frac{(1-x)^{b-1}}{\beta(1, b)}; 0 < x < 1$$

$$\lambda(x) = b_1(1-x)^{b_1 - 1}$$

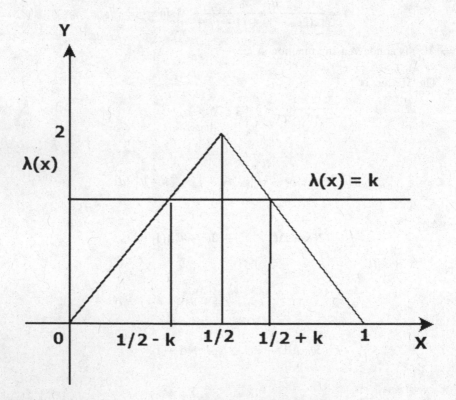

Fig. 7.13 Graph of $\lambda(x)$

$$\lambda'(x) = b_1(b_1 - 1)(1 - x)^{b_1 - 2}(-1) < 0$$

$\lambda(x)$ is nonincreasing function in x then $\lambda(x) > k \Leftrightarrow x < k$. The MP test of size α is given as

$$\phi(X) = \begin{cases} 1 \; ; \; X < k \\ 0 \; ; \; \text{otherwise} \end{cases}$$

$$E_{H_0}\phi(x) = \alpha \Rightarrow k = \alpha$$

Hence our UMP test of size α is given as

$$\phi(X) = \begin{cases} 1 \; ; \; X < \alpha \\ 0 \; ; \; \text{otherwise} \end{cases}$$

This test is UMP because it does not depend on H_1.
See Graph of $\lambda(x)$ for $\beta(1, b_1)$, (Fig. 7.13).

Example 7.4.13 Let P_0, P_1, P_2 be the probability distributions assigning to the integers 1, 2, 3, 4, 5 and 6 the following probabilities

	1	2	3	4	5	6
P_0	0.03	0.02	0.02	0.01	0	0.92
P_1	0.06	0.05	0.08	0.02	0.01	0.78
P_2	0.09	0.05	0.12	0	0.02	0.72
$\frac{P_1}{P_0}$	2	2.5	4	2	∞	0.85
$\frac{P_2}{P_0}$	3	2.5	6	0	∞	0.78

Determine whether there exists a α-level test of $H_0 : P = P_0$ which is UMP against alternative P_1 and P_2 when
   (i) $\alpha = 0.01$ (ii) $\alpha = 0.05$ (iii) $\alpha = 0.07$
   (A) $H_0 : P = P_0$ against $H_1 : P = P_1$

$$\frac{P_1}{P_0} = \begin{cases} \infty & ; \; X = 5 \\ 4 & ; \; X = 3 \\ 2.5 & ; \; X = 2 \\ 2 & ; \; X = 1 \text{ or } 4 \\ 0.85 & ; \; X = 6 \end{cases}$$

(i) $\alpha = 0.01$
   Define

$$\phi_1(X) = \begin{cases} 1 \; ; \; X = 5 \\ \gamma \; ; \; X = 3 \\ 0 \; ; \; \text{otherwise} \end{cases}$$

$$E_{P_0}\phi_1(x) = P(x = 5) + \gamma P(x = 3) = 0.01$$

$$\Rightarrow 0 + \gamma(0.02) = 0.01 \Rightarrow \gamma = 0.5$$

The MP test is

$$\phi_1(X) = \begin{cases} 1 & ; X = 5 \\ 0.5 & ; X = 3 \\ 0 & ; \text{otherwise} \end{cases}$$

Power $= E_{P_1}\phi_1(x) = 0.05$
   (ii) $\alpha = 0.05$
Define a MP test

$$\phi_2(X) = \begin{cases} 1 ; X = 5, 3, 2 \text{ and } 4 \\ 0 ; \text{otherwise} \end{cases}$$

Power $= E_{P_1}\phi_2(x) = 0.16$
   Define one more MP test of the same size

$$\phi_3(X) = \begin{cases} 1 ; X = 5, 3, 2 \\ \gamma ; X = 1 \\ 0 ; \text{otherwise} \end{cases}$$

$$E_{P_0}\phi_3(x) = 0.02 + 0.02 + \gamma(0.03) = 0.05 \Rightarrow \gamma = \frac{1}{3}$$

Hence our test is

$$\phi_3(X) = \begin{cases} 1 ; X = 5, 3, 2 \\ \frac{1}{3} ; X = 1 \\ 0 ; \text{otherwise} \end{cases}$$

Power $= E_{P_1}\phi_3(x) = 0.16$
   Note: In this case, we have two MP tests, i.e., ϕ_2 and ϕ_3.
   (iii) $\alpha = 0.07$
Define a MP test

$$\phi_4(X) = \begin{cases} 1 ; X = 5, 3, 2, 1 \\ 0 ; \text{otherwise} \end{cases}$$

One can easily see that $E_{P_0}\phi_4(x) = 0.07$
   Power $= E_{P_1}\phi(x) = 0.20$
   Define one more randomized test

$$\phi_5(X) = \begin{cases} 1 ; X = 5, 3, 2 \\ \gamma ; X = 1, 4 \\ 0 ; \text{otherwise} \end{cases}$$

$$E_{P_0}\phi(x) = 0.07 \Rightarrow \gamma = \frac{3}{4}$$

The MP test $\phi_5(X)$ is given as

$$\phi_5(X) = \begin{cases} 1 & ; \ X = 5, 3, 2 \\ 0.75 & ; \ X = 1, 4 \\ 0 & ; \ \text{otherwise} \end{cases}$$

Power $= E_{P_1}\phi_5(x) = 0.20$
Define one more MP test

$$\phi_6(X) = \begin{cases} 1 & ; \ X = 5, 3, 2, 4 \\ \gamma & ; \ X = 1 \\ 0 & ; \ \text{otherwise} \end{cases}$$

$$E_{P_0}\phi_6(x) = 0.07 \Rightarrow \gamma = \frac{2}{3}$$

The MP test $\phi_6(X)$ is given as

$$\phi_5(X) = \begin{cases} 1 & ; \ X = 5, 3, 2, 4 \\ \frac{2}{3} & ; \ X = 1 \\ 0 & ; \ \text{otherwise} \end{cases}$$

Power $= E_{P_1}\phi_6(x) = 0.20$
One should note that ϕ_4, ϕ_5 and ϕ_6 are MP tests of size 0.07.
(B) To test $H_0 : P = P_0$ against $H_1 : P = P_2$

$$\frac{P_2}{P_0} = \begin{cases} \infty & ; \ X = 5 \\ 6 & ; \ X = 3 \\ 3 & ; \ X = 1 \\ 2.5 & ; \ X = 2 \\ 0.78 & ; \ X = 6 \\ 0 & ; \ X = 4 \end{cases}$$

(i) $\alpha = 0.01$
Define

$$\phi_7(X) = \begin{cases} 1 & ; \ X = 5 \\ \gamma & ; \ X = 3 \\ 0 & ; \ \text{otherwise} \end{cases}$$

$$E_{P_0}\phi_7(x) = P(x = 5) + \gamma P(x = 3) = 0.01 \Rightarrow \gamma = 0.5$$

The MP test of size 0.01 is given as

$$\phi_7(X) = \begin{cases} 1 & ; X = 5 \\ 0.5 & ; X = 3 \\ 0 & ; \text{otherwise} \end{cases}$$

Power $= E_{H_1}\phi_7(x) = 0.08$
(ii) $\alpha = 0.05$
The MP test of is given as

$$\phi_8(X) = \begin{cases} 1 & ; X = 5, 3, 1 \\ 0 & ; \text{otherwise} \end{cases}$$

Power $= E_{H_1}\phi_8(x) = 0.23$
(iii) $\alpha = 0.07$
The MP test of is given as

$$\phi_9(X) = \begin{cases} 1 & ; X = 5, 3, 1, 2 \\ 0 & ; \text{otherwise} \end{cases}$$

Power $= E_{P_2}\phi_9(x) = 0.28$
(C) To find UMP for $H_0 : P = P_0$ against $H_1 : P = P_1$ or P_2

$$\max\left\{\frac{P_1}{P_0}, \frac{P_2}{P_0}\right\} = \begin{cases} \infty & ; X = 5 \\ \max(4, 6) & ; X = 3 \\ \max(2, 3) & ; X = 1 \\ \max(2.5, 2.5) & ; X = 2 \\ \max(2, 0) & ; X = 4 \\ \max(0.85, 0.78) & ; X = 6 \end{cases}$$

$$\max\left\{\frac{P_1}{P_0}, \frac{P_2}{P_0}\right\} = \begin{cases} \infty & ; X = 5 \\ 6 & ; X = 3 \\ 3 & ; X = 1 \\ 2.5 & ; X = 2 \\ 2 & ; X = 4 \\ 0.85 & ; X = 6 \end{cases}$$

(i) $\alpha = 0.01$
To find UMP test

$$\phi_{10}(X) = \begin{cases} 1 & ; X = 5 \\ \gamma & ; X = 3 \\ 0 & ; \text{otherwise} \end{cases}$$

$$E\phi_{10}(x) = 0.01 \Rightarrow \gamma = \frac{1}{7}$$

The UMP test becomes

$$\phi_{10}(X) = \begin{cases} 1 \ ; \ X = 5 \\ \frac{1}{2} \ ; \ X = 3 \\ 0 \ ; \ \text{otherwise} \end{cases}$$

Power $= E_{P_1}\phi_{10}(x) = 0.05$ and $E_{P_2}\phi_{10}(x) = 0.08$

Power of ϕ_1 and ϕ_{10} is same for $\alpha = 0.01$ under P_1.

Similarly Power of ϕ_7 and ϕ_{10} is same for $\alpha = 0.01$ under P_2.

Further, one can get class of UMP test such as $\phi_1^*(X) = a\phi_1 + (1-a)\phi_{10}$ and $\phi_2^*(X) = b\phi_7 + (1-b)\phi_{10}, 0 \le a, b \le 1$,

(ii) $\alpha = 0.05$

The MP test is given as

$$\phi_{11}(X) = \begin{cases} 1 \ ; \ X = 5, 3, 1 \\ 0 \ ; \ \text{otherwise} \end{cases}$$

Since $E_{P_0}\phi_{11}(x) = 0.05$

Power $(P_1) = E_{P_1}\phi_{11}(x) = 0.01 + 0.08 + 0.06 = 0.15$

Power $(P_2) = E_{P_2}\phi_{11}(x) = 0.02 + 0.12 + 0.09 = 0.23$

Define one more MP test as

$$\phi_{12}(X) = \begin{cases} 1 \ ; \ X = 5, 3, 2 \\ \gamma \ ; \ X = 1 \\ 0 \ ; \ \text{otherwise} \end{cases}$$

$$E_{P_0}\phi_{12}(x) = 0.05 \Rightarrow \gamma = \frac{1}{3}$$

The MP test becomes

$$\phi_{12}(X) = \begin{cases} 1 \ ; \ X = 5, 3, 2 \\ \frac{1}{3} \ ; \ X = 1 \\ 0 \ ; \ \text{otherwise} \end{cases}$$

Power $(P_1) = E_{P_1}\phi_{12}(x) = (0.01 + 0.08 + 0.05) + \frac{1}{3}(0.06) = 0.14 + 0.02 = 0.16$

Power $(P_2) = E_{P_2}\phi_{12}(x) = (0.02 + 0.12 + 0.05) + \frac{1}{3}(0.09) = 0.19 + 0.03 = 0.22$

The test depends on P_1 and P_2

Hence UMP test does not exist for $P = P_1$ or P_2.

(iii) $\alpha = 0.07$

The MP test is given as

$$\phi_{13}(X) = \begin{cases} 1 \ ; \ X = 5, 3, 2, 1 \\ 0 \ ; \ \text{otherwise} \end{cases}$$

Since $E_{P_0}\phi_{13}(x) = 0.07$

Power $(P_1) = E_{P_1}\phi_{13}(x) = 0.20$

Power $(P_2) = E_{P_2}\phi_{13}(x) = 0.28$

Power of $\phi_{13}, \phi_4, \phi_5$ and ϕ_6 is same for $P = P_1$

Similarly Power of ϕ_{13} and ϕ_9 is same under $P = P_2$

Hence the MP test ϕ_{13} does not depend on H_1.

Therefore ϕ_{13} is UMP test.

Example 7.4.14 Let the distribution of X be given by

X: 0   1   2   3

P(X): θ 2θ $0.9 - 2\theta$ $0.1 - \theta$

where $0 < \theta < 1$.

For testing $H_0 : \theta = 0.05$ against $H_1 : \theta > 0.05$ at $\alpha = 0.05$, determine which of the following tests (if any) is UMP?

(i) $\phi(0) = 1, \phi(1) = \phi(2) = \phi(3) = 0$

(ii) $\phi(1) = 0.5, \phi(0) = \phi(2) = \phi(3) = 0$

(iii) $\phi(3) = 1, \phi(0) = \phi(1) = \phi(2) = 0$

Now, $H_0 : \theta = 0.05$ against $H_1 : \theta > 0.05$

X	0	1	2	3
P_0	0.05	0.1	0.8	0.05
P_1	θ_1	$2\theta_1$	$0.9 - 2\theta_1$	$0.1 - \theta_1$
$\frac{P_1}{P_0}$	$20\theta_1$	$20\theta_1$	$1.125 - \frac{5}{2}\theta_1$	$2 - 20\theta_1$

Since $\theta < 0.10 \Rightarrow 20\theta < 2$

$$\max\left(\frac{P_1}{P_0}\right) = \begin{cases} 2 & ; X = 0 \\ 2 & ; X = 1 \\ 0.875 & ; X = 2 \\ 0 & ; X = 3 \end{cases}$$

The MP test is given by

(i)

$$\phi_1(X) = \begin{cases} 1 ; X = 0 \\ 0 ; \text{otherwise} \end{cases}$$

We can easily see that $E\phi_1(x) = 0.05$

Power $= E_{H_1}\phi_1(X) = \theta_1$

(ii)

$$\phi_2(X) = \begin{cases} 0.5 ; X = 1 \\ 0 \quad ; \text{otherwise} \end{cases}$$

$E\phi_2(X) = (0.5)(0.1) = 0.05$

Power $= E_{H_1} \phi_2(X) = (0.5)(2\theta_1) = \theta_1$

(iii)

$$\phi_3(X) = \begin{cases} 1 \; ; \; X = 3 \\ 0 \; ; \; \text{otherwise} \end{cases}$$

$E_{H_0} \phi_3(x) = 0.05$

Power $= E_{H_1} \phi_3(x) = 0.1 - \theta_1$

We can conclude that ϕ_1 and ϕ_2 are UMP test.

Example 7.4.15 Let $X_1, X_2, \ldots, X_n$ are iid rvs from $\cup(0, \theta)$. Find the UMP test for testing

(A) $H_0 : \theta = \theta_0$ against $H_1 : \theta > \theta_0$
(B) $H_0 : \theta = \theta_0$ against $H_1 : \theta < \theta_0$
(C) $H_0 : \theta = \theta_0$ against $H_1 : \theta \neq \theta_0$

(A) $H_0 : \theta = \theta_0$ against $H_1 : \theta > \theta_0$

$$f(x_1, x_2, \ldots, x_n | \theta) = \theta^{-n} \mathbf{I}(\theta - x_{(n)})$$

where

$$\mathbf{I}(\theta - x_{(n)}) = \begin{cases} 1 \; ; \; X_{(n)} < \theta \\ 0 \; ; \; \text{otherwise} \end{cases}$$

There are three cases

(i) $0 < X_{(n)} < \theta_0$ (ii) $\theta_0 < X_{(n)} < \theta_1$ (iii) $\theta_1 < X_{(n)} < \infty$

$$\lambda(x) = \begin{cases} (\frac{\theta_0}{\theta_1})^n \frac{\mathbf{I}(\theta_1 - X_{(n)})}{\mathbf{I}(\theta_0 - X_{(n)})} \; ; \; 0 < X_{(n)} < \theta_0 \\ \frac{\theta_1^{-n} \mathbf{I}(\theta_1 - X_{(n)})}{0} \; ; \; \theta_0 \leq X_{(n)} < \theta_1 \\ \frac{0}{0} \; ; \; X_{(n)} \geq \theta_1 \end{cases}$$

$\lambda(x)$ is nondecreasing in $X_{(n)}$.

Hence $\lambda(x) > k \Leftrightarrow X_{(n)} > k$.

$$f_{X_{(n)}}(x) = \begin{cases} \frac{nx^{n-1}}{\theta^n} \; ; \; 0 < x < \theta \\ 0 \; \; \; ; \; \text{otherwise} \end{cases}$$

The MP test is defined as

(a) Define ϕ_1 as

$$\phi_1(x) = \begin{cases} 1 \; ; \; X_{(n)} > k \\ \alpha \; ; \; \text{otherwise} \end{cases}$$

$$E_{H_0} \phi_1(x) = \alpha \Rightarrow P[X_{(n)} > k] + \alpha P[X_{(n)} \leq k] = \alpha$$

If $k > \theta_0 \Rightarrow P_{H_0}[X_{(n)} > k] = 0$ and $k \leq \theta_0 \Rightarrow P_{H_0}[X_{(n)} \leq \theta_0] = 1$

We can write ϕ_1 as

$$\phi_1(x) = \begin{cases} 1 \; ; \; X_{(n)} > \theta_0 \\ \alpha \; ; \; X_{(n)} \le \theta_0 \end{cases}$$

Hence,

$$E\phi_1(x) = 1 \times P[X_{(n)} > \theta_0] + \alpha P[X_{(n)} \le \theta_0] = \alpha$$

The test $\phi_1(x)$ is UMP as it does not depend on H_1.

(b) Define ϕ_2 as

$$\phi_2(X) = \begin{cases} 1 \; ; \; X_{(n)} > k \\ 0 \; ; \; X_{(n)} \le k \end{cases}$$

$$P[X_{(n)} > k] = \alpha \Rightarrow \int_k^{\theta_0} n \frac{x^{n-1}}{\theta_0^n} dx = \alpha$$

$$\Rightarrow 1 - \frac{k^n}{\theta_0^n} = \alpha$$

$$k = \theta_0 (1 - \alpha)^{\frac{1}{n}}$$

This test does not depend on H_1. Hence ϕ_2 is a UMP test.

We can write ϕ_2 as

$$\phi_2(X) = \begin{cases} 1 \; ; \; X_{(n)} > \theta_0 (1 - \alpha)^{\frac{1}{n}} \\ 0 \; ; \; X_{(n)} \le \theta_0 (1 - \alpha)^{\frac{1}{n}} \end{cases}$$

We can get class of UMP test by the linear combination of ϕ_1 and ϕ_2.

Hence $\phi_a^* = a\phi_1 + (1 - a)\phi_2$ is also a class of UMP tests, where $a \in [0, 1]$.

Power of ϕ_1

$$E_{H_1}\phi_1(x) = \int_{\theta_0}^{\theta_1} n \frac{x^{n-1}}{\theta^n} dx + \alpha \int_0^{\theta_0} n \frac{x^{n-1}}{\theta_1^n} dx$$

$$= 1 - \left(\frac{\theta_0}{\theta_1}\right)^n (1 - \alpha)$$

Power of ϕ_2

$$E_{H_1}\phi_2(x) = \int_{\theta_0(1-\alpha)^{\frac{1}{n}}}^{\theta_1} n\frac{x^{n-1}}{\theta_1^n}dx = \frac{\theta_1^n - \theta_0^n(1-\alpha)^{\frac{1}{n}}}{\theta_1^n}$$

$$= 1 - \left(\frac{\theta_0}{\theta_1}\right)^n(1-\alpha)$$

One should note that ϕ_a^* also has the same power.

(B) $H_0 : \theta = \theta_0$ against $H_1 : \theta = \theta_1 < \theta_0$

There are three cases

(i) $0 < X_{(n)} < \theta_1$ (ii) $\theta_1 < X_{(n)} < \theta_0$ (iii) $\theta_0 < X_{(n)} < \infty$

$$\lambda(x) = \begin{cases} (\frac{\theta_0}{\theta_1})^n & ; \ 0 < X_{(n)} < \theta_1 \\ \frac{0}{\theta_1^{-n}\mathbf{I}(X_{(n)}-\theta_0)} & ; \ \theta_1 \le X_{(n)} < \theta_0 \\ \frac{0}{0} & ; \ \theta_0 \le X_{(n)} < \infty \end{cases}$$

$\lambda(x)$ is nonincreasing in $X_{(n)}$.

Hence $\lambda(x) > k \Leftrightarrow X_{(n)} < k$.

The MP test is given as

$$\phi_3(X) = \begin{cases} 1 \ ; \ X_{(n)} < k \\ 0 \ ; \ X_{(n)} \ge k \end{cases}$$

When $k > \theta_0 \Rightarrow P[X_{(n)} < k] = 1$

If $k < \theta_0$ then

$$E\phi_3(x) = \alpha \Rightarrow P_{H_0}[X_{(n)} < k] = \int_0^k \frac{nx^{n-1}}{\theta_0^n}dx \Rightarrow k = \theta_0(\alpha)^{\frac{1}{n}}$$

This test does not depend on H_1; it is UMP.

The UMP test is given as

$$\phi_3(X) = \begin{cases} 1 \ ; \ X_{(n)} < \theta_0(\alpha)^{\frac{1}{n}} \\ 0 \ ; \ X_{(n)} \ge \theta_0(\alpha)^{\frac{1}{n}} \end{cases}$$

Power of $\phi_3(x)$

$$E_{H_1}\phi_3(x) = \int_0^{\theta_0(\alpha)^{\frac{1}{n}}} \frac{nx^{n-1}}{\theta_0^n}dx = \left(\frac{\theta_0}{\theta_1}\right)^n \alpha$$

(C) $H_0 : \theta = \theta_0$ against $H_1 : \theta \neq \theta_0$

We can write one more UMP test as

$$\phi_4(x) = \begin{cases} 1 \; ; \; X_{(n)} < \theta_0(\alpha)^{\frac{1}{n}} \text{ or } X_{(n)} > \theta_0 \\ 0 \; ; \; X_{(n)} > \theta_0(\alpha)^{\frac{1}{n}} \end{cases}$$

We will verify the size of $\phi_4(X)$

$$E_{H_1}\phi_4(X) = P[X_{(n)} < \theta_0(\alpha)^{\frac{1}{n}}] + P[X_{(n)} > \theta_0]$$

$$= \int\limits_0^{\theta_0(\alpha)^{\frac{1}{n}}} \frac{nx^{n-1}}{\theta_0^n} dx + 0 = \alpha$$

Power of $\phi_4(X)$ if $\theta_1 < \theta_0$

$$E_{H_1}\phi_4(x) = P[X_{(n)} < \theta_0(\alpha)^{\frac{1}{n}}] + P[X_{(n)} > \theta_0]$$

$$= \int\limits_0^{\theta_0(\alpha)^{\frac{1}{n}}} \frac{nx^{n-1}}{\theta_1^n} dx = \left(\frac{\theta_0}{\theta_1}\right)^n \alpha$$

Power of $\phi_4(X)$ if $\theta_1 > \theta_0$

$$E_{H_1}\phi_4(x) = P[X_{(n)} < \theta_0(\alpha)^{\frac{1}{n}}] + P_{H_1}[X_{(n)} > \theta_0]$$

$$= \int\limits_0^{\theta_0(\alpha)^{\frac{1}{n}}} \frac{nx^{n-1}}{\theta_1^n} dx + \int\limits_{\theta_0}^{\theta_1} \frac{nx^{n-1}}{\theta_1^n} dx$$

$$= 1 - \left(\frac{\theta_0}{\theta_1}\right)^n (1 - \alpha)$$

Power of ϕ_1, ϕ_2 and ϕ_4 is same for $\theta_1 > \theta_0$ and Power of ϕ_3 and ϕ_4 is same for $\theta_1 < \theta_0$.

Example 7.4.16 Let $X_1, X_2, \ldots, X_n$ be a random sample of size n from the pmf

$$P[X = x] = \frac{1}{N}; \quad x = 1, 2, \ldots, N. \ (N \geq 1)$$

Find the UMP test for testing

(A) $H_0 : N = N_0$ against $H_0 : N > N_0$
(B) $H_0 : N = N_0$ against $H_0 : N < N_0$
(C) $H_0 : N = N_0$ against $H_0 : N \neq N_0$

We will find a test for (A)
(A) $H_0 : N = N_0$ against $H_0 : N > N_0$

$$f(x_1, x_2, \ldots, x_n | N) = N^{-n} \mathbf{I}(N - x_{(n)})$$

where

$$\mathbf{I}(N - x_{(n)}) = \begin{cases} 1 \; ; \; X_{(n)} < N \\ 0 \; ; \; \text{otherwise} \end{cases}$$

There are three cases
(a) $0 < X_{(n)} \leq N_0$ (b) $N_0 < X_{(n)} \leq N_1$ (c) $N_1 < X_{(n)} < \infty$

$$\lambda(x) = \begin{cases} (\frac{N_0}{N_1})^n & ; \; X_{(n)} \leq N_0 \\ \frac{N_1^{-n} \mathbf{I}(N_1 - X_{(n)})}{0} & ; \; N_0 < X_{(n)} \leq N_1 \\ \frac{0}{0} & ; \; X_{(n)} > N_1 \end{cases}$$

$\lambda(x)$ is nondecreasing in $X_{(n)}$.
Hence $\lambda(x) > k \Leftrightarrow X_{(n)} > k$.
Let $X_{(n)} = Y$

$$P[Y = y] = \left(\frac{y}{N}\right)^n - \left(\frac{y-1}{N}\right)^n ; \quad y = 1, 2, \ldots, N$$

Define a MP test
(i)

$$\phi_1(X) = \begin{cases} 1 \; ; \; X_{(n)} > k \\ 0 \; ; \; \text{otherwise} \end{cases}$$

To find k such that $E_{H_0} \phi(X) = \alpha$

$$\Rightarrow P_{H_0}[X_{(n)} > k] = \alpha \tag{7.4.7}$$

Note that if $k > N_0 \Rightarrow P[X_{(n)} > k] = 0$ and if $k \leq N_0 \Rightarrow P[X_{(n)} \leq N_0] = 1$

From (7.4.7), under H_0,

$$P[X_{(n)} > k] = P[Y > k] = \sum_{y=k+1}^{N_0} \left(\frac{y}{N_0}\right)^n - \left(\frac{y-1}{N_0}\right)^n = \alpha$$

$$= \sum_{y=k+1}^{N_0} \left[y^n - (y-1)^n\right] = \alpha N_0^n$$

$$\Rightarrow (k+1)^n - k^n + (k+2)^n - (k+1)^n + \cdots + N_0^n - (N_0-1)^n = \alpha N_0^n$$

$$\Rightarrow N_0^n - k^n = \alpha N_0^n$$

$$\Rightarrow k = N_0(1-\alpha)^{\frac{1}{n}}$$

This test does not depend on H_1. Hence it is a UMP test.

The UMP test ϕ_1 is given as

$$\phi_1(X) = \begin{cases} 1 \; ; \; X_{(n)} > N_0(1-\alpha)^{\frac{1}{n}} \\ 0 \; ; \; \text{otherwise} \end{cases}$$

(ii) Define one more MP test as

$$\phi_2(X) = \begin{cases} 1 \; ; \; X_{(n)} > N_0 \\ \alpha \; ; \; X_{(n)} \leq N_0 \end{cases}$$

$$E_{H_0}\phi_2(X) = P[X_{(n)} > N_0] + \alpha P[X_{(n)} \leq N_0] = 0 + \alpha = \alpha$$

Power of $\phi_1(X) = E_{H_1}\phi_1(X)$

$$= \sum_{y=N_0(1-\alpha)^{\frac{1}{n}}+1}^{N_1} \left[\left(\frac{y}{N_1}\right)^n - \left(\frac{y-1}{N_1}\right)^n\right]$$

Let $N_1' = N_0(1-\alpha)^{\frac{1}{n}} + 1$

$$= \sum_{y=N_1'}^{N_1} \left[\left(\frac{y}{N_1}\right)^n - \left(\frac{y-1}{N_1}\right)^n\right]$$

$$= \frac{1}{N_1^n}\left[N_1^n - (N_1'^n - 1)\right] = \frac{1}{N_1^n}\left[N_1^n - N_0^n(1-\alpha)\right] = 1 - \left(\frac{N_0}{N_1}\right)^n(1-\alpha)$$

Power of $\phi_2(X)$

$$E_{H_1}\phi_2(X) = \sum_{y=N_0+1}^{N_1}\left[\left(\frac{y}{N_1}\right)^n - \left(\frac{y-1}{N_1}\right)^n\right] + \alpha\sum_{y=1}^{N_0}\left[\left(\frac{y}{N_1}\right)^n - \left(\frac{y-1}{N_1}\right)^n\right]$$

$$= \frac{1}{N_1^n}\left[N_1^n - N_0^n\right] + \frac{\alpha}{N_1^n}N_0^n$$

$$= 1 - \left(\frac{N_0}{N_1}\right)^n(1-\alpha)$$

(B) $H_0 : N = N_0$ against $H_1 : N < N_0$

There are three cases

(i) $0 < X_{(n)} \le N_1$ (ii) $N_1 < X_{(n)} \le N_0$ (iii) $N_0 < X_{(n)} < \infty$.

$$\lambda(x) = \begin{cases} (\frac{N_0}{N_1})^n & ; \ 0 < X_{(n)} \le N_1 \\ \frac{0}{N_0^{-n}I(X_{(n)}-N_0)} & ; \ N_1 < X_{(n)} \le N_0 \\ \frac{0}{0} & ; \ N_0 < X_{(n)} < \infty \end{cases}$$

$\lambda(x)$ is nonincreasing in $X_{(n)}$.

Then, $\lambda(x) > k \Leftrightarrow X_{(n)} < k$.

The MP test is given as

$$\phi_3(X) = \begin{cases} 1 \ ; \ X_{(n)} \le k \\ 0 \ ; \ \text{otherwise} \end{cases}$$

$$E_{H_0}\phi_3(X) = \alpha = \sum_{y=1}^{k}\left[\left(\frac{y}{N_0}\right)^n - \left(\frac{y-1}{N_0}\right)^n\right] = \frac{k^n}{N_0^n}$$

$$\Rightarrow k = N_0\alpha^{\frac{1}{n}}$$

This test does not depend on H_1. Hence it is a UMP test.

The UMP test ϕ_3 is written as

$$\phi_3(X) = \begin{cases} 1 \ ; \ X_{(n)} \le N_0\alpha^{\frac{1}{n}} \\ 0 \ ; \ \text{otherwise} \end{cases}$$

Power of $\phi_3(X) = E_{H_1}\phi_3(X)$, let $N_0\alpha^{\frac{1}{n}} = N_0'$

$$= \sum_{y=1}^{N_0'}\left[\left(\frac{y}{N_1}\right)^n - \left(\frac{y-1}{N_1}\right)^n\right] = \left(\frac{N_0'}{N_1}\right)^n\alpha$$

(C) $H_0 : N = N_0$ against $H_0 : N \neq N_0$

We can write one more UMP test as

$$\phi_4(X) = \begin{cases} 1 \; ; \; X_{(n)} \leq N_0\alpha^{\frac{1}{n}} \text{ or } X_{(n)} > N_0 \\ 0 \; ; \text{ otherwise} \end{cases}$$

$$E_{H_0}\phi_4(X) = P[X_{(n)} \leq N_0\alpha^{\frac{1}{n}}] + P[X_{(n)} > N_0]$$

$$= \sum_{y=1}^{N_0'}\left[\left(\frac{y}{N_0}\right)^n - \left(\frac{y-1}{N_0}\right)^n\right] + 0 = \left(\frac{N_0}{N_0}\right)^n \alpha = \alpha,$$

where $N_0' = N_0\alpha^{\frac{1}{n}}$.

Power of $\phi_4(X)$ if $N_1 < N_0$

$$E_{H_1}\phi_4(X) = P[X_{(n)} \leq N_0\alpha^{\frac{1}{n}}] + P[X_{(n)} > N_0]$$

$$= \sum_{y=1}^{N_0'}\left[\left(\frac{y}{N_1}\right)^n - \left(\frac{y-1}{N_1}\right)^n\right] + 0 = \left(\frac{N_0}{N_1}\right)^n \alpha$$

Power of $\phi_4(X)$ if $N_1 > N_0$

$$E_{H_1}\phi(X) = \sum_{y=N_0+1}^{N_1}\left[\left(\frac{y}{N_1}\right)^n - \left(\frac{y-1}{N_1}\right)^n\right] + \sum_{y=1}^{N_0'}\left[\left(\frac{y}{N_1}\right)^n - \left(\frac{y-1}{N_1}\right)^n\right]$$

$$= \frac{N_1^n - N_0^n}{N_1^n} + \left(\frac{N_0}{N_1}\right)^n \alpha$$

$$= 1 - \left(\frac{N_0}{N_1}\right)^n (1 - \alpha)$$

Power of ϕ_1, ϕ_2 and ϕ_4 is same for $N_1 > N_0$ and Power of ϕ_3 and ϕ_4 is same for $N_1 < N_0$.

Example 7.4.17 Let X be a rv under H_0 against H_1 as follows:

$H_0 : X \sim f_0(x)$, where

$$f_0(x) = \frac{1}{\sqrt{2\pi}} \exp\left[-\frac{x^2}{2}\right] \; ; \; -\infty < x < \infty$$

against $H_1 : X \sim f_1(x)$, where

$$f_1(x) = \frac{1}{2} e^{-|x|} \; ; \; -\infty < x < \infty$$

Test H_0 against H_1 based on a single observation.

$$\lambda(x) = \frac{f_1(x)}{f_0(x)} = \frac{2^{-1} \exp^{-|x|}}{(2\pi)^{-\frac{1}{2}} \exp\left[-\frac{x^2}{2}\right]}$$

By NP lemma, $\lambda(x) > k$

$$\Rightarrow \frac{x^2}{2} - |x| > k$$

$$\Rightarrow [|x| - 1]^2 > k$$

$$\Rightarrow |x| > k + 1 \text{ or } |x| < 1 - k$$

Hence we define a MP test as follows:

$$\phi(X) = \begin{cases} 1 \; ; \; |x| > k + 1 \text{ or } |x| < 1 - k \\ 0 \; ; \text{ otherwise} \end{cases}$$

$E\phi(X) = \alpha$

$$\Rightarrow P[|x| > 1 + k] + P[|x| < 1 - k] = \alpha$$

$$\Rightarrow 1 - P[|x| > 1 + k] - P[|x| < 1 - k] = 1 - \alpha$$

$$\Rightarrow P[|x| \leq k + 1] - P[|x| < 1 - k] = 1 - \alpha$$

$$\Rightarrow \int\limits_{1-k}^{1+k} \frac{2e^{-\frac{x^2}{2}}}{\sqrt{2\pi}} = 1 - \alpha$$

$$\Rightarrow \int\limits_{1-k}^{1+k} \frac{e^{-\frac{x^2}{2}}}{\sqrt{2\pi}} = \frac{1 - \alpha}{2}$$

$$\Rightarrow \Phi(1+k) - \Phi(1-k) = \frac{1-\alpha}{2}$$

Given α, k has to be found out by trial and error.

To calculate value of k we can use following R progran.

```
# To solve the equation phi(1+k)-phi(1-k) = (1-alpha)/2
  # Defining function.
    f <- function(k)
    {
      alpha <- 0.05;
      l = pnorm(1+k,0,1)-pnorm(1-k,0,1)-((1-alpha)/2);
      return(l)
    }
  # To solve function using uniroot.
    x = uniroot(f,c(0,5))
  # OUTPUT
    print(x$root)

  # To calculate manually
    k <- c(seq(from = 0, to = 0.9, by = 0.1),0.9951,seq(from = 1, to = 1.5, by = 0.1))
    phi1 <- pnorm(1+k,0,1); phi2 <- pnorm(1-k,0,1);
    phi <- phi1-phi2
    output <- data.frame(k,phi1,phi2,phi)
    alpha <- 0.05; al <- (1-alpha)/2;
    a <- min(which(phi>al))
    print(c("function has value greater than (1-alpha)/2 at k=",k[a]))
```

```
OUTPUT
Using uniroot function in R
 0.995046
Calculating manually
     k        phi1        phi2        phi
1  0.0000    0.8413447   0.8413447   0.00000000
2  0.1000    0.8643339   0.8159399   0.04839406
3  0.2000    0.8849303   0.7881446   0.09678573
4  0.3000    0.9031995   0.7580363   0.14516317
5  0.4000    0.9192433   0.7257469   0.19349646
6  0.5000    0.9331928   0.6914625   0.24173034
7  0.6000    0.9452007   0.6554217   0.28977897
8  0.7000    0.9554345   0.6179114   0.33752312
9  0.8000    0.9640697   0.5792597   0.38480997
10 0.9000    0.9712834   0.5398278   0.43145560
11 0.9951    0.9769840   0.5019548   0.47502920
12 1.0000    0.9772499   0.5000000   0.47724987
13 1.1000    0.9821356   0.4601722   0.52196342
 "function has value greater than (1-alpha)/2 at k="  0.9951.
```

Example 7.4.18 Let $X_1, X_2, \ldots, X_n$ are iid rvs from $f(x|\theta)$,

$$f(x|\theta) = \begin{cases} \frac{\theta}{x^2} \; ; \; 0 < \theta \leq x \\ 0 \; \; ; \text{otherwise} \end{cases}$$

Find the MP test of size α for testing

(A) $H_0 : \theta = \theta_0$ against $H_1 : \theta = \theta_1 > \theta_0$
(B) $H_0 : \theta = \theta_0$ against $H_1 : \theta = \theta_1 < \theta_0$
(C) $H_0 : \theta = \theta_0$ against $H_1 : \theta \neq \theta_0$

(A) $H_0 : \theta = \theta_0$ against $H_1 : \theta = \theta_1 > \theta_0$

$$f(x_1, x_2, \ldots, x_n | \theta) = \frac{\theta^n \mathbf{I}(x_{(1)} - \theta)}{\prod_{i=1}^{n} x_i^2}$$

There are three cases: (i) $0 < x_{(1)} < \theta_0$ (ii) $\theta_0 \leq x_{(1)} < \theta_1$ (iii) $\theta_1 \leq x_{(1)} < \infty$

$$\lambda(x) = \begin{cases} \frac{0}{0} & ; \ 0 < x_{(1)} < \theta_0 \\ \frac{0}{\theta_0^{-n} \mathbf{I}(x_{(1)} - \theta)(\prod_{i=1}^{n} x_i^2)^{-1}} & ; \ \theta_0 \leq x_{(1)} < \theta_1 \\ \left(\frac{\theta_1^n}{\theta_0^n}\right) \frac{\mathbf{I}(x_{(1)} - \theta_1)}{\mathbf{I}(x_{(1)} - \theta_0)} & ; \ \theta_1 \leq x_{(1)} < \infty \end{cases}$$

$\lambda(x)$ is nondecreasing in $x_{(1)}$.

Hence, $\lambda(x) > k \Leftrightarrow x_{(1)} > k$.

$$f(x_{(1)}) = \begin{cases} \frac{n\theta^n}{x^{n+1}} & ; \ \theta < x_{(1)} < \infty \\ 0 & ; \ \text{otherwise} \end{cases}$$

The MP test is defined as

$$\phi_1(x) = \begin{cases} 1 & ; \ X_{(1)} > k \\ 0 & ; \ \text{otherwise} \end{cases}$$

$$\mathrm{E}\phi_1(X) = \alpha = \int_k^\infty \frac{n\theta_0^n}{x^{n+1}} dx = \alpha$$

$$\Rightarrow k = \frac{\theta_0}{\alpha^{\frac{1}{n}}}$$

This test does not depend on H_1. Hence, it is a UMP test.

The UMP test ϕ_1 is written as

$$\phi_1(X) = \begin{cases} 1 & ; \ X_{(1)} > \theta_0 \alpha^{-\frac{1}{n}} \\ 0 & ; \ \text{otherwise} \end{cases}$$

Note that $X_{(1)} > \theta_0 \alpha^{-\frac{1}{n}}$ and $X_{(1)} > \theta_1 \Rightarrow X_{(1)} > \max(\theta_1, \theta_0 \alpha^{-\frac{1}{n}})$

Power of $\phi_1(x)$

$$\mathrm{E}_{H_1}\phi_1(x) = \int_{\theta_1}^\infty \frac{n\theta_1^n}{x^{n+1}} dx = 1$$

(B) $H_0 : \theta = \theta_0$ against $H_1 : \theta < \theta_0$

There are three cases: (i) $0 < x_{(1)} < \theta_1$ (ii) $\theta_1 \leq x_{(1)} < \theta_0$ (iii) $\theta_0 \leq x_{(1)} < \infty$

$$\lambda(x) = \begin{cases} \frac{0}{0} & ; 0 < x_{(1)} < \theta_1 \\ \frac{\theta_1^{-n} I(x_{(1)} - \theta_1)(\prod_{i=1}^n x_i^2)^{-1}}{0} & ; \theta_1 \leq x_{(1)} < \theta_0 \\ (\frac{\theta_1}{\theta_0})^n \frac{I(x_{(1)} - \theta_1)}{I(x_{(1)} - \theta_0)} & ; \theta_0 \leq x_{(1)} < \infty \end{cases}$$

Now, $\lambda(x)$ is nonincreasing in $X_{(1)}$.

Hence, $\lambda(x) > k \Leftrightarrow X_{(1)} < k$.

The MP test is defined as

$$\phi_3(X) = \begin{cases} 1 ; X_{(1)} \leq k \\ 0 ; \text{otherwise} \end{cases}$$

$$E_{H_0} \phi_3(X) = \alpha = \int_{\theta_0}^{k} \frac{n\theta_0^n}{x^{n+1}} dx = \alpha$$

$$\Rightarrow k = \frac{\theta_0}{(1 - \alpha)^{\frac{1}{n}}}$$

Hence, we can write the MP test as

$$\phi_3(X) = \begin{cases} 1 ; X_{(1)} < \theta_0(1 - \alpha)^{-\frac{1}{n}} \\ 0 ; \text{otherwise} \end{cases}$$

This test does not depend on H_1. Hence, it is a UMP test.

(C) In this case, NP lemma cannot be used to get UMP test for testing $H_0 : \theta = \theta_0$ against $H_1 : \theta \neq \theta_0$.

7.5   Families with Monotone Likelihood Ratio

If we wish to test $H_0 : \theta \leq \theta_0$ against $H_1 : \theta > \theta_0$ then it is not possible to find UMP test. Because the MP test of $H_0 : \theta \leq \theta_0$ against $H_1 : \theta > \theta_0$ depends on H_1, i.e., on θ_1. Here, we consider a special case of distributions that is large enough to include the one parameter exponential family, for which a UMP test of one-sided hypothesis exists.

When the alternative hypothesis is composite, i.e., $H_1 : \theta \in \Theta_1$, then the power can be different for different alternatives. For each particular alternative θ_1, a test is MP test of size α for an alternative θ_1 if the test is most powerful for the simple alternative $H_1 : \theta = \theta_1$

If a particular test function $\phi^*(x)$ is the MP test of size α for all alternatives $\theta \in \Theta_1$, then we say that $\phi^*(x)$ is a uniformly most powerful (UMP) test of size α.

Definition 7.5.1 Let $\{f_\theta : \theta \in \Theta\}$ be a family of pdf(pmf). We say that $\{f_\theta\}$ has a monotone likelihood ratio (MLR) in statistics $T(x)$ if for $\theta_2 > \theta_1$, whenever f_{θ_1} and f_{θ_2} are distinct, i.e., $f(x|\theta_1) \neq f(x|\theta_2) \; \forall \; x$, the ratio $\frac{f(x|\theta_2)}{f(x|\theta_1)}$ is a non decreasing function of $T(x)$ for the set of values of x for which at least one of $f(x|\theta_1)$ and $f(x|\theta_2)$ is greater than zero.

Definition 7.5.2 A class of tests ϕ_α is defined as

$$\phi_\alpha = \{\phi \in D| \sup_{\theta \in \Theta_0} E_\theta \phi(x) \leq \alpha\}$$

Theorem 7.5.1 *The one-parameter exponential family*

$$f(x|\theta) = \exp\{Q(\theta)T(x) + S(x) + D(\theta)\}$$

where $Q(\theta)$ is nondecreasing, has MLR in $T(x)$.

Proof Let $\theta_2 > \theta_1$

$$\lambda(x) = \frac{f(x|\theta_2)}{f(x|\theta_1)}$$

$$\lambda(x) = \frac{\exp\{Q(\theta_2)T(x) + S(x) + D(\theta_2)\}}{\exp\{Q(\theta_1)T(x) + S(x) + D(\theta_1)\}}$$

$$= \exp[D(\theta_2) - D(\theta_1)] \exp[T(x)\{Q(\theta_2) - Q(\theta_1)\}]$$

Differentiate $\lambda(x)$ with respect to x,

$$\lambda'(x) = [Q(\theta_2) - Q(\theta_1)]T'(x) \exp[\{Q(\theta_2) - Q(\theta_1)\}T(x)] \exp[D(\theta_2) - D(\theta_1)],$$

where $T'(x)$ is derivative of $T(x)$.

Given that $Q(\theta)$ is nondecreasing

$$\Rightarrow [Q(\theta_2) - Q(\theta_1)] > 0 \quad \text{for} \quad \theta_2 > \theta_1$$

$$\Rightarrow \lambda'(x) > 0$$

Hence $\lambda(x)$ is nondecreasing and $f(x|\theta)$ has MLR property in $T(x)$.

Theorem 7.5.2 *Let $X_1, X_2, \ldots, X_n$ be iid rvs from one-parameter exponential family. Then a UMP test exist for testing $H_0 : \theta = \theta_0$ against $H_1 : \theta = \theta_1 > \theta_0$.*

Proof By Theorem 1, $\lambda(x)$ is nondecreasing for $\theta_1 > \theta_0$. It has MLR property in $T(x)$.

$$\lambda(x) = \exp\left[\sum_{i=1}^{n^*} T(x_i)\{Q(\theta_1) - Q(\theta_0)\} + n\{D(\theta_1) - D(\theta_0)\} \right]$$

By NP lemma, $\lambda(x) > k \Leftrightarrow \sum_{i=1}^{n} T(x_i) > k$.

The MP test is given by

$$\phi_1(x) = \begin{cases} 1 \; ; \; \sum_{i=1}^{n} T(x_i) > k \\ \gamma \; ; \; \sum_{i=1}^{n} T(x_i) = k \\ 0 \; ; \text{otherwise} \end{cases}$$

Since ϕ_1 does not depend on any specific values of θ_1. Hence ϕ_1 is UMP test of level α.

Remark: 1. If we are testing $H_0 : \theta = \theta_0$ against $H_1 : \theta = \theta_1 < \theta_0$, similarly we get UMP test of size α as

$$\phi_2(x) = \begin{cases} 1 \; ; \; \sum_{i=1}^{n} T(x_i) < k \\ \gamma \; ; \; \sum_{i=1}^{n} T(x_i) = k \\ 0 \; ; \text{otherwise} \end{cases}$$

2. Theorem 7.5.1 includes Binomial, Poisson, normal, gamma etc.
3. One should note that $\cup(0, \theta)$, which does not belong to exponential family has an MLR property.

Theorem 7.5.3 *Let the rv X has pdf(pmf) $f(x|\theta)$, where $f(x|\theta)$ has an MLR in $T(x)$. Consider the one-sided testing problem, $H_0 : \theta \leq \theta_0$ against $H_1 : \theta > \theta_0, \theta_0 \in \Theta$, any test of the form*

$$\phi(x) = \begin{cases} 1 \; ; \; T(x) > t_0 \\ \gamma \; ; \; T(x) = t_0 \\ 0 \; ; \; T(x) < t_0 \end{cases} \tag{7.5.1}$$

has nondecreasing power function and is UMP of its size α provided that $\alpha > 0$.

Moreover, for every $0 \leq \alpha \leq 1$ and every $\theta_0 \in \Theta$, there exist a $t_0, -\infty < t_0 < \infty$ and $0 \leq \gamma \leq 1$, such that the test described in (7.5.1) is UMP of its size α for testing H_0 against H_1.

Proof Let θ_1, $\theta_2 \in \Theta$, $\theta_1 < \theta_2$.

Consider the testing problem $H_0 : \theta = \theta_1$ against $H_1 : \theta = \theta_2$.

By using NP lemma, MP test of size α is given as

$$\phi(x) = \begin{cases} 1 \; ; \; \lambda(x) > k \\ \gamma \; ; \; \lambda(x) = k \\ 0 \; ; \; \lambda(x) < k \end{cases} \tag{7.5.2}$$

with $0 \leq k < \infty$ and $E_{H_0}\phi(x) = \alpha > 0$.

Next, for $k = \infty$, the test

$$\phi(x) = \begin{cases} 1 \; ; \; f(x|\theta_1) = 0 \\ 0 \; ; \; f(x|\theta_1) > 0 \end{cases} \tag{7.5.3}$$

is MP of size 0.

Now $f(x|\theta)$ has MLR in $T(x)$. It implies that $\lambda(x)$ is nondecreasing function in $T(x)$.

Hence $\lambda(x) > k \Leftrightarrow T(x) > k$, k is chosen such that $E_{\theta_1}\phi(x) = \alpha > 0$. Let $k = t_0$

$$\phi(x) = \begin{cases} 1 \; ; \; T(x) > t_0 \\ \gamma \; ; \; T(x) = t_0 \\ 0 \; ; \; T(x) < t_0 \end{cases} \tag{7.5.4}$$

Now we shall show that the test given in (7.5.4) has a nondecreasing power function.

Consider a test $\phi^* = \alpha \Rightarrow E_{H_0}\phi^*(x) = E_{H_1}\phi^*(x) = \alpha$

Power of test (7.5.4) is at least α

$$\Rightarrow E_{H_1}\phi(x) = E_{\theta_2}\phi(x) \geq \alpha = E_{\theta_2}\phi^*(x)$$

But $\alpha = E_{\theta_1}\phi(x)$

$$\Rightarrow E_{\theta_2}\phi(x) \geq E_{\theta_1}\phi(x)$$

$\Rightarrow$ Power function of the test ϕ, i.e., $E_{H_1}\phi(x)$ is nondecreasing function of θ, $\theta_2 > \theta_1$, provided that its size $E_{\theta_1}\phi(x) > 0$.

Let $\theta_1 = \theta_0$ and $\theta_2 > \theta_0$, the testing problem can be written as

$$H_0 : \theta = \theta_0 \quad \text{against} \quad H_1 : \theta > \theta_0 \tag{7.5.5}$$

The corresponding class of level α tests becomes

$$\{\phi | E_{\theta_0}\phi(x) \leq \alpha\} \tag{7.5.6}$$

in which we shall find out a UMP test for testing problem given in (7.5.5). The test in (7.5.4) is UMP of size α in the class (7.5.6) since it does not depend on H_1.

Now, consider the testing problem

$H_0 : \theta \leq \theta_0$ against $H_1 : \theta > \theta_0$.

The class of level α tests for testing this problem would be

$$\{\phi| \sup_{\theta \leq \theta_0} E\phi(x) = \alpha\} = \{\phi|E_{H_0}\phi(x) \leq \alpha, \forall \theta \leq \theta_0\} \qquad (7.5.7)$$

The test ϕ in (7.5.1) belongs to the class given in (7.5.7) since its power function is nondecreasing function of θ and its size $\alpha > 0$. Further, class of tests in (7.5.7) is contained in (7.5.6) because the number of restrictions in (7.5.7) is more than that of in (7.5.6). Therefore, the UMP of size α test ϕ in the larger class becomes UMP of size α test in the smaller class because it belongs to a smaller class.

Hence, provided $\alpha > 0$, the test in (7.5.4) is UMP of size α for testing $H_0 : \theta \leq \theta_0$ against $H_1 : \theta > \theta_0$.

From (7.5.1), we can write as

$$\sup_{\theta \leq \theta_0} E\phi(x) = \alpha \Rightarrow E_{\theta_0}\phi(x) = \alpha$$

$$\Rightarrow P[T > t_0] + \gamma P[T = t_0] = \alpha$$

$$\Rightarrow 1 - P[T > t_0] - \gamma P[T = t_0] = 1 - \alpha$$

$$\Rightarrow P[T \leq t_0] - \gamma P[T = t_0] = 1 - \alpha \qquad (7.5.8)$$

Note that $P[T \leq t_0]$ is a distribution function. It is nondecreasing and right continuous function of t_0.

If $P[T \leq t_0] = 1 - \alpha$ then $\gamma = 0$

If $\gamma > 0$ then

$$P[T < t_0] + (1 - \gamma)P[T = t_0] = 1 - \alpha$$

$$P[T < t_0] \leq 1 - \alpha \qquad (7.5.9)$$

and

$$P[T \leq t_0] = (1 - \alpha) + \gamma P[T = t_0]$$

$$(1 - \alpha) < P[T \leq t_0] \qquad (7.5.10)$$

From (7.5.9) and (7.5.10),

$$P[T < t_0] \le (1 - \alpha) < P[T \le t_0]$$

Hence, from (7.5.8), for $\gamma = \gamma_0$

$$\gamma_0 = \frac{P[T \le t_0] - (1 - \alpha)}{P[T = t_0]}; \quad 0 < \gamma_0 \le 1$$

Next, consider the case $\alpha = 0$.

Define the support of $f(x|\theta)$ under H_0 and H_1,

$$S_0 = \{x | f(x|\theta_0) > 0\} = \{x | a < x < b\}$$

$$S_1 = \{x | f(x|\theta_1) > 0\} = \{x | c < x < d\}$$

then $a < c$. Without loss of generality assume that $b \le d$. Consider a test ϕ of the form as:

$$\phi(x) = \begin{cases} 1 \; ; \; T(x) > T(b) \\ 0 \; ; \; \text{otherwise} \end{cases} \tag{7.5.11}$$

$$\sup_{\theta \le \theta_0} E_{H_0} \phi(x) = 0 \Rightarrow E_{\theta_0} \phi(x) = 0$$

Consider any other test ϕ_1 of size 0.

Then

$$E_{\theta_0} \phi_1(x) = 0 \Rightarrow \int_{S_0} \phi_1(x) f(x|\theta) dx = 0$$

$$\Rightarrow \phi_1(x) = 0 \quad \text{on} \quad S_0$$

Next, consider the power of the test ϕ at any $\theta > \theta_0$,

$$E\phi(x) = \int_{x \ge b} \phi(x) f(x|\theta) dx \ge \int_{x \ge b} \phi_1(x) f(x|\theta) dx$$

$$= \int_{x \ge b} f(x|\theta) dx \ge \int_{x \ge b} \phi_1(x) f(x|\theta) dx$$

Since $0 \le \phi_1(x) \le 1$.

Now,

$$\int\limits_{x \geq b} \phi_1(x) f(x|\theta) dx = \int\limits_{S_0} \phi_1(x) f(x|\theta) dx + \int\limits_{S_0{}^c} \phi_1(x) f(x|\theta) dx$$

Now $\phi_1(x) = 0$ on S_0

$$= \int\limits_{S_0{}^c} \phi_1(x) f(x|\theta) dx$$

$$= \mathrm{E}_{H_1} \phi_1(x)$$

Hence $\mathrm{E}_{H_1} \phi(x) \geq \mathrm{E}_{H_1} \phi_1(x)$.
   It implies that $\phi(x)$ is UMP of size 0.

Theorem 7.5.4 *For one-parameter exponential family, there exist a UMP test of the hypothesis $H_0 : \theta \leq \theta_1$ or $\theta \geq \theta_2$ against $H_1 : \theta_1 < \theta < \theta_2$.*
   The test function is given as

$$\phi(x) = \begin{cases} 1 \ ; \ c_1 < T(x) < c_2 \\ \gamma_i \ ; \ T(x) = c_i (i = 1, 2) \\ 0 \ ; \ T(x) < c_1 \ or \ T(x) > c_2 \end{cases}$$

where the c's and γ's are given by
(i) $\mathrm{E}_{\theta_1} \phi(x) = \mathrm{E}_{\theta_2} \phi(x) = \alpha$
(ii) The test minimizes $\mathrm{E}_\theta \phi(x)$ subject to (i) for all $\theta < \theta_1$ or $\theta > \theta_2$
(iii) For $0 < \alpha < 1$, the power function of this test has a maximum at a point θ_0 between θ_1 and θ_2 and decreases strictly as θ tends away from θ_0 in either direction, unless there exist two values t_1, t_2 such that

$$P_\theta[T(x) = t_1] + P_\theta[T(x) = t_2] = 1 \ \forall \ \theta$$

Example 7.5.1 Which of the following distributions possesses an MLR property.
   (i) Binomial (n, p) (ii) Cauchy $(1, \theta)$ (iii) Gamma $(p, \frac{1}{\sigma})$

$$(i) f(x|p) = \binom{n}{x} p^x q^{n-x}; \quad x = 0, 1, 2, \ldots, n, \ 0 < p < 1, \quad q = 1 - p$$

$$= \binom{n}{x} \left(\frac{p}{q}\right)^x q^n$$

$$= \binom{n}{x} \exp[x \ln \frac{p}{q} + n \ln q]$$

$$= \binom{n}{x} \exp[Q(p)T(x) + nD(p)]$$

$$= \exp[Q(p)T(x) + nD(p) + H(x)]$$

where

$$Q(p) = \ln \frac{p}{q}, \ T(x) = x, \ D(p) = \ln q, \ H(x) = \ln \binom{n}{x}$$

$\frac{dQ(p)}{dp} > 0$, i.e., $Q(p)$ is strictly increasing. This family has MLR property.

$$(ii) \ \ f(x|\sigma) = \frac{e^{-\frac{x}{\sigma}} x^{p-1}}{\sigma^p \Gamma(p)}; x > 0, \sigma > 0, p > 0$$

$$f(x|\sigma) = \exp\left[-\frac{x}{\sigma} + (p-1)\ln x - p \ln \sigma - \ln \Gamma(p)\right]$$

For p known

$$= \exp[Q(\sigma)T(x) + H(x) + D(p)]$$

where

$$Q(\sigma) = -\frac{1}{\sigma}, \ T(x) = x, \ H(x) = (p-1)\ln x - \ln p, \ D(p) = -p \ln \sigma$$

Since $\frac{dQ(\sigma)}{d\sigma} > 0 \ \forall \ \sigma$.

This belongs to exponential family. Hence this family has MLR property.
(iii)

$$\frac{f(x|\theta_2)}{f(x|\theta_1)} = \frac{1 + (x - \theta_1)^2}{1 + (x - \theta_2)^2} \to 1 \text{ as } x \to +\infty \text{ or } -\infty \qquad (7.5.12)$$

Hence, $C(1, \theta)$ does not have an MLR property.

Example 7.5.2 Let the rv X have hypergeometric pmf:

$$P[X = x|M] = \frac{\binom{M}{x}\binom{N-M}{n-x}}{\binom{N}{n}}; \quad x = 0, 1, 2, \ldots, M$$

Find UMP test to test $H_0 : M \leq M_0$ against $H_1 : M > M_0$

$$\lambda(x) = \frac{P[X = x|M+1]}{P[X = x|M]} = \frac{\binom{M+1}{x}\binom{N-M-1}{n-x}}{\binom{M}{x}\binom{N-M}{n-x}}$$

$$= \frac{M+1}{N-M} \times \frac{N-M-n+x}{M+1-x} > 0$$

We see that $P[X = x|M]$ has MLR in x.

From Theorem 7.5.3, there exist a UMP test of size α and is given as

$$\phi(x) = \begin{cases} 1 \; ; \; x > k \\ \gamma \; ; \; x = k \\ 0 \; ; \; x < k \end{cases}$$

k and γ are determined from $E_{H_0}\phi(x) = \alpha$.

Example 7.5.3 Let $X_1, X_2, \ldots, X_n$ be iid rvs with $N(\mu, 1)$. Find UMP test for H_0 : $\mu \leq \mu_0$ or $\mu \geq \mu_1$ against $H_1 : \mu_0 < \mu < \mu_1$.

From the Theorem 7.5.4, the UMP test is given as

$$\phi(x) = \begin{cases} 1 \; ; \; c_1 < \sum_{i=1}^{n} x_i < c_2 \\ 0 \; ; \; \text{otherwise} \end{cases}$$

Determine c_1 and c_2 such that

$$E_{\mu_0}\phi(x) = E_{\mu_1}\phi(x) = \alpha$$

$$\Rightarrow P_{\mu_0}\left[c_1 < \sum x_i < c_2\right] = P_{\mu_1}\left[c_1 < \sum X_i < c_2\right]$$

if $X_i(i = 1, 2, \ldots, n) \sim N(\mu_0, 1)$ then $\sum_{i=1}^{n} X_i \sim N(n\mu_0, n)$.
Similarly if $X_i(i = 1, 2, \ldots, n) \sim N(\mu_1, 1)$ then $\sum_{i=1}^{n} X_i \sim N(n\mu_1, n)$
Let $Z_0 = \frac{\sum x_i - n\mu_0}{\sqrt{n}}$ and $Z_1 = \frac{\sum x_i - n\mu_1}{\sqrt{n}}$

$$P_{\mu_0}\left[\frac{c_1 - n\mu_0}{\sqrt{n}} < Z_0 < \frac{c_2 - n\mu_0}{\sqrt{n}}\right] = P_{\mu_1}\left[\frac{c_1 - n\mu_1}{\sqrt{n}} < Z_1 < \frac{c_2 - n\mu_1}{\sqrt{n}}\right] = \alpha$$

$Z_i \sim N(0, 1); i = 0, 1$.
Given α, n, μ_0 and μ_1,

$$\Phi\left[\frac{c_2 - n\mu_0}{\sqrt{n}}\right] - \Phi\left[\frac{c_1 - n\mu_0}{\sqrt{n}}\right] = \alpha \tag{7.5.13}$$

and

$$\Phi\left[\frac{c_2 - n\mu_1}{\sqrt{n}}\right] - \Phi\left[\frac{c_1 - n\mu_1}{\sqrt{n}}\right] = \alpha, \tag{7.5.14}$$

where Φ is the df of Z.

We can solve Eqs. (7.5.13) and (7.5.14) simultaneously to get the values c_1 and c_2 using R

```
# Given data
  n <- 10; mu0 <- 0.2; mu1 <- 0.3; alpha <- 0.05;
# To find c2 such that Phi((c2-nmu0)/sqrt(n)) > alpha
  x <- seq(-4,4,0.1) # possible values for c2
  # Standardization
    z1 <- (x-(n*mu0))/sqrt(n);
    z2 <- (x-(n*mu1))/sqrt(n);
  # To find cumulative probability
    cdf1 <- pnorm(z1,0,1);
    cdf2 <- pnorm(z2,0,1);
  # To find c2
    a <- min(which(cdf1 > alpha))
    b <- min(which(cdf2 > alpha))
    c2 <- max(x[a],x[b])
  # To find value of c1
    Eqn2 <- alpha-0.1
    while(Eqn2 < alpha)
    {
        z3 <- (c2-(n*mu0))/sqrt(n);
        Eqn1 <- pnorm(z3,0,1)-alpha;
        c1 <- qnorm(Eqn1,(n*mu0),sqrt(n))
        z4 <- (c2-(n*mu1))/sqrt(n);
        z5 <- (c1-(n*mu1))/sqrt(n);
        Eqn2 <- pnorm(z4,0,1)-pnorm(z5,0,1);
        if(Eqn3 <= alpha) { c1_pre <- c1; c2_pre <- c2 }
        c2 <- c2+0.1
    }
# Check
  c1 <- c1_pre; c2 <- c2_pre
  z1 <- (c2-(n*mu0))/sqrt(n);
  z2 <- (c1-(n*mu0))/sqrt(n);
  z3 <- (c2-(n*mu1))/sqrt(n);
  z4 <- (c1-(n*mu1))/sqrt(n);
  Eqn1 <- pnorm(z1,0,1)-pnorm(z2,0,1)
  Eqn2 <- pnorm(z3,0,1)-pnorm(z4,0,1)
# OUTPUT
print(c("c1 =",c1));
print(c("c2 =",c2));
print("CHECK")
print("calculated alpha for equation one"); Eqn1
print("calculated alpha for equation two"); Eqn2
# RESULT
  # OUTPUT
  "c1 ="           "2.29843477913486"
  "c2 =" "2.7"
  "CHECK"
  "calculated alpha for equation one" =  0.05
  "calculated alpha for equation two" =  0.04999609
```

Remark: The UMP test for testing $H_0 : \theta_1 \leq \theta \leq \theta_2$ against $H_1 : \theta = \theta_0$ for one-parameter exponential family does not exist.

Example 7.5.4 Let $X_1, X_2, \ldots, X_n$ be a random sample from $(0, \sigma^2)$. Find the UMP test for

(i) $H_0 : \sigma = \sigma_0$ against $H_1 : \sigma > \sigma_0$
(ii) $H_0 : \sigma = \sigma_0$ against $H_1 : \sigma < \sigma_0$

By using the Theorem 7.5.3,
(i) The UMP test is given as

$$\phi_1(x) = \begin{cases} 1 \ ; & \sum_{i=1}^{n} x_i^2 > c_1 \\ 0 \ ; & \text{otherwise} \end{cases}$$

Under H_0, $\frac{\sum_{i=1}^{n} x_i^2}{\sigma_0^2} \sim \chi_n^2$

if $\alpha = E\phi_1(x) = P[\sum_{i=1}^{n} x_i^2 > c_1]$
then $c_1 = \sigma_0^2 \chi_{n,\alpha}^2$
The UMP test is

$$\phi_1(x) = \begin{cases} 1 \ ; & \sum_{i=1}^{n} x_i^2 > \sigma_0^2 \chi_{n,\alpha}^2 \\ 0 \ ; & \text{otherwise} \end{cases}$$

(ii) Similarly as in (i), we can write UMP test:

$$\phi_2(x) = \begin{cases} 1 \ ; & \sum_{i=1}^{n} x_i^2 < \sigma_0^2 \chi_{n,1-\alpha}^2 \\ 0 \ ; & \text{otherwise} \end{cases}$$

Note: ϕ_1 and ϕ_2 are not UMP for testing $H_0 : \sigma = \sigma_0$ against $H_1 : \sigma \neq \sigma_0$.

Example 7.5.5 Let $X_1, X_2, \ldots, X_n$ be a random sample from $N(\theta, 1)$, where θ is unknown. Show that there is no uniformly most powerful test of $H_0 : \theta = \theta_0$ against $H_1 : \theta \neq \theta_0$

By NP lemma,

$$\lambda(x) = \exp\left[-\frac{1}{2}\sum(x_i - \theta_1)^2 + \frac{1}{2}\sum(x_i - \theta_0)^2\right] \geq k$$

$$\Rightarrow \sum(x_i - \theta_0)^2 - \sum(x_i - \theta_1)^2 \geq k$$

$$\Rightarrow n(\theta_0^2 - \theta_1^2) + 2\sum x_i(\theta_1 - \theta_0) \geq k$$

$$\Rightarrow \sum x_i \geq \frac{k}{2(\theta_1 - \theta_0)} + \frac{n(\theta_0 + \theta_1)}{2} \quad \text{if} \quad \theta_1 > \theta_0$$

$$\Rightarrow \sum x_i \leq \frac{n}{2}(\theta_0 + \theta_1) + \frac{k}{2(\theta_1 - \theta_0)} \quad \text{if} \quad \theta_1 < \theta_0$$

Consider $H_1 : \theta = \theta_1 = \theta_0 + 1$

$$\Rightarrow \sum x_i \geq \frac{k}{2} + \frac{n(2\theta_0 + 1)}{2} \quad \text{if} \quad \theta_1 > \theta_0$$

and if $H_1 : \theta = \theta_1 = \theta_0 - 1$

$$\Rightarrow \sum x_i \leq \frac{n}{2}(2\theta_0 - 1) - \frac{k}{2} \quad \text{if} \quad \theta_1 < \theta_0$$

Thus a best critical region for testing the simple hypothesis against an alternative hypothesis $H_1 : \theta = \theta_1 = \theta_0 + 1$ will not serve as a BCR for testing $H_1 : \theta = \theta_1 = \theta_0 - 1$.

Hence, there is no uniformly MP test for $H_1 : \theta \neq \theta_0$.

Example 7.5.6 Let $X_1, X_2, \ldots, X_n$ be a random sample from $f(x|\theta), \theta \in \Theta$, where

$$f(x|\theta) = a(\theta)h(x); \quad -\infty < x < \theta,$$

Show that this family has MLR.

Let $\theta_1 < \theta_2$

$$\lambda(x) = \frac{f(x_1, x_2, \ldots, x_n|\theta_2)}{f(x_1, x_2, \ldots, x_m|\theta_1)}$$

$$= \frac{[a(\theta_2)]^n \prod_{i=1}^{n} h(x_i)\mathbf{I}(\theta_2 - X_{(n)})}{[a(\theta_1)]^n \prod_{i=1}^{n} h(x_i)\mathbf{I}(\theta_1 - X_{(n)})}$$

There are three cases

(i) $X_{(n)} < \theta_1 < \theta_2$ (ii) $\theta_1 \leq X_{(n)} < \theta_2$
(iii) $\theta_1 < \theta_2 \leq x_{(n)}$

$$\lambda(x) = \begin{cases} \frac{[a(\theta_2)]^n}{[a(\theta_1)]^n} & ; \ -\infty < X_{(n)} < \theta_1 \\ \frac{[a(\theta_2)]^n \prod_{i=1}^{n} h(x_i)}{0} & ; \ \theta_1 \leq X_{(n)} < \theta_2 \\ \frac{0}{0} & ; \ \theta_2 \leq X_{(n)} < \infty \end{cases}$$

$\lambda(x)$ is increasing in $X_{(n)}$. Therefore this family has MLR property.

Example 7.5.7 Show that the double exponential family (known as Laplace distribution) of distribution

$$f(x|a, \theta) = \frac{1}{2\theta} \exp\left[-\frac{|x - a|}{\theta}\right]$$

has monotone likelihood ratio, when a is unknown and θ is known.

$H_0 : a = a_1$ against $H_1 : a = a_2 > a_1$

$$\lambda(x) = \frac{f(x|a_2)}{f(x|a_1)} = \exp\left\{\frac{1}{\theta}\left[|x - a_1| - |x - a_2|\right]\right\}$$

There are three cases

(i) $(x - a_i) \leq 0$ for $i = 1, 2$

$$\lambda(x) = \exp\left\{\frac{(a_1 - a_2)}{\theta}\right\}$$

(ii) $(x - a_1) > 0$ and $(x - a_2) < 0$

$$\lambda(x) = \exp\left\{\frac{(2x - a_1 - a_2)}{\theta}\right\}$$

One can observe $\lambda'(x) > 0$ for $a_1 < a_2$ and $\lambda(x) \to \exp\{\frac{a_2 - a_1}{\theta}\}$ as $x \uparrow a_2$.
$\lambda(x)$ is nondecreasing in x.
(iii) $(x - a_1) < 0$ and $(x - a_2) > 0$
$\Rightarrow x < a_1$ and $x > a_2$
$\Rightarrow a_2 < x < a_1$, which is not possible
(iv) $(x - a_i) > 0;\ i = 1, 2$.

$$h(x) = \exp\left\{\frac{(a_2 - a_1)}{\theta}\right\}$$

From (i), (ii) and (iv), we can see that $x \uparrow a_1$ in $(-\infty, a_1), x \uparrow a_2$ in $(-\infty, a_2)$ and $x \uparrow \infty$ in (a_2, ∞), $\lambda(x)$ equals at $\exp\left\{\frac{(a_1 - a_2)}{\theta}\right\}$ and increases to $\exp\left\{\frac{(a_2 - a_1)}{\theta}\right\}$ and becomes constant at $\exp\left\{\frac{(a_2 - a_1)}{\theta}\right\}$. We can conclude that $\lambda(x)$ is nondecreasing in x and this family possesses MLR property.

Example 7.5.8 Consider the following problem from Lehman (1986).

Let X be length of life of an electron tube. Assume that X has an exponential distribution with mean 2θ. Hence, pdf of X is

$$f(x|\theta) = \frac{1}{2\theta}\exp\left(-\frac{x}{2\theta}\right);\ x > 0$$

Let n such tubes be put on test simultaneously, i.e., we draw an independent sample $X_1, X_2, \ldots, X_n$ from the exponential population. Let X's be ordered and denoted by $Y_1 \leq Y_2 \leq \cdots \leq Y_n$, where Y_1 be the life of a tube which gets fused first, Y_2 be the life of a tube which gets fused next to it and so on. We may continue this experiment till we get rth tube fused. This process is known as inverse sampling. Same model arises in life testing applications where n bulbs are put on testing and this number n is held fixed by replacing each burned out bulb with a new one and denoting Y_1 as the time at which first burn out bulb is replaced, Y_2 as the time at which second bulb is replaced, etc., each measured from some fixed time.

Obtain a UMP test size α for $H_0 : \theta \geq \theta_0$ against $H_1 : \theta < \theta_0$

On the basis of $Y_1, Y_2, \ldots, Y_r$.

The joint distribution of $(Y_1, Y_2, \ldots, Y_r)$ is

$$f(y_1, y_2, \ldots, y_r | \theta) = \frac{n!}{(n-r)!} \prod_{i=1}^{r} f(y_i)[1 - F(y_r)]^{n-r}$$

$$f(y_i) = \frac{1}{2\theta} e^{-\frac{y_i}{2\theta}}, \quad F(y) = 1 - e^{-\frac{y}{2\theta}}$$

$$f(y_1, y_2, \ldots, y_r | \theta) = \frac{n!}{(n-r)!} \frac{1}{(2\theta)^r} \exp\left[-\frac{\sum_{i=1}^{r} y_i}{2\theta}\right] \exp\left[-\frac{y_r(n-r)}{2\theta}\right]$$

$$f(y_1, y_2, \ldots, y_r | \theta) = \frac{n!}{(n-r)!} \frac{1}{(2\theta)^r} \exp\left\{-\frac{1}{2\theta}\left[\sum_{i=1}^{r} y_i + y_r(n-r)\right]\right\}$$

This belongs to exponential family. By using Theorem 7.5.3, we can give a UMP test for testing $H_0 : \theta \leq \theta_0$ against $H_1 : \theta > \theta_0$,

$$\phi(T) = \begin{cases} 1 ; & T > t_0 \\ 0 ; & \text{otherwise} \end{cases} \tag{7.5.15}$$

where $T = \sum_{i=1}^{r} y_i + y_r(n-r)$.

We have to find a distribution of T,

Note that

$$\frac{1}{\theta} \sum_{i=1}^{r} (n-i+1)(y_i - y_{i-1}) = \frac{1}{\theta}\left\{\sum_{i=1}^{r} y_i + (n-r)y_r\right\}$$

The joint distribution of (y_i, y_{i-1}) is

$$f(y_i, y_{i-1}) = \frac{n!}{(i-2)!(n-i)!}\left[1 - \exp\left(-\frac{y_{i-1}}{2\theta}\right)\right]^{i-2}$$

$$\exp\left[-\frac{y_i}{2\theta}(n-i)\right] \frac{1}{2\theta} \exp\left[-\frac{y_i}{2\theta}\right] \frac{1}{2\theta} \exp\left[-\frac{y_{i-1}}{2\theta}\right],$$

Let $C = \frac{n!}{(i-2)!(n-i)!}$

$$f(y_i, y_{i-1}) = \frac{C}{(2\theta)^2}\left[1 - \exp\left(-\frac{y_{i-1}}{2\theta}\right)\right]^{i-2} \exp\left[-\frac{y_i(n-i+1)}{2\theta}\right] \exp\left[-\frac{y_{i-1}}{2\theta}\right]$$

Let $U_i = Y_i - Y_{i-1} \Rightarrow Y_i = U_i + Y_{i-1}$

$$f(y_i, y_{i-1}) = \frac{C}{(2\theta)^2} \left[1 - \exp\left(-\frac{y_{i-1}}{2\theta}\right)\right]^{i-2} \exp\left[-\frac{(u_i + y_{i-1})(n - i + 1)}{2\theta}\right] \exp\left[-\frac{y_{i-1}}{2\theta}\right]$$

Let $W = 1 - \exp\left[-\frac{y_{i-1}}{2\theta}\right], dw = \frac{\exp[-\frac{y_{i-1}}{2\theta}]}{2\theta} dy_{i-1}$

$$f_{u_i}(u) = \frac{C}{(2\theta)} \exp\left[-\frac{u_i(n - i + 1)}{2\theta}\right] \int_0^1 w^{i-2}(1 - w)^{n-i+1} dw$$

$$= \frac{C}{(2\theta)} \exp\left[-\frac{u_i(n - i + 1)}{2\theta}\right] \beta(i - 1, n - i + 2)$$

Now

$$C\beta(i - 1, n - i + 2) = \frac{n!}{(i - 2)!(n - i)!} \frac{\Gamma(i - 1)\Gamma(n - i + 2)}{\Gamma(n + 1)} = n - i + 1$$

$$f_{u_i}(u) = \frac{n - i + 1}{(2\theta)} \exp\left[-\frac{u(n - i + 1)}{2\theta}\right]$$

if $v_i = \frac{(n-i+1)u_i}{\theta}$ then $f(v_i) = \frac{e^{-\frac{v_i}{2}}}{2}$

i.e., $v_i \sim \chi_2^2$ then $\sum_{i=1}^r v_i \sim \chi_{2r}^2$

$$\sum v_i = \frac{1}{\theta} \sum (n - i + 1)u_i \sim \chi_{2r}^2$$

$$= \frac{1}{\theta} \sum_{i=1}^r y_i + (n - r)y_r \sim \chi_{2r}^2 \tag{7.5.16}$$

Hence $T = \sum_{i=1}^r y_i + (n - r)y_r \sim G(r, \frac{1}{2\theta})$ We can write the UMP test as given in (7.5.15) of size α as

$$\phi(T) = \begin{cases} 1 \; ; \; T > \theta_0 \chi_{2r,\alpha}^2 \\ 0 \; ; \; \text{otherwise} \end{cases}$$

For example, $r = 4$

if $\theta_0 = 3$ then $\chi_{8,0.05}^2 = 15.5073$ then $t_0 = 46.5219$

Our UMP test is

$$\phi(T) = \begin{cases} 1 \; ; \; T > 46.5219 \\ 0 \; ; \; \text{otherwise} \end{cases}$$

Example 7.5.9 Find UMP test for the logistic distribution with location parameter θ w̌ ṱṣḷ ṟ̃ẖ , ḷ ⩽ ₥ₕ ₐₚₐₙₙₛₗ ₕ₁ ' ₕ ⩽ ₕ₀ ᵗᵒᵣ ₐ ₛ₁ᵤₐ ₒₜ ₐₗ

$$f(x|\theta) = \frac{\exp[-(x-\theta)]}{\{1 + \exp[-(x-\theta)]\}^2}; \quad -\infty < x < \infty, -\infty < \theta < \infty$$

Let $\theta_1 < \theta_2$ and $\lambda(x) = \frac{f(x|\theta_2)}{f(x|\theta_1)}$

$$\lambda(x) = \exp[\theta_2 - \theta_1]\left[\frac{1 + \exp[-(x-\theta_1)]}{1 + \exp[-(x-\theta_2)]}\right]^2$$

Let $x < y \Rightarrow \lambda(x) < \lambda(y)$, i.e., to prove $\lambda(x) - \lambda(y) < 0$

$$\left[\frac{1 + \exp[-(x-\theta_1)]}{1 + \exp[-(x-\theta_2)]}\right] < \left[\frac{1 + \exp[-(y-\theta_1)]}{1 + \exp[-(y-\theta_2)]}\right]$$

$$\Rightarrow [1 + e^{-(x-\theta_1)}][1 + e^{-(y-\theta_2)}] < [1 + e^{-(x-\theta_2)}][1 + e^{-(y-\theta_1)}]$$

$$\Rightarrow 1 + e^{-(y-\theta_2)} + e^{-(x-\theta_1)} + e^{-(x-\theta_1)}e^{-(y-\theta_2)} < 1 + e^{-(y-\theta_1)} + e^{-(x-\theta_2)} + e^{-(x-\theta_2)}e^{-(y-\theta_1)}$$

$$\Rightarrow e^{-(y-\theta_2)} - e^{-(x-\theta_2)} + e^{-(x-\theta_1)} - e^{-(y-\theta_1)} + e^{-(x-\theta_1)}e^{-(y-\theta_2)} - e^{-(x-\theta_2)}e^{-(y-\theta_1)} < 0$$

$$\Rightarrow e^{-(y-\theta_2)} - e^{-(x-\theta_2)} + e^{-(x-\theta_1)} - e^{-(y-\theta_1)} < 0$$

(the other term is zero)

$$\Rightarrow e^{\theta_2}[e^{-y} - e^{-x}] + e^{\theta_1}[e^{-x} - e^{-y}] < 0$$

$$\Rightarrow (e^{-x} - e^{-y})(e^{\theta_1} - e^{\theta_2}) < 0,$$

which is always true because

$$\theta_1 < \theta_2 \Rightarrow e^{\theta_1} < e^{\theta_2} \Rightarrow (e^{\theta_1} - e^{\theta_2}) < 0$$

$y > x \Rightarrow -y < -x$

$$e^{-y} < e^{-x} \Rightarrow (e^{-x} - e^{-y}) > 0$$

Hence $\lambda(x) > \lambda(y)$ if $\theta_1 < \theta_2$.
Therefore, logistic distribution has MLR property.

By using the Theorem 7.5.3, we can write UMP test as

$$\phi(x) = \begin{cases} 1 \; ; \; x > x_0 \\ 0 \; ; \; \text{otherwise,} \end{cases}$$

x_0 is determined by $E_{H_0} \phi(x) = \alpha$.
Hence,

$$\int_{x_0}^{\infty} \frac{e^{-(x-\theta_0)}}{[1 + e^{-(x-\theta_0)}]^2} = \alpha$$

Let $w = 1 + e^{-(x-\theta_0)} \Rightarrow dw = -e^{-(x-\theta_0)} dx$
Now, $x = x_0 \Rightarrow w = 1 + e^{-(x-\theta_0)}$ and $x = \infty \Rightarrow w = 1$

$$\int_{1}^{1+e^{-(x-\theta_0)}} \frac{dw}{w^2} = \alpha$$

$$\left[1 + e^{-(x-\theta_0)} \right]^{-1} = 1 - \alpha$$

$$1 + e^{-(x-\theta_0)} = \frac{1}{1-\alpha}$$

$$e^{-(x-\theta_0)} = \frac{\alpha}{1-\alpha}$$

$$x_0 = \theta_0 - \log \frac{\alpha}{1-\alpha}$$

Therefore, UMP test for testing $H_0 : \theta \le \theta_0$ against $H_1 : \theta > \theta_0$ would be

$$\phi(x) = \begin{cases} 1 \; ; \; x > \theta_0 - \log \frac{\alpha}{1-\alpha} \\ 0 \; ; \; \text{otherwise,} \end{cases} \qquad (7.5.17)$$

Example 7.5.10 Let the rv X has the following pdf $f(x|\theta)$:

$$f(x|\theta) = \frac{\theta}{(\theta + x)^2}; \quad x > 0, \theta > 0$$

Obtain UMP test for testing $H_0 : \theta \le \theta_0$ against $H_1 : \theta > \theta_0$

$$\lambda(x) = \frac{\theta_1}{\theta_0} \left(\frac{\theta_0 + x}{\theta_1 + x} \right)^2$$

$$\log \lambda(x) = \log \frac{\theta_1}{\theta_0} + 2[\log(\theta_0 + x) - \log(\theta_1 + x)]$$

$$\frac{d \log \lambda(x)}{dx} = 2\left[\frac{1}{\theta_0 + x} - \frac{1}{\theta_1 + x}\right] = \frac{2(\theta_1 - \theta_0)}{(\theta_0 + x)(\theta_1 + x)} > 0$$

$\Rightarrow \lambda'(x) > 0 \ \forall \ x \Rightarrow$ It has MLR property.

We can write UMP test as

$$\phi(x) = \begin{cases} 1 \; ; \; x > k \\ 0 \; ; \; \text{otherwise}, \end{cases}$$

$$E\phi(x) = \alpha \Rightarrow \int_k^\infty \frac{\theta_0}{(\theta_0 + x)^2} dx = \alpha$$

$$\Rightarrow k = \theta_0 \left(\frac{1}{\alpha} - 1\right) = \frac{\theta_0(1 - \alpha)}{\alpha}$$

Hence (7.5.17) becomes

$$\phi(x) = \begin{cases} 1 \; ; \text{if } x > \frac{\theta_0(1-\alpha)}{\alpha} \\ 0 \; ; \text{otherwise} \end{cases}$$

7.6 Exercise 7

1. There are two density $P_1(X)$, $P_2(X)$ to describe a particular experiment which has record space $X = \{0, 1, 2, 3, 4, 5\}$

Hypothesis	X	0	1	2	3	4	5
S_1	$P_1(X)$	0.30	0.20	0.05	0.10	0.15	0.20
S_2	$P_2(X)$	0.05	0.15	0.20	0.40	0.10	0.10

Two decision rules are proposed.

Rule 1: If $0 \leq X \leq 3$ decide for S_2, otherwise decide for S_1

Rule 2: If $X < 4$ decide for S_1, if $X \geq 4$ decide for S_2

By calculating the liabilities to error for those two decision rules find which is better. Can you give a reason for your choice? Can you improve on this choice by reducing the liability to error of the first kind?

2. The identification of a cell as Type A or Type B is long and expensive, but a recent series of experiments have shown that the quantity (in certain units) of an easily quantifiable 1 chemical is well described by a $N(15, 3)$ random variable for type A and by a $N(20, 6)$ random variable for type B. In future it is decided to classify a cell as type A if it contains not more than 12 units of the chemical, and as type B otherwise. Obtain suitable measures of the liabilities to misclassify.

Where should the point of division be drawn if it is required to set the probability of misclassifying a type A cell equal to 0.06? What then is the probability of the other kind misclassification?

3. A dispute has arisen between two archeologists over dating of a specimen. A claims that it is 5000 years old and B that it is 10,000 years old. It is known that such specimens emit a certain type of radioactive particle, the number of particles emitted in any one minute being described by a Poisson (1) random variable if A's claim is true and by a Poisson (5) random variable if B's claim is true.

An arbiter suggests that, after a minute counting, he should decide in favor of A if not more than 2 particles are observed and decide for B otherwise. Investigate the liabilities to error with this decision rule.

What is the minimum number of complete minutes for which counting should be recorded if the probability of deciding for A when in fact B is correct is to be less than 0.06? What is the corresponding probability of deciding for B when in fact A is correct?

4. A person claims to have telepathic power in the sense that he can say which of the two colored cards is being observed by his partner with probability 0.7 rather than 0.6, which would be the appropriate value for guessing of colors randomly presented to the partner. As a preliminary test he is asked to state the colors on 8 such cards randomly presented to his partner. It is decided to accept him for further tests if he scores at least 7 successes, and otherwise to dismiss him. Evaluate the appropriate measures of the probabilities to unjustified acceptance and to wrongful dismissal in such a test.

A person who has passed this preliminary test is now subjected to longer series of 500 cards (again randomly presented). It is agreed to accept him for even more tests if he scores at least 300 successes, and otherwise to dismiss him. Find the appropriate measures of liability for this series. (Use the normal approximation to the binomial.) How many successes should have been demanded if the probability of unjustified acceptance was to kept to 0.06?

5. For a process which produces components at independent operations and with lifetimes varying according to the density function $p(x) = \theta e^{-\theta x}$ ($x \geq 0, \theta > 0$). Show that the probability that two components having lifetimes greater than a is $e^{-2\theta a}$, and the probability that the total lifetime of two components is greater than a is $(1 + \theta a)e^{-\theta a}$.

For process A, θ is known to be 2 for process B, θ is known to be 3. Components from these processes are not easily distinguishable and unfortunately a large batch of unlabeled components have been discovered.Compare the following rules, for deciding, on the results of life testing two components, which process the batch has come from.

Rule 1: If both lifetimes are greater than $\frac{1}{2}$ decide that it was process A, otherwise decide for process B.

Rule 2: If the total lifetime of the two components is greater than 2 decide for A, otherwise decide for B.

What critical value should replace $\frac{1}{2}$ in Rule 1 if the probability of deciding for process B when in fact A was used to be 0.1? What then is the probability of deciding for process A when in fact B was used?

6. A sample of size 1 is taken from a population distribution $P(\lambda)$. To test $H_0 : \lambda = 2$ against $H_1 : \lambda = 3$, consider the nonrandomized test

$$\phi(x) = \begin{cases} 1 \; ; \; x > 3 \\ 0 \; ; \; x \leq 3, \end{cases}$$

Find the probabilities of type I and type II errors and the power of the test against $\lambda = 2$. If it is required to achieve a size equal to 0.05, how should one modify the test ϕ? Plot the power function for $H_1 : \lambda > 3$.

7. A sample of size 1 is taken from exponential pdf with parameter θ, i.e., $X \sim G(1, \theta)$. To test $H_0 : \theta = 2$ against $H_1 : \theta > 2$, the test to be used is the nonrandomized test

$$\phi(x) = \begin{cases} 1 \; ; \; x > 2 \\ 0 \; ; \; x \leq 2, \end{cases}$$

Find the size of the test. What is the power function? Plot the power functions.

8. Let X_1, X_2 be iid observations from

$$f(x, \theta) = \frac{1}{\theta} e^{-\frac{x}{\theta}}; \quad 0 < x < \infty, \theta > 0$$

Consider the acceptance region as $w = \{(x_1, x_2) | x_1 + x_2 < 6\}$ for testing $H_0 : \theta = 2$ against $H_1 : \theta = 4$. Determine type I and type II errors.

9. Let X_1, X_2 be random sample drawn from

$$f(x, \theta) = \theta x^{\theta-1}; \quad 0 < x < 1$$

If we test $H_0 : \theta = 1$ against $H_1 : \theta = 2$ with the critical region $w = \{(x_1, x_2) | (x_1 x_2) \geq \frac{1}{2}\}$.

Find the size and power of the test.

10. Let $X_1, X_2, \ldots, X_{10}$ be a random sample from $N(\mu, \sigma^2)$. Find a MP test of the hypothesis $H_0 : \mu = 0, \sigma^2 = 1$ against the alternative hypothesis $H_1 : \mu = 1, \sigma^2 = 4$.

11. Let $X_1, X_2, \ldots, X_n$ be a random sample from a normal distribution with mean μ and variance 100. It is designed to test $H_0 : \mu = 75$ against $H_1 : \mu = 78$. Find test of level of significance 0.05 and with power equal to 0.90 approximately.

12. Let

$$f_\theta(x) = \frac{1}{\pi[1 + (x - \theta)^2]}, \quad -\infty < x < \infty$$

Using a single observation find a most powerful test of size 0.10 to test the hypothesis $H_0 : \theta = 2$ against $H_1 : \theta = 4$. Use R.

13. Let $X_1, X_2, \ldots, X_{10}$ be a random sample of size 10 from a $N(0, \sigma^2)$. Find a best critical region of size $\alpha = 0.05$ for testing $H_0 : \sigma^2 = 1$ against $H_1 : \sigma^2 = 2$. In this a best critical region against alternative $H_1 : \sigma^2 > 1$.

14. Consider the two independent normal distributions $N(\mu_1, 400)$ and $N(\mu_2, 225)$. Find a UMP test to test the hypothesis $H_0 : \mu_1 - \mu_2 = 0$ against the alternative $H_1 : \mu_1 - \mu_2 > 0$ such that the power at two points $\beta(\mu_1 - \mu_2 = 0) = 0.05$ and $\beta(\mu_1 - \mu_2 = 10) = 0.90$ approximately.

15. Find the Neyman–Pearson size α test of $H_0 : \theta = \theta_0$ against $H_1 : \theta = \theta_1(\theta_1 < \theta_0)$ based on a sample size 1 from the pdf

$$f_0(x) = 2\theta x + 2(1 - \theta)(1 - x), \quad 0 < x < 1\, \theta \in [0, 1].$$

(Take $\alpha = 0.02, 0.10, \theta_0 = 4$ and $\theta_1 = 2$)

16. Find the Neyman–Pearson size α test of $H_0 : \beta = 1$ against $H_1 : \beta = \beta_1(>1)$ based on sample size 1 from

$$f(x, \beta) = \begin{cases} \beta x^{\beta-1} \; ; \; 0 < x < 1 \\ 0 \qquad ; \text{otherwise} \end{cases}$$

17. Let X be an observation in $U(0, 1)$. Find an MP size α test of $H_0 : X \sim f(x) = 4x$ if $0 < x < \frac{1}{2}$, and $= 4 - 4x$ if $\frac{1}{2} \le x < 1$ against $H_1 : X \sim f(x) = 1$ if $0 < x < 1$. Find the power of your test.

18. Let $X_1, X_2, \ldots, X_n$ be a random sample with common pdf

$$f_0(x) = \frac{1}{2\theta} \exp -\frac{|x|}{\theta}, \quad x \in R\, \theta > 0$$

Find a size α MP test for testing $H_0 : \theta = \theta_0$ versus $H_1 : \theta = \theta_1(>\theta_0)$

19. Let $X \sim f_i$, $i = 0, 1$ where

(i)

x	1	2	3	4	5
$f_0(x)$	$\frac{1}{5}$	$\frac{1}{5}$	$\frac{1}{5}$	$\frac{1}{5}$	$\frac{1}{5}$
$f_1(x)$	$\frac{1}{6}$	$\frac{1}{4}$	$\frac{1}{6}$	$\frac{1}{4}$	$\frac{1}{6}$

(ii)

x	1	2	3	4	5
$f_0(x)$	$\frac{1}{4}$	$\frac{1}{4}$	0	$\frac{1}{2}$	0
$f_1(x)$	$\frac{1}{5}$	$\frac{1}{5}$	$\frac{1}{5}$	0	$\frac{2}{5}$

(Take $\alpha = 0.02, 0.06$).

(a) Find the form of the MP test of its size.

(b) Find the size and the power of your test for various values of the cut off point.

20. Let X have the binomial distribution $B(n, p)$ and consider the hypothesis $H_0 : p = p_0$ against $H_1 : p = p_1 > p_0$ at level of significance α. Determine the boundary values of the UMP unbiased test for $n = 10$, $\alpha = 0.1, p_0 = 0.2$ and $\alpha = 0.05, p_0 = 0.4$ and in each case graph the power functions of both the unbiased and the equal tails test.

21. Let $\frac{T_n}{\theta}$ have a χ^2 distribution with n degrees of freedom. For testing $H_0 : \theta = 1$ at level of significance $\alpha = 0.05$, find n so large that the power of the UMP unbiased test is ≥ 0.90 against both $\theta \geq 2$ and $\theta \leq \frac{1}{2}$. How large does n have to be if the test is not required to be unbiased? (see Definition 8.1.1).

22. Let $X_1, X_2, \ldots, X_n$ be iid $N(5, 1)$. Draw a sample of size 10.
    Test (i) $H_0 : \mu = 5$ against $H_1 : \mu \neq 5$, assume $\alpha = 0.5$
    (ii) $H_0 : \mu = 5$ against $H_1 : \mu > 5$
    (iii) $H_0 : \mu = 5$ against $H_1 : \mu < 5$
    Draw the power curve of all three test on the same graph paper and comment.

23. Let $X_1, X_2, \ldots, X_n$ be iid $N(0, \theta)$ with $\theta = 6$. Draw a sample of size 10.
    Test (i) $H_0 : \theta = 6$ against $H_1 : \theta \neq 6$, assume $\alpha = 0.5$
    (ii) $H_0 : \theta = 6$ against $H_1 : \theta > 6$
    (iii) $H_0 : \theta = 6$ against $H_1 : \theta < 6$
    Plot the power function of all the three test on the same graph and comment.

24. Suppose a certain type of 40 W bulb has been standardized so that the mean life of the bulb is 1500 h and the standard deviation is 200 h. A random sample of 25 of these bulbs from lot having mean θ was tested and found to have a mean life of 1380 h.

(a) Test at 1 percent significance level the hypothesis $H_0 : \theta = 1500$ against the alternative $H_1 : \theta < 1500$

(b) Compute the power of the test at $\theta = 1450, 1400, 1300, 1200, 1150,$ and plot the power function.

25. A sample of size 9 from a population which is normally distributed with mean 1260 is as follows:

1268, 1271, 1259, 1266, 1257, 1263, 1272, 1260, 1256.

(a) Test at 5 percent level of significance the hypothesis $\sigma^2 = 40$ against the alternative $\sigma^2 < 40$

(b) Compute the power of the test when $\sigma^2 = 36, 32, 28, 24,$ and plot the power function.

26. A sample of size 5 is observed from a binomial distribution $B(20, p)$. Find a UMP test for resting the hypothesis $H_0 : p = \frac{1}{2}$ against the alternative $H_1 : p > \frac{1}{2}$ at 5 percent level significance.

27. A sample of size 10 is obtained from a Poisson distribution with parameter m. Construct a test of level of significance $\alpha = 0.01$ to test $H_0 : m = 3$ against the alternative $H_1 : m > 3$.

28. Let X be the number of successes in n independent trials with probability p of successes, and let $\phi(X)$ be the UMP test for testing $p \leq p_0$ against $p > p_0$ at level of significance α.

(i) For $n = 6, p_0 = 0.25$ and the levels $\alpha = 0.05, 0.1, 0.2$ determine k and γ and find the power of the test against $P_1 = 0.3, 0.4, 0.5, 0.6, 0.7$.

(ii) If $p_0 = 0.2$ and $\alpha = 0.05$, and it is desired to have power $\beta \geq 0.9$ against $p_1 = 0.4$, determine the necessary sample size (a) by using tables of the binomial distribution, (b) by using the normal approximation.

(iii) Use the normal approximation to determine the sample size required when $\alpha = 0.05, \beta = 0.9, p_0 = 0.01, p_1 = 0.02$.

29. Let $X_1, X_2, \ldots, X_n$ be independently distributed with density

$$f(x) = (2\theta)^{-1} \exp\left[-\frac{x}{2\theta}\right], \quad x > 0, \theta > 0$$

and let $Y_1 \leq Y_2 \leq \cdots \leq Y_n$ be the ordered X's.

Assume that Y_1 becomes available first, then Y_2, etc., and that observation is continued until Y_r has been observed. On the basis of $Y_1, \ldots, Y_r$ it is desired to test $H : \theta > \theta_0 = 1000$ at level $\alpha = 0.05$ against $H_1 : \theta < \theta_0$

(i) Determine the rejection region when $r = 4$, and find the power of the test against $\theta_1 = 500$.

(ii) Find the value of r required to get power $\beta \geq 0.95$ against the alternative.

30. When a Poisson process is observed for a time interval of length r, the number X of events occurring has the Poisson distribution $P(\lambda r)$. Under an alternative scheme, the process is observed until r events have occurred, and the time T of observation is then a random variable such that $2\lambda T$ has a χ^2 distribution with 2r degrees of freedom. For testing $H : \lambda \leq \lambda_0$ at level α one can, under either design, obtain a specified power β against an alternative λ_1 by choosing T and r sufficiently.

(i) The ratio of the time of observation required for this purpose under the first design to the expected time required under the second is λ_1

(ii) Determine for which values of each of the two designs is preferable when $\lambda_0 = 1$, $\lambda_1 = 2$, $\alpha = 0.05$, $\beta = 0.9$.

31. Let $X_1, X_2, \ldots, X_n$ are iid rvs from $U(\theta, \theta + 1)$. Find a UMP test of size α to test $H_0 : \theta \leq \theta_0$ against $H_1 : \theta > \theta_0$.
   Further, test $H_0 : \theta \leq 2$ against $H_1 : \theta > 2$ for the following data. Let $\alpha = 0.05$.
   2.69 2.72 2.60 2.61 2.65 2.55 2.65
   2.02 2.32 2.04 2.99 2.98 2.39 2.63
   2.00 2.04 2.62 2.78 2.19 2.05

32. Let $X_1, X_2, \ldots, X_n$ are iid rvs from $U(\theta, 2\theta)$. Find a UMP test of size α to test $H_0 : \theta \leq \theta_0$ against $H_1 : \theta > \theta_0$.
   Further, test $H_0 : \theta \leq 4$ against $H_1 : \theta > 4$ for the following data. Let $\alpha = 0.05$.
   4.51 5.64 3.88 5.88 3.50 5.45 5.84
   5.54 3.52 5.38 4.16 4.14 3.75 3.96
   4.36 5.96 4.66 5.15 5.67 3.46

33. Let $X_1, X_2, \ldots, X_n$ are iid rvs from the following distribution as

$$f(x|\lambda) = \begin{cases} \frac{\lambda}{(1+x)^{\lambda+1}} & ; x > 0 \\ 0 & ; \text{otherwise} \end{cases}$$

Obtain a UMP test for testing $H_0 : \lambda \leq \lambda_0$ against $H_1 : \lambda > \lambda_0$.
   Further test $H_0 : \lambda \leq 1$ against $H_1 : \lambda > 1$ for the following data. Let $\alpha = 0.05$.
   1.10 0.32 0.14 0.23 0.20 0.05 1.48
   0.86 0.35 0.39 0.23 2.18 0.32 5.11
   7.77

34. Let the rv X_1 has exponential distribution with mean $\frac{1}{\theta}$ and the rv X_2 has $g(x_2|\theta)$,

$$g(x_2|\theta) = \theta x_2^{\theta-1}; \quad 0 < x < 1, \ \theta > 0$$

Obtain a MP test of size α for testing $H_0 : \theta = \theta_0$ against $H_1 : \theta_1 > \theta_0$. Can it be a UMP test?

35. Let the rv X_1 is $B(n, p)$ and the rv X_2 is $NB(r, \theta)$. Obtain a MP test of size α for testing $H_0 : \theta = 0.2$ against $H_1 : \theta_1 = 0.3$. Can it be a UMP test? Assume n and r are known. If $n = 5$ and $r = 3$, test the same hypothesis for the following data
   2, 1,11, 01, 5, 18, 12, 5.

36. Let the rvs $X_1, X_2, \ldots, X_n$ are $N(\mu_i, \sigma^2)$, $i = 1, 2, \ldots, n$, μ_i is known. Obtain a MP test of size α to test $H_0 : \sigma^2 = \sigma_0^2$ against $H_1 : \sigma^2 = \sigma_1^2 > \sigma_0^2$.

37. Let the rvs $X_1, X_2, \ldots, X_n$ are $N(\mu, \sigma_i^2)$, $i = 1, 2, \ldots, n$, σ_i are known. Obtain a MP test of size α to test $H_0 : \mu = \mu_0$ against $H_1 : \mu = \mu_1 < \mu_0$.

38. Let the rvs X_i has $P(i\lambda)$, $i = 1, 2, \ldots, n$. Obtain a MP test of size α to test $H_0 : \lambda = \lambda_0$ against $H_1 : \lambda = \lambda_1 < \lambda_0$.

39. Let the rvs $X_1, X_2, \ldots, X_n$ be iid rvs with $\cup(-k\theta, k\theta)$, $\theta > 0$, k is known. Obtain a MP test of size α for testing $H_0 : \theta = \theta_0$ against $H_1 : \theta_1 \neq \theta_0$.

40. Let the rvs $X_1, X_2, \ldots, X_n$ be iid rvs with (i) $\cup(-\theta, 0)$, $\theta > 0$ (ii) $\cup(0, \theta^2)$, $\theta > 0$ (iii) $\cup(\theta, \theta^2)$ $\theta > 0$.

   Obtain UMP test of size α for testing $H_0 : \theta = \theta_0$ against $H_1 : \theta \neq \theta_0$.

References

Gibbson JD, Pratt JW (1975) P value: interpretation and methodology. Am Stat 29(1):20–25

Lehman EL (1986) Testing statistical hypotheses, 2nd edn. Wiley, New York

Chapter 8
Unbiased and Other Tests

In earlier chapters, we have discussed the most powerful and UMP tests. Many times, we cannot get UMP tests for testing $H_0 : \theta = \theta_0$ against $H_1 : \theta \neq \theta_0$. Then how to get UMP tests? In the beginning, we will consider the extension of Neyman–Pearson Lemma to the cases where f_0 and f_1 may take negative values. They may not be necessarily densities but may satisfy some other conditions. We, therefore, discuss the maximization of $\int \phi f \, dx$ for some integrable function f over a class of critical functions ϕ, satisfying several other conditions. This extension is also known as generalized Neyman–Pearson Lemma. We will only state this Lemma with a brief proof. The detailed proof is given in Lehman (1986). Further, we will consider unbiased and other tests.

8.1 Generalized NP Lemma and UMPU Test

Theorem 8.1.1 *Suppose we have $(m+1)$ functions $g_0(x), g_1(x), \ldots, g_m(x)$ which are integrable and let $0 \le \phi(x) \le 1$ such that*

$$\int \phi(x) g_i(x) dx = c_i, i = 1, 2, \ldots, m \tag{8.1.1}$$

where c_i's are known constants.

Let $\phi_0(x)$ be a function such that

$$\phi_0(x) = \begin{cases} 1 \; ; \; g_0(x) > \sum_{i=1}^{m} k_i g_i(x) \\ 0 \; ; \text{otherwise} \end{cases}$$

then $\int \phi_0(x) g_0(x) dx \ge \int \phi(x) g_0(x) dx$
where $\phi_0(x)$ and $\phi(x)$ satisfy (8.1.1).

© Springer Science+Business Media Singapore 2016
U.J. Dixit, *Examples in Parametric Inference with R*,
DOI 10.1007/978-981-10-0889-4_8

Proof Consider

$$[\phi_0(x) - \phi(x)]\left[g_0(x) - \sum_{i=1}^{m} k_i g_i(x)\right] \geq 0 \qquad (8.1.2)$$

One can easily see that if $\phi_0(x) = 1$ then (8.1.1) is nonnegative.
Similarly if $\phi_0(x) = 0$ then also (8.1.1) is nonnegative.
From (8.1.2)

$$[\phi_0(x) - \phi(x)]\left[g_0(x) - \sum_{i=1}^{m} k_i g_i(x)\right] \geq 0$$

$$\Rightarrow \int [\phi_0(x) - \phi(x)] g_0(x) dx \geq \sum_{i=1}^{m} k_i \left[\int \phi_0(x) g_i(x) dx - \int \phi(x) g_i(x) dx\right]$$

From (8.1.1), $\sum_{i=1}^{m} k_i [c_i - c_i] = 0$

$$\Rightarrow \int [\phi_0(x) - \phi(x)] g_0(x) dx \geq 0$$

$$\Rightarrow \int \phi_0(x) g_0(x) dx \geq \int \phi(x) g_0(x) dx$$

Remark We can see the difference between NP lemma and its extension

　(i)　There is an equality in (8.1.1).
　(ii)　The functions $g_0(x), g_1(x), \ldots, g_m(x)$ need not be pdf.
　(iii)　k_i's need not be nonnegative.

Definition 8.1.1 A test $\phi(x)$ is called an unbiased test of size α if $E_{H_0}\phi(x) \leq \alpha$ or $\beta_\phi(\theta) \leq \alpha, \theta \in \Theta_0$ or $\sup_{\theta \in \Theta_0} \beta_\phi(\theta) = \alpha$, and $E_{H_1}\phi(x) \geq \alpha$ or $\beta_\phi(\theta) \geq \alpha, \theta \in \Theta_1$,

Construction of UMP unbiased (UMPU) test
Assume that $f(x|\theta)$ involves single parameter. In this case, we will find an UMPU test for testing $H_0 : \theta = \theta_0$ against $H_1 : \theta \neq \theta_0$ for a size of α.
From the Definition 8.1.1,

$$\beta_\phi(\theta) \leq \alpha, \theta \in \Theta_0 \qquad (i)$$

and

$$\beta_\phi(\theta) \geq \alpha, \theta \neq \theta_0 \qquad (ii)$$

Suppose that the power function $E_{H_1}\phi(x)$ is a continuous function of θ.
From (i) and (ii), $E_{H_0}\phi(x)$ has minimum at $\theta = \theta_0$.

Therefore, we want to find ϕ such that

$$\beta_\phi(\theta_0) = \alpha \qquad \text{(iii)}$$

and

$$\beta_\phi(\theta) > \alpha \text{ for } \theta \neq \theta_0 \qquad \text{(iv)}$$

If $\beta_\phi(\theta)$ is differentiable, then

$$\frac{d\beta_\phi(\theta)}{d\theta}|_{\theta=\theta_0} = 0 \Rightarrow \beta'_\phi(\theta_0) = 0 \qquad \text{(v)}$$

Class of test satisfying (i) and (ii) is a subset of a class of tests satisfying (i) and (v). Now we have to find a MP test satisfying (i) and (v). If the test is independent of θ, then it is UMPU.

Now our problem reduces to find a test ϕ such that

$$E_{H_0}\phi(X) = \int \phi(x)f(x|\theta_0)dx = \alpha \qquad \text{(vi)}$$

and if regularity conditions are satisfied then

$$\frac{dE_\theta\phi(x)}{d\theta}|_{\theta=\theta_0} = \frac{d\beta_\phi(\theta)}{d\theta}|_{\theta=\theta_0} = \int \frac{d}{d\theta}\phi(x)f(x|\theta_0)dx|_{\theta=\theta_0} = 0 \qquad \text{(vii)}$$

Now $\phi(x)$ maximizes the power $\int \phi(x)f(x|\theta_1)dx$ such that to find $\phi_0(x)$ that satisfies (vi) and (vii).

i.e., $\int \phi_0(x)f(x|\theta_1)dx \geq \int \phi(x)f(x|\theta_1)dx \; \forall \; \theta$

Using Theorem 8.1.1, i.e., GNP lemma, $g_0(x) = f(x|\theta_1)$, $g_1(x) = f(x|\theta_0)$ and $g_2(x) = \frac{df(x|\theta)}{d\theta}|_{\theta=\theta_0}$

$$\phi_0(x) = \begin{cases} 1 \; ; \; f(x|\theta_1) > k_1 f(x|\theta_0) + k_2[\frac{df(x|\theta)}{d\theta}]_{\theta=\theta_0} \\ 0 \; ; \; \text{otherwise} \end{cases}$$

$$\phi_0(x) = \begin{cases} 1 \; ; \; \frac{f(x|\theta_1)}{f(x|\theta_0)} > k_1 + k_2[\frac{d\log f(x|\theta)}{d\theta}]_{\theta=\theta_0} \\ 0 \; ; \; \text{otherwise} \end{cases} \qquad (8.1.3)$$

where k_1 and k_2 are such that $E_{H_0}\phi_0(x) = \alpha$ and $\frac{dE\phi_0(x)}{d\theta}|_{\theta=\theta_0} = 0$

Theorem 8.1.2 *Let the rv X has pdf(pmf) $f(x|\theta), \theta \in \Theta$. Assume that $f(x|\theta)$ belongs to a one parameter exponential family.*

$$f(x|\theta) = A(x)\exp[\theta T(x) + D(\theta)], \; x \in R, \theta \in \Theta$$

Then prove the test if

(i) U is continuous

$$\phi_0(x) = \begin{cases} 1 \; ; \; u < c_1 \text{ or } u > c_2 \\ 0 \; ; \; \text{otherwise} \end{cases} \tag{8.1.4}$$

(ii) U is discrete

$$\phi_0(x) = \begin{cases} 1 \; ; \; u < c_1 \text{ or } u > c_2 \\ \gamma_i \; ; \; u = c_i, i = 1, 2 \\ 0 \; ; \; \text{otherwise} \end{cases} \tag{8.1.5}$$

is UMPU of size α for testing $H_0 : \theta = \theta_0$ against $H_1 : \theta \neq \theta_0$, where $u = \sum_{i=1}^{n} T(x_i)$

Proof

$$f(x_1, x_2, \ldots, x_n | \theta) = \prod_{i=1}^{n} A(x_i) \exp\left[\theta \sum_{i=1}^{n} T(x_i) + nD(\theta)\right]$$

Using GNP and (8.1.3),

$$\phi_0(x) = \begin{cases} 1 \; ; \; if \; \frac{f_1(x|\theta_1)}{f_0(x|\theta_0)} > k_1 + k_2[\frac{d \log f(x|\theta)}{d\theta}]_{\theta=\theta_0} \\ 0 \; ; \; \text{otherwise} \end{cases} \tag{8.1.6}$$

$$\frac{f_1(x_1, x_2, \ldots, x_n | \theta_1)}{f_0(x_1, x_2, \ldots, x_n | \theta_0)} = \frac{\exp[\theta_1 \sum_{i=1}^{n} T(x_i) + nD(\theta_1)]}{\exp[\theta_0 \sum_{i=1}^{n} T(x_i) + nD(\theta_0)]}$$

$$= \exp\left[(\theta_1 - \theta_0) \sum_{i=1}^{n} T(x_i) + n\{D(\theta_1) - D(\theta_0)\}\right] \tag{8.1.7}$$

Next,

$$\log f(x|\theta) = \sum_{i=1}^{n} \log A(x_i) + \theta \sum_{i=1}^{n} T(x_i) + nD(\theta)$$

$$\frac{d \log f(x|\theta)}{d\theta} = \sum_{i=1}^{n} T(x_i) + nD'(\theta) \tag{8.1.8}$$

From (8.1.6), (8.1.7), and (8,1,8)

$$\phi_0(x) = \begin{cases} 1 \; ; & \exp\left[(\theta_1 - \theta_0) \sum_{i=1}^{n} T(x_i) + n\{D(\theta_1) - D(\theta_0)\}\right] > k_1 + k_2 \left[\sum_{i=1}^{n} T(x_i) + nD'(\theta_0)\right] \\ 0 \; ; & \exp\left[(\theta_1 - \theta_0) \sum_{i=1}^{n} T(x_i) + n\{D(\theta_1) - D(\theta_0)\}\right] \le k_1 + k_2 \left[\sum_{i=1}^{n} T(x_i) + nD'(\theta_0)\right] \end{cases}$$

$$= \begin{cases} 1 \; ; & \exp\left[(\theta_1 - \theta_0) \sum_{i=1}^{n} T(x_i)\right] > k_1^* + k_2^* \sum_{i=1}^{n} T(x_i) \\ 0 \; ; & \exp\left[(\theta_1 - \theta_0) \sum_{i=1}^{n} T(x_i)\right] \le k_1^* + k_2^* \sum_{i=1}^{n} T(x_i) \end{cases}$$

where

$$k_1^* = \frac{k_1}{n[D(\theta_1) - D(\theta_0)]} + \frac{k_2 n D'(\theta_0)}{n[D(\theta_1) - D(\theta_0)]}, k_2^* = \frac{k_2}{n[D(\theta_1) - D(\theta_0)]}$$

Let $h[\sum_{i=1}^{n} T(x_i)] = \exp[(\theta_1 - \theta_0) \sum T(x_i)] - k_2^* \sum_{i=1}^{n} T(x_i)$
Then

$$\phi_0(x) = \begin{cases} 1 \; ; & h\left[\sum_{i=1}^{n} T(x_i)\right] > k_1^* \\ 0 \; ; & h\left[\sum_{i=1}^{n} T(x_i)\right] \le k_1^* \end{cases}$$

Nature of $h\left[\sum_{i=1}^{n} T(x_i)\right]$
Let $U(x) = \sum_{i=1}^{n} T(x_i)$

$$h(u) = \exp[(\theta_1 - \theta_0)u(x)] - k_2^* u(x)$$

$$h'(u) = (\theta_1 - \theta_0) \exp[(\theta_1 - \theta_0)u(x)] - k_2^*$$

$$h''(u) = (\theta_1 - \theta_0)^2 \exp[(\theta_1 - \theta_0)u(x)] > 0; \quad \text{(see Fig. 8.1)}$$

This implies that h(u) is convex in U. If $h(u) > k_1^* \Rightarrow u < c_1$ or $u > c_2$ (Fig. 8.1).

Fig. 8.1 Graph of $h(u)$

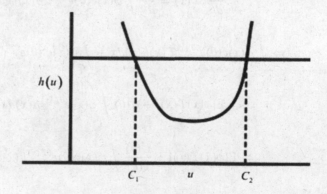

The UMPU test is given as

$$\phi_0(x) = \begin{cases} 1 & ; \ u < c_1 \ \text{or} \ u > c_2 \\ 0 & ; \ \text{otherwise} \end{cases}$$

If U is discrete

$$\phi_0(x) = \begin{cases} 1 & ; \ u < c_1 \ \text{or} \ u > c_2 \\ \gamma_i & ; \ u = c_i, \ i = 1, 2 \\ 0 & ; \ \text{otherwise} \end{cases}$$

$\gamma_i (i = 1, 2)$ can be determined such that $E\phi_0(x) = \alpha$ and $\frac{d}{d\theta}[P(u < c_1) + P(u > c_2)]_{\theta=\theta_0} = 0$.

Conclusion: For one parameter exponential family, we have seen how to obtain UMPU test for testing $H_0 : \theta = \theta_0$ against $H_1 : \theta \neq \theta_0$.

These conditions are as follows:

(i) $E\phi(x) = \alpha$
(ii) $\frac{d}{d\theta}E\phi(x)|_{\theta=\theta_0} = 0$

These conditions can be put in different form. Hence, we consider the following theorem:

Theorem 8.1.3 *Let* $f(x|\theta) = c(\theta)e^{\theta T(x)}h(x); \ x \in R, \theta \in \Theta$.
Prove that

$$E\phi(x)T(x) = \alpha ET(x), \tag{8.1.9}$$

where $\phi(x)$ is defined in (8.1.4) or (8.1.5)

Proof The test $\phi(x)$ defined in (8.1.4) or (8.1.5) satisfies (i) and (ii) defined in the conclusion.

$$\frac{d}{d\theta}E\phi(x) = \frac{d}{d\theta}\int \phi(x)c(\theta)e^{\theta T(x)}h(x)|_{\theta=\theta_0} = 0$$

$$\Rightarrow \int \phi(x)c(\theta)e^{\theta T(x)}T(x)h(x)dx + \int \phi(x)c'(\theta)e^{\theta T(x)}h(x)dx|_{\theta=\theta_0} = 0$$

$$\Rightarrow E[\phi(x)T(x)] + c'(\theta)\int \phi(x)e^{\theta T(x)}h(x)dx|_{\theta=\theta_0} = 0$$

$$\Rightarrow E[\phi(x)T(x)] + \frac{c'(\theta)}{c(\theta)}\int \phi(x)c(\theta)e^{\theta T(x)}h(x)dx|_{\theta=\theta_0} = 0$$

$$\Rightarrow E[\phi(x)T(x)] + \frac{c'(\theta)}{c(\theta)}E\phi(x) = 0 \tag{8.1.10}$$

This is true for all ϕ, which are unbiased.

Let $\phi(x) = \alpha$

$$\alpha E[T(x)] + \frac{c'(\theta)}{c(\theta)} \alpha = 0$$

Then

$$E[T(x)] = -\frac{c'(\theta)}{c(\theta)} \qquad (8.1.11)$$

From (8.1.10) and (8.1.11),

$$E[\phi(x)T(x)] = E[\phi(x)]E[T(x)]$$

Then we get the result as in (8.1.9),

$$E[\phi(x)T(x)] = \alpha E[T(x)]$$

Remark 1 We can find the constants c_1 and c_2 from (i) $E[\phi(x)] = \alpha$ and
(ii) $E[\phi(x)T(x)] = \alpha E[T(x)]$.

Remark 2 A simplification of the test is possible if for $\theta = \theta_0$, the distribution of T
is symmetric about some point a.
i.e.,

$$P_{\theta_0}[T < a - u] = P_{\theta_0}[T > a + u] \quad \forall u$$

Then any test which is symmetric about a and satisfies $E_{H_0}\phi(x) = \alpha$, i.e., it satisfies
(8.1.9).

  Let $\psi(t)$ be symmetric about a and $E\psi(t) = \alpha$, then we have to show that $\psi(t)$ is
unbiased, i.e., it satisfies (8.1.9).

$$\begin{aligned}
E_{H_0}\psi(t)T(x) &= E_{H_0}[(T - a)\psi(t) + a\psi(t)] \\
&= E_{H_0}(T - a)\psi(t) + aE_{H_0}\psi(t) \\
&= 0 + aE_{H_0}\psi(t)
\end{aligned}$$

(As T is symmetric about a then $ET = a$.)

$$= a\alpha = \alpha ET(x)$$

Hence it satisfies (8.1.9).
Therefore $\psi(t)$ is unbiased.

Remark 3 c_i's and γ_i's can be found out such that
$E_{H_0}\psi(t) = \alpha$

$$\Rightarrow P_{H_0}[T < c_1] + \gamma_1 P_{H_0}[T = c_1] = \frac{\alpha}{2}$$

and

$$\Rightarrow P_{H_0}[T > c_2] + \gamma_2 P_{H_0}[T = c_2] = \frac{\alpha}{2}$$

Further $\gamma_1 = \gamma_2$ and $a = \frac{c_1 + c_2}{2}$

Example 8.1.1 Let the rvs $X_1, X_2, \ldots X_n$ are iid rvs $N(\theta, 1)$. Find UMPU test for testing $H_0 : \theta = \theta_0$ against $H_1 : \theta \neq \theta_0$

$$f(x_1, x_2, \ldots, x_n | \theta) = (2\pi)^{-\frac{n}{2}} \exp\left[-\frac{1}{2}\sum_{i=1}^{n}(x_i - \theta)^2\right]$$

This belongs to a one parameter exponential family. Hence, from Theorem 8.1.2,
$U = \sum T(x_i) = \sum x_i$
The UMPU test is

$$\phi(x) = \begin{cases} 1 \; ; \; \sum x_i < c_1 \text{ or } \sum x_i > c_2 \\ 0 \; ; \text{ otherwise} \end{cases}$$

OR

$$\phi(x) = \begin{cases} 1 \; ; \; \bar{x} < c_1 \text{ or } \bar{x} > c_2 \\ 0 \; ; \text{ otherwise} \end{cases}$$

$\bar{X}$ is distributed as $N(\theta, \frac{1}{n})$
$E_{H_0}\phi(x) = \alpha$

$$P_{H_0}(\bar{X} < c_1) + P(\bar{X} > c_2) = \alpha$$

$$\Rightarrow \int_{-\infty}^{c_1} \frac{\sqrt{n}}{\sqrt{2\pi}} \exp\left[-\frac{n(\bar{x} - \theta_0)^2}{2}\right] d\bar{x} + \int_{c_2}^{-\infty} \frac{\sqrt{n}}{\sqrt{2\pi}} \exp\left[-\frac{n(\bar{x} - \theta_0)^2}{2}\right] d\bar{x} = \alpha$$

Next, $\frac{dE\phi(x)}{d\theta}\big|_{\theta=\theta_0} = 0$

$$\rightarrow \frac{d}{d\theta}\left[\int_{-\infty}^{c_1} \frac{\sqrt{n}}{\sqrt{2\pi}} \exp\left[-\frac{n(\bar{r}-\theta_0)^2}{2}\right] dr + \int_{c_2}^{-\infty} \frac{\sqrt{n}}{\sqrt{2\pi}} \exp\left[-\frac{n(\bar{r}-\theta_0)^2}{2}\right] dr\right]\Big|_{\theta=\theta_0} = 0$$

$$\Rightarrow \frac{d}{d\theta}\left[\int_{-\infty}^{\sqrt{n}(c_1-\theta)} \frac{e^{-\frac{t^2}{2}}}{\sqrt{2\pi}} dt + \int_{\sqrt{n}(c_2-\theta)}^{\infty} \frac{e^{-\frac{t^2}{2}}}{\sqrt{2\pi}} dt\right]\Big|_{\theta=\theta_0} = 0$$

$$\Rightarrow -\frac{\sqrt{n}}{\sqrt{2\pi}} \exp\left[-\frac{n(c_1-\theta_0)^2}{2}\right] + \frac{\sqrt{n}}{\sqrt{2\pi}} \exp\left[-\frac{n(c_2-\theta_0)^2}{2}\right] = 0$$

$$\Rightarrow \exp\left[-\frac{n(c_1-\theta_0)^2}{2}\right] = \exp\left[-\frac{n(c_2-\theta_0)^2}{2}\right]$$

$$\Rightarrow (c_1-\theta_0)^2 = (c_2-\theta_0)^2$$

$$\Rightarrow (c_1-\theta_0)^2 - (c_2-\theta_0)^2 = 0$$

$$\Rightarrow (c_1-c_2)(c_1+c_2-2\theta_0) = 0$$

$$\Rightarrow c_1 = c_2 \text{ and } c_1+c_2-2\theta_0 = 0$$

$$\Rightarrow c_1 = 2\theta_0 - c_2 \text{ or } c_2 = 2\theta_0 - c_1$$

Now $E\phi(x) = \alpha$ and $\sqrt{n}(c_2-\theta_0) = \sqrt{n}(2\theta_0-c_1-\theta_0) = \sqrt{n}(\theta_0-c_1)$
Hence,

$$\int_{-\infty}^{\sqrt{n}(c_1-\theta_0)} \frac{e^{-\frac{t^2}{2}}}{\sqrt{2\pi}} dt + \int_{\sqrt{n}(\theta_0-c_1)}^{\infty} \frac{e^{-\frac{t^2}{2}}}{\sqrt{2\pi}} dt = \alpha$$

$$\Rightarrow 2 \int_{-\infty}^{\sqrt{n}(c_1-\theta_0)} \frac{e^{-\frac{t^2}{2}}}{\sqrt{2\pi}} dt = \alpha$$

$$\Rightarrow \int_{-\infty}^{\sqrt{n}(c_1-\theta_0)} \frac{e^{-\frac{t^2}{2}}}{\sqrt{2\pi}} dt = \frac{\alpha}{2}$$

$$\Rightarrow \int_{-\sqrt{n}(c_1-\theta_0)}^{\infty} \frac{e^{-\frac{t^2}{2}}}{\sqrt{2\pi}} dt = \frac{\alpha}{2}$$

$$\Rightarrow -\sqrt{n}(c_1 - \theta_0) = Z_{\frac{\alpha}{2}}$$

$$\Rightarrow c_1 = \theta_0 - \frac{Z_{\frac{\alpha}{2}}}{\sqrt{n}} \text{ and } c_2 = \theta_0 + \frac{Z_{\frac{\alpha}{2}}}{\sqrt{n}}$$

Hence, the UMPU test is

$$\phi(x) = \begin{cases} 1 \; ; \; \bar{x} < \theta_0 - \frac{Z_{\frac{\alpha}{2}}}{\sqrt{n}} or \; \bar{x} > \theta_0 + \frac{Z_{\frac{\alpha}{2}}}{\sqrt{n}} \\ 0 \; ; \; \text{otherwise} \end{cases}$$

Example 8.1.2 Let the rvs $X_1, X_2, \ldots X_n$ are iid rvs $N(0, \sigma^2)$. Find UMPU test for testing $H_0 : \sigma = \sigma_0$ against $H_1 : \sigma \neq \sigma_0$

$$f(x_1, x_2, \ldots, x_n | \sigma^2) = \left(\frac{1}{\sigma\sqrt{2\pi}}\right)^n \exp\left[-\frac{\sum_{i=1}^n x_i^2}{2\sigma^2}\right]; \; x_i \in R, \; \sigma > 0$$

This belongs to a one parameter exponential family.
Hence, from Theorem 8.1.2, $\sum T(x_i) = \sum x_i^2$,
The UMPU test is

$$\phi(x) = \begin{cases} 1 \; ; \; \sum x_i^2 < c_1 \; \text{ or } \; \sum x_i^2 > c_2 \\ 0 \; ; \; \text{otherwise} \end{cases}$$

c_1 and c_2 are such that $E_{H_0}\phi(x) = \alpha$ and $\frac{dE\phi(x)}{d\sigma}|_{\sigma=\sigma_0} = 0$. Let $\sum x_i^2 = t$.
Now $\frac{\sum x_i^2}{\sigma_0^2} \sim \chi_n^2$, $E_{H_0}\phi(x) = \alpha$

$$\Rightarrow \int_0^{\frac{c_1}{\sigma_0^2}} \frac{e^{-\frac{t}{2}} t^{\frac{n}{2}-1}}{2^{\frac{n}{2}} \Gamma\frac{n}{2}} dt + \int_{\frac{c_2}{\sigma_0^2}}^{\infty} \frac{e^{-\frac{t}{2}} t^{\frac{n}{2}-1}}{2^{\frac{n}{2}} \Gamma\frac{n}{2}} dt = \alpha \qquad (8.1.12)$$

and $\frac{dE\phi(x)}{d\sigma}|_{\sigma=\sigma_0} = 0$

$$-\frac{2c_1}{\sigma_0^3} \frac{e^{-\frac{c_1}{2\sigma_0^2}} (\frac{c_1}{\sigma_0^2})^{\frac{n}{2}-1}}{2^{\frac{n}{2}} \Gamma\frac{n}{2}} + \frac{2c_2}{\sigma_0^3} \frac{e^{-\frac{c_2}{2\sigma_0^2}} (\frac{c_2}{\sigma_0^2})^{\frac{n}{2}-1}}{2^{\frac{n}{2}} \Gamma\frac{n}{2}} = 0$$

$$\Rightarrow -c_1 e^{-\frac{c_1}{2\sigma_0^2}} (c_1)^{\frac{n}{2}-1} + c_2 e^{-\frac{c_2}{2\sigma_0^2}} (c_2)^{\frac{n}{2}-1} = 0$$

$$\Rightarrow \exp\left[-\frac{c_1}{2\sigma_0^2} + \frac{c_2}{2\sigma_0^2}\right] = \left(\frac{c_2}{c_1}\right)^{\frac{n}{2}}$$

$$\Rightarrow -\frac{1}{2\sigma_0^2}[c_1 - c_2] = \frac{n}{2} \log \frac{c_2}{c_1}$$

$$\Rightarrow \log c_2 - \log c_1 = \frac{c_2 - c_1}{n\sigma_0^2} \tag{8.1.13}$$

$$\Rightarrow n\sigma_0^2 = \frac{c_2 - c_1}{\log c_2 - \log c_1} \tag{8.1.14}$$

c_1 and c_2 satisfying (8.1.12) and (8.1.14) are found by trial and error method. Let

$$\int_{-\infty}^{\frac{c_1}{\sigma_0^2}} f(t)dt = \alpha_1 \quad \text{and} \quad \int_{\frac{c_2}{\sigma_0^2}}^{\infty} f(t)dt = \alpha_2, \tag{8.1.15}$$

and $\alpha = \alpha_1 + \alpha_2$,
Then,

$$\frac{c_1}{\sigma_0^2} = \chi^2_{n,1-\alpha_1}, \ \frac{c_2}{\sigma_0^2} = \chi^2_{n,\alpha_2} \quad \text{and} \quad \frac{c_2 - c_1}{\log c_2 - \log c_1} = n\sigma_0^2,$$

Find c_1 and c_2 from these equations and start with $\alpha_1 = \frac{\alpha}{2}$.

Example 8.1.3 Consider $X \sim B(10, p)$ and assume $\alpha = 0.1$. To test $H_0 : p = 0.2$ against $H_1 : p \neq 0.2$. The UMPU test is given by

$$\phi(x) = \begin{cases} 1 & ; \ x < c_1 \text{ or } x > c_2 \\ \gamma_1 & ; \ x = c_1 \\ \gamma_2 & ; \ x = c_2 \\ 0 & ; \ \text{otherwise} \end{cases}$$

To find c_1, c_2, γ_1 and γ_2, we will use the following equations

$$E_{H_0} \phi(x) = \alpha$$

$$\Rightarrow \sum_{x=0}^{c_1-1} \binom{10}{x}(0.2)^x(0.8)^{10-x} + \sum_{c_2+1}^{10} \binom{10}{x}(0.2)^x(0.8)^{10-x}$$

$$+ \gamma_1 \binom{10}{c_1}(0.2)^{c_1}(0.8)^{10-c_1} + \gamma_2 \binom{10}{c_2}(0.2)^{c_2}(0.8)^{10-c_2} = 0.10 \quad (8.1.16)$$

Next,

$$E\phi(x)T(x) = \alpha ET(x) = \alpha n p_0$$

$$\sum_{x=0}^{c_1-1} \binom{9}{x-1}(0.2)^{x-1}(0.8)^{10-x} + \sum_{c_2+1}^{10} \binom{9}{x-1}(0.2)^{x-1}(0.8)^{10-x}$$

$$+ \gamma_1 \binom{9}{c_1-1}(0.2)^{c_1-1}(0.8)^{10-c_1} + \gamma_2 \binom{9}{c_2-1}(0.2)^{c_2-1}(0.8)^{10-c_2} = 0.10$$

$$\Rightarrow \sum_{y=0}^{c_1-2} \binom{9}{y}(0.2)^y(0.8)^{9-y} + \sum_{c_2}^{9} \binom{9}{y}(0.2)^y(0.8)^{9-y} + \gamma_1 \binom{9}{c_1-1}(0.2)^{c_1-1}(0.8)^{10-c_1}$$

$$+ \gamma_2 \binom{9}{c_2-1}(0.2)^{c_2-1}(0.8)^{10-c_2} = 0.10 \qquad\qquad (8.1.17)$$

	B(10,0.2)		B(9,0.2)
X	P(X=x)	X	P(X=x)
0	0.107374	0	0.134218
1	0.268435	1	0.301990
2	0.301990	2	0.301990
3	0.201327	3	0.176161
4	0.088080	4	0.066060
5	0.026424	5	0.016515
6	0.005505	6	0.002753
7	0.000786	7	0.000295
8	0.000074	8	0.000018
9	0.000004	9	0.000001
10	0.000000		

from (8.1.16) and (8.1.17), $c_1 = 0$ and $c_2 = 4$
From (8.1.16),

$$0 + 0.03279 + 0.107374\gamma_1 + 0.08808\gamma_2 = 0.10 \qquad (8.1.18)$$

From (8.1.17),

$$0 + 0.08564 + (0)\gamma_1 + 0.17616\gamma_2 = 0.10 \qquad (8.1.19)$$

Solving (8,1,18) and (8,1,19) simultaneously,
$\gamma_1 = 0.5591$ and $\gamma_2 - 0.08152$
The UMPU test is

$$\phi(x) = \begin{cases} 1 & ; \ x < 0 \text{ or } x > 4, \\ 0.5591 & ; \ x = 0, \\ 0.08152 & ; \ x = 4, \\ 0 & ; \text{ otherwise} \end{cases}$$

Example 8.1.4 Let the rvs $X_1, X_2, \ldots X_n$ are iid rvs with $B(k, p)$. Find UMPU test for testing $H_0 : p = p_0$ against $H_1 : p \neq p_0$

$$f(x_1, x_2, \ldots, x_n | p) = \prod_{i=1}^{n} \binom{k}{x_i} p^t q^{nk-t}; \ T = \sum_{i=1}^{n} x_i, \ q = 1 - p$$

$$= \prod_{i=1}^{n} \binom{k}{x_i} \left(\frac{p}{q}\right)^t q^{nk};$$

This belongs to one parameter exponential family because $\theta = \frac{p}{q}$, $\sum T(x_i) = \sum x_i$ (see, Theorem 8.1.2)
The UMPU test is

$$\phi(x) = \begin{cases} 1 & ; \ T < c_1 \text{ or } T > c_2, \\ \gamma_i & ; \ T = c_i, i = 1, 2 \\ 0 & ; \text{ otherwise} \end{cases}$$

c_i's and γ_i's $(i = 1, 2)$ are such that $E\phi(x) = \alpha$ and $\frac{dE\phi(x)}{dp} = 0$ or $E\phi(x)T(x) = \alpha ET(x)$
Now T is distributed as $B(nk, p)$.
$E_{H_0}\phi(x) = \alpha$,

$$\Rightarrow \sum_{r=0}^{c_1-1} \binom{nk}{r} p_0^r q_0^{nk-r} + \sum_{r=c_2+1}^{nk} \binom{nk}{r} p_0^r q_0^{nk-r} + \gamma_1 \binom{nk}{c_1} p_0^{c_1} q_0^{nk-c_1}$$

$$+ \gamma_2 \binom{nk}{c_2} p_0^{c_2} q_0^{nk-c_2} = \alpha \qquad (8.1.20)$$

Consider

$$A(c) = \sum_{r=c}^{n} \binom{n}{r} p^r q^{n-r}$$

$$\frac{dA(c)}{dp} = \sum_{r=c}^{n} \binom{n}{r} r p^{r-1} q^{n-r} - \sum_{r=c}^{n} \binom{n}{r} (n-r) p^r q^{n-r-1}$$

$$= n \sum_{r=c}^{n} \left[\binom{n-1}{r-1} p^{r-1} q^{n-r} - \binom{n-1}{r} p^r q^{n-r-1} \right]$$

Let

$$D_r = \binom{n-1}{r} p^r q^{n-r-1}$$

$$\frac{dA(c)}{dp} = n \sum_{r=c}^{n} [D_{r-1} - D_r]$$

$$= n \sum_{r=c}^{n} [D_{r-1} - D_r]$$

$$= n[D_{c-1} - D_n] \tag{8.1.21}$$

Note that $D_n = 0$ because $\binom{n-1}{n} = 0$

$$\frac{dA(c)}{dp} = n D_{c-1}$$

$$= n \binom{n-1}{c-1} p^{c-1} q^{n-c} \tag{8.1.22}$$

$$1 - A(c) = \sum_{r=0}^{c-1} \binom{n}{r} p^r q^{n-r}$$

$$\frac{d[1 - A(c)]}{dp} = -\frac{dA(c)}{dp} = -n \binom{n-1}{c-1} p^{c-1} q^{n-c} \tag{8.1.23}$$

From (8.1.20), use (8.1.23)

$$\frac{dE\phi(x)}{dp} = -nk \binom{nk-1}{c_1-1} p_0^{c_1-1} q_0^{nk-c_1} + nk \binom{nk-1}{c_2} p_0^{c_2} q_0^{nk-c_2-1}$$

$$+ \gamma_1 \binom{nk}{c_1} \{c_1 p_0^{c_1-1} q_0^{nk-c_1} - (nk-c_1) p_0^{c_1} q_0^{nk-c_1-1}\}$$

$$+ \gamma_2 \binom{nk}{c_2} \{c_2 p_0^{c_2-1} q_0^{nk-c_2} - (nk-c_2) p_0^{c_2} q_0^{nk-c_2-1}\} = 0 \tag{8.1.24}$$

To get c_1, c_2, γ_1 and γ_2 from (8.1.20) and (8.1.24), unique solution is not possible. A program in R is written for (8.1.20) and (8.1.24) to get c_1, c_2, γ_1 and γ_2 for $n = 10, k = 5, H_0 : p = 0.95$

```
library('rootSolve')
biump=function(n,k,p0,alpha){ eq=function(gamma){
c(f1=pbinom(c1-1,n*k,p0)+1-pbinom(c2,n*k,p0)+gamma[1]*dbinom(c1,n*k,p0)+gamma[2]*dbinom(c2,n*k,p0)-alpha,
f2=-n*k*dbinom(c1-1,n*k-1,p0)+n*k*dbinom(c2,n*k-1,p0)+
gamma[1]*choose(n*k,c1)*(c1*p0^(c1-1)*(1-p0)^(n*k-c1)-(n*k-c1)*p0^c1*(1-p0)^(n*k-c1-1))
+gamma[2]*choose(n*k,c2)*(c2*p0^(c2-1)*(1-p0)^(n*k-c2)-(n*k-c2)*p0^c2*(1-p0)^(n*k-c2-1)))}
for (c1 in 1:(n*k-1)){ for (c2 in (c1+1):(n*k))
kk=multiroot(f=eq,c(0,0))$root if (kk[1]>0 & kk[1]<=1 & kk[2]>0 &
kk[2]<=1)
{print(c('gamma1,gamma2=',kk));print(c('c1=',c1,'c2=',c2));break}} }
biump(n=10,k=2,p0=0.95,alpha=0.05)
biump(n=10,k=3,p0=0.95,alpha=0.05)
```

```
biump(n=10,k=4,p0=0.95,alpha=0.05)
```

```
biump(n=10,k=2,p0=0.95,alpha=0.01)
biump(n=10,k=3,p0=0.95,alpha=0.01)
biump(n=10,k=4,p0=0.95,alpha=0.01)
biump(n=10,k=5,p0=0.95,alpha=0.01)
```

The UMPU test is for $n = 10$, $p = 0.95$, $\alpha = 0.05$, $k = 2$

$$\phi(x) = \begin{cases} 1 & ; \ T < 16 \ \text{or} \ T > 20, \\ 0.6896 & ; \ T = 16 \\ 0.1067 & ; \ T = 20 \\ 0 & ; \ \text{otherwise} \end{cases}$$

Now using the condition

$$E\phi(x)T(x) = \alpha E T(x)$$

$$E\phi(x)T(x) = \sum_{r=0}^{c_1-1} r\binom{nk}{r} p_0^r q_0^{nk-r} + \sum_{r=c_2+1}^{nk} r\binom{nk}{r} p_0^r q_0^{nk-r} + \gamma_1 c_1 \binom{nk}{c_1} p_0^{c_1} q_0^{nk-c_1}$$

$$+ \gamma_2 c_2 \binom{nk}{c_2} p_0^{c_2} q_0^{nk-c_2} = \alpha nk p_0 = \alpha E T(x)$$

$$\Rightarrow \sum_{r=0}^{c_1} \binom{nk-1}{r-1} p_0^{r-1} q_0^{nk-r} + \sum_{r=c_2+1}^{nk} \binom{nk-1}{r-1} p_0^{r-1} q_0^{nk-r} + \gamma_1 \binom{nk-1}{c_1-1} p_0^{c_1-1} q_0^{nk-c_1}$$

$$+ \gamma_2 \binom{nk-1}{c_2-1} p_0^{c_2-1} q_0^{nk-c_2} = \alpha \tag{8.1.25}$$

From (8.1.24) and (8.1.25), $c_1, c_2 \, \gamma_1$ and γ_2 can be obtained. One has to use binomial tables.

As $n > 30$ and p tends to $\frac{1}{2}$, the distribution of

$$\frac{T - nkp_0}{\sqrt{nkp_0q_0}} \to N(0, 1).$$

Using normal tables, one can find c_1 and c_2. In this situation $\gamma_1 = \gamma_2 = 0$.

Now for sample sizes which are not too small and values of p_0 are not too close to 0 or 1, the distribution of T is approximately symmetric.

In this case, much simpler equal tails test gives a good approximation to the unbiased test.

c_1 and c_2 are determined so that

$$\sum_{r=0}^{c_1-1} \binom{nk}{r} p_0^r q_0^{nk-r} + \gamma_1 \binom{nk}{c_1} p_0^{c_1} q_0^{nk-c_1} = \frac{\alpha}{2} \tag{8.1.26}$$

$$\sum_{r=c_2+1}^{nk} \binom{nk}{r} p_0{}^r q_0{}^{nk-r} + \gamma_2 \binom{nk}{c_2} p_0{}^{c_2} q_0{}^{nk-c_2} = \frac{\alpha}{2} \qquad (8.1.27)$$

Similarly, program is written in R for (8.1.26) and (8.1.27) to get c_1, c_2, γ_1, and γ_2.

```
# Given data
 n = 10; p = 0.35; alpha = 0.05; k = 3; q = 1-p; m = n*k
# First equation
  # To find c1
    a <- seq(from=1,to=m,by=1) # Declaring possible values for c1.
    cdf = pbinom(a,m,p)-dbinom(a,m,p) # P(T < c1)
    ind = min(which(cdf < (alpha/2))) # Gives value of c1 such that P(T < c1) < alpha
    c1 = a[ind]
  # To find gamma1
    gam1 = -0.1 # Declaring gamma variable.
    while(gam1 < 0 || gam1 > 1)
      {
        gam1 = ((alpha/2)-pbinom(c1,m,p)+dbinom(c1,m,p))/dbinom(c1,m,p)
        c1_pre = c1; gam1_pre = gam1;
        c1 = c1+1;
      }
# Second equation
  b <- seq(from=0,to=(m-1),by=1) # Declaring possible values for c2.
  cdf = 1-pbinom(b,m,p) # P( T > c2)
  # To find c2 such that P(T > c2) < alpha/2
  if(cdf[m] < (alpha/2))
    {
      ind = min(which(cdf < (alpha/2))); c2 = a[ind];
    }
  if(cdf[m] > (alpha/2)) { c2 = m}
  # To find gamma2
  gam2 = -0.1 # Declaring gamma variable.
  while(gam2 < 0 || gam2 > 1)
    {
      gam2 = ((alpha/2)-1+pbinom(c2,m,p))/dbinom(c2,m,p)
      c2_pre = c2; gam2_pre = gam2;
      c2 = c2-1;
    }
# Assignment
  c1 = c1_pre; gamma1 = gam1_pre
  c2 = c2_pre; gamma2 = gam2_pre
# To check value
  alpha1 = pbinom(c1,m,p)-dbinom(c1,m,p)+(gamma1*dbinom(c1,m,p))
  alpha2 =  1-pbinom(c2,m,p)+(gamma2*dbinom(c2,m,p))
# OUTPUT
print(c("c1=",c1));
print(c("c2=",c2));
print(c("gamma1=",gamma1));
print(c("gamma2=",gamma2));
print("Check")
print(c("First equation",alpha1));
print(c("Second equation",alpha2));
# RESULT
 # OUTPUT
 "c1=" "6"
 "c2=" "16"
 "gamma1="              "0.0493338989789363"
 "gamma2="              "0.713138558357069"
 "Check"
 "First equation" "0.025"
 "Second equation" "0.025"
```

The UMPU test is for $n = 10$, $p = 0.35$, $\alpha = 0.05$, $k = 3$

$$\phi(x) = \begin{cases} 1 & ; \ T < 6 \ \text{or} \ T > 16, \\ 0.049 & ; \ T = 6, \\ 0.7131 & ; \ T = 16, \\ 0 & ; \ \text{otherwise} \end{cases}$$

Example 8.1.5 Let $X \sim P(\lambda)$ and assume $\alpha = 0.05$ To test $H_0 : \lambda = 2$ against $H_1 : \lambda \neq 2$

The UMPU test is given by

$$\phi(x) = \begin{cases} 1 & ; \ x < c_1 \ \text{or} \ x > c_2, \\ \gamma_1 & ; \ x = c_1, \\ \gamma_2 & ; \ x = c_2, \\ 0 & ; \ \text{otherwise} \end{cases}$$

To find c_1, c_2, γ_1 and γ_2, we will use the following equations:

$$E_{H_0}\phi(x) = \alpha$$

$$\Rightarrow \sum_{r=0}^{c_1-1} \frac{e^{-\lambda_0}(\lambda_0)^r}{r!} + \gamma_1 \frac{e^{-\lambda_0}(\lambda_0)^{c_1}}{c_1!} + \sum_{r=c_2+1}^{\infty} \frac{e^{-\lambda_0}(\lambda_0)^r}{r!} + \gamma_2 \frac{e^{-\lambda_0}(\lambda_0)^{c_2}}{c_2!} = \alpha$$

$$\sum_{r=0}^{c_1-1} \frac{e^{-2}2^r}{r!} + \gamma_1 \frac{e^{-2}2^{c_1}}{c_1!} + \sum_{r=c_2+1}^{\infty} \frac{e^{-2}2^r}{r!} + \gamma_2 \frac{e^{-2}2^{c_2}}{c_2!} = 0.05 \qquad (8.1.28)$$

Next $E_{H_0}[\phi(x)T(x)] = \alpha E_{H_0} T(x)$

$$\Rightarrow \sum_{s=0}^{c_1-2} \frac{e^{-\lambda_0}\lambda_0^s}{s!} + \gamma_1 \frac{e^{-\lambda_0}\lambda_0^{c_1}}{c_1!} + \sum_{s=c_2}^{\infty} \frac{e^{-\lambda_0}-\lambda_0^s}{s!} + \gamma_2 \frac{e^{-\lambda_0}\lambda_0^{c_2-1}}{c_2-1!} = \alpha$$

$$\sum_{s=0}^{c_1-1} \frac{e^{-2}2^s}{s!} + \gamma_1 \frac{e^{-2}2^{c_1}}{c_1!} + \sum_{s=c_2}^{\infty} \frac{e^{-2}2^s}{s!} + \gamma_2 \frac{e^{-2}2^{c_2-1}}{c_2-1!} = 0.05 \qquad (8.1.29)$$

Consider $c_1 = 0$ and $c_2 = 6$
From (8.1.28),

$$0 + (0.13534)\gamma_1 + (1 - 0.99547) + (0.01203)\gamma_2 = 0.05$$

$$(0.13534)\gamma_1 + (0.01203)\gamma_2 = 0.04547 \qquad (8.1.30)$$

	Poisson(2)	
X	P(X=x)	F(x)
0	0.135335	0.13534
1	0.270671	0.40601
2	0.270671	0.67668
3	0.180447	0.85712
4	0.090224	0.94735
5	0.036089	0.98344
6	0.012030	0.99547
7	0.003437	0.99890
8	0.000859	0.99976
9	0.000191	0.99995
10	0.000038	0.99999
11	0.000007	1.00000
12	0.000001	1.00000

From (8.1.29),

$$0 + (0.01656) + 0 + \gamma_2(0.03089) = 0.05 \qquad (8.1.31)$$

$\Rightarrow \gamma_2 = 0.9266$
From (8.1.30), $\gamma_1 = 0.2536$ The UMPU test is

$$\phi(x) = \begin{cases} 1 & ; \ x < 0 \ \text{or} \ x > 6, \\ 0.2536 & ; \ x = 0, \\ 0.9266 & ; \ x = 6, \\ 0 & ; \ \text{otherwise} \end{cases}$$

Example 8.1.6 Let the rvs $X_1, X_2, \ldots X_n$ are iid rvs with $P(\lambda)$. Find UMPU test for testing $H_0 : \lambda = \lambda_0$ against $H_1 : \lambda \neq \lambda_0$

$$f(x_1, x_2, \ldots, x_n | \lambda) = \frac{e^{-n\lambda} \lambda^t}{\prod_{i=1}^{n} x_i!}; \ x_i = 0, 1, 2 \ldots$$

This belongs to one parameter exponential family.
In this case $\sum_{i=1}^{n} T(x_i) = \sum_{i=1}^{n} x_i = T$. Now $t \sim P(n\lambda)$.
The UMPU test is

$$\phi(x) = \begin{cases} 1 & ; \ T < c_1 \ \text{or} \ T > c_2 \\ \gamma_i & ; \ T = c_i, i = 1, 2, \\ 0 & ; \ \text{otherwise} \end{cases} \qquad (8.1.32)$$

c_1, c_2, γ_1 and γ_2 are such that $E_{H_0}\phi(x) = \alpha$ and $ET(x)\phi(x) = \alpha ET(x)$
From $E_{H_0}\phi(x) = \alpha$

$$\Rightarrow \sum_{r=0}^{c_1-1} \frac{e^{-n\lambda_0}(n\lambda_0)^r}{r!} + \gamma_1 \frac{e^{-n\lambda_0}(n\lambda_0)^{c_1}}{c_1!} + \sum_{r=c_2+1}^{\infty} \frac{e^{-n\lambda_0}(n\lambda_0)^r}{r!}$$

$$+ \gamma_2 \frac{e^{-n\lambda_0}(n\lambda_0)^{c_2}}{c_2!} = \alpha \tag{8.1.33}$$

Consider

$$\Rightarrow \sum_{r=0}^{c_1-1} \frac{e^{-n\lambda_0}(n\lambda_0)^r}{r!} + \sum_{r=c_1}^{c_2} \frac{e^{-n\lambda_0}(n\lambda_0)^r}{r!} + \sum_{r=c_2+1}^{\infty} \frac{e^{-n\lambda_0}(n\lambda_0)^r}{r!} = 1 \tag{8.1.34}$$

$$\Rightarrow 1 - \sum_{c_1}^{c_2} \frac{e^{-n\lambda_0}(n\lambda_0)^r}{r!} = \sum_{r=0}^{c_1-1} \frac{e^{-n\lambda_0}(n\lambda_0)^r}{r!} + \sum_{r=c_2+1}^{\infty} \frac{e^{-n\lambda_0}(n\lambda_0)^r}{r!}$$

(8.1.33) becomes

$$\Rightarrow 1 - \sum_{r=c_1}^{c_2} \frac{e^{-n\lambda_0}(n\lambda_0)^r}{r!} + \gamma_1 \frac{e^{-n\lambda_0}(n\lambda_0)^{c_1}}{c_1!} + \gamma_2 \frac{e^{-n\lambda_0}(n\lambda_0)^{c_2}}{c_2!} = \alpha$$

$$\Rightarrow \sum_{r=c_1}^{c_2} \frac{e^{-n\lambda_0}(n\lambda_0)^r}{r!} - \gamma_1 \frac{e^{-n\lambda_0}(n\lambda_0)^{c_1}}{c_1!} - \gamma_2 \frac{e^{-n\lambda_0}(n\lambda_0)^{c_2}}{c_2!} = 1 - \alpha \tag{8.1.35}$$

From $ET(x)\phi(x) = \alpha ET(x)$

$$\Rightarrow \sum_{r=0}^{c_1-1} r\frac{e^{-n\lambda_0}(n\lambda_0)^r}{r!} + \sum_{r=c_2+1}^{\infty} r\frac{e^{-n\lambda_0}(n\lambda_0)^r}{r!} + c_1\gamma_1 \frac{e^{-n\lambda_0}(n\lambda_0)^{c_1}}{c_1!}$$

$$+ c_2\gamma_2 \frac{e^{-n\lambda_0}(n\lambda_0)^{c_2}}{c_2!} = \alpha n\lambda_0 \tag{8.1.36}$$

$$\Rightarrow \sum_{r=1}^{c_1-1} \frac{e^{-n\lambda_0}(n\lambda_0)^{r-1}}{(r-1)!} + \sum_{r=c_2+1}^{\infty} \frac{e^{-n\lambda_0}(n\lambda_0)^{r-1}}{(r-1)!} + \gamma_1 \frac{e^{-n\lambda_0}(n\lambda_0)^{c_1-1}}{(c_1-1)!}$$

$$+ \gamma_2 \frac{e^{-n\lambda_0}(n\lambda_0)^{c_2-1}}{(c_2-1)}! = \alpha$$

$$\Rightarrow \sum_{s=0}^{c_1-2} \frac{e^{-n\lambda_0}(n\lambda_0)^s}{s!} + \sum_{s=c_2}^{\infty} \frac{e^{-n\lambda_0}(n\lambda_0)^s}{s!} + \gamma_1 e^{-n\lambda_0} \frac{(n\lambda_0)^{c_1-1}}{(c_1-1)!}$$

$$+ \gamma_2 \frac{e^{-n\lambda_0}(n\lambda_0)^{c_2-1}}{(c_2-1)!} = \alpha \tag{8.1.37}$$

Now

$$\sum_{s=0}^{c_1-2} \frac{e^{-n\lambda_0}(n\lambda_0)^s}{s!} + \sum_{s=c_1-1}^{c_2-1} \frac{e^{-n\lambda_0}(n\lambda_0)^s}{s!} + \sum_{s=c_2}^{\infty} \frac{e^{-n\lambda_0}(n\lambda_0)^s}{s!} = 1 \quad (8.1.38)$$

Hence,

$$\sum_{s=0}^{c_1-2} e^{-n\lambda_0}\frac{(n\lambda_0)^s}{s!} + \sum_{s=c_2}^{\infty} \frac{e^{-n\lambda_0}(n\lambda_0)^s}{s!} = 1 - \sum_{s=c_1-1}^{c_2-1} \frac{e^{-n\lambda_0}(n\lambda_0)^s}{s!}$$

$$\Rightarrow 1 - \sum_{s=c_1-1}^{c_2-1} \frac{e^{-n\lambda_0}(n\lambda_0)^s}{s!} + \gamma_1\frac{e^{-n\lambda_0}(n\lambda_0)^{c_1-1}}{(c_1-1)!} + \gamma_2\frac{e^{-n\lambda_0}(n\lambda_0)^{c_2-1}}{(c_2-1)!} = \alpha$$

$$\Rightarrow \sum_{s=c_1-1}^{c_2-1} \frac{e^{-n\lambda_0}(n\lambda_0)^s}{s!} - \gamma_1\frac{e^{-n\lambda_0}(n\lambda_0)^{c_1-1}}{(c_1-1)!} - \gamma_2\frac{e^{-n\lambda_0}(n\lambda_0)^{c_2-1}}{(c_2-1)!} = 1 - \alpha$$

$$(8.1.39)$$

We have to solve the Eqs. (8.1.35) and (8.1.39) to get c_1, c_2, γ_1, and γ_2, but it is difficult to solve.

A program in R is written as c_1, c_2, γ_1 and γ_2 for $n = 10$, $\lambda = 8.2$, $\alpha = 0.05$

```
library('rootSolve')
poiump=function(n,lambda0,alpha,m){
eq=function(gamma){
c(f1=1-(ppois(c2,n*lambda0)-ppois(c1-1,n*lambda0))+gamma[1]*dpois(c1,n*lambda0)+gamma[2]*dpois(c2,n*lambda0)-alpha,
f2=(ppois(c2-1,n*lambda0)-ppois(c1-2,n*lambda0))-
gamma[1]*dpois(c1-1,n*lambda0)-gamma[2]*dpois(c2-1,n*lambda0)-1+alpha)}
for (c1 in 1:(m-1)){
for (c2 in (c1+1):(m)){
kk=multiroot(f=eq,c(0,0))$root
if (kk[1]>0 & kk[1]<=1 & kk[2]>0 & kk[2]<=1)
(print(c('gamma1,gamma2=',kk));print(c('c1=',c1,'c2=',c2));break}}
}
poiump(n=10,lambda0=8.2,alpha=0.05,m=100)
```

The UMPU test is

$$\phi(x) = \begin{cases} 1 & ; \ T < 65 \ \text{or} \ T > 100 \\ 0.3709 & ; \ T = 65, \\ 0.1007 & ; \ T = 100, \\ 0 & ; \ \text{otherwise} \end{cases}$$

As an approximation, we can use equal tail test,

$$\sum_{r=0}^{c_1-1} \frac{e^{-n\lambda_0}(n\lambda_0)^r}{r!} + \gamma_1\frac{e^{-n\lambda_0}(n\lambda_0)^{c_1}}{c_1!} = \frac{\alpha}{2} \quad (8.1.40)$$

$$\sum_{c_2+1}^{\infty} \frac{e^{-n\lambda_0}(n\lambda_0)^r}{r!} + \gamma_2 \frac{e^{-n\lambda_0}(n\lambda_0)^{c_2}}{c_2!} = \frac{\alpha}{2} \qquad (8.1.41)$$

Similarly, program is written in R for (8.1.40) and (8.1.41) to get c_1, c_2, γ_1 and γ_2.

```
find_c1 = function(a)
  {
    cdf = ppois(a,m)-dpois(a,m) # P(T < c1)
    if(cdf < (alpha/2)) { return(a)}
    if(cdf >= (alpha/2)) {print("some error")}
  }
find_gamma1 = function(a)
  {
    gam1 = ((alpha/2)-ppois(a,m)+dpois(a,m))/dpois(a,m);
    return(gam1)
  }
# Given data
  n = 10; lambda = 8.2; alpha = 0.05; m = n*lambda
# First equation
  a = 1; g1 = -0.1
  while(g1 < 0 || g1 > 1)
    {
      c1 = find_c1(a);
      g1 = find_gamma1(a);
      a = a+1;
    }
# Second equation
  b = 0 # Declaring possible values for c2.
  cdf = alpha
  while(cdf >= (alpha/2))
    {
    cdf = 1-ppois(b,m) # P( T > c2)
    b_pre = b
    b = b + 1
    }
  c2 = b_pre;
# To find gamma2
  gam2 = -0.1 # Declaring gamma variable.
  while(gam2 < 0 || gam2 > 1)
    {
      gam2 = ((alpha/2)-1+ppois(c2,m))/dpois(c2,m)
      c2_pre = c2; gam2_pre = gam2;
      c2 = c2+1;
    }
```

```
# Assignment
  gamma1 = g1
  c2 = c2_pre; gamma2 = gam2_pre
# To check value
  alpha1 = ppois(c1,m)-dpois(c1,m)+(gamma1*dpois(c1,m))
  alpha2 =  1-ppois(c2,m)+(gamma2*dpois(c2,m))
# OUTPUT
print(c("c1=",c1));
print(c("c2=",c2));
print(c("gamma1=",gamma1));
print(c("gamma2=",gamma2));
print("Check")
print(c("First equation",alpha1));
print(c("Second equation",alpha2));
# RESULT
  "c1=" "65"
  "c2=" "100"
  "gamma1="              "0.223352054419766"
  "gamma2="              "0.274038885245226"
  "Check"
  "First equation" "0.025"
  "Second equation" "0.025"
```

The UMPU test is for $n = 10$, $\lambda = 8.2$, $\alpha = 0.05$

$$\phi(x) = \begin{cases} 1 & ; \ T < 65 \text{ or } T > 100 \\ 0.22234 & ; T = 65, \\ 0.2740 & ; T = 100, \\ 0 & ; \text{otherwise} \end{cases}$$

For large n, we can use normal approximations

$$\frac{\sum X_i - n\lambda_0}{\sqrt{n\lambda_0}} \to N(0,1)$$

Hence

$$P_{H_0}(T < c_1) = \frac{\alpha}{2} \quad \text{and} \quad P(T > c_2) = \frac{\alpha}{2}$$

$$P\left(Z < \frac{c_1 - n\lambda_0}{\sqrt{n\lambda_0}}\right) = \frac{\alpha}{2} \quad \text{and} \quad P\left(Z > \frac{c_2 - n\lambda_0}{\sqrt{n\lambda_0}}\right) = \frac{\alpha}{2}$$

$$P\left(Z < \frac{c_1 - n\lambda_0}{\sqrt{n\lambda_0}}\right) = \frac{\alpha}{2} \quad \text{and} \quad P(Z < \frac{c_2 - n\lambda_0}{\sqrt{n\lambda_0}}) = 1 - \frac{\alpha}{2}$$

$$\rightarrow \frac{c_1 - n\lambda_0}{\sqrt{n\lambda_0}} = -Z_{\frac{\alpha}{2}} \quad \text{and} \quad \frac{c_2 - n\lambda_0}{\sqrt{n\lambda_0}} = Z_{\frac{\alpha}{2}}$$

$$\Rightarrow c_1 = n\lambda_0 - \sqrt{n\lambda_0}Z_{\frac{\alpha}{2}} \quad \text{and} \quad c_2 = n\lambda_0 + \sqrt{n\lambda_0}Z_{\frac{\alpha}{2}}$$

In this case, γ_1 and γ_2 is equal to zero.
Approximate UMPU test is

$$\phi(x) = \begin{cases} 1 \; ; \; T < n\lambda_0 - \sqrt{n\lambda_0}Z_{\frac{\alpha}{2}} \text{ or } T > n\lambda_0 + \sqrt{n\lambda_0}Z_{\frac{\alpha}{2}} \\ 0 \; ; \; \text{otherwise} \end{cases}$$

8.2 Locally Most Powerful Test (LMPT)

Sometimes, when UMP test does not exist, i.e., there is no single critical region which is the best for all alternatives, we may find regions which are best for alternatives close to null hypothesis and hope that such regions will do well for distant alternatives.

A locally most powerful test is the one which is most powerful in the neighborhood of the null hypothesis.

Let $H_0 : \theta = \theta_0$ against $H_1 : \theta > \theta_0$, i.e., $\theta = \theta_0 + \delta, \delta > 0$

Let ϕ is locally most powerful, then $E_{H_0}\phi(x) = \beta_\phi(\theta_0) = \alpha$ and $\beta_\phi(\theta) > \beta_{\phi^*}(\theta_0) \; \forall \; \theta$, where $\theta_0 \le \theta < \theta_0 + \delta \; \forall \; \delta > 0$

We have to maximize $\beta(\theta)$ in the interval $\theta_0 \le \theta < \theta_0 + \delta$.

Construction of LMP test

Expand $\beta_\phi(\theta)$ around θ_0 by Taylor Series expansions.

$$\beta_\phi(\theta) = \beta_\phi(\theta_0) + (\theta - \theta_0)\beta'_\phi(\theta_0) + o(\delta^2)$$

Maximizing $\beta_\phi(\theta)$ is equivalent to maximizing $\beta'_\phi(\theta)$, where

$$\beta'_\phi(\theta_0) = \frac{d}{d\theta} \int \phi(x) f(x|\theta) dx |_{\theta=\theta_0}$$

Assuming that differentiation under the integral sign holds.

Hence, we have to find ϕ such that it maximizes $\int \frac{d}{d\theta} f(x|\theta)\phi(x)dx$ subject to $\int \phi(x) f(x|\theta_0)dx = \alpha$.

Using extension of NP lemma, the test is given by

$$\phi(x) = \begin{cases} 1 \; ; \; \frac{df(x|\theta)}{d\theta}|_{\theta=\theta_0} > kf(x|\theta_0) \\ 0 \; ; \; \text{otherwise} \end{cases}$$

This test can be written as

$$\phi(x) = \begin{cases} 1 \; ; \; \frac{d \log f(x|\theta)}{d\theta}|_{\theta=\theta_0} > k \\ 0 \; ; \; \text{otherwise} \end{cases} \tag{8.2.1}$$

LMP test in a sample of size n

Suppose we have a random sample $X_1, X_2 < \dots, X_n$ from the pdf $f(x|\theta)$. Then LMPT for testing $H_0 : \theta = \theta_0$ against $H_1 : \theta > \theta_0$ is given as

$$\phi(x) = \begin{cases} 1 \; ; \; \sum \frac{d \log f(x|\theta)}{d\theta}|_{\theta=\theta_0} > k \\ 0 \; ; \; \text{otherwise} \end{cases} \tag{8.2.2}$$

Note: If $f(x|\theta)$ is such that

$$\mathrm{E}\left[\frac{d \log f(x|\theta)}{d\theta}\right]_{\theta=\theta_0} = 0$$

and

$$\mathrm{V}\left[\frac{d \log f(x|\theta)}{d\theta}\right]_{\theta=\theta_0} = \mathbf{I}(\theta_0),$$

where $\mathbf{I}(\theta_0)$ is a Fisher's Information.
Then for large n,

$$[n\mathbf{I}(\theta_0)]^{-\frac{1}{2}} \sum_{i=1}^{n} \frac{d \log f(x|\theta)}{d\theta}|_{\theta=\theta_0} \sim N(0, 1).$$

Hence, if Z_α is upper $\alpha\%$ value of $N(0, 1)$, then an approximate value of $K = Z_\alpha \sqrt{n\mathbf{I}(\theta_0)}$

Example 8.2.1 Let the rv X is $N(\theta, 1 + a\theta^2)$, $a > 0$ and known. Find the LMP test for testing $H_0 : \theta = 0$ against $H_1 : \theta > 0$

$$f(x|\theta) = \frac{1}{\sqrt{2\pi}\sqrt{1 + a\theta^2}} \exp\left[-\frac{1}{2}\frac{(x - \theta)^2}{(1 + a\theta^2)}\right]$$

$$\log f(x|\theta) = -\frac{1}{2}\log(1 + a\theta^2) - \frac{1}{2}\log 2\pi - \frac{(x - \theta)^2}{2(1 + a\theta^2)}$$

$$\frac{d \log f(x|\theta)}{d\theta} = \frac{-2a\theta}{2(1 + a\theta^2)} + \frac{(x - \theta)}{1 + a\theta^2} + \frac{a\theta(x - \theta)^2}{[1 + a\theta^2]^2}$$

Under H_0, i.e., $\theta = 0$, $\frac{d \log f(x|\theta)}{d\theta} = x$

In a sample of size n,

$$\sum_{i=1}^{n} \frac{d \log f(x_i|\theta)}{d\theta} = \sum_{i=1}^{n} x_i$$

The LMPT is

$$\phi(x) = \begin{cases} 1 \; ; \; \bar{X} > k \\ 0 \; ; \text{otherwise} \end{cases}$$

Now under H_0, $\bar{X} \sim N(0, \frac{1}{n})$

$$E\phi(x) = P(\bar{X} > k) = \alpha$$
$$= P(\sqrt{n}\bar{X} > k\sqrt{n}) = \alpha$$

Hence, $Z_\alpha = k\sqrt{n} \Rightarrow k = \frac{Z_\alpha}{\sqrt{n}}$

The LMPT is

$$\phi(x) = \begin{cases} 1 \; ; \; \bar{X} > \frac{Z_\alpha}{\sqrt{n}} \\ 0 \; ; \text{otherwise} \end{cases}$$

Construction of Locally Most Powerful Unbiased Test

If UMPU test does not exist, then we can find a test which is most powerfully unbiased in the neighborhood of θ_0. And then we have to take the alternatives very close to θ_0 and maximize the power locally.

We have to test $H_0 : \theta = \theta_0$ against $H_1 : \theta \neq \theta_0$

Determine a test ϕ such that

$$E_{H_0}\phi(x) = \alpha \; and \; E_{H_1}\phi(x) \geq \alpha \; \text{for} \; \theta \neq \theta_0 \tag{8.2.3}$$

It maximizes $E_\theta\phi(x)$ when $|\theta - \theta_0| < \delta$, when δ is very small.

Expand $E_\theta\phi(x)$ around θ_0

$$E_\theta\phi(x) = E_{\theta_0}\phi(x) + (\theta - \theta_0)\frac{dE\phi(x)}{d\theta}|_{\theta=\theta_0} + \frac{(\theta - \theta_0)^2}{2} \frac{d^2E\phi(x)}{d\theta^2}|_{\theta=\theta_0} + o(\delta^2)$$

From (8.2.3), $E_\theta\phi(x)$ has minimum at $\theta = \theta_0$. It implies that $\frac{dE\phi(x)}{d\theta}|_{\theta=\theta_0} = 0$

Maximizing $E\phi(x)$ when $|\theta - \theta_0| < \delta$ is equivalent to maximizing $\frac{d^2E\phi(x)}{d\theta^2}|_{\theta=\theta_0}$ subject to $E_{\theta_0}\phi(x) = \alpha$ and $\frac{dE\phi(x)}{d\theta}|_{\theta=\theta_0} = 0$

i.e., Maximize $\int \phi(x)\frac{d^2 f(x|\theta)}{d\theta^2}dx|_{\theta=\theta_0}$ subject to $\int \phi(x)f(x|\theta_0) = \alpha$

and $\int \phi(x)\frac{df(x|\theta)}{d\theta}dx|_{\theta=\theta_0} = 0$

Using Extension of NP lemma with $c_1 = \alpha, c_2 = 0, g_0 = \frac{d^2 f(x|\theta)}{d\theta^2}|_{\theta=\theta_0}, g_1 = f(x|\theta_0)$

and $g_2 = \frac{df(x|\theta)}{d\theta}|_{\theta=\theta_0}$

Hence,

$$\phi(x) = \begin{cases} 1 \; ; \; \frac{d^2 f(x|\theta)}{d\theta^2}|_{\theta=\theta_0} > k_1 f(x|\theta_0) + k_2 \frac{df(x|\theta)}{d\theta}|_{\theta=\theta_0} \\ 0 \; ; \; \text{otherwise} \end{cases}$$

Consider

$$\frac{d^2 f(x|\theta)}{d\theta^2}|_{\theta=\theta_0} > k_1 f(x|\theta_0) + k_2 \frac{df(x|\theta)}{d\theta}|_{\theta=\theta_0}$$

$$\Rightarrow \frac{1}{f(x|\theta_0)} \frac{d^2 f(x|\theta)}{d\theta^2}|_{\theta=\theta_0} > k_1 + \frac{k_2}{f(x|\theta_0)} \frac{df(x|\theta)}{d\theta}|_{\theta=\theta_0} \qquad (8.2.4)$$

Next,

$$\frac{df}{d\theta} = \frac{1}{f} \frac{df}{d\theta} f = \frac{d\log f}{d\theta} f$$

$$\frac{d^2 f}{d\theta^2} = \frac{d^2 \log f}{d\theta^2} f + \frac{d\log f}{d\theta} \frac{df}{d\theta}$$

$$= \frac{d^2 \log f}{d\theta^2} f + \frac{d\log f}{d\theta} \left(\frac{1}{f} \frac{df}{d\theta} f \right)$$

$$= \frac{d^2 \log f}{d\theta^2} f + \left(\frac{d\log f}{d\theta} \right)^2 f$$

$$\frac{1}{f(x|\theta)} \frac{d^2 f}{d\theta^2} = \frac{d^2 \log f(x|\theta)}{d\theta^2} + \left[\frac{d\log f(x|\theta)}{d\theta} \right]^2$$

(8.2.4) will become

$$\frac{d^2 \log f(x|\theta)}{d\theta^2}|_{\theta=\theta_0} + \left(\frac{d\log f(x|\theta)}{d\theta} \right)^2 |_{\theta=\theta_0} > k_1 + k_2 \frac{d\log f(x|\theta)}{d\theta}|_{\theta=\theta_0}$$

The LMPUT is

$$\phi(x) = \begin{cases} 1 \; ; \; \frac{d^2 \log f(x|\theta)}{d\theta^2}|_{\theta=\theta_0} + (\frac{d\log f(x|\theta)}{d\theta})^2|_{\theta=\theta_0} > k_1 + k_2 \frac{d\log f(x|\theta)}{d\theta}|_{\theta=\theta_0} \\ 0 \; ; \; \text{otherwise} \end{cases} \quad (8.2.5)$$

k_1 and k_2 are such that $E_{H_0} \phi(x) = \alpha$ and $\frac{dE\phi(x)}{d\theta}|_{\theta=\theta_0} = 0$

Example 8.2.2 Let $X_1, X_2, \ldots, X_n$ be iid with Cauchy distribution $C(\theta)$. Obtain LMPT for testing $H_0 : \theta = 0$ against $H_1 : \theta > 0$.

$$f(x|\theta) = \frac{1}{\pi} \frac{1}{1 + (x - \theta)^2} \; ; \quad -\infty < x < \infty, \; \theta > 0$$

$$\log f(x|\theta) = -\log \pi - \log[1 + (x - \theta)^2]$$

$$\frac{d \log f(x|\theta)}{d\theta} - \frac{2(x-\theta)}{[1+(x-\theta)^2]}$$

For a sample of size n,

$$\sum_{i=1}^{n} \frac{d \log f(x_i|\theta)}{d\theta}\Big|_{\theta=0} = 2 \sum_{i=1}^{n} \frac{x_i}{1+x_i^2}.$$

Using (8.2.2), LMPT is given as

$$\phi(x) = \begin{cases} 1 ; & \sum_{i=1}^{n} \frac{x_i}{1+x_i^2} > k \\ 0 ; & \text{otherwise} \end{cases}$$

k is chosen such that

$$E\phi(x) = \alpha \Rightarrow P\left[\sum_{i=1}^{n} \frac{x_i}{1+x_i^2} > k\right] = \alpha \qquad (8.2.6)$$

It is difficult to obtain the distribution of $\sum_{i=1}^{n} \frac{x_i}{1+x_i^2}$.

By applying CLT, the distribution of $\sum_{i=1}^{n} Y_i \sim AN(n\mu, n\sigma^2)$, where $Y_i = \frac{X_i}{1+X_i^2}$ and $EY_i = \mu$ and $V(Y_i) = \sigma^2$; $i = 1, 2, \ldots, n$

Now

$$EY_i^r = \int_{-\infty}^{\infty} \left(\frac{x_i}{1+x_i^2}\right)^r \frac{1}{\pi} \frac{dx_i}{1+x_i^2}$$

$$= \int_{-\infty}^{\infty} \frac{x_i^r}{\pi[1+x_i^2]^{r+1}} dx_i$$

If r is odd then $EY_i^r = 0$. Let r is even and $x_i = \tan\theta \Rightarrow dx_i = \sec^2\theta d\theta$
If $x_i = 0 \Rightarrow \tan\theta = 0 \Rightarrow \theta = 0$ and $x_i = \infty \Rightarrow \tan\theta = \infty \Rightarrow \theta = \frac{\pi}{2}$

$$E(Y_i^r) = \frac{2}{\pi} \int_0^{\frac{\pi}{2}} \frac{(\tan\theta)^r \sec^2(\theta)}{[\sec^2(\theta)]^{r+1}} d\theta$$

$$= \frac{2}{\pi} \int_0^{\frac{\pi}{2}} \frac{(\tan\theta)^r}{[\sec^2(\theta)]^r} d\theta$$

$$= \frac{2}{\pi} \int_0^{\frac{\pi}{2}} \sin^r(\theta) \cos^r(\theta) d\theta$$

We know that

$$\int_0^{\frac{\pi}{2}} \sin^{2m-1}(\theta) \cos^{2n-1}(\theta) d\theta = \frac{\beta(m, n)}{2}$$

$$= \frac{2}{\pi} \frac{\beta(\frac{r+1}{2}, \frac{r+1}{2})}{2} = \frac{\beta(\frac{r+1}{2}, \frac{r+1}{2})}{\pi}$$

Put r = 2, $E(Y_i^2) = \frac{1}{8}$ Hence, $E(Y_i) = 0$, $V(Y_i) = \frac{1}{8}$
From (8.2.6)

$$P\left[\sum Y_i > k\right] = P\left[\sqrt{\frac{8}{n}} \sum Y_i > \sqrt{\frac{8}{n}} k\right] = \alpha$$

$$Z_\alpha = \sqrt{\frac{8}{n}} k \Rightarrow k => \sqrt{\frac{n}{8}} Z_\alpha$$

The LMPT is

$$\phi(x) = \begin{cases} 1 \; ; \; \sum_{i=1}^n Y_i > \sqrt{\frac{n}{8}} Z_\alpha \\ 0 \; ; \; \text{otherwise} \end{cases}$$

Example 8.2.3 Let $X_1, X_2, \ldots, X_n$ be iid rv with Cauchy distribution $C(\theta)$. Obtain LMPU test for testing $H_0 : \theta = 0$ against $H_1 : \theta \neq 0$.

$$\sum_{i=1}^n \frac{d \log f(x_i|\theta)}{d\theta} \Big|_{\theta=0} = 2 \sum_{i=1}^n \frac{x_i}{(1 + x_i^2)}$$

$$\sum_{i=1}^n \frac{d^2 \log f(x_i|\theta)}{d\theta^2} \Big|_{\theta=0} = 4 \sum_{i=1}^n \frac{x_i^2}{(1 + x_i^2)^2} - 2 \sum_{i=1}^n \frac{1}{1 + x_i^2}$$

$$= 2 \sum_{i=1}^n \frac{x_i^2 - 1}{(1 + x_i^2)^2}$$

From (8.2.5), the LMPU test is

$$\phi(x) = \begin{cases} 1 \; ; \; 2\sum_{i=1}^n \frac{x_i^2-1}{(1+x_i^2)^2} + 4(\sum_{i=1}^n \frac{x_i}{1+x_i^2})^2 > k_1 + k_2 \sum_{i=1}^n \frac{2x_i}{1+x_i^2} \\ 0 \; ; \; \text{otherwise} \end{cases}$$

k_1 and k_2 are such that $E_{H_0}\phi(x) = \alpha$ and $\frac{dE\phi(x)}{d\theta}\Big|_{\theta=0} = 0$

Remark If UMPU test exists then it is LMPU but converse is not true.

Example 8.2.4 Let $X_1, X_2, \ldots, X_n$ be iid rvs with $N(\theta, 1)$. Show that UMPU test for testing $H_0 : \theta = 0$ against $H_1 : \theta \neq 0$ is also LMPU test.
We have already given an UMPU test in the Example 8.1.1.

The UMPU test is

$$\phi(x) = \begin{cases} 1 ; & \bar{x} < c_1 \text{ or } \bar{x} > c_2 \\ 0 ; & \text{otherwise} \end{cases}$$

Now we will find the LMPU test for testing $H_0 : \theta = 0$ against $H_1 : \theta \neq 0$

$$\log f(x_1, x_2, \ldots, x_n | \theta) = const - \frac{1}{2} \sum_{i=1}^{n} (x_i - \theta)^2$$

$$\frac{d \log f}{d\theta} = \sum_{i=1}^{n} (x_i - \theta)$$

$$\frac{d \log f}{d\theta} |_{\theta=0} = \sum_{i=1}^{n} x_i$$

$$\frac{d^2 \log f}{d\theta^2} |_{\theta=0} = -n$$

The LMPU test is given as

$$\phi(x) = \begin{cases} 1 ; & -n + (\sum_{i=1}^{n} x_i)^2 > k_1 + k_2 (\sum_{i=1}^{n} x_i) \\ 0 ; & \text{otherwise} \end{cases}$$

$$\Rightarrow \phi(x) = \begin{cases} 1 ; & n^2 \bar{x}^2 - n > k_1 + k_2 n \bar{x} \\ 0 ; & \text{otherwise} \end{cases}$$

Consider $n^2 \bar{x}^2 - n - k_1 - k_2 n \bar{x} > 0 \Rightarrow \bar{x}^2 - \frac{k_2 \bar{x}}{n} - \frac{k_1}{n^2} - \frac{1}{n} > 0$

$$\text{Let } t = \bar{x} \Rightarrow g(t) = t^2 - \frac{k_2 t}{n} - \left(\frac{k_1}{n^2} + \frac{1}{n} \right) > 0$$

$$\Rightarrow g(t) = t^2 + A_1 t + A_2 > 0,$$

where $A_1 = -\frac{k_2}{n}$ and $A_2 = -\left(\frac{k_1}{n^2} + \frac{1}{n} \right)$
$g'(t) = 2t - \frac{k_2}{n}$ $g''(t) = 2 > 0$
g(t) is convex in $t \Rightarrow t < c_1$ or $t > c_2$ Hence,

$$\phi(x) = \begin{cases} 1 ; & \bar{x} < c_1 \text{ or } \bar{x} > c_2 \\ 0 ; & \text{otherwise} \end{cases}$$

which is same as UMPU test.

Example 8.2.5 Let $X_1, X_2, \ldots, X_n$ be iid rvs with $N(0, \sigma^2)$. Find LMPU test for testing $H_0 : \sigma^2 = \sigma_0^2$ against $H_1 : \sigma^2 \neq \sigma_0^2$

Let $\sigma^2 = \theta$

$$f(x_1, x_2, \ldots, x_n | \sigma^2) = -\frac{n}{\sqrt{2\pi\theta}} \exp\left[-\frac{\sum x_i^2}{2\theta}\right]$$

$$\log f(x_1, x_2, \ldots, x_n | \sigma^2) = -\frac{n}{2} \log \theta - \frac{\sum x_i^2}{2\theta}$$

$$\frac{d \log f}{d\theta}\Big|_{\theta=\theta_0} = -\frac{n}{2\theta_0} + \frac{\sum x_i^2}{2\theta_0^2}$$

$$\frac{d^2 \log f}{d\theta^2}\Big|_{\theta=\theta_0} = \frac{n}{2\theta_0^2} - \frac{\sum x_i^2}{\theta_0^3}$$

$$\phi(x) = \begin{cases} 1 ; & \left(-\frac{\sum x_i^2}{\theta_0^3} + \frac{n}{2\theta_0^2}\right) + \left(\frac{\sum x_i^2}{2\theta_0} - \frac{n}{2\theta_0}\right)^2 > k_1 + k_2 \left(\frac{\sum x_i^2}{2\theta_0^2} - \frac{n}{2\theta_0}\right) \\ 0 ; & \text{otherwise} \end{cases}$$

Let $\frac{\sum x_i^2}{\theta_0} = v$

$$\phi(x) = \begin{cases} 1 ; & \left(-\frac{v}{\theta_0^2} + \frac{n}{2\theta_0^2}\right) + \left(\frac{v}{2\theta_0} - \frac{n}{2\theta_0}\right)^2 > k_1 + k_2 \left(\frac{v}{2\theta_0} - \frac{n}{2\theta_0}\right) \\ 0 ; & \text{otherwise} \end{cases}$$

Let

$$g(v) = \left(\frac{v-n}{2\theta_0}\right)^2 + \frac{n}{2\theta_0^2} - \frac{v}{\theta_0^2} - k_1 - \frac{k_2}{2\theta_0}(v-n)$$

$$g'(v) = 2\frac{(v-n)}{4\theta_0^2} - \frac{1}{\theta_0^2} - \frac{k_2}{2\theta_0}$$

$$g''(v) = \frac{1}{2\theta_0^2} > 0$$

g(v) is convex, then LMPU test is given as

$$\phi(x) = \begin{cases} 1 ; & \frac{\sum x_i^2}{\theta_0} < c_1 \text{ or } \frac{\sum x_i^2}{\theta_0} > c_2 \\ 0 ; & \text{otherwise} \end{cases}$$

c_1 and c_2 are such that $\mathrm{E}_{H_0}\phi(x) = \alpha$ and $\frac{d\mathrm{E}\phi(x)}{d\theta}\Big|_{\theta=\theta_0} = 0$
Further note that $\frac{\sum x_i^2}{\theta_0} \sim \chi_n^2 \Rightarrow c_1 = \chi_{n,1-\alpha}^2$ and $c_2 = \chi_{n,\alpha}^2$.

8.3 Similar Test

In many testing problems, the hypothesis is for a single parameter but the distribution of the observable random variables depends on more than one parameter. Therefore, we will obtain UMPU test for composite alternatives when pdf(pmf) involves more than one parameter.

Let $f(x|\theta)$ be a family of pdf with parameter θ, $\theta \in \Theta$, where Θ is a parametric space.

Suppose $H_0 : \theta_0 \in \Theta_0$, where Θ_0 contains more than one point, $H_1 : \theta \in \Theta - \Theta_0$

Let $\phi(x)$ be the test of level of significance α.

$$E_{H_0}\phi(x) \leq \alpha \; \forall \; \theta \in \Theta_0$$

$$E_{H_1}\phi(x) > \alpha \; \forall \; \theta \in \Theta - \Theta_0$$

Definition 8.3.1 A test $\phi(x)$ of significance level α is called a similar test for testing $H_0 : \theta \in \wedge$, where $\wedge$ is a subset of Θ if

$$E_{H_0}\phi(x) = \alpha \; \forall \; \theta \in \wedge,$$

where a set $\wedge$ in the parametric space Θ is called the boundary set of subsets of Θ_0 and $\Theta - \Theta_0$.

Further if $\theta \in \wedge$, then there are points in Θ_0 and $\Theta - \Theta_0$, which are arbitrarily close to θ.

For example, consider $N(\mu, \sigma^2)$ $\Theta = \{-\infty < \mu < \infty, \sigma > 0\}$

If $\Theta_0 : \mu = 0$ and $\Theta - \Theta_0 : \mu \neq 0$ then $\wedge : \mu = 0$.

Hence, $\wedge$ is called boundary of Θ_0 and $\Theta - \Theta_0$.

Theorem 8.3.1 *Let $\phi(x)$ be an unbiased test of level of significance α for testing $H_0 : \theta \in \Theta_0$ against $H_1 : \theta \in \Theta - \Theta_0$. Suppose that $E_{H_1}\phi(x)$ is a continuous function of θ then $\phi(x)$ is a similar test for testing $H_0 : \theta \in \wedge$ where $\wedge$ is a boundary of Θ_0 and $\Theta - \Theta_0$.*

Proof Let θ_0 be a point in $\wedge$. Assume that there exists a sequence θ_{1n} in Θ_0 such that

$$\lim_{n \to \infty} \theta_{1n} = \theta_0$$

Since $E_{H_1}\phi(x)$ is a continuous function of θ.

$$\lim_{n \to \infty} E_{\theta_{1n}}\phi(x) = E_\theta \phi(x)$$

$$\Rightarrow \beta_\phi(\theta_{1n}) \to \beta_\phi(\theta_0)$$

$$Since \ \beta_\phi(\theta_{1n}) \le \alpha \ \forall \, n, \theta_{1n} \in \Theta_0$$

$$\Rightarrow \beta_\phi(\theta_0) \le \alpha \tag{8.3.1}$$

There exist a sequence θ_{2n} in $\Theta - \Theta_0$ such that $\lim_{n\to\infty} \theta_{2n} = \theta_0$
Since ϕ is unbiased

$$E_{\theta_{2n}}\phi(x) \ge \alpha \Rightarrow \beta_\phi(\theta_{2n}) \ge \alpha$$

$$\Rightarrow E_{\theta_0}\phi(x) \ge \alpha \Rightarrow \beta_\phi(\theta_0) \ge \alpha \tag{8.3.2}$$

From (8.3.1) and (8.3.2), $E_{\theta_0}\phi(x) = \alpha$

$$\Rightarrow \beta_\phi(\theta_0) = \alpha \ \text{for} \ \theta_0 \in \wedge$$

$\Rightarrow \phi(x)$ is a similar test for testing $\underline{\theta} \in \wedge$.

Note: Thus, if $\beta_\phi(\theta)$ is continuous in θ for any ϕ- an unbiased test of size α of H_0 against H_1 is also α-similar for the pdf(pmf) of $\wedge$, i.e., for $\{f(x|\theta) : \theta \in \wedge\}$. We can find MP similar test of $H_0 : \theta \in \wedge$ against H_1 and if this test is unbiased of size α, then necessarily it is MP in the smaller class.

Definition 8.3.2 A test ϕ is UMP among all α-similar test on the boundary $\wedge$, is said to be a UMP α-similar test.

Remark 1 Let C_1 be the class of all unbiased test for testing $H_0 : \theta \in \Theta_0$ against $H_1 : \theta \in \Theta - \Theta_0$. Since $E_\theta\phi(x)$ is a continuous function. Similarly, let C_2 be the class of all similar test for testing $H_0 : \theta \in \wedge$ against $H_1 : \theta \in \Theta - \Theta_0$ then $C_1 \subseteq C_2$.

Remark 2 It is frequently easier to find a UMP α-similar test. Moreover, tests that are UMP similar on the boundary are often UMP unbiased.

Theorem 8.3.2 *Let The power function of the test ϕ of $H_0 : \theta \in \Theta_0$ against $H_1 : \theta \in \Theta - \Theta_0$ be continuous in θ, then a UMP α-similar test is UMP unbiased provided that its size is α for testing H_0 against H_1.*

Proof Let ϕ_0 be a UMP α-similar. Then

$$E_\theta\phi_0(x) \le \alpha \ \ \theta \in \Theta_0 \ \ (i)$$

Consider a trivial test $\phi(x) = \alpha$

$$\beta_{\phi_0}(\theta) \ge \beta_\phi(\theta) \ \ (Because \ \phi_0 \ is \ UMP)$$

$$\beta_{\phi_0}(\theta) \ge \alpha \ \ \theta \in \Theta - \Theta_0 \ \ (ii)$$

From (i) and (ii), ϕ_0 is also unbiased.

Remark 3 Hence, we see that class of all unbiased size α test is a subclass of the class of all α-similar tests

Remark 4 The continuity of power function $\beta_\phi(\theta)$ is not always easy to check. If the family $f(x|\theta)$ belongs to exponential family, then the power function is continuous.

In the following example, we will show that UMP α-similar test is unbiased.

Example 8.3.1 Let $X_1, X_2, \ldots, X_n$ be iid random sample from $N(\theta, 1)$, we wish to test $H_0 : \theta \leq 0$ against $H_1 : \theta > 0$

Since this family of densities has an MLR in $T = \sum_{i=1}^{n} x_i$. We can give a UMP test as

$$\phi(t) = \begin{cases} 1 \; ; T > k \\ 0 \; ; \text{otherwise} \end{cases}$$

Now we will find similar test

$$\Theta_0 = \{\theta \leq 0\}, \Theta - \Theta_0 = \{\theta > 0\}, \wedge = \{\theta = 0\}$$

Distribution of $T \sim N(n\theta, n)$, which is a one parameter exponential family, then the power function of any test ϕ based on T is continuous in θ. It follows that any unbiased size α test has the property, $\beta_\phi(\theta) = \alpha$ of similarity over $\wedge$.

Now we have to find a UMP test of $H_0 : \theta \in \wedge$ against $H_1 : \theta > 0$
By NP lemma,

$$\lambda(t) = \exp\left[\frac{t^2}{2n} - \frac{(t - n\theta)^2}{2n}\right] > k \Leftrightarrow T > k$$

The UMP test is given as

$$\phi(t) = \begin{cases} 1 \; ; T > k \\ 0 \; ; \text{otherwise} \end{cases}$$

where k is determined as

$$E_{H_0}\phi(t) = \alpha \Rightarrow P[T > k] = P\left[\frac{T}{\sqrt{n}} > \frac{k}{\sqrt{n}}\right] = \alpha$$

$$\frac{k}{\sqrt{n}} = Z_\alpha \Rightarrow k = \sqrt{n}Z_\alpha$$

Since ϕ is independent of H_1 as long as $\theta > 0$, we see that the test

$$\phi(t) = \begin{cases} 1 \; ; T > \sqrt{n}Z_\alpha \\ 0 \; ; \text{otherwise} \end{cases}$$

is UMP α-similar.
We need only to check that ϕ is of right size for testing H_0 against H_1.

We have $H_0 : \theta \leq 0$

$$E_{H_0}\phi(t) = P_\theta[T > \sqrt{n}Z_\alpha]$$

$$= P_\theta\left[\frac{T - n\theta}{\sqrt{n}} > \frac{\sqrt{n}Z_\alpha - n\theta}{\sqrt{n}}\right]$$

$$= P_\theta\left[\frac{T - n\theta}{\sqrt{n}} > Z_\alpha - \sqrt{n}\theta\right]$$

Since $\theta < 0 \Rightarrow Z_\alpha - \sqrt{n}\theta > 0 \Rightarrow Z_\alpha - \sqrt{n}\theta > Z_\alpha$
Hence

$$P\left[\frac{T - n\theta}{\sqrt{n}} > Z_\alpha - \sqrt{n}\theta\right] \leq P\left[\frac{T - n\theta}{\sqrt{n}} > Z_\alpha\right] = \alpha$$

$$\Rightarrow E_{H_0}\phi(t) \leq \alpha$$

Therefore, ϕ is UMP unbiased.

8.4 Neyman Structure Tests

In this case, we shall restrict ourselves to test which are similar for testing $H_0 : \theta \in \wedge$ against $H_1 : \theta \in \Theta - \Theta_0$, where $\wedge$ is a boundary of Θ_0 and $\Theta - \Theta_0$.

Definition 8.4.1 Let $T(x)$ be a sufficient statistics for $f(x|\theta)$, whenever $\theta \in \wedge$, then a test $\phi(X)$ is called a Neyman Structure test for testing $\theta \in \wedge$ if

$$E_{H_0}[\phi(X)|T(X) = t] = \alpha \ \forall \ t \ \text{and} \ \theta \in \wedge.$$

Theorem 8.4.1 *Every test $\phi(X)$ having Neyman Structure for $\theta \in \wedge$ is a similar test for $\theta \in \wedge$*

Proof Let $\phi(X)$ be Neyman Structure test and $T(X)$ be a sufficient statistics for $\theta \in \wedge$

$$E_{H_0}[\phi(X)|T(x)] = \alpha \qquad\qquad (i)$$

$$E_T\left\{E_{H_0}[\phi(X)|T(x)]\right\} = E_T(\alpha) = \alpha \ \forall \ t \ \text{and} \ \forall \ \theta \in \wedge.$$

$\Rightarrow$ It is a similar test.

One should note that a complete family is always boundedly complete but the converse is not true.
Note: See Theorem 1.5.2 and Example 1.5.10

Theorem 8.4.2 *Let $T(X)$ be boundedly complete sufficient statistics for $\theta \in \wedge$, then every similar test for testing $\theta \in \wedge$ has a Neyman Structure.*

Proof Let $\phi(X)$ be a similar test for testing $\theta \in \wedge$. Hence $E_{H_0}\phi(X) = \alpha \ \forall \ \theta \in \wedge$. We have to prove that $E[\phi(X)|T(x) = t] = \alpha$.

Let $\psi(X) = \phi(X) - \alpha$

$$E[\psi(X)] = 0 \ \forall \ \theta \in \wedge$$

Since $E[\psi(X)|T(X) = t]$ is independent of θ (Because $T(X)$ is sufficient)

$$E_T E[\psi(X)|T(x) = t] = 0 \ \forall \ \theta \in \wedge$$

Let $\psi^*(T) = E[\psi(X)|T(x) = t] \Rightarrow E\psi^*(T) = 0$

Now $\phi(X)$ lies between 0 and 1. $\psi(X)$ lies between $-\alpha$ and $(1 - \alpha)$. It implies that $\psi(X)$ is bounded. Hence, its expectation also lies between $-\alpha$ and $(1 - \alpha)$. $T(X)$ is boundedly complete which imply

$$\psi^*[T(x)] = 0 \ \forall \ T(X) \tag{i}$$

$$\begin{aligned} \psi^*[T(x)] &= E[\psi(X)|T(x) = t] \\ &= E[\phi(X) - \alpha|T(x) = t] \\ &= E[\phi(X)|T(x) = t] - \alpha \end{aligned}$$

From (i)

$$E[\phi(X)|T(x) = t] - \alpha = 0$$

$$\Rightarrow E[\phi(X)|T(x) = t] = \alpha$$

This implies that $\phi(X)$ has Neyman Structure.

Remark Let C_1=Class of unbiased test.
C_2= class of similar test $\forall \ \theta \in \wedge$
C_3=Class of Neyman Structure test $\forall \ \theta \in \wedge$.
$C_1 \subseteq C_2, C_3 \subseteq C_2$, and $C_2 \subseteq C_3 \Rightarrow$ every similar test is Neyman Structure provided that sufficient Statistics is bounded complete.
$\Rightarrow C_2 = C_3$; see Fig. 8.2.

Steps to Obtain Neyman Structure Test

(i) Find $\wedge$: boundary of Θ_0 and $\Theta - \Theta_0$
   Then test $H_0 : \theta \in \wedge$ against $H_1 : \theta \in \Theta - \Theta_0$
(ii) Find sufficient Statistics T(X) on $\wedge$

Fig. 8.2 Graphical presentation of relation between unbiased, similar and Neyman structure test

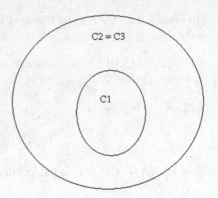

(iii) Show that T(X) is boundedly complete

(iv) Obtain the conditional pdf of $f(x|T(x))$ under H_0, i.e., for $\theta \in \wedge$ and $f(x|T(X))$ under H_1, i.e., $\theta \in \Theta - \Theta_0$.

(v) Obtain most powerful test using the theorem stated in Chap. 7.

Remark Suppose if we want to find UMPU test of size α for $H_0 : \theta \in \wedge$ against $H_1 : \theta \in \Theta - \Theta_0$, then using above steps find UMP Neyman Structure Test of size α for $H_0 : \theta \in \wedge$ against $H_1 : \theta \in \Theta - \Theta_0$.

If there exists a boundedly complete sufficient statistics, then this test is also UMP similar test for $\theta \in \wedge$.

If this test has to be unbiased, it should satisfy

$$E_\theta \phi(X) \leq \alpha \ \forall \ \theta \in \Theta_0$$

and $E_\theta \phi(X) \geq \alpha \ \forall \ \theta \in \Theta - \Theta_0$

Example 8.4.1 Let $X_1, X_2, \ldots, X_{k_1}$ and $Y_1, Y_2, \ldots, Y_{k_2}$ are iid rvs $B(n_1, p_1)$ and $B(n_2, p_2)$, respectively, where n_1 and n_2 are known. Find Neyman Structure test for testing $H_0 : p_1 = p_2$ against $H_1 : p_1 > p_2$.

In this case $\wedge : p_1 = p_2$

$$f(x, y|p_1, p_2) = \prod_{i=1}^{k_1} \binom{n_1}{x_i} p_1^{t_1} q_1^{n_1 k_1 - t_1} \prod_{i=1}^{k_2} \binom{n_2}{y_i} p_2^{t_2} q_1^{n_2 k_2 - t_2},$$

where, $q_1 = 1 - p_1, q_2 = 1 - p_2, T_1 = \sum_{i=1}^{k_1} x_i, T_2 = \sum_{j=1}^{k_2} y_j$
Now $T_1 \sim B(n_1 k_1, p_1)$ and $T_2 \sim B(n_2 k_2, p_2)$.
Then under H_0, $T_1 + T_2 \sim B(n_1 k_1 + n_2 k_2, p_1)$
$T_1 + T_2$ is sufficient and complete under H_0.

$$f_{T_0}(x\ y|T_1 + T_2 = t) = \frac{\prod_{i=1}^{k_1}\binom{n_1}{x_i}\prod_{j=1}^{k_2}\binom{n_2}{y_j}}{\binom{n_1k_1+n_2k_2}{t}}\ x_i = 0, 1, 2, \ldots, n_1, \ y_j = 0, 1, 2, \ldots, n_2,$$

$$i = 1, 2, \ldots, k_1, j = 1, 2, \ldots, k_2$$

Now to find the distribution of $T_1 + T_2 = t$ under H_1

$$f(t_1, t_2|p_1, p_2) = \binom{n_1k_1}{t_1}\binom{n_2k_2}{t_2}p_1^{t_1}q_1^{n_1k_1-t_1}p_2^{t_2}q_2^{n_2k_2-t_2},$$

Let $T_1 + T_2 = T$ and $T_2 = T - T_1$

$$f(t_1, t_2|p_1, p_2) = \binom{n_1k_1}{t_1}\binom{n_2k_2}{t-t_1}p_1^{t_1}q_1^{n_1k_1-t_1}p_2^{t-t_1}q_2^{n_2k_2-t+t_1},$$

$$= \binom{n_1k_1}{t_1}\binom{n_2k_2}{t-t_1}\left(\frac{p_1}{p_2}\right)^{t_1}\left(\frac{q_2}{q_1}\right)^{t_1}\left(\frac{p_2}{q_2}\right)^{t}q_1^{n_1k_1}q_2^{n_2k_2},$$

$$P[T = t] = q_1^{n_1k_1}q_2^{n_2k_2}\left(\frac{p_2}{q_2}\right)^{t}\sum_{t_1=0}^{\min(n_1k_1,t)}\binom{n_1k_1}{t_1}\binom{n_2k_2}{t-t_1}\left(\frac{p_1q_2}{p_2q_1}\right)^{t_1}$$

$$f(x, y|p_1, p_2) = \prod_{i=1}^{k_1}\binom{n_1}{x_i}\prod_{j=1}^{k_2}\binom{n_2}{y_j}p_1^{t_1}q_1^{n_1k_1-t_1}p_2^{t_2}q_2^{n_2k_2-t_2},$$

$$f(x, y|p_1, p_2) = \prod_{i=1}^{k_1}\binom{n_1}{x_i}\prod_{j=1}^{k_2}\binom{n_2}{y_j}p_1^{t_1}q_1^{n_1k_1-t_1}p_2^{t-t_1}q_2^{n_2k_2-t+t_1},$$

$$= \prod_{i=1}^{k_1}\binom{n_1}{x_i}\prod_{j=1}^{k_2}\binom{n_2}{y_j}q_1^{n_1k_1}q_2^{n_2k_2}\left(\frac{p_2}{q_2}\right)^{t}\left(\frac{p_1q_2}{p_2q_1}\right)^{t_1},$$

$$f_{H_1}(x, y|T_1 + T_2 = t) = \frac{\prod_{i=1}^{k_1}\binom{n_1}{x_i}\prod_{j=1}^{k_2}\binom{n_2}{y_j}(\frac{p_1q_2}{p_2q_1})^{t_1}}{\sum_{t_1=0}^{\min(n_1k_1,t)}\binom{n_1k_1}{t_1}\binom{n_2k_2}{t-t_1}(\frac{p_1q_2}{p_2q_1})^{t_1}},$$

$$\lambda(x, y|T) = \frac{f_{H_1}(x, y|T_1 + T_2 = t)}{f_{H_0}(x, y|T_1 + T_2 = t)}$$

$$= \frac{\binom{n_1k_1+n_2k_2}{t}(\frac{p_1q_2}{p_2q_1})^{t_1}}{\sum_{t_1=0}^{\min(n_1k_1,t)}\binom{n_1k_1}{t_1}\binom{n_2k_2}{t-t_1}(\frac{p_1q_2}{p_2q_1})^{t_1}}$$

Hence

$$
\phi(X, Y | T_1 + T_2 = t) = \begin{cases} 1 \; ; \; (\frac{p_1 q_2}{p_2 q_1})^{t_1} > k(t, p_1, p_2) \\ \gamma \; ; \; (\frac{p_1 q_2}{p_2 q_1})^{t_1} = k(t, p_1, p_2) \\ 0 \; ; \; (\frac{p_1 q_2}{p_2 q_1})^{t_1} < k(t, p_1, p_2) \end{cases}
$$

$$
\Rightarrow \phi(X, Y | T_1 + T_2 = t) = \begin{cases} 1 \; ; \; T_1 > k(t, p_1, p_2) \\ \gamma \; ; \; T_1 = k(t, p_1, p_2) \\ 0 \; ; \; T_1 < k(t, p_1, p_2) \end{cases}
$$

To find the distribution of T_1 given $T_1 + T_2 = t$ under H_0

$$
P_{H_0}[T_1 = t_1 | T_1 + T_2 = t] = \frac{P[T_1 = t_1, T_2 = t - t_1]}{P[T_1 + T_2 = t]}
$$

$$
= \frac{\binom{n_1 k_1}{t_1}\binom{n_2 k_2}{t - t_1}}{\binom{n_1 k_1 + n_2 k_2}{t}}, t_1 = 0, 1, 2, \ldots \min(n_1 k_1, t)
$$

$$
E_{H_0}[\phi(X, Y | T_1 + T_2 = t)] = \alpha
$$

$$
\Rightarrow P[T_1 > k(t, p_1, p_2) | T_1 + T_2 = t] + \gamma P[T_1 = k(t, p_1, p_2) | T_1 + T_2 = t] = \alpha
$$

$$
\Rightarrow \sum_{t_1 = k+1}^{\min(n_1 k_1, t)} \frac{\binom{n_1 k_1}{t_1}\binom{n_2 k_2}{t - t_1}}{\binom{n_1 k_1 + n_2 k_2}{t}} + \gamma \frac{\binom{n_1 k_1}{k}\binom{n_2 k_2}{t - k}}{\binom{n_1 k_1 + n_2 k_2}{t}} = \alpha
$$

This is a conditional test as it depends on t.

One should note that this test does not depend on p_1 and p_2. Hence this test is UMP similar for $p_1 > p_2$.

We have written a program in R to calculate k and γ.

```
# Given data
  x = c(1,1,2,3,2,2,1,1,0,2);
  y = c(3,3,3,2,3,2,1,3,2,3,1,3,3,1,2)
  alpha = 0.05; n1 = 4; n2 = 5;
  k1 = length(x); k2 = length(y); m <- n1*k1; N <- n1*k1+n2*k2
# To find k such that first term is < alpha
  t1 = sum(x); t2 <- sum(y); T = t1+t2;
  a = seq(from=0,to=min(m,T)-1,by=1); # possible values for k
  la = length(a)
# to find cumulative probability, i.e., P(t1 > k)
  cpk <- rep(0,la) # declaring variable to find cumulative probability.
  for(i in 1:la)
    {
      for(j in (a[i]+1):min(m,T))
        {
          cpk[i] = cpk[i] + dhyper(j,m,N-m,T);
        }
    }
```

```
ind = min(which(cpk < alpha))   # gives cumulative probability < alpha
# To find gamma
{cdflit(k) = ind-1, b <- dhyper(k,m,N-m,T);
gamma = (alpha-cpk[k+1])/b
# To check the answer
check <- cpk[k+1]+(gamma*dhyper(k,m,N-m,T))
# OUTPUT
print(c("k=",k))
print(c("gamma =",gamma))
print(c("check=",check))
# RESULT
"k=" "22"
"gamma ="                  "0.918666332250108"
"check="                   "0.05"
```

Example 8.4.2 Let $X_1, X_2, \ldots, X_n$ be a random sample from $N(\mu, \sigma^2)$, where μ is unknown. Find Neyman Structure test or UMP similar test for testing $H_0 : \sigma^2 = \sigma_0^2$ against $H_1 : \sigma^2 > \sigma_0^2$, where μ is unknown.

Now $\bar{x}$ and s^2 are jointly sufficient for (μ, σ^2), where $s^2 = \sum_{i=1}^{n}(x_i - \bar{x})^2$. Further, $\bar{x}$ and s^2 are independent random variables.

$$f(\bar{x}, s^2) = \frac{n}{\sigma\sqrt{2\pi}} \exp\left[-\frac{n(\bar{x}-\mu)^2}{2\sigma^2}\right] \frac{e^{-\frac{s^2}{2\sigma^2}}(\frac{s^2}{\sigma^2})^{\frac{n-1}{2}-1}}{2^{\frac{n-1}{2}}\Gamma(\frac{n-1}{2})} \frac{1}{\sigma^2}$$

$$= \frac{c \exp[-\frac{n}{2\sigma^2}(\bar{x}-\mu)^2](s^2)^{\frac{n-1}{2}-1}\exp[-\frac{s^2}{2\sigma^2}]}{\sigma^n}$$

where c is constant.
Under $H_0 : \bar{x} \sim N(\mu, \frac{\sigma_0^2}{n})$

$$f_{H_0}(\bar{x}, s^2|\bar{x}) = \frac{c_1}{\sigma_0^{n-1}} \exp\left[-\frac{s^2}{2\sigma_0^2}\right](s^2)^{\frac{n-1}{2}-1},$$

where c_1 is constant.
Under $H_1, \bar{x} \sim N(\mu, \frac{\sigma^2}{n}), \quad \sigma^2 > \sigma_0^2$

$$f_{H_1}(\bar{x}, s^2|\bar{x}) = \frac{c\sigma^{-n}\exp[-\frac{1}{2\sigma^2}\{n(\bar{x}-\mu)^2+s^2\}](s^2)^{\frac{n-1}{2}-1}}{c_1\sigma^{-1}\exp[-\frac{1}{2\sigma^2}\{n(\bar{x}-\mu)^2\}]}$$

$$= c_2\sigma^{-(n-1)}(s^2)^{\frac{n-1}{2}-1}\exp\left[-\frac{s^2}{2\sigma^2}\right],$$

$$\lambda(\bar{x}, s^2|\bar{X}) = \frac{f_{H_1}}{f_{H_0}} = c_2\exp\left[-\frac{s^2}{2}\left(\frac{1}{\sigma^2}-\frac{1}{\sigma_0^2}\right)\right],$$

where c_2 is function of (σ^2, σ_0^2)

$$\Rightarrow \lambda(\bar{x}, s^2 | \bar{x}) > k$$

$$\Rightarrow s^2 > k(\sigma, \sigma_0)$$

Neyman Structure Test is given as

$$\phi(\bar{x}, s^2 | \bar{x}) = \begin{cases} 1 \; ; \; s^2 > k(\sigma, \sigma_0) \\ 0 \; ; \; s^2 \leq k(\sigma, \sigma_0) \end{cases}$$

$k(\sigma, \sigma_0)$ is determined as

$$P_{H_0}[s^2 > k(\sigma, \sigma_0) | \bar{x}] = \alpha$$

$$\Rightarrow P_{H_0}\left[\frac{s^2}{\sigma_0^2} > k | \bar{x}\right] = \alpha$$

Now though it is a conditional probability but one can write it out as

$$\Rightarrow P_{H_0}\left[\frac{s^2}{\sigma_0^2} > k\right] = \alpha,$$

because $\bar{x}$ and s^2 are independent.

$$\frac{s^2}{\sigma_0^2} \sim \chi_{n-1}^2 \Rightarrow k = \chi_{n-1,\alpha}^2$$

The test is

$$\Rightarrow \phi(\bar{x}, s^2 | \bar{x}) = \begin{cases} 1 \; ; \; \frac{s^2}{\sigma_0^2} > \chi_{n-1,\alpha}^2 \\ 0 \; ; \; \text{otherwise} \end{cases}$$

Note: This test is UMP similar test.

Example 8.4.3 Let $X_1, X_2, \ldots, X_n$ be a random sample from $N(\mu, \sigma^2)$, where σ^2 is unknown. Find Neyman Structure test for testing $H_0 : \mu = \mu_0$, σ^2 unknown against $H_1 : \mu > \mu_0$

Now $\bar{x}$ and s^2 are jointly sufficient for (μ, σ^2)

$$f(\bar{x}, s^2) = \frac{c}{\sigma^n} \exp\left[-\frac{1}{2\sigma^2}\{n(\bar{x} - \mu)^2 + s^2\}\right](s^2)^{\frac{n-1}{2}-1},$$

where c is constant, $s^2 = \sum_{i=1}^{n}(x_i - \bar{x})^2$.

$$\frac{\sqrt{n}(\bar{x} - \mu_0)}{\psi} \sim N(0, 1) \text{ and } \frac{s^2}{\sigma^2} \sim \chi^2_{n-1}$$

$$\Rightarrow \frac{n(\bar{x} - \mu_0)^2 + s^2}{\sigma^2} \sim \chi^2_n$$

Let $w = n(\bar{x} - \mu_0)^2 + s^2$ is a sufficient statistics for $N(\mu, \sigma^2)$.

$$f_{H_0}(\bar{x}, s^2 | w) = \frac{c\sigma^{-n} \exp(-\frac{w}{2\sigma^2})[w - n(\bar{x} - \mu_0)^2]^{\frac{n-1}{2} - 1}}{c_1 \sigma^{-n} \exp(-\frac{w}{2\sigma^2}) w^{\frac{n}{2} - 1}},$$

$$= \frac{c_2 [w - n(\bar{x} - \mu_0)^2]^{\frac{n-1}{2} - 1}}{w^{\frac{n}{2} - 1}}$$

Now to find $f_{H_1}(\bar{x}, s^2 | w)$

Let $\mu_1 = \mu - \mu_0$ and $(\bar{x} - \mu_0) \sim N(\mu_1, \frac{\sigma^2}{n})$

$\Rightarrow \sqrt{n}(\bar{x} - \mu_0) \sim N(\sqrt{n}\mu_1, \sigma^2)$,

Let $v = \sqrt{n}(\bar{x} - \mu_0)$

Consider

$$w = s^2 + [n(\bar{x} - \mu_0)]^2$$
$$= s^2 + v^2$$

$$f(s^2, v) = \frac{(\frac{s^2}{\sigma^2})^{\frac{n-1}{2} - 1} \exp(-\frac{s^2}{2\sigma^2})}{2^{\frac{n-1}{2}} \Gamma(\frac{n-1}{2}) \sigma^2} \frac{\exp[-\frac{1}{2\sigma^2}(v - \sqrt{n}\mu_1)^2]}{\sigma\sqrt{2\pi}}$$

$$= \frac{c}{\sigma^n} (s^2)^{\frac{n-1}{2} - 1} \exp\left[-\frac{1}{2\sigma^2} \{s^2 + (v - \sqrt{n}\mu_1)^2\} \right]$$

We have to find a joint distribution of s^2 and w.

Since $w = s^2 + v^2 \Rightarrow v = \sqrt{w - s^2}$ and $s^2 = s^2$

$$J = \frac{\partial(v, s^2)}{\partial(w, s^2)} = \begin{pmatrix} \frac{\partial v}{\partial w} & \frac{\partial v}{\partial s^2} \\ \frac{\partial s^2}{\partial w} & \frac{\partial s^2}{\partial s^2} \end{pmatrix} = \begin{pmatrix} \frac{1}{2}(w - s^2)^{-\frac{1}{2}} & -\frac{1}{2}(w - s^2)^{-\frac{1}{2}} \\ 0 & 1 \end{pmatrix}$$

$$= \frac{1}{2}(w - s^2)^{-\frac{1}{2}}$$

$$|J| = \frac{1}{2}(w - s^2)^{-\frac{1}{2}}$$

$$f(s^2, w) = \frac{c}{\sigma^n}(s^2)^{\frac{n-1}{2} - 1} \exp\left[-\frac{1}{2\sigma^2}\{s^2 + (v - \sqrt{n}\mu_1)^2\} \right] |J|$$

$$f(s^2, w) = \frac{c}{\sigma^n}(s^2)^{\frac{n-1}{2}-1}\exp\left[-\frac{1}{2\sigma^2}\{s^2 + w - s^2 - 2\sqrt{n}\mu_1\sqrt{w-s^2} + n\mu_1^2\}\right]|J|$$

$$= \frac{c}{\sigma^n}(s^2)^{\frac{n-1}{2}-1}\exp\left[-\frac{1}{2\sigma^2}\{w - 2\sqrt{n}(w-s^2)^{\frac{1}{2}}\mu_1 + n\mu_1^2\}\right]\frac{1}{2}\left(w-s^2\right)^{-\frac{1}{2}}$$

$$v^2 > 0 \Rightarrow w - s^2 > 0 \Rightarrow s^2 < w$$

$$f(w) = \int_0^w \frac{c}{\sigma^n}(s^2)^{\frac{n-1}{2}-1}\exp\left[-\frac{1}{2\sigma^2}\{w - 2\sqrt{n}(w-s^2)^{\frac{1}{2}}\mu_1 + n\mu_1^2\}\right]\frac{1}{2}\left(w-s^2\right)^{-\frac{1}{2}}ds^2$$

Let $\frac{s^2}{w} = u \Rightarrow ds^2 = wdu$

$$= \frac{c}{\sigma^n}\int_0^1 (wu)^{\frac{n-1}{2}-1}\exp\left[-\frac{w}{2\sigma^2}\right]\exp[-\frac{\sqrt{n}w^{\frac{1}{2}}(1-u)^{\frac{1}{2}}\mu_1}{\sigma^2} - \frac{n\mu_1^2}{2\sigma^2}]w^{-\frac{1}{2}}(1-u)^{-\frac{1}{2}}wdu$$

$$= \frac{c}{\sigma^n}w^{\frac{n}{2}-1}\exp\left[-\frac{w}{2\sigma^2}\right]\int_0^1 u^{\frac{n-1}{2}-1}(1-u)^{-\frac{1}{2}}\exp[-\frac{\sqrt{n}w^{\frac{1}{2}}(1-u)^{\frac{1}{2}}\mu_1}{\sigma^2} - \frac{n\mu_1^2}{2\sigma^2}]du$$

$$= \frac{c}{\sigma^n}w^{\frac{n}{2}-1}\exp\left[-\frac{w}{2\sigma^2}\right]g(w, \mu_1, \sigma^2)$$

$$f_{H_1}(\bar{x}, s^2|w) = \frac{c\sigma^{-n}\exp[-\frac{1}{2\sigma^2}\{n(\bar{x}-\mu)^2 + s^2\}][w - n(\bar{x}-\mu_0)^2]^{\frac{n-1}{2}-1}}{c\sigma^{-n}w^{\frac{n}{2}-1}\exp[-\frac{w}{2\sigma^2}]g(w, \mu_1, \sigma^2)}$$

Consider

$$\exp\left[-\frac{1}{2\sigma^2}\{n(\bar{x}-\mu_0+\mu_0-\mu)^2 + s^2\}\right]$$

$$= \exp\left[-\frac{1}{2\sigma^2}\{n(\bar{x}-\mu_0)^2 + s^2 - 2n(\bar{x}-\mu_0)\mu_1 + n\mu_1^2\}\right]$$

$$= \exp\left[-\frac{1}{2\sigma^2}\{w - 2n\bar{x}\mu_1 + 2n\mu_0\mu_1 + n\mu_1^2\}\right]$$

$$= \exp(-\frac{w}{2\sigma^2})\exp[-\frac{1}{2\sigma^2}\{-2n\bar{x}\mu_1 + 2n\mu_0\mu_1 + n\mu_1^2\}]$$

$$f_{H_1}(\bar{x}, s^2|w) = \frac{[w - n(\bar{x}-\mu_0)^2]^{\frac{n-1}{2}-1}\exp[-\frac{1}{2\sigma^2}\{-2n\bar{x}\mu_1 + 2n\mu_0\mu_1 + n\mu_1^2\}]}{w^{\frac{n}{2}-1}g(w, \mu_1, \sigma^2)}$$

$$\lambda(\bar{x}, s^2|w) = \frac{f_{H_1}(\bar{x}, s^2|W)}{f_{H_0}(\bar{X}, s^2|W)} = const.\exp[-\frac{1}{2\sigma^2}(-2n\bar{x}\mu_1)] \qquad (8.4.1)$$

$$= const. \exp\left[\frac{n\bar{x}\mu_1}{\sigma^2}\right] \tag{8.4.2}$$

$$= const. \exp\left[\frac{n\bar{x}(\mu - \mu_0)}{\sigma^2}\right] \tag{8.4.3}$$

The Neyman Structure test is

$$\phi(\bar{x}, s^2|w) = \begin{cases} 1 \; ; & \exp[\frac{n\bar{x}(\mu - \mu_0)}{\sigma^2}] > k \\ 0 \; ; & \text{otherwise} \end{cases}$$

$$\Rightarrow \phi(\bar{x}, s^2|w) = \begin{cases} 1 \; ; & \frac{\sqrt{n}(\bar{x} - \mu_0)}{s} > k(\mu, \mu_0, \sigma^2) \\ 0 \; ; & \text{otherwise} \end{cases} \tag{8.4.4}$$

Let $t = \frac{\sqrt{n}(\bar{X} - \mu_0)}{s}$

Consider, $h(t) = \frac{\sqrt{n}(\bar{x} - \mu_0)}{\sqrt{s^2 + n(\bar{x} - \mu_0)^2}} = \frac{t}{\sqrt{1 + t^2}}$

$h'(t) > 0 \Rightarrow h(t)$ is increasing in t

$$\frac{n(\bar{x} - \mu_0)}{\sqrt{w}} > k \Rightarrow \frac{\sqrt{n}(\bar{x} - \mu_0)}{s} > k$$

From (8.4.4)

$$\frac{\sqrt{n}(\bar{x} - \mu_0)}{\sqrt{w}} > k(\mu, \mu_0, \sigma^2)$$

$$\Rightarrow P\left[\frac{\sqrt{n}(\bar{x} - \mu_0)}{\sqrt{w}} > k|w\right] = P\left[\frac{\sqrt{n}(\bar{x} - \mu_0)}{s} > k\right] = \alpha \Rightarrow k = t_{n-1, \alpha}$$

Hence, Neyman Structure test is

$$\phi(\bar{x}, s^2|w) = \begin{cases} 1 \; ; & [\frac{\sqrt{n}(\bar{x} - \mu_0)}{s}] > t_{n-1, \alpha} \\ 0 \; ; & \text{otherwise} \end{cases}$$

Example 8.4.4 Let $X_1, X_2, \ldots, X_m$ and $Y_1, Y_2, \ldots, Y_n$ are iid rvs as $P(\lambda_1)$ and $P(\lambda_2)$, respectively. Find (i) Neyman Structure test (ii) UMPU test (iii) UMP similar test for testing $H_0 : \lambda_1 = \lambda_2$ against $H_1 : \lambda_1 > \lambda_2$.

Let $T_1 = \sum_{i=1}^{m} x_i$ and $T_2 = \sum_{j=1}^{n} y_j$
T_1 is sufficient for λ_1 and T_2 is sufficient for λ_2. Under H_0, the distribution of $T_1 + T_2$ is $P((m + n)\lambda)$.

Moreover, under H_0, $T_1 + T_2$ is complete sufficient statistics.

$$f(t_1, t_2) = \frac{e^{-m\lambda_1}(m\lambda_1)^{t_1}}{t_1!} \frac{e^{-n\lambda_2}(n\lambda_2)^{t_2}}{t_2!}$$

$$f_{H_0}(t_1, t_2 | T_1 + T_2 = t) = \frac{f(T_1 = t_1, T_2 = t - t_2)}{f(t)}$$

$$= \frac{e^{-m\lambda}(m\lambda)^{t_1}}{t_1!} \frac{e^{-n\lambda}(n\lambda)^{t_2}}{t_2!} \frac{t!}{e^{-\lambda(m+n)}[\lambda(m+n)]^t}$$

$$= \binom{t}{t_1} \left(\frac{m}{m+n}\right)^{t_1} \left(\frac{n}{m+n}\right)^{t-t_1} ; t_1 = 0, 1, 2, \ldots, t$$

Under H_1, $T_1 + T_2$ is distributed as $P[m\lambda_1 + n\lambda_2]$

$$f_{H_1}(t_1, t_2 | T_1 + T_2 = t) = \binom{t}{t_1} \left(\frac{m\lambda_1}{m\lambda_1 + n\lambda_2}\right)^{t_1} \left(\frac{n\lambda_2}{m\lambda_1 + n\lambda_2}\right)^{t-t_1} ; t_1 = 0, 1, 2, \ldots, t$$

$$\lambda(t_1, t_2 | T) = \frac{f_{H_1}(t_1, t_2 | T_1 + T_2 = t)}{f_{H_0}(t_1, t_2 | T_1 + T_2 = t)} = \left(\frac{\lambda_1}{m\lambda_1 + n\lambda_2}\right)^{t_1} \left(\frac{\lambda_2}{m\lambda_1 + n\lambda_2}\right)^{t-t_1} (m+n)^t$$

$$\lambda(t_1, t_2 | T > k(t) \Rightarrow (\tfrac{\lambda_1}{\lambda_2})^{t_1} > k(\lambda_1, \lambda_2, t) \Rightarrow t_1 > k(\lambda_1, \lambda_2, t)$$

The Neyman Structure test is given as

$$\phi(t_1, t_2 | T = t) = \begin{cases} 1 ; t_1 > k(\lambda_1, \lambda_2, t) \\ \gamma ; t_1 = k(\lambda_1, \lambda_2, t) \\ 0 ; t_1 < k(\lambda_1, \lambda_2, t) \end{cases}$$

The distribution of T_1 given $T_1 + T_2$ under H_0 as

$$P[T_1 = t_1 | T_1 + T_2 = t] = \binom{t}{t_1} \left(\frac{m}{m+n}\right)^{t_1} \left(\frac{n}{m+n}\right)^{t-t_1} ; t_1 = 0, 1, 2, \ldots, t$$

Now $E_{H_0}\phi(t_1, t_2 | T) = \alpha$

$$P_{H_0}[T_1 > k(\lambda_1, \lambda_2, t) | T = t] + \gamma P_{H_0}[T_1 = k(\lambda_1, \lambda_2, t) | T_1 + T_2 = t] = \alpha$$

$$\sum_{t_1 = k+1}^{t} \binom{t}{t_1} \left(\frac{m}{m+n}\right)^{t_1} \left(\frac{n}{m+n}\right)^{t-t_1} + \gamma \binom{t}{k} \left(\frac{m}{m+n}\right)^{t} \left(\frac{n}{m+n}\right)^{t-k} = \alpha$$

One can find k and γ according to the example as stated in UMP tests. This test is also UMPU and UMP similar test.

```
# Given data
x = c(4,8,5,6,3,3,11,10,8,4,2,1);
y = c(3,?,?,6,6,6,0,0,0,...,...,0);
alpha = 0.05
m = length(x); n = length(y); p = m/(m+n); q = 1-p;
# To find k such that cumulative probability is < alpha
t1 = sum(x); t2 <- sum(y); T = t1+t2;
a = seq(from=0,to=(T-1),by=1); # possible values for k
la = length(a)
# to find cumulative probability, i.e., P(t1 > k)
cpk <- rep(0,la) # declaring variable to find cumulative probability.
for(i in 1:la)
  {
    for(j in (a[i]+1):T)
      {
        cpk[i] = cpk[i] + dbinom(j,T,p);
      }
  }
ind = min(which(cpk < alpha))  # gives cumulative probability < alpha
# To find gamma
k = ind-1; b <- dbinom(k,T,p);
gamma = (alpha-cpk[k+1])/b
# To check the answer
check <- cpk[k+1]+(gamma*dbinom(k,T,p))
# OUTPUT
print(c("k=",k))
print(c("gamma =",gamma))
print(c("check=",check))
# RESULT
"k=" "64"
"gamma ="           "0.238699811290301"
"check=" "0.05"
```

Example 8.4.5 Let $X_1, X_2, \ldots, X_m$ and $Y_1, Y_2, \ldots, Y_n$ are iid rvs with $N(\mu_1, \sigma_1^2)$ and $N(\mu_2, \sigma_2^2)$, respectively, where μ_1 and μ_2 are unknown.

Find Neyman Structure Test for testing $H_0 : \sigma_1^2 = \sigma_2^2$ against $H_1 : \sigma_1^2 > \sigma_2^2$, where μ_1 and μ_2 are unknown.

$$\bar{x} \sim N\left(\mu_1, \frac{\sigma_1^2}{m}\right), \bar{y} \sim N\left(\mu_2, \frac{\sigma_2^2}{n}\right)$$

$$\frac{s_1^2}{\sigma_1^2} \sim \chi_{m-1}^2 \text{ and } \frac{s_2^2}{\sigma_2^2} \sim \chi_{n-1}^2$$

$$f(\bar{x}, \bar{y}, s_1^2, s_2^2) = \frac{\sqrt{m}}{\sigma_1 \sqrt{2\pi}} \exp\left[-\frac{m}{2\sigma_1^2}(\bar{x}-\mu_1)^2\right] \frac{\sqrt{n}}{\sigma_2 \sqrt{2\pi}} \exp\left[-\frac{n}{2\sigma_2^2}(\bar{y}-\mu_2)^2\right]$$

$$\times \frac{(s_1^2)^{\frac{m-1}{2}-1} e^{-\frac{s_1^2}{2\sigma_1^2}}}{2^{\frac{m-1}{2}} \Gamma(\frac{m-1}{2})\sigma_1^{m-1}} \frac{(s_2^2)^{\frac{n-1}{2}-1} e^{-\frac{s_2^2}{2\sigma_2^2}}}{2^{\frac{n-1}{2}} \Gamma(\frac{n-1}{2})\sigma_2^{n-1}}$$

$$= \frac{c}{\sigma_1^m \sigma_2^n} \exp\left[-\frac{m}{2\sigma_1^2}(\bar{x}-\mu_1)^2 - \frac{n}{2\sigma_2^2}(\bar{y}-\mu_2)^2 - \frac{s_1^2}{2\sigma_1^2} - \frac{s_2^2}{2\sigma_2^2}\right] (s_1^2)^{\frac{m-1}{2}-1} (s_2^2)^{\frac{n-1}{2}-1} \quad (8.4.5)$$

where c is a constant.

Under H_0, $(\bar{x}, \bar{y}, s_1^2 + s_2^2)$ is sufficient and complete statistic.

$\frac{s_1^2+s_2^2}{\sigma^2} \sim \chi_{m+n-2}^2$

$$f(\bar{x}, \bar{y}, s_1^2, s_2^2) = \frac{c}{\sigma_1^m \sigma_2^n} \exp\left[-\frac{m(\bar{x}-\mu_1)^2}{2\sigma_1^2} - \frac{n(\bar{y}-\mu_2)^2}{2\sigma_2^2} + \frac{s_1^2+s_2^2}{2\sigma^2}\right](s_1^2+s_2^2)^{\frac{m+n-2}{2}-1} \quad (8.4.6)$$

Now to find $f_{H_0}(\bar{x}, \bar{y}, s_1^2, s_2^2 | \bar{x}, \bar{y}, s_1^2 + s_2^2)$,

$$f_{H_0}(\bar{x}, \bar{y}, s_1^2, s_2^2 | \bar{x}, \bar{y}, s_1^2 + s_2^2) = \frac{f(\bar{x}, \bar{y}, s_1^2, s_2^2)}{f(\bar{x}, \bar{y}, s_1^2 + s_2^2)}$$

$$= \frac{(s_1^2)^{\frac{m-1}{2}-1}(s^2 - s_1^2)^{\frac{n-1}{2}-1}}{(s^2)^{\frac{m+n-2}{2}-1}}, \quad (8.4.7)$$

where $s^2 = s_1^2 + s_2^2$
　To find $f_{H_1}(\bar{x}, \bar{y}, s^2)$

$$f_{H_1}(\bar{x}, \bar{y}, s^2) = \frac{c}{\sigma_1^m \sigma_2^n}(s_1^2)^{\frac{m-1}{2}-1}(s^2 - s_1^2)^{\frac{n-1}{2}-1}$$

$$\times \exp[-\frac{m(\bar{x}-\mu_1)^2}{2\sigma_1^2} - \frac{n(\bar{y}-\mu_2)^2}{2\sigma_2^2} - \frac{s_1^2}{2\sigma_1^2} - \frac{(s^2-s_1^2)}{2\sigma_2^2}]$$

$$f_{H_1}(\bar{x}, \bar{y}, s^2) = \frac{c}{\sigma_1^m \sigma_2^n} \exp\left[-\frac{m(\bar{x}-\mu_1)^2}{2\sigma_1^2} - \frac{n(\bar{y}-\mu_2)^2}{2\sigma_2^2} - \frac{s^2}{2\sigma_2^2}\right]$$

$$\times \int_0^{s^2} \exp\left[-\frac{s_1^2}{2}\left(\frac{1}{\sigma_1^2} - \frac{1}{\sigma_2^2}\right)\right](s_1^2)^{\frac{m-1}{2}-1}(s^2 - s_1^2)^{\frac{n-1}{2}-1}ds_1^2$$

Let $u = \frac{s_1^2}{s^2} \Rightarrow us^2 = s_1^2 \Rightarrow dus^2 = ds_1^2$
　Consider

$$\int_0^{s^2} \exp\left[-\frac{s_1^2}{2}\left(\frac{1}{\sigma_1^2} - \frac{1}{\sigma_2^2}\right)\right](s_1^2)^{\frac{m-1}{2}-1}(s^2 - s_1^2)^{\frac{n-1}{2}-1}ds_1^2$$

$$= \int_0^1 \exp\left[-\frac{us^2}{2}\left(\frac{1}{\sigma_1^2} - \frac{1}{\sigma_2^2}\right)\right]u^{\frac{m-1}{2}-1}(1-u)^{\frac{n-1}{2}-1}(s^2)^{\frac{m+n-2}{2}-1}du$$

$$= (s^2)^{\frac{m+n-2}{2}-1}g(s^2, \sigma_1^2, \sigma_2^2)$$

Hence

$$f_{H_1}(\bar{x}, \bar{y}, s^2) = \frac{c}{\sigma_1^m \sigma_2^n} \exp\left[-\frac{m(\bar{x} - \mu_1)^2}{2\sigma_1^2} - \frac{n(\bar{y} - \mu_2)^2}{2\sigma_2^2} - \frac{s^2}{2\sigma_2^2}\right] g(s^2, \sigma_1^2, \sigma_2^2)(s^2)^{\frac{m+n-2}{2}-1}$$

$$f_{H_1}(\bar{x}, \bar{y}, s_1^2, s_2^2) = \frac{c}{\sigma_1^m \sigma_2^n} \exp\left[-\frac{m(\bar{x} - \mu_1)^2}{2\sigma_1^2} - \frac{n(\bar{y} - \mu_2)^2}{2\sigma_2^2} - \frac{s_1^2}{2\sigma_2^2} - \frac{(s^2 - s_1^2)}{2\sigma_2^2}\right]$$
$$\times (s_1^2)^{\frac{m-1}{2}-1}(s^2 - s_1^2)^{\frac{n-1}{2}-1}$$

$$f_{H_1}(\bar{x}, \bar{y}, s_1^2, s_2^2 | \bar{x}, \bar{y}, s^2) = \frac{\exp\left[-\frac{s_1^2}{2}\left(\frac{1}{\sigma_1^2} - \frac{1}{\sigma_2^2}\right)\right](s_1^2)^{\frac{m-1}{2}-1}(s^2 - s_1^2)^{\frac{n-1}{2}-1}}{(s^2)^{\frac{m+n-2}{2}-1} g(s^2, \sigma_1^2, \sigma_2^2)}$$

$$\lambda(\bar{x}, \bar{y}, s_1^2, s_2^2 | \bar{x}, \bar{y}, s^2) = \frac{f_{H_1}(\bar{x}, \bar{y}, s_1^2, s_2^2 | \bar{x}, \bar{y}, s^2)}{f_{H_0}(\bar{x}, \bar{y}, s_1^2, s_2^2 | \bar{x}, \bar{y}, s^2)} = \frac{\exp\left[-\frac{s_1^2}{2}\left(\frac{1}{\sigma_1^2} - \frac{1}{\sigma_2^2}\right)\right]}{g(s^2, \sigma_1^2, \sigma_2^2)}$$

We define a Neyman Structure test as

$$\phi(\bar{x}, \bar{y}, s_1^2, s_2^2 | \bar{x}, \bar{y}, s^2) = \begin{cases} 1 ; & \frac{\exp\left[-\frac{s_1^2}{2}\left(\frac{1}{\sigma_1^2} - \frac{1}{\sigma_2^2}\right)\right]}{g(s^2, \sigma_1^2, \sigma_2^2)} > k(s^2, \sigma_1^2, \sigma_2^2) \\ 0 ; & \text{otherwise} \end{cases}$$

$$\phi(\bar{x}, \bar{y}, s_1^2, s_2^2 | \bar{x}, \bar{y}, s^2) = \begin{cases} 1 ; & s_1^2 > k(s^2, \sigma_1^2, \sigma_2^2) \\ 0 ; & \text{otherwise} \end{cases}$$

$$\Rightarrow \phi(\bar{x}, \bar{y}, s_1^2, s_2^2 | \bar{x}, \bar{y}, s^2) = \begin{cases} 1 ; & \frac{s_1^2}{s_1^2 + s_2^2} > k(s^2, \sigma_1^2, \sigma_2^2) \\ 0 ; & \text{otherwise} \end{cases}$$

Let $w = \frac{s_1^2}{s_2^2} \Rightarrow \frac{s_1^2}{s_1^2 + s_2^2} = \frac{w}{1+w}$

$g(w) = \frac{w}{1+w} \Rightarrow g'(w) = \frac{1}{(w+1)^2} > 0$

g(w) is increasing in w.
   Then

$$\phi(\bar{x}, \bar{y}, s_1^2, s_2^2 | \bar{x}, \bar{y}, s^2) = \begin{cases} 1 ; & w > k(s^2, \sigma_1^2, \sigma_2^2) \\ 0 ; & \text{otherwise} \end{cases}$$

$$\phi(\bar{x}, \bar{y}, s_1^2, s_2^2 | \bar{x}, \bar{y}, s^2) = \begin{cases} 1 ; & \frac{s_1^2}{s_2^2} > k(s^2) \\ 0 ; & \text{otherwise} \end{cases} \tag{8.4.8}$$

We have to find the distribution of $\frac{s_1^2}{s_2^2}$ given S^2 under H_0.

Let $w = \frac{s_1^2}{s_2^2}$, $v = s_1^2 + s_2^2$

$\frac{s_1^2}{\sigma^2} \sim \chi_{m-1}^2$ and $\frac{s_2^2}{\sigma^2} \sim \chi_{n-1}^2$

$$f(s_1^2, s_2^2) = c(s_1^2)^{\frac{m-1}{2}-1}(s_2^2)^{\frac{n-1}{2}-1}e^{-\frac{(s_1^2+s_2^2)}{2\sigma^2}}$$

Now $s_1^2 = ws_2^2$, $v = s_1^2 + s_2^2 \Rightarrow s_2^2 = \frac{v}{1+w}$, $s_1^2 = \frac{wv}{1+w}$

$$J = \begin{pmatrix} \frac{\partial s_1^2}{\partial w} & \frac{\partial s_1^2}{\partial v} \\ \frac{\partial s_2^2}{\partial w} & \frac{\partial s_2^2}{\partial v} \end{pmatrix} = \begin{pmatrix} \frac{v}{(1+w)^2} & \frac{w}{1+w} \\ -\frac{v}{(1+w)^2} & \frac{1}{1+w} \end{pmatrix} = \frac{v}{(1+w)^2}$$

$$\begin{aligned} f(w, v) &= f(s_1^2, s_2^2)|J| \\ &= c\left(\frac{wv}{1+w}\right)^{\frac{m-1}{2}-1}\left(\frac{v}{1+w}\right)^{\frac{n-1}{2}-1}\left(\frac{v}{1+w}\right)e^{-\frac{v}{2\sigma^2}} \\ &= c\frac{w^{\frac{m-1}{2}-1}}{(1+w)^{\frac{m+n}{2}-2}}v^{\frac{m+n}{2}-2}e^{-\frac{v}{2\sigma^2}} \\ &= f_1(w)f_2(w) \end{aligned}$$

w and v are independent

Hence, $f(w|v) = f(w)$

therefore, $\frac{s_1^2/(m-1)}{s_2^2/(n-1)} \sim F_{m-1,n-1}$

From (8.4.8), $P\left[\frac{s_1^2/(m-1)}{s_2^2/(n-1)} > k\right] = \alpha \Rightarrow k = F_{m-1,n-1,\alpha}$

Hence Neyman Structure Test is

$$\phi(\bar{x}, \bar{y}, s_1^2, s_2^2) = \begin{cases} 1 \; ; & \frac{s_1^2/(m-1)}{s_2^2/(n-1)} > F_{m-1,n-1,\alpha} \\ 0 \; ; & \text{otherwise} \end{cases} \tag{8.4.9}$$

Example 8.4.6 Let $X_1, X_2, \ldots, X_{n_1}$ and $Y_1, Y_2, \ldots, Y_{n_2}$ are iid rvs with $N(\mu_1, \sigma^2)$ and $N(\mu_2, \sigma^2)$, respectively, where σ^2 is unknown. Find Neyman Structure test for testing $H_0 : \mu_1 = \mu_2$ against $H_1 : \mu_1 > \mu_2$, where σ^2 is unknown.

Note that $(\bar{x}, \bar{y}, s^2)$ is sufficient and complete for (μ_1, μ_2, σ^2), where $s^2 = \sum(x_i - \bar{x})^2 + \sum(y_i - \bar{y})^2$

$$f(\bar{x}, \bar{y}, s^2) = \frac{c}{\sigma^{n_1+n_2}} \exp\left[-\frac{1}{2\sigma^2}\{n_1(\bar{x} - \mu_1) + n_2(\bar{y} - \mu_2)\}\right]e^{-\frac{s^2}{2\sigma^2}}(s^2)^{\frac{n_1+n_2-2}{2}-1},$$

Let $n = n_1 + n_2$

Under H_0, i.e., $\mu_1 = \mu_2$, $(n_1\bar{x} + n_2\bar{y}, \sum_{i=1}^{n_1} x_i^2 + \sum_{j=1}^{n_2} y_j^2)$ are jointly sufficient and complete for (μ_1, μ_2, σ^2).

Let $z = \frac{n_1\bar{x}+n_2\bar{y}}{n}$, $w = \sum_{i=1}^{n_1} x_i^2 + \sum_{j=1}^{n_2} y_j^2$ are any functions of sufficient statistics.

Let $s^2 = s_1^2 + s_2^2 = \sum_{i=1}^{n_1} x_i^2 - n_1\bar{x}^2 + \sum_{j=1}^{n_2} y_j^2 - n_2\bar{y}^2$

we want to find the distribution of z, w and $\bar{y}$.

$s^2 = w - n_1\bar{x}^2 - n_2\bar{y}^2 \Rightarrow w = s^2 + n_1\bar{x}^2 + n_2\bar{y}^2$

$$J = \frac{\partial(\bar{x}, \bar{y}, s^2)}{\partial(z, w, y)} = \begin{pmatrix} \frac{\partial \bar{x}}{\partial z} & \frac{\partial \bar{x}}{\partial w} & \frac{\partial \bar{x}}{\partial \bar{y}} \\ \frac{\partial \bar{y}}{\partial z} & \frac{\partial \bar{y}}{\partial w} & \frac{\partial \bar{y}}{\partial \bar{y}} \\ \frac{\partial s^2}{\partial z} & \frac{\partial s^2}{\partial w} & \frac{\partial s^2}{\partial \bar{y}} \end{pmatrix} = \begin{pmatrix} \frac{n}{n_1} & \frac{1}{2n_1\bar{x}} & 0 \\ \frac{1}{n_2} & \frac{1}{2n_2 y} & 1 \\ 0 & 1 & 0 \end{pmatrix}$$

$|J| = \frac{n}{n_1}$

Consider

$$n_1(\bar{x} - \mu_1)^2 + n_2(\bar{y} - \mu_2)^2 + s^2 = n_1\bar{x}^2 - 2n_1\mu_1\bar{x} + n_1\mu_1^2 + n_2\bar{y}^2 - 2n_2\mu_2\bar{y} + n_2\mu_2^2 + s^2$$

Since $w = s^2 + n_1\bar{x}^2 + n_2\bar{y}^2$

$$= w - 2n_1\mu_1\bar{x} - 2n_2\mu_2\bar{y} + n_1\mu_1^2 + n_2\mu_2^2$$

Next, since $z = \frac{n_1\bar{x} + n_2\bar{y}}{n} \Rightarrow \bar{x} = \frac{nz - n_2\bar{y}}{n_1}$

$$s^2 = w - n_1\bar{x}^2 - n_2\bar{y}^2 = w - n_1\left(\frac{nz - n_2\bar{y}}{n_1}\right)^2 - n_2\bar{y}^2,$$

Further,

$$w - 2n_1\bar{x}\mu_1 - 2n_2\mu_2\bar{y} + n_1\mu_1^2 + n_2\mu_2^2$$

$$= w - 2n_1\mu_1\left(\frac{nz - n_2\bar{y}}{n_1}\right) - 2n_2\mu_2\bar{y} + n_1\mu_1^2 + n_2\mu_2^2$$

$$= w - 2\mu_1[nz - n_2\bar{y}] - 2n_2\mu_2\bar{y} + n_1\mu_1^2 + n_2\mu_2^2$$

$$= w - 2n\mu_1 z + 2\mu_1 n_2\bar{y} - 2n_2\mu_2\bar{y} + n_1\mu_1^2 + n_2\mu_2^2$$

$$= w - 2n\mu_1 z - 2n_2\bar{y}(\mu_2 - \mu_1) + n_1\mu_1^2 + n_2\mu_2^2$$

$$f(z, w, \bar{y}) = f(\bar{x}, \bar{y}, s^2)|J|$$

$$= c\exp\left[-\frac{1}{2\sigma^2}\{w - 2n\mu_1 z - 2n_2\bar{y}(\mu_2 - \mu_1) + n_1\mu_1^2 + n_2\mu_2^2\}\right]$$

$$\times \left[w - n_2\bar{y}^2 - n_1\{\frac{nz - n_2\bar{y}}{n_1}\}^2\right]^{\frac{n-2}{2}-1}$$

Let $g(w, z|\mu_1, \mu_2, \sigma^2) = \int f(z, w, \bar{y})d\bar{y}$

Under H_0

$$f_{H_0}\left(\bar{x}, \bar{y}, s^2|z, w\right) = \frac{c\exp\left[-\frac{1}{2\sigma^2}\{w - 2n\mu z + n\mu^2\}\right]\left[w - n_2\bar{y}^2 - n_1\{\frac{nz - n_2\bar{y}}{n_1}\}^2\right]^{\frac{n-2}{2}-1}}{g\left(w, z|\mu, \sigma^2\right)}$$

$$f_{H_1}(\bar{x}, \bar{y}, s^2 | z, w) = c \exp[-\frac{1}{2\sigma^2}\{w - 2n\mu_1 z - 2n_2\bar{y}(\mu_2 - \mu_1) + n_1\mu_1^2 + n_2\mu_2^2\}]$$

$$\times \frac{[w - n_2\bar{y}^2 - n_1\{\frac{nz - n_2\bar{y}}{n_1}\}^2]^{\frac{n-2}{2}-1}}{g(w, z | \mu_1, \mu_2, \sigma^2)}$$

$$\lambda(\bar{x}, \bar{y}, s^2 | z, w) = \frac{f_{H_1}(\bar{x}, \bar{y}, s^2 | z, w)}{f_{H_0}(\bar{x}, \bar{y}, s^2 | z, w)}$$

$$\Rightarrow \exp\left[\frac{n_2\bar{y}(\mu_2 - \mu_1)}{\sigma^2}\right] > k(z, w, \mu_1, \mu_2, \sigma^2)$$

$$\Rightarrow \bar{y} > k(z, w, \mu_1, \mu_2, \sigma^2)$$

The Neyman Structure test is given as

$$\Rightarrow \phi(\bar{x}, \bar{y}, s^2 | z, w) = \begin{cases} 1 \; ; \; \bar{y} > k(\mu_1, \mu_2, z, w, \sigma^2) \\ 0 \; ; \; \text{otherwise} \end{cases} \qquad (8.4.10)$$

Consider $\bar{y} > k \Rightarrow n\bar{y} > k$

$$\Rightarrow (n_1 + n_2)\bar{y} - nz > k$$

$$\Rightarrow n_1\bar{y} + n_2\bar{y} - n_1\bar{x} - n_2\bar{y} > k$$

$$\Rightarrow n_1(\bar{y} - \bar{x}) > k \Rightarrow (\bar{y} - \bar{x}) > k$$

Next,

$$s_1^2 + s_2^2 = \sum x_i^2 - n_1\bar{x}^2 + \sum y_i^2 - n_2\bar{y}^2$$

$$= \sum x_i^2 + \sum y_i^2 - n_1\bar{x}^2 - n_2\bar{y}^2$$

$$s^2 = w - n_1\bar{x}^2 - n_2\bar{y}^2$$

$$s^2 = w - nz^2 + nz^2 - n_1\bar{x}^2 - n_2\bar{y}^2$$

Further,

$$
nz^2 - n_1\bar{x}^2 - n_2\bar{y}^2 = n\left(\frac{n_1x + n_2\bar{y}}{n}\right)^2 - n_1\bar{x}^2 - n_2\bar{y}^2
$$

$$
= \frac{1}{n}\left[n_1^2\bar{x}^2 + n_2^2\bar{y}^2 + 2n_1n_2\bar{x}\bar{y}\right] - n_1\bar{x}^2 - n_2\bar{y}^2
$$

$$
= \frac{1}{n}\left[n_1^2\bar{x}^2 + n_2^2\bar{y}^2 + 2n_1n_2\bar{x}\bar{y} - nn_1\bar{x}^2 - nn_2\bar{y}^2\right]
$$

$$
= \frac{1}{n}\left[n_1^2\bar{x}^2 + n_2^2\bar{y}^2 + 2n_1n_2\bar{x}\bar{y} - (n_1 + n_2)n_1\bar{x}^2 - (n_1 + n_2)n_2\bar{y}^2\right]
$$

$$
= \frac{1}{n}\left[2n_1n_2\bar{x}\bar{y} - n_1n_2\bar{x}^2 - n_1n_2\bar{y}^2\right]
$$

$$
= \frac{-n_1n_2}{n}\left[\bar{y}^2 + \bar{x}^2 - 2\bar{x}\bar{y}\right] = -\frac{n_1n_2}{n}(\bar{y} - \bar{x})^2
$$

$$
s_1^2 + s_2^2 = w - nz^2 - \frac{n_1n_2}{n}(\bar{y} - \bar{x})^2
$$

(8.4.10) will be

$$
\phi(\bar{x}, \bar{y}, s^2 | w, z) = \begin{cases} 1 ; & \bar{y} - \bar{x} > k \\ 0 ; & \text{otherwise} \end{cases}
$$

$$
\Rightarrow \phi(\bar{x}, \bar{y}, s^2 | w, z) = \begin{cases} 1 ; & \frac{\bar{y} - \bar{x}}{\sqrt{w - nz^2}} > k \\ 0 ; & \text{otherwise} \end{cases}
$$

Now

$$
\frac{\bar{y} - \bar{x}}{\sqrt{w - nz^2}} = \frac{(\bar{y} - \bar{x})}{\sqrt{s^2 + \frac{n_1n_2}{n}(\bar{y} - \bar{x})^2}}
$$

Let $\frac{\bar{y} - \bar{x}}{s} = v$, $h(v) = \frac{v}{\sqrt{1 + \frac{n_1n_2}{n}v^2}} \Rightarrow h'(v) > 0 \Rightarrow h(v)$ is an increasing function in v.

Hence $v > k$,

The Neyman Structure test is

$$
\phi(\bar{x}, \bar{y}, s^2 | w, z) = \begin{cases} 1 ; & \frac{\bar{y} - \bar{x}}{s} > k \\ 0 ; & \text{otherwise} \end{cases} \tag{8.4.11}
$$

To find the conditional distribution of $\frac{\bar{y} - \bar{x}}{s}$ given (z, w) under H_0

$$
f(\bar{x}, \bar{y}, s^2) = c\exp\left[-\frac{1}{2\sigma^2}\{n_1(\bar{x} - \mu)^2 + n_2(\bar{y} - \mu)^2 + s^2\}\right](s^2)^{\frac{n-2}{2}-1}
$$

$t = \frac{\bar{y} - \bar{x}}{s}$, $w = \sum x_i^2 + \sum y_i^2 = s^2 + nz^2 + \frac{n_1n_2}{n}(\bar{y} - \bar{x})^2$

$$w = s^2 \left[1 + \frac{n_1 n_2}{n} t^2\right] + nz^2 \tag{8.4.12}$$

Now $n_1(\bar{x} - \mu)^2 + n_2(\bar{y} - \mu)^2 + s^2$

$$= \sum x_i^2 + \sum y_i^2 - 2n_1\mu\bar{x} - 2n_2\mu\bar{y} + (n_1 + n_2)\mu^2$$

$$= \sum x_i^2 + \sum y_i^2 - 2\mu(n_1\bar{x} + n_2\bar{y}) + (n_1 + n_2)\mu^2$$

$$= \sum x_i^2 + \sum y_i^2 - 2n\mu z + (n_1 + n_2)\mu^2$$

$$= w - 2n\mu z + n\mu^2 \tag{8.4.13}$$

Next,

$$w - nz^2 = s^2 \left[1 + \frac{n_1 n_2}{n} t^2\right]$$

$$\frac{w - nz^2}{1 + \frac{n_1 n_2}{n} t^2} = s^2 \tag{8.4.14}$$

$$f(\bar{x}, \bar{y}, s^2) = c \exp\left[-\frac{1}{2\sigma^2}\{w - 2n\mu z + n\mu^2\}\right] \left[\frac{w - nz^2}{1 + \frac{n_1 n_2}{n} t^2}\right]^{\frac{n-2}{2} - 1}$$

$$= c \exp\left[-\frac{1}{2\sigma^2}\{w - 2n\mu z + n\mu^2\}\right] (w - nz^2)^{\frac{n-2}{2} - 1} \left[\frac{1}{1 + \frac{n_1 n_2}{n} t^2}\right]^{\frac{n-2}{2} - 1}$$

$$= g_1(w, z) g_2(t)$$

Hence, (w, z) and t are independent.

Therefore, the conditional distribution of $\frac{\bar{y} - \bar{x}}{s}$ given (w, z) is the distribution of $\frac{\bar{y} - \bar{x}}{s}$.

Now $\bar{y} - \bar{x} \sim N(\mu_2 - \mu_1, \sigma^2\{\frac{1}{n_1} + \frac{1}{n_2}\})$

under H_0 $\bar{y} - \bar{x} \sim N(0, \sigma_1^2)$, where $\sigma_1^2 = \sigma^2\{\frac{1}{n_1} + \frac{1}{n_2}\}$

$$\Rightarrow \left(\sqrt{\frac{1}{n_1} + \frac{1}{n_2}}\right)^{-1} (\bar{y} - \bar{x}) \sim N(0, \sigma^2),$$

$$\hat{\sigma}^2 = \frac{s_1^2 + s_2^2}{n_1 + n_2 - 2} = \frac{\sum x_i^2 - n_1\bar{x}^2 + \sum y_i^2 - n_2\bar{y}^2}{n_1 + n_2 - 2}$$

Hence

$$\frac{\left(\sqrt{\frac{1}{n_1} + \frac{1}{n_2}}\right)^{-1} (\bar{y} - \bar{x})}{\sqrt{\frac{s_1^2 + s_2^2}{n_1 + n_2 - 2}}} \sim t_{n_1 + n_2 - 2}$$

From (8.4.11),

$$E[\phi(\bar{x}, \bar{y}, s^2 | z, w)] = P\left[\frac{\bar{y} - \bar{x}}{s} > k\right] = \alpha$$

$$\Rightarrow k = t_{n_1 + n_2 - 2, \alpha}$$

Hence Neyman Structure test is

$$\phi(\bar{x}, \bar{y}, s^2 | w, z) = \begin{cases} 1 \; ; \; \frac{\bar{y} - \bar{x}}{s} > t_{n_1 + n_2 - 2, \alpha} \\ 0 \; ; \; \text{otherwise} \end{cases} \tag{8.4.15}$$

8.5 Likelihood Ratio Tests

In earlier sections, we had obtained UMPU, LMPU, similar and Neyman Structure tests for some distributions. Perhaps, they do not exist for other distributions. We also had seen that computations of UMP unbiased tests in multiparameter case are usually complicated. Since these are α Similar tests having Neyman Structure.

In this section, we consider a classical procedure for constructing tests that has some intuitive appeal and that frequently, though not necessarily, leads to optimal tests. The procedure also leads to tests that have some desirable large-sample properties. Neyman and Pearson (1928) suggested a simple method for testing a general testing problem. Consider a random sample $X_1, X_2, \ldots, X_n$ from $f(x|\theta), \theta \in \Theta$ and we have to test

$$H_0 : \theta \in \Theta_0 \; against \; H_1 : \theta \in \Theta_1 \tag{8.5.1}$$

The likelihood ratio test for testing (8.5.1) is defined as

$$\lambda(x) = \frac{\sup_{\theta \in \Theta_0} L(\theta | x)}{\sup_{\theta \in \Theta} L(\theta | x)} \tag{8.5.2}$$

where $L(\theta|x)$ is the likelihood function of x.

Definition 8.5.1 Let $L(\theta|x)$ be a likelihood for a random sample having the joint pdf(pmf) $f(x|\theta)$ for $\theta \in \Theta$. The likelihood ratio is defined to be

$$\lambda(x) = \frac{\sup_{\theta \in \Theta_0} L(\theta | x)}{\sup_{\theta \in \Theta} L(\theta | x)}.$$

The numerator of the likelihood ratio $\lambda(x)$ is the best explanation of X that the null hypothesis H_0 can provide and the denominator is the best possible explanation of X. H_0 is rejected if there is a much better explanation of X than the best one provided by H_0. Further, it implies that smaller values of λ leads to the rejection of H_0 and larger values of λ leads to acceptance of H_0. Therefore, the critical region would be of the type $\lambda(x) < k$. Note that $0 \le \lambda \le 1$. The constant k is determined by

$$\sup_{\theta \in \Theta_0} P[\lambda(x) < k] = \alpha.$$

If the distribution of $\lambda(x)$ is continuous, then the size α is exactly attained and no randomization on the boundary is required. Similarly, if the distribution of $\lambda(x)$ is discrete, the size may not attain α and then we require randomization. We will see the following theorems without proof.

Theorem 8.5.1 *For $0 \le \alpha \le 1$ nonrandomized Neyman–Pearson and likelihood ratio test of a simple hypothesis against a simple alternative exist, they are equivalent.*

Theorem 8.5.2 *For testing $\theta \in \Theta_0$ against $\theta \in \Theta_1$, the likelihood ratio test is a function of every sufficient statistics for θ.*

Example 8.5.1 Let $X_1, X_2, \ldots, X_n$ be a random sample of size n from a normal distribution with mean μ and variance σ^2. Obtain the likelihood ratio for testing $H_0 : \mu = \mu_0$ against $H_1 : \mu \ne \mu_0$

Case (i) σ^2 is known
Note that there is no UMP test for this problem.
The likelihood function is

$$L(\mu|X_1, X_2, \ldots, X_n) = \prod_{i=1}^{n} \frac{1}{\sigma\sqrt{2\pi}} \exp\left[-\frac{1}{2\sigma^2}(x_i - \mu)^2\right]$$

$$= \left(\frac{1}{\sigma\sqrt{2\pi}}\right)^n \exp\left[-\frac{1}{2\sigma^2} \sum_{i=1}^{n}(x_i - \mu)^2\right]$$

Consider $\sum_{i=1}^{n}(x_i - \mu)^2 = \sum_{i=1}^{n}(x_i - \bar{x} + \bar{x} - \mu)^2 = \sum_{i=1}^{n}(x_i - \bar{x})^2 + n(\bar{x} - \mu)^2$

$$L(\mu|X) = \left(\frac{1}{\sigma\sqrt{2\pi}}\right)^n \exp\left[-\frac{n}{2\sigma^2}(\bar{x} - \mu)^2\right] \exp\left[-\frac{1}{2\sigma^2} \sum_{i=1}^{n}(x_i - \bar{x})^2\right]$$

$$\sup_{\theta \in \Theta_0} L(\mu|x) = \sup_{\mu = \mu_0} L(\mu|x) = L(\mu_0)$$

$$= \left(\frac{1}{\sigma\sqrt{2\pi}}\right)^n \exp\left[-\frac{n}{2\sigma^2}(\bar{x} - \mu_0)^2\right] \exp\left[-\frac{1}{2\sigma^2} \sum_{i=1}^{n}(x_i - \bar{x})^2\right]$$

$$\sup_{\theta \in \Theta} L(\mu|x) = L(\hat{\mu}|x), \ \hat{\mu} \ is \ the \ mle \ of \ \mu.$$

$\hat{\mu}$ is the mle of μ

Hence, mle of $\mu = \hat{\mu} = \bar{X}$

$$L(\hat{\mu}|x) = \left(\frac{1}{\sigma\sqrt{2\pi}}\right)^n \exp\left[-\frac{1}{2\sigma^2}\sum_{i=1}^{n}(x_i - \bar{x})^2\right]$$

From (8.5.2)

$$\lambda(x) = \exp\left[-\frac{n}{2\sigma^2}(\bar{x} - \mu_0)^2\right]$$

The likelihood ratio is just a function of $\frac{n(\bar{x}-\mu_0)^2}{\sigma^2}$ and will be small when the quantity is large.
The LR test is given as

$$\phi(x) = \begin{cases} 1 \; ; & \left|\frac{\bar{x}-\mu_0}{\frac{\sigma}{\sqrt{n}}}\right| > k \\ 0 \; ; & \text{otherwise} \end{cases}$$

Since $\left(\frac{\bar{x}-\mu_0}{\frac{\sigma}{\sqrt{n}}}\right)$ is $N(0, 1)$. Then k can be obtained as $P\left[\left(\frac{\bar{x}-\mu_0}{\frac{\sigma}{\sqrt{n}}}\right) > k\right] = \frac{\alpha}{2}$.
Therefore, $k = Z_{\frac{\alpha}{2}} = \frac{\alpha}{2}$th quantile of $N(0, 1)$
Then the test is given as

$$\phi(x) = \begin{cases} 1 \; ; & \left|\frac{\bar{x}-\mu_0}{\frac{\sigma}{\sqrt{n}}}\right| > Z_{\frac{\alpha}{2}} \\ 0 \; ; & \text{otherwise} \end{cases}$$

Note: The reader should see Example 8.1.1.

This LR test is also UMPU.
Case (ii) σ^2 unknown
We have to test $H_0 : \mu = \mu_0$ against $H_1 : \mu \neq \mu_0$
From (8.5.1),

$$\sup_{\mu=\mu_0,\sigma^2>0} L(\mu|x) = \sup_{\sigma^2>0}\left(\frac{1}{\sigma\sqrt{2\pi}}\right)^n \exp\left[-\frac{1}{2\sigma^2}\sum_{i=1}^{n}(x_i - \mu_0)^2\right]$$

MLE of $\sigma^2 = \frac{1}{n}\sum_{i=1}^{n}(x_i - \mu_0)^2$
Let $s_0^2 = \sum_{i=1}^{n}(x_i - \mu_0)^2 \Rightarrow \hat{\sigma}^2 = \frac{s_0^2}{n}$
Under H_0,

$$\sup_{\sigma^2>0} L(\mu|x) = \sup_{\sigma^2>0}\left(\frac{\sqrt{n}}{s_0\sqrt{2\pi}}\right)^n \exp\left(-\frac{n}{2}\right) \tag{8.5.3}$$

Further, MLE of μ and σ^2 is $\hat{\mu}$ and $\hat{\sigma}^2$

$\hat{\mu} = \bar{X}$ and $\hat{\sigma^2} = \frac{s^2}{n}, s^2 = \sum(x_i - \bar{x})^2$

$$\sup_{\mu, \sigma^2} \left(\frac{1}{\sigma\sqrt{2\pi}} \right)^{\frac{n}{2}} \exp\left\{ -\frac{2}{\sigma^2} \sum_{i=1}^{n}(x_i - \mu)^2 \right\} = \left(\frac{\sqrt{n}}{s\sqrt{2\pi}} \right)^n \exp\left(-\frac{n}{2} \right) \quad (8.5.4)$$

From (8.5.2), (8.5.3) and (8.5.4)

$$\lambda(x) = \left(\frac{s}{s_0} \right)^n = \left(\frac{s^2}{s_0^2} \right)^{\frac{n}{2}}$$

$$s_0^2 = \sum_{i=1}^{n}(x_i - \mu_0)^2 = \sum_{i=1}^{n}(x_i - \bar{x} + \bar{x} - \mu_0)^2$$

$$= \sum_{i=1}^{n}(x_i - \bar{x})^2 + n(\bar{x} - \mu_0)^2 = s^2 + n(\bar{x} - \mu_0)^2$$

$$\lambda(x) = \left[\frac{s^2}{s^2 + n(\bar{x} - \mu_0)^2} \right]^{\frac{n}{2}} = \left[\frac{1}{1 + \frac{n(\bar{x} - \mu_0)^2}{s^2}} \right]^{\frac{n}{2}}$$

CR is $\lambda(x) < k \Rightarrow \frac{n(\bar{x} - \mu_0)^2}{s^2} > k \Rightarrow \left| \frac{(\bar{x} - \mu_0)}{\frac{s}{\sqrt{n}}} \right| > k$. The likelihood ratio test is

$$\phi(x) = \begin{cases} 1 \; ; \; |\frac{\sqrt{n}(\bar{x} - \mu_0)}{s}| > k \\ 0 \; ; \text{otherwise} \end{cases}$$

Now $\frac{(\bar{x} - \mu_0)}{\frac{\sigma}{\sqrt{n}}} \sim N(0, 1)$ and $\frac{s^2}{\sigma^2} \sim \chi_{n-1}^2$
Hence

$$\frac{\frac{(\bar{x} - \mu_0)}{\frac{\sigma}{\sqrt{n}}}}{\sqrt{\frac{(n-1)s^2}{\sigma^2}{n-1}}} \sim t_{n-1}$$

The distribution t_{n-1} is symmetric about 0,

$$P_{H_0}\left\{ \frac{\sqrt{n}(\bar{x} - \mu_0)}{s} > k \right\} = \frac{\alpha}{2}$$

The likelihood ratio test is given as

$$\phi(x) = \begin{cases} 1 \; ; \; |\frac{\sqrt{n}(\bar{x} - \mu_0)}{s}| > t_{n-1, \frac{\alpha}{2}} \\ 0 \; ; \text{otherwise} \end{cases}$$

Note: 1. If we are testing $H_0 : \mu \le \mu_0$ against $H_1 : \mu > \mu_0$

For σ^2 known

Under H_0, MLE of μ

$$\hat{\mu} = \begin{cases} \mu_0 \; ; \; \bar{x} \ge \mu_0 \\ \bar{x} \; ; \; \bar{x} < \mu_0 \end{cases}$$

This gives

$$\sup_{\mu \le \mu_0} L(\mu|x) = \begin{cases} \left(\frac{1}{\sigma\sqrt{2\pi}}\right)^n \exp\left\{-\frac{1}{2\sigma^2}\sum(x_i - \mu_0)^2\right\} \; ; \; \bar{x} \ge \mu_0 \\ \left(\frac{1}{\sigma\sqrt{2\pi}}\right)^n \exp\left\{-\frac{1}{2\sigma^2}\sum(x_i - \bar{x})^2\right\} \; ; \; \text{otherwise} \end{cases}$$

MLE of μ is $\bar{x}$

$$\sup_{\mu \in \Theta} L(\mu|x) = \left(\frac{1}{\sigma\sqrt{2\pi}}\right)^n \exp\left\{-\frac{1}{2\sigma^2}\sum(x_i - \bar{x})^2\right\}$$

From (8.5.2),

The likelihood ratio test is given as

$$\lambda(x) = \begin{cases} \exp\{-\frac{n}{2\sigma^2}(\bar{x} - \mu_0)^2\} \; ; \; \bar{x} \ge \mu_0 \\ 1 \qquad\qquad\qquad\; ; \; \bar{x} < \mu_0 \end{cases}$$

The LR test is given as

$$\phi(x) = \begin{cases} 1 \; ; \; \frac{\sqrt{n}(\bar{x}-\mu_0)}{\sigma} > k \\ 0 \; ; \; \text{otherwise} \end{cases}$$

Since $\frac{\sqrt{n}(\bar{x}-\mu_0)}{\sigma} \sim N(0, 1)$, hence $k = Z_\alpha$.

2. If we are testing $H_0 : \mu \le \mu_0$ against $H_1 : \mu > \mu_0$.

For σ^2 unknown

Under H_0, MLE of μ and σ^2:

$$\hat{\mu} = \begin{cases} \mu_0 \; ; \; \bar{x} \ge \mu_0 \\ \bar{x} \; ; \; \bar{x} < \mu_0 \end{cases}$$

and

$$\hat{\sigma}^2 = \begin{cases} s_0^2 \qquad\quad\; ; \; \bar{x} \ge \mu_0 \\ \frac{\sum(x_i - \bar{x})^2}{n} \; ; \; \bar{x} < \mu_0 \end{cases}$$

Hence, under H_0,

$$\sup_{(\mu,\sigma^2) \in \Theta} L(\mu, \sigma^2|X) = \begin{cases} (\sqrt{2\pi}s_0)^{-n} \exp\{-\frac{n}{2}\} \; ; \; \bar{x} \ge \mu_0 \\ \left(\frac{\sqrt{2\pi}s}{\sqrt{n}}\right)^{-n} \exp\{-\frac{n}{2}\} \; ; \; \text{otherwise} \end{cases}$$

mle of $\mu = \hat{\mu} = \bar{x}$ and $\hat{\sigma^2} = \frac{s^2}{n}$

$$\sup_{(\mu,\sigma^2)\in\Theta} L(\mu, \sigma^2|x) = \left(\frac{\sqrt{2\pi}s}{\sqrt{n}}\right)^{-n} \exp\left\{-\frac{n}{2}\right\}$$

LR is given as

$$\lambda(x) = \begin{cases} (\frac{s^2}{ns_0^2})^{\frac{n}{2}} ; \ \bar{x} \geq \mu_0 \\ 1 \qquad ; \ \bar{x} < \mu_0 \end{cases}$$

Hence LR test is given as

$$\phi(x) = \begin{cases} 1 ; \ \frac{\sqrt{n}(\bar{x}-\mu_0)}{s} > k \\ 0 ; \ \text{otherwise} \end{cases}$$

Since $\frac{\sqrt{n}(\bar{x}-\mu_0)}{s}$ has t-distribution with $(n-1)$ df.
Therefore, $k = t_{n-1,\alpha}$.
This is also Neyman Structure Test, reader should see Example 8.4.3.

Example 8.5.2 Let $X_1, X_2, \ldots, X_n$ be a random sample from $N(\mu, \sigma^2)$. Obtain a LR test to test $H_0 : \sigma^2 = \sigma_0^2$ against $H_1 : \sigma^2 \neq \sigma_0^2$ with population mean μ is unknown.

$$L(\mu, \sigma^2|X) = \left(\frac{1}{\sigma\sqrt{2\pi}}\right)^n \exp\left[-\frac{1}{2\sigma^2}\sum_{i=1}^{n}(x_i - \mu)^2\right]$$

$$\sup_{(\mu,\sigma_0^2)\in\Theta_0} L(\mu, \sigma^2|X) = \left(\frac{1}{\sigma\sqrt{2\pi}}\right)^n \exp\left[-\frac{s^2}{2\sigma_0^2}\right]$$

ML estimate of μ and σ^2 is $\hat{\mu} = \bar{x}$ and $\hat{\sigma^2} = \frac{\sum(x_i-\bar{x})^2}{n} = \frac{s^2}{n}$

$$\sup_{\mu,\sigma^2} L(\mu, \sigma^2|X) = \left(\frac{\sqrt{n}}{s\sqrt{2\pi}}\right)^n \exp\left(-\frac{n}{2}\right)$$

LR is

$$\lambda(x) = \left(\frac{s}{\sqrt{n}\sigma_0}\right)^n \exp\left[-\frac{1}{2}\left\{\frac{s^2}{\sigma_0^2} - n\right\}\right]$$

The CR is $\lambda(x) < k$

$$\Rightarrow \left(\frac{s}{\sqrt{n}\sigma_0}\right)^n \exp\left[-\frac{1}{2}\left\{\frac{s^2}{\sigma_0^2} - n\right\}\right] < k$$

Let $w = \frac{s^2}{\sigma_0^2} \sim \chi^2_{n-1}$

$$\Rightarrow \left(\frac{w}{n}\right)^{\frac{n}{2}} \exp\left(-\frac{w}{2}\right) < k$$

$\Rightarrow w < k_1$ or $w > k_2$
Under H_0

$$P_{H_0}[w < k_1] + P_{H_0}[w > k_2] = \alpha$$

Distributing the error probability, i.e., α equally in tails, we get $k_1 = \chi^2_{n-1,1-\frac{\alpha}{2}}$ and $k_2 = \chi^2_{n-1,\frac{\alpha}{2}}$

The LR test is as

$$\phi(x) = \begin{cases} 1 ; \frac{s^2}{\sigma^2} \leq \chi^2_{n-1,1-\frac{\alpha}{2}} \text{ or } \frac{s^2}{\sigma^2} \geq \chi^2_{n-1,\frac{\alpha}{2}} \\ 0 ; \text{otherwise} \end{cases}$$

Example 8.5.3 Let $X_1, X_2, \ldots, X_m$ and $Y_1, Y_2, \ldots, Y_n$ be independent rvs from $N(\mu_1, \sigma_1^2)$ and $N(\mu_2, \sigma_2^2)$, respectively. Obtain the likelihood ratio test for $H_0 : \mu_1 = \mu_2$ against $H_1 : \mu_1 \neq \mu_2$ under two conditions.

(i) σ_1^2 and σ_2^2 known
(ii) $\sigma_1^2 = \sigma_2^2 = \sigma^2$ unknown

(i) The likelihood function for $(\mu_1, \mu_2) \in \Theta$ is given as

$$L(\mu_1, \mu_2 | X, Y) = \left(\frac{1}{\sigma_1\sqrt{2\pi}}\right)^m \exp\left[-\frac{1}{2\sigma_1^2}\sum_{i=1}^{m}(x_i - \mu_1)^2\right]$$

$$\times \left(\frac{1}{\sigma_2\sqrt{2\pi}}\right)^n \exp\left[-\frac{1}{2\sigma_2^2}\sum_{i=1}^{n}(y_i - \mu_2)^2\right] \quad (8.5.5)$$

MLE of $\mu_1 = \hat{\mu}_1 = \bar{x}$ and MLE of $\mu_2 = \hat{\mu}_2 = \bar{y}$.

$$\sup_{\mu_1,\mu_2} L(\mu_1, \mu_2 | x, y) = \left(\frac{1}{\sigma_1\sqrt{2\pi}}\right)^m \exp\left[-\frac{1}{2\sigma_1^2}\sum_{i=1}^{m}(x_i - \bar{x})^2\right]$$

$$\times \left(\frac{1}{\sigma_2\sqrt{2\pi}}\right)^n \exp\left[-\frac{1}{2\sigma_2^2}\sum_{i=1}^{n}(y_i - \bar{y})^2\right] \quad (8.5.6)$$

Likelihood function for $\mu \in \Theta_0$ is given as

$$L(\mu|X, Y) = \left(\frac{1}{\sigma_1\sqrt{2\pi}}\right)^m \left(\frac{1}{\sigma_1\sqrt{2\pi}}\right)^n \exp\left[-\frac{1}{2\sigma_1^2}\sum_{i=1}^m (x_i - \mu)^2\right] \exp\left[-\frac{1}{2\sigma_2^2}\sum_{i=1}^n (y_i - \mu)^2\right]$$

$$\frac{\partial \log L(\mu|x, y)}{\partial \mu} = 0$$

$$\Rightarrow \hat{\mu} = MLE \ of \ \mu = \frac{\frac{m\bar{x}}{\sigma_1^2} + \frac{n\bar{y}}{\sigma_2^2}}{\frac{m}{\sigma_1^2} + \frac{n}{\sigma_2^2}} \tag{8.5.7}$$

$$\sup_{\mu \in \Theta_0} L(\mu|X, Y) = \left(\frac{1}{\sigma_1\sqrt{2\pi}}\right)^m \left(\frac{1}{\sigma_1\sqrt{2\pi}}\right)^n \exp\left[-\frac{1}{2\sigma_1^2}\sum_{i=1}^m (x_i - \hat{\mu})^2 - \frac{1}{2\sigma_2^2}\sum_{i=1}^n (y_i - \hat{\mu})^2\right]$$

$$= \left(\frac{1}{\sigma_1\sqrt{2\pi}}\right)^m \left(\frac{1}{\sigma_1\sqrt{2\pi}}\right)^n \exp\left[-\frac{1}{2\sigma_1^2}\left\{\sum_{i=1}^m (x_i - \bar{x})^2 + m(\bar{x} - \hat{\mu})^2\right\}\right]$$

$$\times \exp\left[-\frac{1}{2\sigma_2^2}\left\{\sum_{i=1}^n (y_i - \bar{y})^2 + n(\bar{y} - \mu)^2\right\}\right] \tag{8.5.8}$$

$$\lambda(x, y) = \frac{\sup_{\mu \in \Theta_0} L(\mu|X, Y)}{\sup_{\mu_1, \mu_2} L(\mu_1, \mu_2|x, y)} \tag{8.5.9}$$

From (8.5.6) and (8.5.8),

$$\lambda(x, y) = \exp\left[-\frac{m(\bar{x} - \hat{\mu})^2}{2\sigma_1^2} - \frac{n(\bar{y} - \hat{\mu})^2}{2\sigma_2^2}\right] \tag{8.5.10}$$

From (8.5.7)

$$\hat{\mu} = \frac{\frac{\bar{x}}{\left(\frac{\sigma_1^2}{m}\right)} + \frac{\bar{y}}{\left(\frac{\sigma_2^2}{n}\right)}}{\frac{m}{\sigma_1^2} + \frac{n}{\sigma_2^2}}$$

Let $a_1 = \frac{\sigma_1^2}{m}, a_2 = \frac{\sigma_2^2}{n}$

$$= \frac{\frac{\bar{x}}{a_1} + \frac{\bar{y}}{a_2}}{\frac{1}{a_1} + \frac{1}{a_2}} = \frac{a_2\bar{x} + a_1\bar{y}}{a_2 + a_1}$$

$$\hat{\mu} = \frac{\frac{\sigma_2^2}{n}\bar{x} + \frac{\sigma_1^2}{m}\bar{y}}{\frac{\sigma_2^2}{n} + \frac{\sigma_1^2}{m}}$$

Consider $\bar{x} - \hat{\mu}$

$$= \bar{x} - \frac{\frac{\sigma_2^2}{n}\bar{x} + \frac{\sigma_1^2}{m}\bar{y}}{\frac{\sigma_2^2}{n} + \frac{\sigma_1^2}{m}}$$

$$= \frac{\bar{x}[\frac{\sigma_2^2}{n} + \frac{\sigma_1^2}{m}] - \frac{\sigma_2^2\bar{x}}{n} - \frac{\sigma_1^2\bar{y}}{m}}{\frac{\sigma_2^2}{n} + \frac{\sigma_1^2}{m}}$$

$$\bar{x} - \hat{\mu} = \frac{\frac{\sigma_1^2}{m}(\bar{x} - \bar{y})}{\frac{\sigma_2^2}{n} + \frac{\sigma_1^2}{m}} \qquad (8.5.11)$$

$$\bar{y} - \hat{\mu} = \frac{\frac{\sigma_2^2}{n}(\bar{y} - \bar{x})}{\frac{\sigma_1^2}{m} + \frac{\sigma_2^2}{n}} \qquad (8.5.12)$$

From (8.5.10),

$$\lambda(x, y) = \exp\left[-\frac{m}{2\sigma_1^2}\left\{\frac{\frac{\sigma_1^2}{m}(\bar{x} - \bar{y})}{\frac{\sigma_1^2}{m} + \frac{\sigma_2^2}{n}}\right\}^2 - \frac{n}{2\sigma_2^2}\left\{\frac{\frac{\sigma_2^2}{n}(\bar{y} - \bar{x})}{\frac{\sigma_1^2}{m} + \frac{\sigma_2^2}{n}}\right\}^2\right]$$

$$= \exp\left[-\frac{1}{2}\frac{(\bar{x} - \bar{y})^2}{(\frac{\sigma_1^2}{m} + \frac{\sigma_2^2}{n})^2}\right]$$

$$\lambda(x, y) < k \Leftrightarrow \left(\frac{(\bar{x} - \bar{y})^2}{\frac{\sigma_1^2}{m} + \frac{\sigma_2^2}{n}}\right)^2 > k$$

$$\Leftrightarrow \left|\frac{(\bar{x} - \bar{y})}{\sqrt{\frac{\sigma_1^2}{m} + \frac{\sigma_2^2}{n}}}\right| > k \qquad (8.5.13)$$

Since, $\bar{x} \sim N(\mu_1, \frac{\sigma_1^2}{m}), \bar{y} \sim N(\mu_2, \frac{\sigma_2^2}{n})$ and $\bar{X}$ and $\bar{Y}$ are independent, then $\bar{x} - \bar{y} \sim N(\mu_1 - \mu_2, \frac{\sigma_1^2}{m} + \frac{\sigma_2^2}{n})$

$$\frac{(\bar{x} - \bar{y}) - (\mu_1 - \mu_2)}{\sqrt{\frac{\sigma_1^2}{m} + \frac{\sigma_2^2}{n}}} \sim N(0, 1) \tag{8.5.14}$$

Under H_0,

$$\frac{\bar{x} - \bar{y}}{\sqrt{\frac{\sigma_1^2}{m} + \frac{\sigma_2^2}{n}}} \sim N(0, 1)$$

LR test is given as

$$\phi(x, y) = \begin{cases} 1 \; ; & \frac{|\bar{x} - \bar{y}|}{\sqrt{\frac{\sigma_1^2}{m} + \frac{\sigma_2^2}{n}}} > k \\ 0 \; ; & \text{otherwise} \end{cases}$$

$$E_{H_0}\phi(x, y) = \alpha \Rightarrow P\left[\frac{\bar{x} - \bar{y}}{\sqrt{\frac{\sigma_1^2}{m} + \frac{\sigma_2^2}{n}}} > k\right] = \frac{\alpha}{2} \Rightarrow k = Z_{\frac{\alpha}{2}}$$

LR test will become

$$\phi(x, y) = \begin{cases} 1 \; ; & \frac{|\bar{x} - \bar{y}|}{\sqrt{\frac{\sigma_1^2}{m} + \frac{\sigma_2^2}{n}}} > Z_{\frac{\alpha}{2}} \\ 0 \; ; & \text{otherwise} \end{cases}$$

(ii) The maximum likelihood estimate of μ_1, μ_2 and σ^2 will be $\hat{\mu}_1 = \bar{x}$, $\hat{\mu}_2 = \bar{y}$
$\hat{\sigma}^2 = \frac{1}{m+n}[\sum_{i=1}^{m}(x_i - \bar{x})^2 + \sum_{i=1}^{n}(y_i - \bar{y})^2] = \frac{s_1^2 + s_2^2}{m+n}$
From (8.5.5) and substituting these ML estimates,
we get,

$$L(\mu_1, \mu_2, \sigma^2 | x, y) = \left[\frac{m+n}{\sqrt{2\pi(s_1^2 + s_2^2)}}\right]^{m+n} \exp\left[-\frac{1}{2}(m+n)\right]$$

$$\sup_{\mu_1, \mu_2 \in \Theta} L(\mu_1, \mu_2, \sigma^2 | x, y) = \left[\frac{m+n}{\sqrt{2\pi(s_1^2 + s_2^2)}}\right]^{m+n} \exp\left[-\frac{1}{2}(m+n)\right]$$

Under H_0 :

$$L(\mu, \sigma^2 | x, y) = \left(\frac{1}{\sigma\sqrt{2\pi}}\right)^{m+n} \exp\left[-\frac{1}{2\sigma^2}\left\{\sum_{i=1}^{m}(x_i - \mu)^2 + \sum_{i=1}^{n}(y_i - \mu)^2\right\}\right]$$

ML estimate of μ and σ^2 is $\hat{\mu} = \frac{m\bar{x} + n\bar{y}}{m+n}$

$$\hat{\sigma}^2 = \frac{1}{m+n}\left[\sum_{i=1}^{m}(x_i - \mu)^2 + \sum_{i=1}^{n}(y_i - \mu)^2\right]$$

$$= \frac{1}{m+n}\left[\sum_{i=1}^{m}(x_i - \bar{x})^2 + m(\bar{x} - \hat{\mu})^2 + \sum_{i=1}^{n}(y_i - \bar{y})^2 + n(\bar{y} - \hat{\mu})^2\right]$$

Now

$$(\bar{x} - \hat{\mu})^2 = \left[\bar{x} - \frac{m\bar{x} + n\bar{y}}{m+n}\right]^2 = \left[\frac{n(\bar{x} - \bar{y})}{m+n}\right]^2$$

$$(\bar{y} - \hat{\mu})^2 = \left[\bar{y} - \frac{m\bar{x} + n\bar{y}}{m+n}\right]^2 = \left[\frac{n(\bar{y} - \bar{x})}{m+n}\right]^2$$

$$\hat{\sigma}^2 = \frac{1}{m+n}\left[s_1^2 + s_2^2 + \frac{mn^2(\bar{x} - \bar{y})^2}{(m+n)^2} + \frac{nm^2(\bar{y} - \bar{x})^2}{(m+n)^2}\right] \qquad (8.5.15)$$

$$\hat{\sigma}^2 = \frac{1}{m+n}\left[s_1^2 + s_2^2 + \frac{mn(\bar{x} - \bar{y})^2}{m+n}\right] \qquad (8.5.16)$$

$$\sup_{(\mu,\sigma^2)\in\Theta_0} L(\mu, \sigma^2 | x, y) = \left[\frac{m+n}{\sqrt{2\pi[s_1^2 + s_2^2 + \frac{mn}{m+n}(\bar{x} - \bar{y})^2]}}\right]^{m+n} e^{-(\frac{m+n}{2})} \qquad (8.5.17)$$

$$\lambda(x, y) = \frac{\sup_{(\mu,\sigma^2)\in\Theta_0} L(\mu, \sigma^2 | x, y)}{\sup_{(\mu_1,\mu_2,\sigma^2)\in\Theta_0} L(\mu_1, \mu_2, \sigma^2 | x, y)}$$

$$= \left[\frac{s_1^2 + s_2^2}{s_1^2 + s_2^2 + \frac{mn}{m+n}(\bar{x} - \bar{y})^2}\right]^{(\frac{m+n}{2})}$$

$$= \left[1 + \frac{mn(\bar{x} - \bar{y})^2}{(m+n)(s_1^2 + s_2^2)}\right]^{-(\frac{m+n}{2})} \qquad (8.5.18)$$

$\bar{x} \sim N(\mu_1, \frac{\sigma^2}{m})$ and $\bar{y} \sim N(\mu_2, \frac{\sigma^2}{n})$. $\bar{x} - \bar{y} \sim N(\mu_1 - \mu_2, \sigma^2(\frac{1}{m} + \frac{1}{n}))$

$$= \left[\frac{(\bar{x} - \bar{y}) - (\mu_1 - \mu_2)}{\sigma\sqrt{\frac{1}{m} + \frac{1}{n}}} \right] \sim N(0, 1)$$

$$\frac{s_1^2}{\sigma^2} = \frac{\sum(x_i - \bar{x})^2}{\sigma^2} \sim \chi_{m-1}^2$$

$$\frac{s_2^2}{\sigma^2} = \frac{\sum(y_i - \bar{y})^2}{\sigma^2} \sim \chi_{n-1}^2$$

$$\frac{s_1^2 + s_2^2}{\sigma^2} = \chi_{m+n-2}^2$$

under $H_0 : \mu_1 = \mu_2$

$$t = \frac{\frac{(\bar{x} - \bar{y})}{[\sigma\sqrt{\frac{1}{m} + \frac{1}{n}}]}}{\sqrt{\frac{s_1^2 + s_2^2}{\sigma^2(m+n-2)}}} \sim t_{m+n-2}$$

$$= \frac{\sqrt{\frac{mn}{m+n}}(\bar{x} - \bar{y})}{\sqrt{\frac{s_1^2 + s_2^2}{\sigma^2(m+n-2)}}} \sim t_{m+n-2}$$

$$t^2 = \frac{mn(\bar{x} - \bar{y})^2}{(m+n)(s_1^2 + s_2^2)}(m + n - 2)$$

$$\frac{t^2}{m+n-2} = \frac{mn(\bar{x} - \bar{y})^2}{(m+n)(s_1^2 + s_2^2)} \tag{8.5.19}$$

From (8.5.18)

$$\lambda^{\frac{2}{m+n}}(x, y) = \frac{1}{1 + \frac{mn(\bar{x} - \bar{y})^2}{(m+n)(s_1^2 + s_2^2)}} = \left[1 + \frac{mn(\bar{x} - \bar{y})^2}{(m+n)(s_1^2 + s_2^2)} \right]^{-1}$$

Using (8.5.19),

$$\lambda^{\frac{2}{m+n}}(x, y) = \left[1 + \frac{t^2}{m+n-2} \right]^{-1} = \frac{m+n-2}{m+n-2+t^2}$$

$$\lambda^{\frac{2}{m+n}}(x,y) < k \Rightarrow \frac{m+n-2}{m+n-2} < k \Rightarrow t^2 > k \Rightarrow |t| > k$$

k is obtained such that

$$P_{H_0}[|T| > k] = \alpha$$

Now k is an upper $\frac{\alpha}{2}$ th quantile of t distribution with df $m+n-2$.

Then LR test is given as

$$\phi(x,y) = \begin{cases} 1 \; ; & \frac{|\bar{x}-\bar{y}|}{s\sqrt{\frac{1}{m}+\frac{1}{n}}} > t_{\frac{\alpha}{2},m+n-2} \\ 0 \; ; & \text{otherwise} \end{cases}$$

where $s^2 = \frac{s_1^2 + s_2^2}{m+n-2}$

Example 8.5.4 Let $X_1, X_2, \ldots, X_n$ and $Y_1, Y_2, \ldots, Y_n$ be iid rvs with $N(\mu_1, \sigma_1^2)$ and $N(\mu_2, \sigma_2^2)$, respectively. Find the likelihood ratio test for testing

(a) $H_0 : \mu_1 \leq \mu_2$ against $H_1 : \mu_1 > \mu_2$
(b) $H_0 : \mu_1 \geq \mu_2$ against $H_1 : \mu_1 < \mu_2$

when the population variance (i) σ_1^2 and σ_2^2 known (ii) $\sigma_1^2 = \sigma_2^2 = \sigma^2$ but unknown.

According to note in Example 8.5.1, one can get the following test:

(i) σ_1^2 and σ_2^2 known

$$(a)\phi(x,y) = \begin{cases} 1 \; ; & \frac{(\bar{x}-\bar{y})}{\sqrt{\frac{\sigma_1^2}{m}+\frac{\sigma_2^2}{n}}} > Z_\alpha \\ 0 \; ; & \text{otherwise} \end{cases}$$

$$(b)\phi(x,y) = \begin{cases} 1 \; ; & \frac{(\bar{x}-\bar{y})}{\sqrt{\frac{\sigma_1^2}{m}+\frac{\sigma_2^2}{n}}} > -Z_\alpha \\ 0 \; ; & \text{otherwise} \end{cases}$$

(ii) $\sigma_1^2 = \sigma_2^2 = \sigma^2$ but unknown.

$$(a)\phi(x,y) = \begin{cases} 1 \; ; & \frac{(\bar{x}-\bar{y})}{s\sqrt{\frac{1}{m}+\frac{1}{n}}} > t_{m+n-2,\alpha} \\ 0 \; ; & \text{otherwise} \end{cases}$$

$$(b)\phi(x,y) = \begin{cases} 1 \; ; & \frac{(\bar{x}-\bar{y})}{s\sqrt{\frac{1}{m}+\frac{1}{n}}} > -t_{m+n-2,\alpha} \\ 0 \; ; & \text{otherwise} \end{cases}$$

where $s^2 = \frac{s_1^2 + s_2^2}{m+n-2}$.

Example 8.5.5 A random sample $X_1, X_2, \ldots, X_n$ is taken from $N(\mu, \sigma^2)$. Find the likelihood ratio test of

(a) $H_0 : \sigma^2 = \sigma_0^2$ against $H_1 : \sigma \neq \sigma_0^2$
(b) $H_0 : \sigma^2 \leq \sigma_0^2$ against $H_1 : \sigma^2 > \sigma_0^2$
(c) $H_0 : \sigma^2 > \sigma_0^2$ against $H_1 : \sigma^2 \leq \sigma_0^2$

under the conditions (i) μ known (ii) μ unknown.

(i) μ known

$$(a) L(\sigma^2, \mu | x) = \left(\frac{1}{\sigma \sqrt{2\pi}} \right) \exp \left[-\frac{1}{2\sigma^2} \sum_{i=1}^{n} (x_i - \mu)^2 \right] \qquad (8.5.20)$$

$$\sup_{\sigma^2 \in \Theta_0} L(\sigma^2 | x) = \left(\frac{1}{\sigma_0 \sqrt{2\pi}} \right)^n \exp \left[-\frac{1}{2\sigma_0^2} \sum_{i=1}^{n} (x_i - \mu)^2 \right]$$

From (8.5.20), mle of $\sigma^2 = \hat{\sigma}^2 = \frac{1}{n} \sum_{i=1}^{n} (x_i - \mu)^2$

$$\sup_{\sigma^2 \in \Theta} L(\sigma^2 | x) = \left[\frac{n}{2\pi \sum_{i=1}^{n} (x_i - \mu)^2} \right]^{\frac{n}{2}} \exp \left[-\frac{n}{2} \right]$$

$$\lambda(x) = \frac{\sup_{\sigma^2 \in \Theta_0} L(\sigma^2 | x)}{\sup_{\sigma^2 \in \Theta} L(\sigma^2 | X)}$$

$$= \left[\frac{\sum_{i=1}^{n} (x_i - \mu)^2}{n \sigma_0^2} \right]^{\frac{n}{2}} \exp \left[-\frac{1}{2\sigma_0^2} \sum_{i=1}^{n} (x_i - \mu)^2 + \frac{n}{2} \right]$$

Let $w = \frac{\sum_{i=1}^{n} (x_i - \mu)^2}{\sigma_0^2} \sim \chi_n^2$

$$\lambda(x) = \left(\frac{w}{n} \right)^{\frac{n}{2}} \exp \left(-\frac{w}{2} - \frac{n}{2} \right) < k$$

$$\Rightarrow w^{\frac{n}{2}} e^{-\frac{w}{2}} < k$$

Let $f(w) = w^{\frac{n}{2}} e^{-\frac{w}{2}}$
Plot the function w verses $f(w)$.
$f(w) < k \Rightarrow w < k_1$ and $w > k_2$; see Fig. 8.3
LR test is given as

$$\phi(x) = \begin{cases} 1 \; ; \; w < k_1 \text{ or } w > k_2 \\ 0 \; ; \text{ otherwise} \end{cases}$$

Fig. 8.3 Graph of $f(w)$

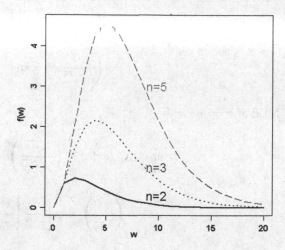

$$E_{H_0}\phi(x) = P[w < k_1] + P[w > k_2] = \alpha$$

One should note that χ^2 is not a symmetric distribution but for mathematical convenience, we consider equal probability $\frac{\alpha}{2}$ on the right and left side of the critical region.

Hence $k_1 = \chi^2_{n,1-\frac{\alpha}{2}}$ and $k_2 = \chi^2_{n,\frac{\alpha}{2}}$

Our LR test is as follows:

$$\phi(x) = \begin{cases} 1 \; ; & \frac{\sum(x_i-\mu_0)^2}{\sigma_0^2} < \chi^2_{n,1-\frac{\alpha}{2}} \text{ or } \frac{\sum(x_i-\mu_0)^2}{\sigma_0^2} > \chi^2_{n,\frac{\alpha}{2}} \\ 0 \; ; & \text{otherwise} \end{cases}$$

(b) Similarly, LR test for testing $H_0 : \sigma^2 \leq \sigma_0^2$ against $H_1 : \sigma^2 > \sigma_0^2$ is given as

$$\phi(x) = \begin{cases} 1 \; ; & \frac{\sum(x_i-\mu_0)^2}{\sigma_0^2} > \chi^2_{n,\alpha} \\ 0 \; ; & \text{otherwise} \end{cases}$$

(c) Further, LR test for testing $H_0 : \sigma^2 \geq \sigma_0^2$ against $H_1 : \sigma^2 < \sigma_0^2$ is given as

$$\phi(x) = \begin{cases} 1 \; ; & \frac{\sum(x_i-\mu_0)^2}{\sigma_0^2} < \chi^2_{n,1-\alpha} \\ 0 \; ; & \text{otherwise} \end{cases}$$

(ii) μ unknown

(a) In this case mle of $\mu = \hat{\mu} = \bar{x}$

$$\sup_{(\mu,\sigma_0^2)\in\Theta_0} L(\mu,\sigma_0^2|x) = \left(\frac{1}{\sigma_0\sqrt{2\pi}}\right)^n \exp\left[-\frac{1}{2\sigma_0^2}\sum(x_i-\bar{x})^2\right]$$

$$= \left(\frac{1}{\sigma_0\sqrt{2\pi}}\right)^n \exp\left[-\frac{s^2}{2\sigma_0^2}\right]$$

MLE of $\sigma^2 = \hat{\sigma}^2 = \frac{s^2}{n}$

$$\sup_{(\mu,\sigma^2)\in\Theta} L(\mu,\sigma^2|x) = \left(\frac{\sqrt{n}}{s\sqrt{2\pi}}\right)^n \exp\left[-\frac{n}{2}\right]$$

$$\lambda(x) = \left(\frac{s^2}{n\sigma_0^2}\right)^{\frac{n}{2}} \exp\left[-\frac{1}{2}\left\{\frac{s^2}{\sigma_0^2}-n\right\}\right]$$

under H_0, $\frac{s^2}{\sigma_0^2} \sim \chi_{n-1}^2$

Let $w = \frac{s^2}{\sigma_0^2}$

$$\lambda(x) = \left(\frac{w}{n}\right)^{\frac{n}{2}} \exp\left[-\frac{1}{2}\{w-n\}\right]$$

$$Now\ \lambda(x) < k \Leftrightarrow w^{\frac{n}{2}}e^{-\frac{w}{2}} < k \Leftrightarrow w < k_1 \text{ and } w > k_2. \tag{8.5.21}$$

LR test is given as

$$\phi(x) = \begin{cases} 1 \text{ ; } w < k_1 \text{ or } w > k_2 \\ 0 \text{ ; otherwise} \end{cases}$$

$$E_{H_0}\phi(x) = \alpha \Rightarrow P_{H_0}(w < k_1) + P_{H_0}(w > k_2) = \alpha$$

$$\Rightarrow k_1 = \chi_{n-1,1-\frac{\alpha}{2}}^2, k_2 = \chi_{n-1,\frac{\alpha}{2}}^2$$

Now LR test is given as

$$\phi(x) = \begin{cases} 1 \text{ ; } \frac{s^2}{\sigma_0^2} \leq \chi_{n-1,1-\frac{\alpha}{2}}^2 \text{ or } \frac{s^2}{\sigma_0^2} > \chi_{n-1,\frac{\alpha}{2}}^2 \\ 0 \text{ ; otherwise} \end{cases}$$

(b) Similarly, LR test for testing $H_0 : \sigma^2 \leq \sigma_0^2$ against $H_1 : \sigma^2 > \sigma_0^2$ is given as

$$\phi(x) = \begin{cases} 1 \text{ ; } \frac{s^2}{\sigma_0^2} \geq \chi_{n-1,\alpha}^2 \\ 0 \text{ ; otherwise} \end{cases}$$

(c) Further, LR test for testing $H_0 : \sigma^2 \geq \sigma_0^2$ against $H_1 : \sigma^2 < \sigma_0^2$

$$\phi(x) = \begin{cases} 1 \; ; \; \frac{s^2}{\sigma_0^2} \leq \chi_{n-1,1-\alpha}^2 \\ 0 \; ; \text{otherwise} \end{cases}$$

Example 8.5.6 Let $X_1, X_2, \ldots, X_m$ and $Y_1, Y_2, \ldots, Y_n$ be two independent sample from (μ_1, σ_1^2) and (μ_2, σ_2^2), respectively. Find out the likelihood ratio test of

(a) $H_0 : \sigma_1^2 = \sigma_2^2$ against $H_1 : \sigma_1^2 \neq \sigma_2^2$
(b) $H_0 : \sigma_1^2 \leq \sigma_2^2$ against $H_1 : \sigma_1^2 > \sigma_2^2$

where μ_1 and μ_2 are unknown.

The likelihood equation for $(\mu_1, \mu_2, \sigma_1^2, \sigma_2^2)$ is given as

$$L(\mu_1, \mu_2, \sigma_1^2, \sigma_2^2) = \left(\frac{1}{2\pi\sigma_1^2}\right)^{\frac{n}{2}} \exp\left[-\frac{1}{2\sigma_1^2}\sum_{i=1}^{n}(x_i - \mu_1)^2\right]\left(\frac{1}{2\pi\sigma_2^2}\right)^{\frac{n}{2}} \exp\left[-\frac{1}{2\sigma_2^2}\sum_{i=1}^{n}(y_i - \mu_2)^2\right]$$

MLE of $(\mu_1, \mu_2, \sigma_1^2, \sigma_2^2)$ is
$\hat{\mu}_1 = \bar{x}, \hat{\mu}_2 = \bar{x}, \hat{\sigma_1^2} = \frac{s_1^2}{m}, \hat{\sigma_2^2} = \frac{s_2^2}{n},$
where $s_1^2 = \sum(x_i - \bar{x})^2, s_2^2 = \sum(y_i - \bar{y})^2$

$$\sup_{\mu_1,\mu_2,\sigma_1^2,\sigma_2^2 \in \Theta} L(\mu_1, \mu_2, \sigma_1^2, \sigma_2^2 | X, Y) = \left(\frac{m}{2\pi s_1^2}\right)^{\frac{m}{2}} \left(\frac{n}{2\pi s_2^2}\right)^{\frac{m}{2}} \exp\left[-\frac{m+n}{2}\right]$$

The likelihood equation for $(\mu_1, \mu_2, \sigma^2) \in \Theta_0$

$$L(\mu_1, \mu_2, \sigma^2 | x, y) = \left(\frac{1}{2\pi\sigma^2}\right)^{\frac{m+n}{2}} \exp\left[-\frac{1}{2\sigma^2}\sum_{i=1}^{m}(x_i - \mu_1)^2 - \frac{1}{2\sigma^2}\sum_{i=1}^{n}(y_i - \mu_2)^2\right]$$

MLE of (μ_1, μ_2, σ^2),
$\hat{\mu}_1 = \bar{x}, \hat{\mu}_2 = \bar{x}, \hat{\sigma}^2 = \frac{1}{m+n}[\frac{s_1^2}{m} + \frac{s_2^2}{n}]$

$$\sup_{\mu_1,\mu_2,\sigma^2 \in \Theta_0} L(\mu_1, \mu_2, \sigma^2 | x, y) = \left[\frac{m+n}{2\pi\left(\frac{s_1^2}{m} + \frac{s_2^2}{n}\right)}\right]^{\frac{m+n}{2}} \exp\left[-\frac{(m+n)}{2}\right]$$

$$\lambda(x, y) = \frac{\sup_{\mu_1,\mu_2,\sigma^2 \in \Theta_0} L(\mu_1, \mu_2, \sigma^2 | x, y)}{\sup_{\mu_1,\mu_2,\sigma_1^2,\sigma_2^2 \in \Theta} L(\mu_1, \mu_2, \sigma_1^2, \sigma_2^2 | x, y)}$$

$$= \frac{(m+n)^{\frac{m+n}{2}} \left(\frac{s_1^2}{m}\right)^{\frac{m}{2}} \left(\frac{s_2^2}{n}\right)^{\frac{n}{2}}}{\left(\frac{s_1^2}{m} + \frac{s_2^2}{n}\right)^{\frac{m+n}{2}}}$$

Define

$$F = \frac{\frac{\sum(x_i - \bar{x})^2}{m-1}}{\frac{\sum(x_i - \bar{x})^2}{n-1}} = \frac{\frac{s_1^2}{m-1}}{\frac{s_2^2}{n-1}} \sim F_{m-1, n-1}$$

Let $\frac{s_1^2}{s_2^2} = \frac{(m-1)F}{(n-1)}$

$$\lambda(x, y) = \frac{(m+n)^{\frac{m+n}{2}} \left(\frac{s_1^2}{m}\right)^{\frac{m}{2}} \left(\frac{s_2^2}{n}\right)^{\frac{n}{2}}}{\left(\frac{\frac{s_1^2}{m}}{\frac{s_2^2}{n}} + 1\right)^{\frac{m+n}{2}} \left(\frac{s_2^2}{n}\right)^{\frac{m+n}{2}}}$$

$$= \frac{(m+n)^{\frac{m+n}{2}} \left(\frac{\frac{s_1^2}{m}}{\frac{s_2^2}{n}}\right)^{\frac{m}{2}}}{\left(\frac{\frac{s_1^2}{m}}{\frac{s_2^2}{n}} + 1\right)^{\frac{m+n}{2}}}$$

$$= \frac{(m+n)^{\frac{m+n}{2}} \left[\frac{n(m-1)F}{m(n-1)}\right]^{\frac{m}{2}}}{\left[1 + \frac{n(m-1)F}{m(n-1)}\right]^{\frac{m+n}{2}}}$$

Now $\lambda(x, y) < k \Leftrightarrow F < k_1$ or $F > k_2$
LR test is given as

$$\phi(x, y) = \begin{cases} 1 \; ; \; \frac{\frac{s_1^2}{m-1}}{\frac{s_2^2}{n-1}} < k_1 \; \text{ or } \; \frac{\frac{s_1^2}{m-1}}{\frac{s_2^2}{n-1}} > k_2 \\ 0 \; ; \; \text{otherwise} \end{cases}$$

Now $\frac{\frac{s_1^2}{m-1}}{\frac{s_2^2}{n-1}} \sim F_{m-1, n-1}$

$$P_{H_0}\left[\frac{\frac{s_1^2}{m-1}}{\frac{s_2^2}{n-1}} < k_1\right] = P_{H_0}\left[\frac{\frac{s_1^2}{m-1}}{\frac{s_2^2}{n-1}} > k_2\right] = \frac{\alpha}{2}$$

$$P_{H_0}[F < k_1] = P_{H_0}[F > k_2] = \frac{\alpha}{2}$$

$k_1 = F_{m-1,n-1,1-\frac{\alpha}{2}}$ and $k_2 = F_{m-1,n-1,\frac{\alpha}{2}}$

LR test is given as

$$\phi(x, y) = \begin{cases} 1 ; & \frac{\frac{s_1^2}{m-1}}{\frac{s_2^2}{n-1}} < F_{m-1,n-1,1-\frac{\alpha}{2}}, \ \frac{\frac{s_1^2}{m-1}}{\frac{s_2^2}{n-1}} < F_{m-1,n-1,\frac{\alpha}{2}} \\ 0 ; & \text{otherwise} \end{cases}$$

(b) Similarly, we can obtain LR test for testing $H_0 : \sigma_1^2 \leq \sigma_2^2$ against $H_1 : \sigma_1^2 > \sigma_2^2$ as

$$\phi(x, y) = \begin{cases} 1 ; & \frac{\frac{s_1^2}{m-1}}{\frac{s_2^2}{n-1}} \geq F_{m-1,n-1,\alpha} \\ 0 ; & \text{otherwise} \end{cases}$$

Further if we are testing $H_0 : \sigma_1^2 \geq \sigma_2^2$ against $H_1 : \sigma_1^2 < \sigma_2^2$, then LR test is

$$\phi(x, y) = \begin{cases} 1 ; & \frac{\frac{s_1^2}{m-1}}{\frac{s_2^2}{n-1}} \leq F_{m-1,n-1,1-\alpha} \\ 0 ; & \text{otherwise} \end{cases}$$

Note: if μ_1 and μ_2 are known then $\frac{\frac{s_1^2}{m-1}}{\frac{s_2^2}{n-1}}$ is replaced by $\frac{\frac{\sum(x_i-\mu_1)^2}{m}}{\frac{\sum(y_i-\mu_2)^2}{n}}$ in all above test.

Example 8.5.7 Let $X_1, X_2, \ldots, X_n$ be a random sample from exponential distribution with mean θ. Test the hypothesis $H_0 : \theta = \theta_0$ against $H_1 : \theta \neq \theta_0$

$$\sup_{\theta \in \Theta_0} L(\theta|X) = \sup_{\theta \in \Theta_0} \theta^{-n} e^{-\frac{t}{\theta}} = \theta_0^{-n} e^{-\frac{t}{\theta_0}}, \ where\ t = \sum_{i=1}^{n} x_i$$

MLE of $\theta = \bar{x} = \frac{t}{n}$

$$\sup_{\theta \in \Theta} L(\theta|x) = \left(\frac{t}{n}\right)^{-n} e^{-n}$$

$$\lambda(x) = \frac{\sup_{\theta \in \Theta_0} L(\theta|x)}{\sup_{\theta \in \Theta} L(\theta|x)} = \frac{\theta_0^{-n} e^{-\frac{t}{\theta_0}}}{(\frac{t}{n})^{-n} e^{-n}} = \left(\frac{t}{n\theta_0}\right)^n \exp\left(-\frac{t}{\theta_0} + n\right)$$

$$\lambda(x) < k \Rightarrow t^n \exp\left(-\frac{t}{\theta_0}\right) < k; \quad \text{see Fig. 8.4}$$

$t < k_1$ or $t > k_2$ (Fig. 8.4).

Fig. 8.4 Graph of $\lambda(x)$

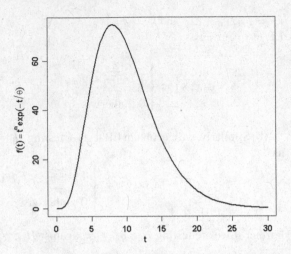

Note that $\frac{2t}{\theta} \sim \chi^2_{2n}$

Let $w = \frac{2t}{\theta} \Rightarrow w < k_1$ or $w > k_2$ LR test is given as

$$\phi(x) = \begin{cases} 1 \; ; \; w < k_1 \text{ or } w > k_2 \\ 0 \; ; \text{ otherwise} \end{cases}$$

$$E_{H_0}\phi(x) = \alpha \Rightarrow P_{H_0}(w < k_1) + P[w > k_2] = \alpha$$

$k_1 = \chi^2_{2n, 1-\frac{\alpha}{2}}$ and $k_2 = \chi^2_{2n, \frac{\alpha}{2}}$

The LR test is given as

$$\phi(x) = \begin{cases} 1 \; ; \; w < \chi^2_{2n, 1-\frac{\alpha}{2}} \text{ or } w > \chi^2_{2n, \frac{\alpha}{2}} \\ 0 \; ; \text{ otherwise} \end{cases}$$

Example 8.5.8 Let $X_1, X_2, \ldots, X_m$ be a random sample from $B(n, p)$. Find the LR test to test $H_0 : p \leq p_0$ against $H_1 : p > p_2$

$$\sup_{p \leq p_0} L(p|x) = \begin{cases} p_0^t(1 - p_0)^{mn-t} \; ; \; p_0 < \frac{\bar{x}}{n}; \bar{x} = \frac{\sum_{i=1}^m x_i}{m} \\ (\frac{\bar{x}}{n})^t(1 - \frac{\bar{x}}{n})^{mn-t} \; ; \; \frac{\bar{x}}{n} \leq p_0 \end{cases}$$

$$\sup_{p \in \Theta} L(p|x) = \left(\frac{\bar{x}}{n}\right)^t \left(1 - \frac{\bar{x}}{n}\right)^{mn-t}$$

$$\lambda(t) = \begin{cases} \frac{p_0^t(1-p_0)^{mn-t}}{(\frac{\bar{x}}{n})^t(1-\frac{\bar{x}}{n})^{mn-t}} \; ; \; p_0 < \frac{\bar{x}}{n} \\ 1 \; ; \; \frac{\bar{x}}{n} < p_0 \end{cases}$$

$\lambda(t) = 1$ if $\bar{x} < np_0$ and $\lambda(t) \leq 1$ if $np_0 < \bar{x}$

$\Rightarrow \lambda(t)$ is decreasing function of t

Hence $\lambda(t) < k \Rightarrow t > k$

LR test is given as

$$\lambda(t) = \begin{cases} 1 \; ; \; t > k \\ \gamma \; ; \; t = k \\ 0 \; ; \; t < k \end{cases}$$

under H_0, $T \sim B(mn, p_0)$

k can be selected as

$$E\phi(t) = \alpha \Rightarrow P_{H_0}[t > k] + \gamma P[t = k] = \alpha$$

This test is same as we have discussed in Chap. 7.

Example 8.5.9 Let $X_1, X_2, \ldots, X_n$ be a random sample from $Poisson(\lambda)$. Obtain the LR test to test $H_0 : \lambda \leq \lambda_0$ against $H_1 : \lambda > \lambda_0$

$$L(\lambda|x) = \frac{e^{-n\lambda}\lambda^t}{\prod_{i=1}^n x_i!}$$

$$\sup_{\lambda \leq \lambda_0} L(\lambda|x) = \begin{cases} \frac{e^{-n\lambda_0}\lambda_0^t}{\prod_{i=1}^n x_i!} \; ; \; \lambda_0 \geq \bar{x} \\ \frac{e^{-n\bar{x}}(n\bar{x})^t}{\prod_{i=1}^n x_i!} \; ; \; \lambda_0 < \bar{x} \end{cases}$$

$$\sup_{\lambda \in \Theta} L(\lambda|x) = \frac{e^{-n\bar{x}}(n\bar{x})^t}{\prod_{i=1}^n x_i!}$$

$$\lambda(t) = \begin{cases} \exp[t - n\lambda_0](\frac{\lambda_0}{t})^t \; ; \; \lambda_0 \geq \bar{x} \\ 1 \qquad\qquad\qquad\; ; \; \lambda_0 < \bar{x} \end{cases}$$

$\lambda(t) \leq 1$ if $\lambda_0 \geq \bar{x}$ and $\lambda(t) = 1$ if $\lambda_0 < \bar{x}$

$\Rightarrow \lambda(t)$ is decreasing function of t if and only if $t > k$

The LR test is given as

$$\lambda(t) = \begin{cases} 1 \; ; \; t > k \\ \gamma \; ; \; t = k \\ 0 \; ; \; \text{otherwise} \end{cases}$$

under H_0, $T = \sum_{i=1}^n X_i \sim P(n\lambda)$

k can be selected as

$$P_{H_0}[t > k] + \gamma P_{H_0}[t = k] = \alpha$$

This test is also same as we have discussed in Chap. 7.

8.6   Exercise 8

1. Let $X_1, X_2, \ldots, X_n$ are iid rvs with $N(\mu, \sigma^2)$, σ^2 is known. Find UMP unbiased test of size α for testing $H_0 : \mu = 0$ against $H_1 : \mu \neq 0$. From the following data, test $H_0 : \mu = 2$ against $H_1 : \mu \neq 2$ for $\alpha = 0.05$
0.81   1.01 2.04 −3.17 0.57 −1.05 −3.83
2.88 −0.44 −2.23 4.09 4.00 −3.63 6.05
1.53
2. Let $X_1, X_2, \ldots, X_n$ are iid rvs from gamma distribution with parameters p(known) and σ unknown.

 (i) Obtain UMP test of size α for testing $H_0 : \sigma \leq \sigma_0$ against $H_1 : \sigma > \sigma_0$.
 (ii) Obtain UMPU test of size α for testing $H_0 : \sigma = \sigma_0$ against $H_1 : \sigma \neq \sigma_0$.

3. Let the rv X is $\beta(a, 1)$. Obtain UMPU test of size α to test $H_0 : a = 1$ against $H_1 : a \neq 1$
4. Let X_1 is $B(n, \theta)$ and X_2 is $NB(r, \theta)$, n and r is known. Obtain UMPU test of size α for testing $H_0 : \theta = \theta_0$ against $H_1 : \theta \neq \theta_0$.
Let $n = 5, r = 4$ and $\alpha = 0.02$
Test $H_0 : \theta = 0.3$ against $H_1 : \theta \neq 0.3$ for the given data as $X_1 = 3$ and $X_2 = 4$.
5. Let X_1 is $\cup(0, \theta_1)$ and X_2 is $\cup(0, \theta_2)$. Obtain UMP test of size α for testing $H_0 : \theta_1 = \theta_2$ against $H_1 : \theta_1 \neq \theta_2$.
6. Let X_1 is $G(1, \frac{1}{\theta_1})$ and X_2 is $G(1, \frac{1}{\theta_2})$. Obtain UMPU test of size α for testing $H_0 : \theta_1 = \theta_2$ against $H_1 : \theta_1 \neq \theta_2$.
7. Let $X_1, X_2, \ldots, X_n$ be a random sample from $f(x|\theta)$, where

$$f(x|\theta) = \begin{cases} \frac{1}{2\theta} \exp[- \mid \frac{x}{\theta} \mid] \; ; \; x \in R \; , \theta > 0 \\ 0 \qquad\qquad\quad ; \text{ otherwise} \end{cases}$$

Obtain UMPU test for testing $H_0 : \theta = \theta_0$ against $H_1 : \theta \neq \theta_0$ of size α.
8. Let X be a rv with $B(n, p)$ and consider the hypothesis $H_0 : p = p_0$ of size α. Determine the boundary values of the UMP unbiased test for $n = 10$, $\alpha = 0.1$, $p_0 = 0.3$ and $\alpha = 0.05$, $p_0 = 0.4$. In each case, plot the graph of the power function of both the unbiased and the equal tails tests. Use R
9. Let $X_1, X_2, \ldots, X_n$ be a sample from $P(\lambda)$. Find UMPU test to test

 (i) $H_0 : \lambda \leq \lambda_0$ against $H_1 : \lambda > \lambda_0$
 (ii) $H_0 : \lambda = \lambda_0$ against $H_1 : \lambda \neq \lambda_0$

Assume that size of test is α.
Test (i) $H_0 : \lambda \leq 3$ against $H_1 : \lambda > 3$
(ii) $H_0 : \lambda = 5$ against $H_1 : \lambda \neq 5$ for the following sample.
2, 1, 4, 3, 5, 0, 2, 2, 1, 3. Use R.
10. Let $\frac{T_n}{\theta}$ have χ_n^2. For testing $H_0 : \theta = 3$, at level of significance $\alpha = 0.05$, find n so large that the power of the test is 0.96 against both $\theta \geq 4$ or $\theta \leq 1$. How large does n have to be if the test is not required to be unbiased? (Use R)

11. Let X be $NB(1, \theta)>$. Find UMPU test of size α to test (i) $H_0 : \theta \leq \theta_0$ against $H_1 : \theta > \theta_0$ and (ii) $H_0 : \theta = \theta_0$ against $H_1 : \theta \neq \theta_0$
If $\theta_0 = 3$, find UMPU test for the sample
1, 3, 1, 4, 1, 1, 1, 2, 4, 2.
12. Find the locally most powerful test for testing $H_0 : \theta = 0$ against $H_1 : \theta > 0$ from the following data if $X \sim N(\theta, 1 + 2\theta^2)$,
-3.11, -1.72, 0.47, -1.32, -0.72, -0.70, 0.36
2.42, -0.98, -2.60, 0.84 ($\alpha = 0.10$, use R).
13. Find the LMPT to test $H_0 : \theta = 4$ against $H_1 : \theta > 4$ from the following data if
(i) $X \sim N(\theta, \theta)$ and (ii) $X \sim N(\theta, \theta^2)$
0.61 2.34 1.15 1.61 1.79 0.01 2.37 0.06
4.11 1.26 1.74 2.70 3.54 3.18 3.41 ($\alpha = 0.03$, use R).
14. Let $X_1, X_2, \ldots, X_n$ be iid rvs with $N(\mu, \sigma^2)$. Obtain Neyman structure test for testing $H_0 : \sigma^2 = 5$ against $H_1 : \sigma^2 > 5$, where μ is unknown, for the following data:
2.09 1.19 2.26 5.02 4.15 1.48 2.41 0.42
-0.32 4.94 -0.29 5.46 9.42 9.34 1.34 ($\alpha = 0.03$, use R).
15. Let the rvs $X_1, X_2, \ldots, X_n$ are iid from $N(0, \sigma^2)$. Find UMPU test for testing $H_0 : \sigma = 3$ against $H_1 : \sigma \neq 3$ for the following data.
0.13 -2.43 -6.19 -0.80 2.62 2.65 0.86
-1.89 0.83 5.01 0.76 -0.29 3.91 3.02
(Let $\alpha = 0.05$, use R)
16. Let $X_1, X_2, \ldots, X_n$ be iid rvs with Cauchy distribution $C(\theta)$. Obtain LMPT for testing $H_0 : \theta = 0$ against $H_1 : \theta > 0$ from the following data.
1.04 0.53 3.56 1.96 2.34 2.46 1.19 2.16 0.92
1.71 4.54 41.95 1.19 6.80 2.09 -5.56 3.28 4.37
5.24 -0.75
(Let $\alpha = 0.05$, use R)
17. Let $X_1, X_2, \ldots, X_n$ be iid rvs with $N(0, \sigma^2)$. Find UMPU test for testing $H_0 : \sigma^2 = 9$ against $H_1 : \sigma^2 \neq 9$ for the data given in problem 15.
18. Let $(X_1, X_2, \ldots, X_{k_1})$ and $(Y_1, Y_2, \ldots, Y_{k_2})$ are iid rvs with $B(n_1, p_1)$ and $B(n_2, p_2)$, respectively, where n_1 and n_2 are known.
Find Neyman Structure test for testing $H_0 : p_1 = p_2$ against $H_1 : p_1 > p_2$ for the following data.
Assume $n_1 = 3$, $n_2 = 4$, $\alpha = 0.05$
X: 3 1 2 1 1 1 1 1 2 2 $-$ $-$
Y: 1 2 2 2 2 1 0 1 3 3 1 3
(Use R)
19. Let $X_1, X_2, \ldots, X_m$ and $Y_1, Y_2, \ldots, Y_n$ are iid rvs as $P(\lambda_1)$ and $P(\lambda_2)$, respectively. Find Neyman Structure test for testing $H_0 : \lambda_1 = \lambda_2$ against $H_1 : \lambda_1 > \lambda_2$ from the following data.
X: 1 2 3 0 2 1 5 0 2 1 $-$ $-$
Y: 1 1 5 2 5 2 2 1 0 6 2 3
(Let $\alpha = 0.02$, use R)
20. Let $X_1, X_2, \ldots, X_n$ be iid rvs with pdf $f(x|\theta)$,

$$f(x|\theta) = \begin{cases} \theta x^{\theta-1} & ; \ 0 < x < 1, \ \theta > 0 \\ 0 & ; \ \text{otherwise} \end{cases}$$

Find UMPU test of size α to test $H_0 : \theta = \theta_0$ against $H_1 : \theta \neq \theta_0$. Further, for $\alpha = 0.02$, find UMPU test for the following sample if $\theta_0 = 5$

0.90, 0.63, 0.51, 0.83, 0.96, 0.87, 0.41, 0.89,
0.71, 0.96, 0.96, 0.99, 0.98, 0.83, 0.92

21. Let thr rv X_1 has exponential distribution with mean $1/\theta$ and the rv X_2 has $g(x_2|\theta)$,

$$g(x_2|\theta) = \theta x_2^{\theta-1} \ ; 0 < x_2 < 1, \ \theta > 0$$

Obtain a UMPU test of size α for testing $H_0 : \theta = \theta_0$ against $H_1 : \theta \neq \theta_0$.
22. Let $X_1, X_2, \ldots, X_n$ be iid rvs with $f(x|\theta)$,

$$f(x|\theta) = \begin{cases} \frac{e^{-\frac{x}{\theta}}}{\theta b} & ; \ 0 < x < c, \ \theta > 0 \\ 0 & ; \ \text{otherwise} \end{cases}$$

where $b = 1 - e^{-\frac{x}{c}}$; $c > 0$. Obtain a Neyman structure test of size α to test $H_0 : c = \infty$ against $H_1 : c < \infty$, θ unknown. (see Dixit and Dixit (2003))
23. Let $X_1, X_2, \ldots, X_n$ be iid rvs with $f(x|\theta)$,

$$f(x|\theta) = \begin{cases} \frac{1}{\theta} \exp\left[-\frac{(x-\mu)}{\theta}\right] & ; \ x > \mu, \ \theta > 0 \\ 0 & ; \ \text{otherwise} \end{cases}$$

Obtain a Neyman structure test of size α to test $H_0 : \mu = 0$ against $H_1 : \mu > 0$, θ unknown.
24. Let $X_1, X_2, \ldots, X_n$ be iid rvs with $\cup(0, \theta)$.
Find LR test for testing $H_0 : \theta = \theta_0$ against $H_1 : \theta \neq \theta_0$.
25. Find LR test for the problem 20.
26. Find LR test for the problem 21.

References

Dixit UJ, Dixit VU (2003) Testing of the parameters of a right truncated exponential distribution. Am J Math Manage Sci 22(3, 4):343–351
Lehman EL (1986) Testing statistical hypotheses, 2nd edn. Wiley, New York
Neyman J, Pearson ES (1928) On the use and interpretation of certain test criteria. Biometrika A20, pp. 175–240, pp. 263–294

Index

© Springer Science+Business Media Singapore 2016
U.J. Dixit, *Examples in Parametric Inference with R*,
DOI 10.1007/978-981-10-0889-4

Printed in the United States
by Bookmasters

Printed in the United States
By Bookmasters